全两册
中英双语

The Importance
of Living

上

生活的
艺术

林语堂 著

越裔 译

湖南文艺出版社
HUNAN LITERATURE AND ART PUBLISHING HOUSE

博集天卷
CS·BOOKY

THE IMPORTANCE OF LIVING

By Lin Yutang

This edition arranged with Curtis Brown Group Ltd.

through Andrew Nurnberg Associates International Limited

著作权合同登记号：图字18-2016-127

图书在版编目（CIP）数据

生活的艺术 = The Importance of Living：全两册：汉英对照/林语堂著；越裔译. — 长沙：湖南文艺出版社，2017.6（2021.8重印）

ISBN 978-7-5404-8081-3

Ⅰ.①生… Ⅱ.①林… ②越… Ⅲ.①人生哲学—汉、英 Ⅳ.①B821

中国版本图书馆CIP数据核字（2017）第095669号

上架建议：名家经典·文化

SHENGHUO DE YISHU: QUAN LIANG CE: HAN-YING DUIZHAO

生活的艺术：全两册：汉英对照

作　　者：林语堂	
译　　者：越　裔	
出 版 人：曾赛丰	
责任编辑：薛　健　刘诗哲	
监　　制：蔡明菲　邢越超	
特约策划：王　维	
特约编辑：蔡文婷	
版权支持：辛　艳　张雪珂	
营销支持：文刀刀　周　茜	
封面设计：棱角视觉	
版式设计：李　洁	

出　　版：湖南文艺出版社
　　　　　（长沙市雨花区东二环一段508号　邮编：410014）
网　　址：www.hnwy.net
印　　刷：三河市兴博印务有限公司
经　　销：新华书店
开　　本：880mm×1270mm　1/32
字　　数：688千字
印　　张：28
版　　次：2017年6月第1版
印　　次：2021年8月第3次印刷
书　　号：ISBN 978-7-5404-8081-3
定　　价：88.00元（全两册）

若有质量问题，请致电质量监督电话：010-59096394
团购电话：010-59320018

生活 | The
的 | Importance
艺术 | of Living

第七章　悠闲的重要
Chapter Seven　The Importance of Loafing

第八章　家庭之乐
Chapter Eight　The Enjoyment of the Home

第九章　生活的享受
Chapter Nine　**The Enjoyment of Living**

第十四章　与上帝的关系
Chapter Thirteen　Relationship to God

自序

本书是一种私人的供状，供认我自己的思想和生活所得的经验。我不想发表客观意见，也不想创立不朽真理。我实在瞧不起自许的客观哲学；我只想表现我个人的观点。我本想题这书的名字为"抒情哲学"，用抒情一词说明这里面所讲的是一些私人的观念。但是这个书名似乎太美，我不敢用，我恐怕目标定得太高，即难于满足读者的期望，况且我的主旨是实事求是的散文，所以用现在的书名较易维持水准，且较自然。让我和草木为友，和土壤相亲，我便已觉得心意满足。我的灵魂很舒服地在泥土里蠕动，觉得很快乐。当一个人悠闲陶醉于土地上时，他的心灵似乎那么轻松，好像是在天堂一般。事实上，他那六尺之躯，何尝离开土壤一寸一分呢？

我颇想用柏拉图的对话方式写这本书。把偶然想到的话说出来，把日常生活中有意义的琐事安插进去，这将是多么自由容易的方式。可是不知什么缘故，我并不如此做。或者是因我恐怕这种文体现在不很流行，没有人喜欢读，而一个作家总是希望自己的作品有人阅读。我所说的对话，它的形式并不是像报纸上的谈话或问答，或分成许多段落的评论；我的意思是指真正有趣的、冗长的、闲逸的谈论，一说就是几页，中间富于迂回曲折，后来在料不到的地方，突然一转，仍旧回到原来的论点，好像一个人因为要使伙伴惊奇，特意翻过一道篱笆回家去一般。我多么喜欢翻篱笆抄小路回家啊！至少会使我的同伴感觉我对于回家的道路和四周的乡野是熟识的……可是我总不敢如此做。

我并不是在创作。我所表现的观念早由许多中西思想家再三思虑过、表现过；我从东方所借来的真理在那边都已陈旧平常了。但它们总是我的观念，它们已经变成自我的一部分。它们所以能在我的生命里生根，是因为它们表现出一些我自己所创造出来的东西，当我第一次见到它们时，我即对它们出于本心的协调了。我喜欢那些思想，并不是因为表现那些思想的是什么伟大人物。老实说，我在读书和写作时都是抄小路走的。我所引用的作家有许多是不见经传的，有些也会使中国文学教授错愕不解。我引用的当中如果有出名人物，那也不过是我在直觉的认可下接受他们的观念，而并不是震于他们的大名。我有一种习惯，最爱购买隐僻无闻的便宜书和断版书，看看是否可以从这些书里发现些什么。如果文学教授们知道了我的思想来源，他们一定会对这么一个俗物显得骇怪。但是在灰烬里拾到一颗小珍珠，比在珠宝店橱窗内看见一粒大珍珠更为快活。

我的思想并不怎样深刻，读过的书也不怎样广博。一个人所读的书太多，便不辨孰是孰非了。我没有读过洛克（Locke，十七世纪英国哲学家）、休谟（Hume，十八世纪苏格兰哲学家）或勃克莱（Berkeley，十七世纪爱尔兰哲学家）的著作，也没有读过大学的哲学课程。从专门技术上讲，我所应用的方法、所受的训练都是错误的，我并不读哲学而只直接拿人生当做课本，这种研究方法是不合惯例的。我的理论根据，大多是从下面所说这些人物方面而来：老妈子黄妈，她具有中国女教的一切良好思想；一个随口骂人的苏州船娘；一个上海的电车售票员；厨子的妻子；动物园中的一只小狮子；纽约中央公园里的一只松鼠；一个发过一句妙论的轮船上的管事；一个在某报天文栏内写文章的记者（已亡故十多年了）；箱子里所收藏的新闻纸；以及任何一个不毁灭我们人生好奇意识的作家，或任何一个不毁灭他自己人生好奇意识的作家……诸如此类，不胜枚举。

我没有受过学院式的哲学训练，所以反而不怕写一本哲学书。观察一切也似乎比较清楚，比较便当，这在正统哲学家看来，不知是不是可算一种补偿。我知道一定有人会说我所用的字句太过于浅俗，说我写得太容易了解，说我太不谨慎，说我在哲学的尊座前说话不低声下气，走路不步伐整齐，态度不惶恐战栗。现代哲学家所最缺乏的似乎是勇气。但我始终徘徊于哲学境界的外面。这倒给我勇气，使我可以根据自己的直觉下判断，思索出自己的观念，创立自己独特的见解，以一种孩子气的厚脸皮，在大庭广众之间把它们直供出来；并且确知在世界另一角落里必有和我同感的人，会表示默契。用这种方法树立观念的人，会常常在惊奇中发现另外一个作家也曾说过相同的话，或有过相同的感觉，只不过是它的表现方法有难易或雅俗之分而已。如此，他便有了一个古代作家替他做证人；他们在精神上成为永久的朋友。

所以我对于这些作家，尤其是对于我精神上的中国朋友，应该表示感谢。当我写这本书时，有一群和蔼可亲的天才和我合作；我希望我们互相亲热。从真实的意义上来说，这些灵魂是与我同在的，我

们之间的精神上的相通，即我所认为是唯一真实的相通方式——两个
时代不同的人有着同样的思想，有着同样的感觉，彼此之间完全了
解。我写这书的时候，他们借着贡献和忠告，给我以特殊的帮助，八
世纪的白居易，十一世纪的苏东坡，以及十六、十七两世纪那许多独
出心裁的人物——浪漫潇洒，富于口才的屠赤水；嬉笑诙谐，独具心
得的袁中郎；多口好奇，独特伟大的李卓吾；感觉敏锐，通晓世故的
张潮；耽于逸乐的李笠翁；乐观风趣的老快乐主义者袁子才；谈笑风
生，热情充溢的金圣叹——这些都是脱略形骸、不拘小节的人，这些
人因为胸蕴太多的独特见解，对事物具有太深的情感，因此不能得到
正统派批评家的称许；这些人太好了，所以不能循规蹈矩，因为太有
道德了，所以在儒家看来便是不"好"的。这些精选出来的同志人数
不多，因此使我享受到更宝贵、更诚挚的快乐。这些人物也许有几个
在本书内不曾述及，可是他们的精神确是同在这部著作里边的。我
想他们在中国总有一天会占到重要的地位，不过是时间问题而已……
还有一些人物，虽然较为晦暗无闻，但是他们恰当的言论也是我所欢
迎的，因为他们将我的意见表示得那么好。我称他们为中国的爱弥
儿（Amiel，瑞士作家，一八二一年至一八八一年）——他们说的话
并不多，但说得总是那么近情，我佩服他们的晓事。此外更有中外古
今的不朽哲人，他们好像是伟大人物的无名祖宗一般，在心灵感动的
当儿，在不知不觉之间说出一些至理名言；最后还有一些更伟大的人
物，我不当他们做我精神上的同志，而当他们是我的先生，他们那清
朗的理解是那么入情入理，又那么超凡入圣，他们的智慧已成自然，
因此表现出来很容易，丝毫不用费力。庄子和陶渊明就是这么一类人
物，他们的精神简朴纯正，非渺小的人所能望其项背。在本书里，我

有时加以相当声明，让他们直接对读者讲话；有时竟代他们说话，虽然表面上好像是我自己的话一般。我和他们的友谊维持得越久，我的思想也就越受他们的影响，我在他们的熏陶下，我的思想就倾向于通俗不拘礼节，无从捉摸，无影无形的类型；正如做父亲的对施予良好的家教所产生的影响一样。我也想以一个现代人的立场说话，而不仅以中国人的立场说话为满足，我不想仅仅替古人做一个虔诚的迻译者，而要把我自己所吸收到我现代脑筋里的东西表现出来。这种方法当然有缺点，但是从大体上说来，确能使这工作比较诚实一些。因此，一切取舍都是根据于我个人的见解。在这本书里我不想把一个诗人或哲学家的思想全盘托出来；假如想要根据本书里所举的少许例证去批判他们的全体，那是不可能的。所以当我结束这篇自序时，必须照例地说，本书如有优点的话，大部分应该归功于我的合作者，至于一切错误、缺点和不正确的见解，当由我自己完全负责。

我要向华尔士先生和夫人（Mr. and Mrs. Walsh）致谢，第一，谢谢他们鼓励我写作本书的念头；第二，谢谢他们坦白有益的批评。我也得感谢韦特先生（Mr. Hugh Wade）帮助我做本书的付印和校对工作，感谢佩弗女士（Miss Lillian Peffer）代我完成书后的索引。

林语堂
作于纽约

Preface

THIS is a personal testimony, a testimony of my own experience of thought and life. It is not intended to be objective and makes no claim to establish eternal truths. In fact I rather despise claims to objectivity in philosophy; the point of view is the thing. I should have liked to call it "A Lyrical Philosophy," using the word "lyrical" in the sense of being a highly personal and individual outlook. But that would be too beautiful a name and I must forego it, for fear of aiming too high and leading the reader to expect too much, and because the main ingredient of my thought is matter-of-fact prose, a level easier to maintain because more natural. Very much contented am I to lie low, to cling to the soil, to be of kin to the sod. My soul squirms comfortably in the soil and sand and is happy. Sometimes when one is drunk with this earth, one's spirit seems so light that he thinks he is in heaven. But actually he seldom rises six feet above the ground.

I should have liked also to write the entire book in the form of a dialogue like Plato's. It is such a convenient form for personal, inadvertent disclosures, for bringing in the significant trivialities of our daily life, above all for idle rambling about the pastures of sweet, silent thought. But somehow I have not done so. I do not know why. A fear, perhaps, that this form of literature being so little in vogue today, no one probably would read it, and a writer after all wants to be read. And when I say dialogue, I do not mean answers and questions like newspaper interviews, or those leaders chopped up into short paragraphs; I mean really good, long, leisurely discourses extending several pages at a stretch, with many detours, and coming back to the original point of discussion by a short cut at the most unexpected spot, like a man returning home by climbing over a hedge, to the surprise of his walking companion. Oh, how I love to reach home by climbing over the back fence, and to travel on bypaths! At least my companion will grant that I am familiar with the way home and with the surrounding countryside... But I dare not.

I am not original. The ideas expressed here have been thought and expressed by many thinkers of the East and West over and over again; those I borrow from the East are hackneyed truths there. They are, nevertheless, my ideas; they have become a part of my being. If they have taken root in my being, it is because they express something original in me, and when I first encountered them, my heart gave an instinctive assent. I like them as ideas and not because the person who expressed them is of any account. In fact, I have traveled the bypaths in my reading as well as in my writing. Many of the authors quoted are names obscure and may baffle a Chinese professor of literature. If some happen to be well-known, I accept their ideas only as they compel my intuitive approval and not because the authors are well-known. It is my habit to buy cheap editions of old, obscure books and see what I can discover there. If the professors of literature knew the sources of my ideas, they would be astounded at the Philistine. But there is a greater pleasure in picking up a small pearl in an ash-can than in looking at a large one in a jeweler's window.

I am not deep and not well-read. If one is too well-read, then one does not know right is right and wrong is wrong. I have not read Locke or Hume or Berkeley, and have not taken a college course in philosophy. Technically speaking, my method and my training are all wrong, because I do not read philosophy, but only read life at first hand. That is an unconventional way of studying philosophy-the incorrect way. Some of my sources are: Mrs. Huang, an amah in my family who has all the ideas that go into the breeding of a good woman in China; a Soochow boat-woman with her profuse use of expletives; a Shanghai street car conductor; my cook's wife; a lion cub in the zoo; a squirrel in Central Park in New York; a deck steward who made one good remark; that writer of a column on astronomy (dead for some ten years now) ; all news in boxes; and any writer who does not kill our sense of curiosity in life or who has not killed it in himself... how can I enumerate them all?

Thus deprived of academic training in philosophy, I am less scared to write a book about it. Everything seems clearer and simpler for it, if that is any compensation in the eyes of orthodox philosophy. I doubt it. I know there will be complaints that my words are not long enough, that I make things too easy to understand, and finally that I lack cautiousness, that I do not whisper low and trip with mincing steps in the sacred mansions of philosophy, looking properly scared as I ought to do. Courage seems to be the rarest of all virtues in a modern philosopher. But I have always wandered outside the precincts of philosophy and that gives me courage. There is a method of appealing to one's own intuitive judgment, of thinking out one's own ideas and forming one's own independent judgments, and confessing them in public with a childish impudence, and sure enough, some kindred souls in another corner of the world will agree with you. A person forming his ideas in this manner will often be astounded to discover how another writer said exactly the same things and felt exactly the same way, but perhaps expressed the ideas more easily and more gracefully. It is then that he discovers the ancient author and the ancient author bears him witness, and they become forever friends in spirit.

There is therefore the matter of my obligations to these authors, especially my Chinese friends in spirit. I have for my collaborators in writing this book a company of genial souls, who I hope like me as much as I like them. For in a very real sense, these spirits have been with me, in the only form of spiritual communion that I recognize as real—when two men separated by the ages think the same thoughts and sense the same feelings and each perfectly understands the other. In the preparation of this book, a few of my friends have been especially helpful with their contributions and advice: Po Chüyi of the eighth century, Su Tungp'o of the eleventh, and that great company of original spirits of the sixteenth and seventeenth centuries–the romantic and voluble T'u Ch'ihshui, the playful, original Yüan Chunglang, the deep, magnificent Li Chowu, the sensitive and sophisticated Chang Ch'ao, the epicure Li Liweng, the happy and gay old hedonist Yüan Tsets'ai, and the bubbling, joking, effervescent Chin Shengt'an–unconventional souls all, men with too much independent judgment and too much feeling for things to be liked by the orthodox critics, men too good to be"moral"and too moral to be"good"for the Confucianists. The smallness of the select company has made the enjoyment of their presence all the more valued and sincere. Some of these may happen not to be quoted, but they are here with me in this book all the same. Their coming back to their own in China is only a matter of time... There have been others, names less well-known, but no less welcome for their apt remarks, because they express my sentiments so well. I call them my Chinese Amiels–people who don't talk much, but always talk sensibly, and I respect their good sense. There are others again who belong to the illustrious company of"Anons"of all countries and ages, who in an inspired moment said something wiser then they knew, like the unknown fathers of great men. Finally there are greater ones still, whom I look up to more as masters than as companions of the spirit, whose serenity of understanding is so human and yet so divine, and whose wisdom seems to have come entirely without effort because it has become completely natural. Such a one is Chuangtse, and such a one is T'ao Yüanming, whose simplicity of spirit is the despair of smaller men. I have sometimes let these souls speak directly to the reader, making proper acknowledgment, and at other times, I have spoken for them while I seem to be speaking for myself. The older my friendship with them, the more likely is my indebtedness to their ideas to be of the familiar, elusive and invisible type, like parental influence in a good family breeding. It is impossible to put one's finger on a definite point of resemblance. I have also chosen to speak as a modern, sharing the modern life, and not only as a Chinese; to give only what

I have personally absorbed into my modern being, and not merely to act as a respectful translator of the ancients. Such a procedure has its drawbacks, but on the whole, one can do a more sincere job of it. The selections are therefore as highly personal as the rejections. No complete presentation of any one poet or philosopher is attempted here, and it is impossible to judge them through the evidences on these pages. I must therefore conclude by saying as usual that the merits of this book, if any, are largely due to the helpful suggestions of my collaborators, while for the inaccuracies, deficiencies and immaturities of judgment, I alone am responsible.

Again I owe my thanks to Mr. and Mrs. Richard J. Walsh, first, for suggesting the idea of the book, and secondly, for their useful and frank criticism. I must also thank Mr. Hugh Wade for cooperating on preparing the manuscript for the press and on the proofs, and Miss Lillian Peffer for making the Index.

Lin Yutang
New York City

生活	The
的	Importance
艺术	of Living

第一章

醒觉

Chapter One　**The Awakening**

一．人生之研究

　　在下面的文章里我不过是表现中国人的观点。我只表现一种中国最优越最聪慧的哲人们所见到而在他们的文字中发挥过的人生观和事物观。我知道这是一种闲适哲学，是在异于现今时代里的闲适生活中所产生的。我总觉得这种人生观是绝对真实的。人类心性既然相同，则在这个国家里能感动人的东西，自然也会感动别的国家的人类。我将要表现中国诗人和学者们的人生观，这种人生观是经过他们的常识和他们的诗意情绪而估定的。我想显示一些异教徒世界的美，显示一个明知此生有涯，但是短短的生命未始没有它的尊严的民族所看到的人生悲哀、美丽、恐惧和喜乐。

　　中国的哲学家是睁着一只眼做梦的人，是一个用爱和讥评心理来观察人生的人，是一个自私主义和仁爱的宽容心混合起来的人，是一个有时从梦中醒来，有时又睡了过去，在梦中比在醒时更觉得富有生气，因而在他清醒时的生活中也含着梦意的人。他把一只眼睁着，一只眼闭着，看透了他四周所发生的事物和他自己的徒劳，而不过仅仅保留着充分的现实感去走完人生应走的道路。因此，他并没有虚幻的憧憬，所以无所谓醒悟；他从来没有怀着过度的奢望，所以无所谓失望。他的精神就是如此得了解放。

观测了中国的文学和哲学之后，我得到一个结论：中国文化的最高理想人物，是一个对人生有一种建于明慧悟性上的达观者。这种达观产生宽宏的怀抱，能使人带着温和的讥评心理度过一生，丢开功名利禄，乐天知命地过生活。这种达观也产生了自由意识，放荡不羁的爱好，傲骨和漠然的态度。一个人有了这种自由的意识及淡漠的态度，才能深切热烈地享受快乐的人生。

我不必说我的哲学思想是否适用于西方人。我们要了解西方人的生活，就得用西方的眼光，用他们自己的性情，他们自己的物质观念，和他们自己的头脑去观察。无疑美国人能忍受中国人所不能忍受的事物，同样中国人也能忍受美国人所不能忍受的事物。我们生下来就不同样，这已有显著的区别。这也不过是比较的看法。我相信在美国的繁忙生活中，他们也一定有一种企望，想躺在一片绿草地上，在美丽的树荫下什么事也不做，只想悠闲自在地去享受一个下午。"醒转来生活吧"（Wake up and live）这种普遍的呼声，在我看来足以证明有一部分美国人宁愿过梦中的光阴，但是美国人终还不至于那么颓丧。问题只是他对这种闲适生活，要想享受的多少和他要怎样使这种生活实现而已。也许美国人只是在这个忙碌的世界上，对于"闲荡"一词有些感到惭愧；可是我确切知道，一如知道他们也是动物一样，他们有时也喜欢松松肌肉，在沙滩上伸伸懒腰，或是静静地躺着，把一条腿舒服地跷起来，把手臂搁在头下当枕头。如果这样，便跟颜回差不多；颜回也有这种美德，孔子在众弟子中，最器重他。我希望看到的，就是他能对这件事抱诚实的态度；譬如他喜欢这件事，便应向全世界实说他喜欢这件事；应在他闲逸自适地躺在沙滩上，而不是在办公室里时，他的灵魂喊着："人生真美妙啊！"

　　所以现在我们将要看到中国整个民族思想所理解的那种哲学和生活艺术。我认为不论是在好的或坏的方面，世界上没有一样东西是如它想象的。因为我们在这里看见一种完全不同的思想典型，由这种思想典型产生了一种簇新的人生观念，任何一个民族的文化都是他的思想产物，这话绝无疑义。中国民族思想在种族性上和西方文化是那么不同，在历史上又和西方文化那么隔离着；因此我们自然能从这种地方找到一些人生问题的新答案，或者好一些，找到一些探讨人生问题的新方法。或者更好些，找到一些人生问题的新论据。我们要知道这种思想的美点和缺点，至少可以由它过去的历史看出。它有光辉灿烂的艺术，也有微不足道的科学，有博大的常识，也有幼稚的逻辑，有精雅温柔的关于人生的闲谈，却没有学者风味的哲学。很多人本来都知道中国人的思想非常实际而精明，爱好中国艺术的人也都知道中国人的思想是极灵敏的；一部分人则承认中国人的思想也是富有诗意和哲理的。至少大家都知道中国人善于用哲理的眼光去观察事物，这比之中国有伟大的哲学或有几个大哲学家的那种说法，更有意义。一个民族产生过几个大哲学家没什么稀罕，但一个民族能以哲理的眼光去观察事物，那是难能可贵的。无论怎样，中国这个民族显然是比较富于哲理性而少实效性，假如不是这样，一个民族经过了四千年专讲效率生活的"高血压"，那是早已不能继续生存了。四千年专重效能的生活能毁灭任何一个民族。一个重要的结果就是：在西方狂人太多了，只好把他们关在疯人院里；而在中国狂人太罕有了，所以崇拜他们；凡具有中国文学知识的人，会证实这一句话。我所要说明的就在乎此。是的，中国有一种轻逸的，一种近乎愉快的哲学，他们的哲学气质，可以在他们那种智慧而快乐的生活哲学里找到最好的论据。

I. *Approach to Life*

IN what follows I am presenting the Chinese point of view, because I cannot help myself. I am interested only in presenting a view of life and of things as the best and wisest Chinese minds have seen it and expressed it in their folk wisdom and their literature. It is an idle philosophy born of an idle life, evolved in a different age, I am quite aware. But I cannot help feeling that this view of life is essentially true, and since we are alike under the skin, what touches the human heart in one country touches all. I shall have to present a view of life as Chinese poets and scholars evaluated it with their common sense, their realism and their sense of poetry. I shall attempt to reveal some of the beauty of the pagan world, a sense of the pathos and beauty and terror and comedy of life, viewed by a people who have a strong feeling of the limitations of our existence, and yet somehow retain a sense of the dignity of human life.

The Chinese philosopher is one who dreams with one eye open, who views life with love and sweet irony, who mixes his cynicism with a kindly tolerance, and who alternately wakes up from life's dream and then nods again, feeling more alive when he is dreaming than when he is awake, thereby investing his waking life with a dream-world quality. He sees with one eye closed and with one eye opened the futility of much that goes on around him and of his own endeavors, but barely retains enough sense of reality to determine to go through with it. He is seldom disillusioned because he has no illusions, and seldom disappointed because he never had extravagant hopes. In this way his spirit is emancipated.

For, after surveying the field of Chinese literature and philosophy, I come to the conclusion that the highest ideal of Chinese culture has always been a man with a sense of detachment (takuan) toward life based on a sense of wise disenchantment. From this detachment comes high-mindedness (k'uanghuai), a high-mindedness which enables one to go through life with tolerant irony and escape the temptations of fame and wealth and achievement, and eventually makes him take what comes. And from this detachment arise also his sense of freedom, his love of vagabondage and his pride and nonchalance. It is only with this sense of freedom and nonchalance that one eventually arrives at the keen and intense joy of living.

It is useless for me to say whether my philosophy is valid or not for the Westerner. To understand Western life, one would have to look at it as a Westerner born, with his own temperament, his bodily attitudes and his own set of nerves. I have no doubt that American nerves can stand a good many things that Chinese nerves cannot stand, and vice versa. It is good that it should be so—that we should all be born different. And yet it is all a question of relativity. I am quite sure that amidst the hustle and bustle of American life, there is a great deal of wistfulness, of the divine desire to lie on a plot of grass under tall beautiful trees of an idle afternoon and just do nothing. The necessity for such common cries as"Wake up and live"is to me a good sign that a wise portion of American humanity prefer to dream the hours away. The American is after all not as bad as all that. It is only a question whether he will have more or less of that sort of thing, and how he will arrange to make it possible. Perhaps the American is merely ashamed of the word"loafing"in a world where everybody is doing something, but somehow, as sure as I know he is also an animal, he likes sometimes to have his muscles relaxed, to stretch on the sand, or to lie still with one leg comfortably curled up and one arm placed below his head as his pillow. If so, he cannot be very different from Yen Huei, who had exactly that virtue and whom Confucius desperately admired among all his disciples. The only thing I desire to see is that he be honest about it, and that he proclaim to the world that he likes it when he likes it, that it is not when he is working in the office but when he is lying idly on the sand that his soul utters, "Life is beautiful. "

We are, therefore, about to see a philosophy and art of living as the mind of the Chinese people as a whole has understood it. I am inclined to think that, in a good or bad sense, there is nothing like it in the world. For here we come to an entirely new way of looking at life by an entirely different type of mind. It is a truism to say that the culture of any nation is the product of its mind. Consequently, where there is a national mind so racially different and historically isolated from the Western cultural world, we have the right to expect new answers to the problems of life, or what is better, new methods of approach, or, still better, a new posing of the problems themselves. We know some of the virtues and deficiencies of that mind, at least as revealed to us in the historical past. It has a glorious art and a contemptible science, a magnificent common sense and an infantile logic, a fine womanish chatter about life and no scholastic philosophy. It is generally known that the Chinese mind is an intensely practical, hard-headed one, and it is also known to some lovers of Chinese art that it is a profoundly sensitive mind; by a still smaller proportion of people, it is accepted as also a profoundly poetic and philosophical mind. At least the Chinese are noted for taking things philosophically, which is saying more than the statement that the Chinese have a great philosophy or have a few great philosophers. For a nation to have a few philosophers is not so unusual, but for a nation to take things philosophically is terrific. It is evident anyway that the Chinese as a nation are more philosophic than efficient, and that if it were otherwise, no nation could have survived the high blood pressure of an efficient life for four thousand years. Four thousand years of efficient living would ruin any nation. An important consequence is that, while in the West, the insane are so many that they are put in an asylum, in China the insane are so unusual that we worship them, as anybody who has a knowledge of Chinese literature will testify. And that, after all, is what I am driving at. Yes, the Chinese have a light, an almost gay, philosophy, and the best proof of their philosophic temper is to be found in this wise and merry philosophy of living.

二．一个准科学公式

现在就让我们先来研究产生这个生活哲学的中国人的理智构造——伟大的现实主义，不充分的理想主义，很多的幽默感，以及对人生和自然的高度诗意感觉性。

人类似分成两种人：一种是理想主义者，另一种是现实主义者，是造成人类进步的两种动力。人性好似泥土，由理想主义浇灌后即变成了柔软可塑的东西，但是使泥土凝结的还是泥土本身，不然我们早就蒸发而化气了。在一切人类活动里，个人的、社会的，或民族的，理想主义和现实主义这两种力都互相牵制着，而真正的进步便是由这两种成分的适当混合而促成；所谓适当的混合就是将泥土保持着适宜的柔软可塑的状态，半湿半燥，恰到好处。例如英国这个最健全的民族，就是现实主义和理想主义适当地混合起来而成的。有些国家常要发生革命，这是因为它们的泥土吸收了一些不能适当同化的外国思想液汁的缘故，以致泥土不能保持着它们的形式。

模糊而缺乏批判精神的理想主义，是极为可笑的。这种理想主义的成分如果太多，于人类颇为危险，它使人徒然地追求虚幻的理想。在任何一个社会或民族里，如果这种幻想的理想主义成分太多，就会时常发生革命。人类好似一对理想主义的夫妻，对于他们的住所永远感到不满意，每三个月总要搬一次家，他们以为没有一块地方是理想的，而没有到过的地方似乎总是好的。幸而人类也富有一种幽默感，其功用，是在纠正人类的梦想，而引导人们去和现实世界相接触。人类不可没有梦想，可是他也不能笑他自己的梦想，两者也许同样的重要。这是多么伟大的天赋，而中国人就富于这种特质。

幽默感（我在下面一章里将做更详细的讨论）似乎和现实主义或称现实感有密切的联系。说笑话者虽常常残酷地使理想主义者感到幻灭，但是在另一方面，完成了一种极重要的任务，这足以使理想主义者不至于把头碰在现实的墙壁上，而受到一个比幻灭更猛烈的撞击。同时也能缓和暴躁急烈分子的紧张心情，使他可以寿命长一些。如能预先让他知道幻灭的无可避免，或许可以使他在最后的撞击里减少一些痛苦，因为一个幽默家始终是像一个负责者将坏的消息温和地告诉垂死的病人。有时一个幽默家的温和警告会挽救垂死者的生命。假如理想主义和幻灭必须在这世界上并存，那么，我们与其说那个说笑话者是残忍的，还不如说人生是残酷的了。

我时常想到一些机械公式，想把人类进步和历史变迁明确地表示出来。这些公式仿佛如下：

"现实"减"梦想"等于"禽兽"

"现实"加"梦想"等于"心痛"（普通叫做"理想主义"）

"现实"加"幽默"等于"现实主义"（普通叫做"保守主义"）

"梦想"减"幽默"等于"热狂"

"梦想"加"幽默"等于"幻想"

"现实"加"梦想"加"幽默"等于"智慧"

这样看来，智慧或最高型的思想，它的形成就是在现实的支持下，用适当的幽默感把我们的梦想或理想主义调和配合起来。

为尝试制造一些准科学的公式起见，不妨进一步照下列的方法来分析各国的民族性。我用"准科学"这个名字，因为我不相信那种呆板的机械公式真能够把人类活动或人类性格表现出来。把人类的活动归纳到一个呆板的公式里，这已经缺少幽默感，因此也就缺乏智慧。我并不是说现在没有拟这一类的公式；在今日之下，这种准科学正盛行。到一个心理学家竟能衡量人类的"智能"（IQ）或"性格"（PQ）[1]时，这世界可真够可怜，因为有人性的学问都被专家跑来篡夺了。但如果我们认为这些公式不过是拿来表现某些意见的简便图解方法，而不拿科学神圣的名义来做我们的护符，则尚没有什么关系。下面所写的是我替某些民族的特性所拟的公式，这些公式完全是照我个人意思而定，绝对无法证实。任何人都可反对它们，修改它们，或另改为他自己的公式。现在以"现"字代表"现实感"（或现实主义），"梦"字代表"梦想"（或理想主义），"幽"字代表"幽默感"——再加上一个重要的成分——"敏"字代表"敏感性"（Sensitivity）[2]，再以"四"代表"最高"，"三"代表"高"，"二"代表"中"，"一"代表"低"，

[1] 智能测验自有其相当有限的用处，这我并不反对；我所反对的是他们认为这些测验于人类性格可以获得数学式的准确答案，或可以作为始终可靠的衡量。

[2] 用 Sensitively 这个词的意义。

这样我们就可以用准化学公式代表下列的民族性。正如硫酸盐和硫化物，或一氧化碳和二氧化碳的作用各不相同，人类和社会也依它们不同的构造，而有不同的作用。在我看来，人类社会或民族在同样的情形之下，却有不同的行为，确是一桩很有趣的事。我们既然不能模仿化学的形式发明"幽默化物"（Humoride）和"幽默盐"（Humorate）一类的名字，自只可用三份"现实主义"、二份"梦想"、二份"幽默"和一份"敏感性"的方式造成一个英国人[3]。

现三　梦二　幽二　敏一　等于英国人

现二　梦三　幽三　敏三　等于法国人

现三　梦三　幽二　敏二　等于美国人

现三　梦四　幽一　敏二　等于德国人

现二　梦四　幽一　敏一　等于俄国人

现二　梦三　幽一　敏一　等于日本人

现四　梦一　幽三　敏三　等于中国人

我不十分知道意大利人、西班牙人、印度人和其他的民族，所以不敢拟议他们的公式，同时上列公式本身就不很靠得住，每一公式都足以引起

[3] 也许有人要根据一些良好的理由，提议另一个"逻"字，以代表逻辑或合理性，认为是造成人类进步的重要元素。但这个"逻"字，总会和敏感性站在反对的地位。因为敏感性是某一种对事物的直接理解。别人尽可以试定这么一个公式。不过据我个人的意见，合理性在人类活动上占着颇低的地位。

严切的批评。这些公式与其说是权威的，不如说是含有挑拨性的。假使我能得到一些新见识或新印象时，我预备把这些公式逐渐修正，以供我自己应用。它们的价值眼前只限于此，这无非是我智识的进步和我愚昧的缺陷的一个记录而已。

这里也许有值得讨论的地方，我认为中国人和法国人最为相近，这从法国人著书和饮食的方式可以清楚地看出来，同时法国人更丰富的理想主义是由比较轻松的性情所产生，而以爱好抽象观念的形式表现出来（请回想他们在文学、艺术和政治运动上的宣言吧）。以"现四"来代表中国人民，是说中国人是世界上最现实化的民族，"梦一"的低分数则表示他们在生活类型或生活理想上似乎缺乏变迁性。中国人的幽默、敏感性和现实主义，我给了较高的分数，这大概是我同中国人接触密切、印象生动的缘故吧。中国人的敏感性，无须我细为解释，从中国的散文、诗歌和绘画即可以得到很好的说明了……日本人和德国人在缺乏幽默感的方面，极为相像（一般的印象），然而无论哪一个民族的特性，总不能以"零"为代表，甚至中国人的理想主义我也不能以"零"来代表它。其间完全不过是程度高低的问题："完全缺乏这种或那种质素"这一类的言论，对各民族有亲切认识者是不会说的。因此我给日本人和德国人不是"幽零"，而是"幽一"，我觉得这是对的。我相信日本人和德国人所以在现在和过去都遭受到政治上的痛苦，原因就是他们缺乏良好的"幽默感"。一个普鲁士的市政顾问官就喜欢人家称他"顾问官"，并且他多么喜欢他制服上的纽扣和徽章啊！一种对"逻辑上必要性"（常常是神圣的）的信念，一种直趋目标，而不做迂回行动的倾向，常常使人遇事过激。其要点不在你信仰什么东西，而是在你怎样去信仰那种东西，并怎样把信仰变成行动。我给日本人"梦三"是特别指出他们对于皇帝和国家的狂热忠诚，这种狂热的忠诚是他们性格上的幽默成分过低的缘故。因为理想主义在不同的国家里代表着不同的东西。正如所谓幽默感包括着许多不同的东西一般……理想主义和现实主义在今日之下的美国正在做有趣的拔河竞赛，二种成分都有相当高的分数，便产生了美国人特有的那种潜力。美国人的理想主义是什么，这问题还是让美国人自己去研究吧；不过有一点我可以说的，他们对什么东西都很热心。从美国人很容易被高尚的思想或高贵的语词感动这一点来说，他们的理想主义想必大部分是高尚的，但是有一部分也不免是自欺欺人而已。美国人的幽默感比之欧洲大陆民族的幽默感也有些不同，可是我的确觉得这种幽默感（爱好玩意儿和原有广博的常识）是美国民族最大的资产。在未来的重要变迁的时代中，詹姆斯·布赖斯（James Bryce，十九世纪末叶的英国历史学家，著有《神圣罗马帝国》《美利坚合众国》等）所说的那种广博常识，于他们将有很大的用处，我希望这种常识能使他们渡过危急的时期。我以很低的分数给予美国人的敏感性，因为在我的印象中，他们能够忍受非常的事物。一般说来，英国人好像是最健全的民族，他们的"现三梦二"，这是民族稳定性的表现。我以为理想的公式似乎是"现三　梦二　幽三　敏二"，因为理想主义或敏感性的成分过多，也不是好事。我以"敏一"代表英国人的敏感性；如果以为这个分数太低，那是要怪英国人自己的！英国人随时随地要露着那种忧郁的样子，我怎能知道他们究竟有什么感觉呢——欢笑，快乐，愤怒，满足？

　　我们对于作家和诗人也可以用同样的公式，现在试举几个著名的人物来做个例：

莎士比亚[4]——现四　梦四　幽三　敏四

德国诗人海涅（Heine）——现三　梦三　幽四　敏三

英国诗人雪莱（Shelley）——现一　梦四　幽一　敏四

美国诗人爱伦·坡（Poe）——现三　梦四　幽一　敏四

李白——现一　梦三　幽二　敏四

杜甫——现三　梦三　幽二　敏四

苏东坡——现三　梦二　幽四　敏三

　　这不过是我随手所写的几个例子。一切诗人显然有很高的敏感性，否则便不称其为诗人了。我觉得爱伦·坡虽有其不可思议的富于想象力的天赋，却是一个极健全的天才。因为，他不是很喜欢做推论的吗？

　　我给中国民族性所定的公式是：

　　现四　梦一　幽三　敏三

　　我用"敏三"去代表丰富的敏感性，这种丰富的敏感性产生一种对人生的适当艺术观念，使中国人很肯定地感到尘世是美满的，因此对人生感到热诚的爱好。不过敏感性还有更大的意义：它事实上代表一种近乎哲学的艺术观念。因这理由，中国哲学家的人生观就是诗人的人生观，而且中

[4]我很犹豫，不知道应该给莎士比亚"敏四"或"敏三"，最后他的"十四行诗"使我决定了。学校教师在批学生分数时，真没有像我批莎士比亚的分数时那么惶恐不安、战战兢兢。

国的哲学是跟诗歌发生联系，而不比西方的哲学是跟科学发生联系的。这种对欢乐、痛苦和人生百态的丰富敏感性，使轻快哲学有造成的可能；这一点可以在下文中看得很明白。人类对于人生悲剧的意识，是由于青春消逝的悲剧的感觉而来，而对人生的那种微妙的深情，是由于一种对昨开今谢的花朵的深情而产生的。起初受到的是愁苦和失败的感觉，随后即变为那狡猾的哲学家的醒觉和哂笑。

在另一面，我们用"现四"来代表浓厚的现实主义，这种浓厚的现实主义就是指一种安于人生现状的态度，是一种认为"二鸟在林，不如一鸟在手"的态度。所以这种现实主义使艺术家的信念变得更坚固，觉得这似乎朝露的人生，更为美丽。同时，也使艺术家和诗人不至于彻底逃避人生。梦中人说："人生不过是一场梦。"现实主义者回说："一点也不错，让我们在梦境里尽量过着美满的生活吧。"但是这里的醒觉者的现实主义是诗人的现实主义，而不是商人的现实主义；那种趾高气扬，欣欣然走上成功之路的年轻进取者的大笑也将失掉，而变为一个手捻长须，低声缓语的老人的大笑。这种梦中人酷爱和平，没有一个人肯为一个梦而拼命斗争。他会一心一意地和他做梦的同志一起过着合理美满的生活。因此人生的高度紧张生活即松弛下去了。

此外，这种现实感觉的主要功用，是要把人生哲学中一切不必要的东西摒除出去，它好似扼住了人生的颈项，以免幻想的翅膀会把它带到一个似乎是美而实则虚幻的世界里去。况且人生的智慧其实就在摒除那种不必要的东西，而把哲学上的问题化到很简单的地步——家庭的享受（夫妻、子、女）、生活的享受、大自然和文化的享受——同时停止其他不相干的科学训练和智识的追求。这么一来，中国哲学家的人生问题即变得稀少简单，同时人生智慧也即是指一种不耐烦的态度——一种对形而上的哲学，以及与人生没有实际关系的智识的不耐烦态度。也指各种人类活动，不论是去获取智识或是东西，都须立刻受人本身的测验以及对生活目标的服从。其结果即是生活的目的，不是什么形而上的实物，而仅是生活本身。

中国人的哲学因为具有这种现实主义和极端不相信逻辑及智能，就变成了一种对人生本身有直接亲热感觉的东西，而不肯让它归纳到任何一种体系里去。因为中国人的哲学里有健全的现实意识，纯然的动物意识和一种明理的精神，因此反而压倒了理性本身，使呆板的哲学体系无从产生。中国有儒道释三教，每一种教都是宏大的哲学体系，但它们都曾被健全的常识冲淡，因而都变成追求人生幸福的共同问题。中国人对任何一个哲学观念、信仰、派别，都不愿专心地相信，或过分起劲地去研究。孔子的一个朋友对他说，他常常三思而后行，孔子诙谐地回答："再，斯可矣。"一个哲学派别的信徒最多不过是一个哲学的学生，可一个人是生活的学生，或者竟是生活的大师哩。

有了这种文化和哲学，最后结果，就是中国人和西洋人成了一个对照，中国人过着一种比较接近大自然和儿童时代的生活，在这种生活里，本能和情感得以自由行动；是一种不太重视智能的生活，敬重肉体也尊崇精神，具有深沉的智慧、轻松的快活和熟悉世故但很孩子气的天真，这些成了一种奇怪的混合物。所以归纳起来说，这种哲学的特征是：第一，一种以艺术眼光对人生的天赋才能；第二，一种于哲理上有意识地回到简单；第三，一种合理近情的生活理想。最后的产品就是一种对于诗人、农夫和放浪者的崇拜，这是可怪的。

II. A Pseudoscientific Formula

LET us begin with an examination of the Chinese mental make-up which produced this philosophy of living: great realism, inadequate idealism, a high sense of humor, and a high poetic sensitivity to life and nature.

Mankind seems to be divided into idealists and realists, and idealism and realism are the two great forces molding human progress. The clay of humanity is made soft and pliable by the water of idealism, but the stuff that holds it together is after all the clay itself, or we might all evaporate into Ariels. The forces of idealism and realism tug at each other in all human activities, personal, social and national, and real progress is made possible by the proper mixture of these two ingredients, so that the clay is kept in the ideal pliable, plastic condition, half moist and half dry, not hardened and unmanageable, nor dissolving into mud. The soundest nations, like the English, have realism and idealism mixed in proper proportions, like the clay which neither hardens and so gets past the stage for the artist's molding, nor is so wishy-washy that it cannot retain its form...

A vague, uncritical idealism always lends itself to ridicule and too much of it might be a danger to mankind, leading it round in a futile wild-goose chase for imaginary ideals. If there were too many of these visionary idealists in any society or people, revolutions would be the order of the day. Human society would be like an idealistic couple forever getting tired of one place and changing their residence regularly once every three months, for the simple reason that no one place is ideal and the place where one is not seems always better because one is not there. Very fortunately, man is also gifted with a sense of humor, whose function, as I conceive it, is to exercise criticism of man's dreams, and bring them in touch with the world of reality. It is important that man dreams, but it is perhaps equally important that he can laugh at his own dreams. That is a great gift, and the Chinese have plenty of it.

The sense of humor, which I shall discuss at more length in a later chapter, seems to be very closely related to the sense of reality, or realism. If the joker is often cruel in disillusioning the idealist, he nevertheless performs a very important function right there by not letting the idealist bump his head against the stone wall of reality and receive a ruder shock. He also gently eases the tension of the hot-headed enthusiast and makes him live longer. By preparing him for disillusion, there is probably less pain in the final impact, for a humorist is always like a man charged with the duty of breaking a sad news gently to a dying patient. Sometimes the gentle warning from a humorist saves the dying patient's life. If idealism and disillusion must necessarily go together in this world, we must say that life is cruel, rather than the joker who reminds us of life's cruelty.

I have often thought of formulas by which the mechanism of human progress and historical change can be expressed. They seem to be as follows:

Reality - Dreams = Animal Being
Reality + Dreams = A Heartache (usually called Idealism)
Reality + Humor = Realism (also called Conservatism)
Dreams - Humor = Fanaticism
Dreams +Humor = Fantasy
Reality + Dreams + Humor = Wisdom

So then, wisdom, or the highest type of thinking, consists in toning down our dreams or idealism with a good sense of humor, supported by reality itself.

As pure ventures in pseudoscientific formulations, we may proceed to analyze national characters in the following manner. I say"pseudoscientific"because I distrust all dead and mechanical formulas for expressing anything connected with human affairs or human personalities. Putting human affairs in exact formulas shows in itself a lack of the sense of humor and therefore a lack of wisdom. I do not mean that these things are not being done: they are. That is why we get so much pseudoscience today. When a psychologist can measure a man's I. Q. or P. Q., it is a pretty poor world, and specialists have risen to usurp humanized scholarship. But if we recognize that these formulas are no more than handy, graphic ways of expressing certain opinions, and so long as we don't drag in the sacred name of science to help advertise our goods, no harm is done. The following are my formulas for the characters of certain nations, entirely personal and completely incapable of proof or verification. Anyone is free to dispute them and change them or add his own, if he does not claim that he can prove his private opinions by a mass of statistical facts and figures. Let"R"stand for a sense of reality (or realism),"D"for dreams (or idealism), "H"for a sense of humor, and—adding one important ingredient—"S"for sensitivity. And further let"4"stand for "abnormally high, ""3"stand for"high, ""2"for"fair,"and"1"for"low,"and we have the following pseudo-chemical formulas for the following national characters. Human beings and communities behave then differently according to their different compositions, as sulphates and sulphides or carbon monoxide and carbon dioxide behave differently from one another. For me, the interesting thing always is to watch how human communities or nations behave differently under identical conditions. As we cannot invent words like"humoride"and"humorate"after the fashion of chemistry, we may put it thus:"3 grains of Realism, 2 grains of Dreams, 2 grains of Humor and 1 grain of Sensitivity make an Englishman. "

R3D2H2S1 =*The English*
R2D3H3S3 =*The French*
R3D3H2S2 =*The Americans*
R3D4H1S2 =*The Germans*
R2D4H1S1 = *The Russians*
R2D3H1S1 = *The Japanese*
R4D1H3S3= *The Chinese*

I do not know the Italians, the Spanish, the Hindus and others well enough even to essay a formula on the subject, realizing that the above are shaky enough as they are, and in any case are enough to bring down a storm of criticism upon my head. Probably these formulas are more provocative than authoritative. I promise to modify them gradually for my own use as new facts are brought to my knowledge, or new impressions are formed. That is all they are worth today—a record of the progress of my knowledge and the gaps of my ignorance.

Some observations may be necessary. It is easy to see that I regard the Chinese as most closely allied to the French in their sense of humor and sensitivity, as is quite evident from the way the French write their books and eat their food, while the more volatile character of the French comes from their greater idealism, which takes the form of love of abstract ideas (recall the manifestoes of their literary, artistic and political movements)."R4"for Chinese realism makes the Chinese the most realistic people;"D1"accounts for something of a drag in the changes in their pattern or ideal of life. The high figures for Chinese humor and sensitivity, as well as for their realism, are perhaps due to my too close association and the vividness of my impressions. For Chinese sensitivity, little justification is needed; the whole story of

Chinese prose, poetry and painting proclaims it... The Japanese and Germans are very much alike in their comparative lack of humor (such is the general impression of people), yet it is really impossible to give a"zero"for any one characteristic in any one nation, not even for idealism in the Chinese people. It is all a question of degree; such statements as a complete lack of this or that quality are not based on an intimate knowledge of the peoples. For this reason, I give the Japanese and the Germans"H1"instead of"H0"and I intuitively feel that I am right. But I do believe that the Japanese and the Germans suffer politically at present, and have suffered in the past, for lacking a better sense of humor. How a Prussian Geheimrat loves to be called a Geheimrat, and how he loves his buttons and metal pins! A certain belief in"logical necessity"(often"holy"or"sacred"), a tendency to fly too straight at a goal instead of circling around it, often carries one too far. It is not so much what you believe in that matters, as the way in which you believe it and proceed to translate that belief into action. By"D3" for the Japanese I am referring to their fanatic loyalty to their emperor and to the state, made possible by a low mixture of humor. For idealism must stand for different things in different countries, as the so-called sense of humor really comprises a very wide variety of things... There is an interesting tug between idealism and realism in America, both given high figures, and that produces the energy characteristic of the Americans. What American idealism is, I had better leave it to the Americans to find out; but they are always enthusiastic about something or other. A great deal of this idealism is noble, in the sense that the Americans are easily appealed to by noble ideals or noble words; but some of it is mere gullibility. The American sense of humor again means a different thing from the Continental sense of humor, but really I think that, such as it is (the love of fun and an innate, broad common sense), it is the greatest asset of the American nation. In the coming years of critical change, they will have great need of that broad common sense referred to by James Bryce, which I hope will tide them over these critical times. I give American sensitivity a low figure because of my impression that they can stand so many things. There is no use quarreling about this, because we will be quarreling about words... The English seem to be on the whole the soundest race: contrast their "R3D2"with the French"R2D3."I am all for"R3D2."It bespeaks stability. The ideal formula for me would seem to be R3D2H3S2, for too much idealism or too much sensitivity is not a good thing, either. And if I give"S1"for English sensitivity, and if that is too low, who is to blame for it except the English themselves? How can I tell whether the English ever feel anything–joy, happiness, anger, satisfaction–when they are determined to look so glum on all occasions?

We might apply the same formula to writers and poets. To take a few well-known types:

Shakespeare = R4D4H3S4
Heine = R3D3H4S3
Shelley = R1D4H1S4
Poe = R3D4H1S4
LiPo = R1D3H2S4
Tu Fu = R3D3H2S4
Su Tungp'o = R3D2H4S3

These are no more than a few impromptu suggestions. But it is clear that all poets have a high sensitivity, or they wouldn't be poets at all. Poe, I feel, is a very sound genius, in spite of his weird imaginative gift. Doesn't he love"ratiocination"?

So my formula for the Chinese national mind is:

R4D1H3S3

There we start with an"S3", standing for high sensitivity, which guarantees a proper artistic approach to life and answers for the Chinese affirmation that this earthly life is beautiful and the consequent intense love of this life. But it signifies more than that; actually it stands for the artistic approach even to philosophy. It accounts for the fact that the Chinese philosopher's view of life is essentially the poet's view of life, and that, in China, philosophy is married to poetry rather than to science as it is in the West. It will become amply clear from what follows that this high sensitivity to the pleasures and pains and flux and change of the colors of life is the very basis that makes a light philosophy possible. Man's sense of the tragedy of life comes from his sensitive perception of the tragedy of a departing spring, and a delicate tenderness toward life comes from a tenderness toward the withered blossoms that bloomed yesterday. First the sadness and sense of defeat, then the awakening and the laughter of the old rogue-philosopher.

On the other hand, we have "R4" standing for intense realism, which means an attitude of accepting life as it is and of regarding a bird in the hand as better than two in the bush. This realism, therefore, both reinforces and supplements the artist's affirmation that this life is transiently beautiful, and it all but saves the artist and poet from escaping from life altogether. The Dreamer says "Life is but a dream," and the Realist replies, "Quite correct. And let us live this dream as beautifully as we can." But the realism of one awakened is the poet's realism and not that of the business man, and the laughter of the old rogue is no longer the laughter of the young go-getter singing his way to success with his head up and his chin out, but that of an old man running his finger through his flowing beard, and speaking in a soothingly low voice. Such a dreamer loves peace, for no one can fight hard for a dream. He will be more intent to live reasonably and well with his fellow dreamers. Thus is the high tension of life lowered.

But the chief function of this sense of realism is the elimination of all non-essentials in the philosophy of life, holding life down by the neck, as it were, for fear that the wings of imagination may carry it away to an imaginary and possibly beautiful, but unreal, world. And after all, the wisdom of life consists in the elimination of non-essentials, in reducing the problems of philosophy to just a few—the enjoyment of the home (the relationship between man and woman and child), of living, of Nature and of culture—and in showing all the other irrelevant scientific disciplines and futile chases after knowledge to the door. The problems of life for the Chinese philosopher then become amazingly few and simple. It means also an impatience with metaphysics and with the pursuit of knowledge that does not lead to any practical bearing on life itself. And it also means that every human activity, whether the acquiring of knowledge or the acquiring of things, has to be submitted immediately to the test of life itself and of its subserviency to the end of living. Again, and here is a significant result, the end of living is not some metaphysical entity—but just living itself.

Gifted with this realism, and with a profound distrust of logic and of the intellect itself, philosophy for the Chinese becomes a matter of direct and intimate feeling of life itself, and refuses to be encased in any system. For there is a robust sense of reality, a sheer animal sense, a spirit of reasonableness which crushes reason itself and makes the rise of any hard and fast philosophic system impossible. There are the three religions of China, Confucianism, Taoism and Buddhism, all magnificent systems in themselves, and yet robust common sense dilutes them all and reduces them all into the common problem of the pursuit of a happy human life. The mature Chinese is always a person who refuses to think too hard or to believe in any single idea or faith or school of philosophy whole-heartedly. When a friend of Confucius told him that he always thought three times before he acted, Confucius wittily replied,"To think twice is quite enough."A follower of a school of philosophy is but a student of philosophy, but a man is a student, or perhaps a master, of life.

The final product of this culture and philosophy is this: in China, as compared with the West, man lives a life closer to nature and closer to childhood, a life in which the instincts and the emotions are given free play and emphasized against the life of the intellect, with a strange combination of devotion to the flesh and arrogance of the spirit, of profound wisdom and foolish gaiety, of high sophistication and childish na?veté. I would say, therefore, that this philosophy is characterized by: first, a gift for seeing life whole in art; secondly, a conscious return to simplicity in philosophy; and thirdly, an ideal of reasonableness in living. The end product is, strange to say, a worship of the poet, the peasant and the vagabond.

三．以放浪者为理想的人

在我这个有着东方精神也有着西方精神的人看来，人类的尊严是由以下几个事实所造成；也就是人类和动物的区别。第一，他们对于追求智识，有着一种近乎戏弄的好奇心和天赋的才能；第二，他们有一种梦想和崇高的理想主义（常常是模糊的、混杂的，或自满的，但亦有价值）；第三，也是最重要的一点，他们能够利用幽默感去纠正他们的梦想，以一种比较健全的现实主义去抑制他们的理想主义；第四，他们不像动物般对于环境始终如一地机械地反应着，而是有决定自己反应的能力和随意改变环境的自由。这一点就是说人类的性格生来是世界上最不容易服从机械律的；人类的心思永远是捉摸不定，无法测度，而常常想着，怎样去逃避那些发狂的心理学家和未有夫妇同居经验的经济学家所要强置在他身上的机械律或是什么唯物辩证法。所以人类是一种好奇的、梦想的、幽默的、任性的动物。

总之，我对人类尊严的信仰，实是在于我相信人类是世上最伟大的放浪者。人类的尊严应和放浪者的理想发生联系，而绝对不应和一个服从纪律、受统驭的兵士的理想发生联系。这样讲起来，放浪者也许是人类中最显赫最伟大的典型，正如兵士也许是人类中最卑劣的典型一样。读者对于我以前的一部著作《吾国与吾民》（My Country and My People）的一般印象是我好似在赞颂"老猾"。现在我希望读者对这一部著作的一般印象是：我正在竭力称颂放浪汉或是流浪汉，我希望在这一点上我能成功，因为世间的事物，有时看来不能像它们外表那么简单。在这个民主主义和个人自由受着威胁的今日，也许只有放浪者和放浪的精神会解放我们，使我们不至于都变成有纪律的、服从的、受统驭的、一式一样的大队中的一个标明号数的兵士，因而无声无息地湮没。放浪者将成为独裁制度的最后的最厉害的敌人。他将成为人类尊严和个人自由的卫士，也将是最后一个被征服者。现代一切文化都靠他去维持。

造物主也许会晓得当他在地球上创造人类时，他是创造了一个放浪者，虽是一个聪明的，然而总还是放浪者。人类放浪的质素，终究是他的最有希望的质素。这个已造成的放浪者，无疑是聪慧的。但他仍是一个很难于约束，很难于处置的青年，他自以为比事实上的他更伟大，更聪慧，依然喜欢胡闹，喜欢顽皮，喜欢一切自由。虽然如此，但亦有许多美点，所以造物主也许还愿意把希望寄托在他身上，正如一个父亲把他的希望寄托在一个聪慧而又有点顽皮的二十岁儿子的身上一般。我常想，他可也有一天情愿退隐，而把这个宇宙交给他的儿子去管理吗？

以中国人的立场来说，我认为文化须先由巧辩矫饰进步到天真淳朴，有意识地进步到简朴的思想和生活里去，才可称为完全的文化；我以为人类必须从智识的智慧，进步到无智的智慧，须变成一个欢乐的哲学家；也必须先感到人生的悲哀，然后感到人生的快乐，这样才可以称为有智慧的人类。因为我们必须先有哭，才有欢笑，有悲哀而后有醒觉，有醒觉而后有哲学的欢笑，再加上和善与宽容。

我以为这个世界太严肃了，因为太严肃，所以必须有一种智慧和欢乐的哲学以为调剂。如果世间有东西可以用尼采所谓愉快哲学（Gay Science）这个名称的话，那么中国人生活艺术的哲学确实名副其实了。只有有快乐的哲学，才有真正深湛的哲学；西方那些严肃的哲学理论，我想还不曾开始了解人生的真意义哩。在我看来，哲学的唯一效用是叫我们对人生抱一种比一般商人的较轻松较快乐的态度。一个五十岁的商人，本来可以退隐，在我看来不是哲学家。这不是一个偶然发生的念头，而是我一个根深蒂固的观念。只有当人类渲染了这种轻快的精神时，世界才会变得更和平、更合理，而可以使人类居住生活。现代的人们对人生过于严肃而充满着烦扰和纠纷。我们应该费一些工夫，把那些态度，根本地研究一下，方能使人生有享受快乐的可能，并使人们的气质有变为比较合理、比较和平、比较不暴躁的可能。

我也许可以把这种哲学称为中国民族的哲学，而不把它叫做任何一个派别的哲学。这个哲学比孔子和老子的更伟大，因为它是超越这两个哲学家以及他们的哲学的；它由这些思想的泉源里吸收资料，把它们融洽调和成一个整体；它从它们智慧的抽象轮廓里，造出一种实际的生活艺术，使普通一般人都可看得见，触得到，并且能够了解。拿全部的中国文学和哲学观察过后，我深深地觉得那种对人生能够尽量享受和聪慧的醒悟哲学，便是他们的共同福音和教训——就是中国民族思想上最恒久的，最具特性的，最永存的叠句唱词。

III. The Scamp as Ideal

TO me, spiritually a child of the East and the West, man's dignity consists in the following facts which distinguish man from animals. First, that he has a playful curiosity and a natural genius for exploring knowledge; second, that he has dreams and a lofty idealism (often vague, or confused, or cocky, it is true, but nevertheless worthwhile) ; third, and still more important, that he is able to correct his dreams by a sense of humor, and thus restrain his idealism by a more robust and healthy realism; and finally, that he does not react to surroundings mechanically and uniformly as animals do, but possesses the ability and the freedom to determine his own reactions and to change surroundings at his will. This last is the same as saying that human personality is the last thing to be reduced to mechanical laws; somehow the human mind is forever elusive, uncatchable and unpredictable, and manages to wriggle out of mechanistic laws or a materialistic dialectic that crazy psychologists and unmarried economists are trying to impose upon him. Man, therefore, is a curious, dreamy, humorous and wayward creature.

In short, my faith in human dignity consists in the belief that man is the greatest scamp on earth. Human dignity must be associated with the idea of a scamp and not with that of an obedient, disciplined and regimented soldier. The scamp is probably the most glorious type of human being, as the soldier is the lowest type, according to this conception. It seems in my last book, My Country and My People, the net impression of readers was that I was trying to glorify the"old rogue."It is my hope that the net impression of the present one will be that I am doing my best to glorify the scamp or vagabond. I hope I shall succeed. For things are not so simple as they sometimes seem. In this present age of threats to democracy and individual liberty, probably only the scamp and the spirit of the scamp alone will save us from becoming lost as serially numbered units in the masses of disciplined, obedient, regimented and uniformed coolies. The scamp will be the last and most formidable enemy of dictatorships. He will be the champion of human dignity and individual freedom, and will be the last to be conquered. All modern civilization depends entirely upon him.

Probably the Creator knew well that, when He created man upon this earth, He was producing a scamp, a brilliant scamp, it is true, but a scamp nonetheless. The scamp-like qualities of man are, after all, his most hopeful qualities. This scamp that the Creator has produced is undoubtedly a brilliant chap. He is still a very unruly and awkward adolescent, thinking himself greater and wiser than he really is, still full of mischief and naughtiness and love of a free-for-all. Nevertheless, there is so much good in him that the Creator might still be willing to pin on him His hopes, as a father sometimes pins his hopes on a brilliant but somewhat erratic son of twenty. Would He be willing some day to retire and turn over the management of this universe to this erratic son of His? I wonder...

Speaking as a Chinese, I do not think that any civilization can be called complete until it has progressed from sophistication to unsophistication, and made a conscious return to simplicity of thinking and living, and I call no man wise until he has made the progress from the wisdom of knowledge to the wisdom of foolishness, and become a laughing philosopher, feeling first life's tragedy and then life's comedy. For we must weep before we can laugh. Out of sadness comes the awakening and out of the awakening comes the laughter of the philosopher, with kindliness and tolerance to boot.

The world, I believe, is far too serious, and being far too serious, it has need of a wise and merry philosophy. The philosophy of the Chinese art of living can certainly be called the "gay science," if anything can be called by that phrase used by Nietzsche. After all, only a gay philosophy is profound philosophy; the serious philosophies of the West haven't even begun to understand what life is. To me personally, the only function of philosophy is to teach us to take life more lightly and gayly than the average business man does, for no business man who does not retire at fifty, if he can, is in my eyes a philosopher. This is not merely a casual thought, but is a fundamental point of view with me. The world can be made a more peaceful and more reasonable place to live in only when men have imbued themselves in the light gayety of this spirit. The modern man takes life far too seriously, and because he is too serious, the world is full of troubles. We ought, therefore, to take time to examine the origin of that attitude which will make possible a whole-hearted enjoyment of this life and a more reasonable, more peaceful and less hot-headed temperament.

I am perhaps entitled to call this the philosophy of the Chinese people rather than of any one school. It is a philosophy that is greater than Confucius and greater than Laotse, for it transcends these and other ancient philosophers; it draws from these fountain springs of thought and harmonizes them into a whole, and from the abstract outlines of their wisdom, it has created an art of living in the flesh, visible, palpable and understandable by the common man. Surveying Chinese literature, art and philosophy as a whole, it has become quite clear to me that the philosophy of a wise disenchantment and a hearty enjoyment of life is their common message and teaching-the most constant, most characteristic and most persistent refrain of Chinese thought.

第二章

关于人类的观念

Chapter Two　**Views of Mankind**

一. 基督徒、希腊人、中国人

关于人类的观念，世上有好几种：即传统的基督教观念，希腊的异教徒观念和中国人的道教和孔教的观念（因为佛教的观念太悲观了，所以我不把它包括进去）。这些观念，由它们深湛的讽喻意义上说来，并没有什么分别，尤其是在具有高深的生物学和人类智识的现代人，给予它们一种广义的解释后，更不能分其轩轾，可是在它们原来的形式上分别仍是存在的。

依传统的正统基督教观念，人类是完善的、天真的、愚笨的、快乐的，赤裸着身体在伊甸园里生活。后来人类有了智识和智慧，于是堕落了，这就是痛苦的起因。所谓痛苦，主要是由于（一）男人方面的流汗工作；（二）女人方面的生男育女的疼痛。为要显示人类的缺点起见，基督教又引进一种人类的新成分，和原来的天真完美相对照。这种新成分就是魔鬼，它大概是由肉体方面去活动；而人类较高尚的天性由灵魂方面去活动，我不知道"灵魂"在基督教神学里是什么时候发明的，但是这"灵魂"变成了一种实物，而不是一种机能，变成了一种实质，而不是一种状态；它把灵魂不值拯救的禽兽和人类明确地划分了。在这里，逻辑便发生了问题，因为"魔鬼"的来源必须解释，然而当中世纪的神学家，用他们平常的学者逻辑去研讨这个问题时，他们便陷入了进退维谷的境地。他们不能承认"非上帝"的"魔鬼"和上帝并存永生。所以在无可奈何中，他们只得说"魔鬼"一定是一个堕落的天使，但是这又引起了罪恶的来源问题（因为另外总得有一个"魔鬼"来引诱这个天使去堕落啊）。因此，这种理论便不能使人满意，他们也只好随它去了。虽然如此，这理论却产生了神灵和肉体相对的奇怪观念；这个玄妙的观念至今存在，对于我们的人生观和幸福还有着很大的影响[5]。

接踵而至的，便是"赎罪"的理论，这理论依然是由牺牲的观念假借而来；从这个理论推想起来，上帝好像是一个喜欢人间烟火味的神，不愿意无代价赦免人类的罪恶。基督教有了这种理论，人类一下子就可以寻到一个可以赦免一切罪恶的方法，因此人类又找到了获得完美的方法。基督教思想中最奇突的一点就是完美观念。因为基督教是从上古世界的崩溃中所产生，所以有一种着重来世的倾向，拯救问题替代了人生幸福问题，或者替代了简朴生活本身问题。这观念的含义就是人类要怎样才能脱离这个腐败、混乱和灭亡中的世界，而到另一个世界里去。因此，就有了永生的观念。这和《创世记》里上帝不要人类永生的原始说法是矛盾的。根据《创世记》的记载，亚当和夏娃之所以被逐出伊甸园，并不是像一般人所相信的那样为了偷尝善恶树的果子，而是上帝怕他们再度违背命令，去偷吃生命树的果子，因而得到永生：

---

[5] 在现代思想进步的过程中，"魔鬼"是一个被弃掉的东西，这是值得欣幸的事实。我相信今日在一百个不相信有上帝的进步基督徒之中，相信真魔鬼的（除了比喻的意义之外）恐怕不上五人，同时，相信真地狱的也和相信真天堂的同归消灭。

耶和华上帝说：那人已经和我们相似，能知道善恶，现在恐怕他又伸
手去摘生命树的果子吃，就永远活着。

耶和华上帝便打发他出伊甸园，去耕种他身所自出之土。

于是把他赶了出去，又在伊甸园的东边安设四面转动能发火焰的剑，
去把守生命树的道路。

善恶树似乎是在乐园的正中央，生命树却是在靠近东门的地方，据我
们知道在那边智天使还驻守着，以防人类侵犯。

总之，现在还存有一种以为人类是完全堕落的信念，今生的享乐就是
罪恶，以为刻苦就是美德，以为人类除了被一种外来的伟大力量拯救外，
不能自救。罪恶仍是今日通行的基督教教义的根本理论。教士在讲道的时
候，第一步是使人体会到罪恶的存在，以及人类本性的不良（是传教士应
用藏在袖子里的现成药方时的必要条件）。总之，如果你不先使一个人相
信他是罪人，你便不能劝诱他做基督教徒。有人曾说过一句颇为刻薄的话：
"我国的宗教已经成为一种罪恶的反省，体面的人士不敢再走进教堂了。"

希腊的异教世界是一个绝对不同的世界，所以他们对于人类的观念亦
异。最值得注意的就是：希腊人要他们的神成为凡人一般，基督教徒则反
之，要使凡人跟神一样。在奥林匹克那些确是些快乐的、好色的、谈恋
爱、会说谎、好吵架，也会背誓的急性易怒的家伙；正像希腊人那样喜
欢打猎、驾马车、掷标枪——他们也很喜欢结婚，而且生了许多的私生
子。讲到神和人的区别，神不过具有在天上会打雷在地上会培育植物的
能力而已，他们都是永生的，喝花蜜酿的仙酒而不喝酒——不过用来酿
成的果实是差不多的。我们觉得可以和这班人亲近，我们可以背了一个

行李，和阿波罗（Apollo，司日轮，音乐、诗、医疗、预言等之神）或雅典娜（Athene，司智慧，学术技艺，战争之神）一同去行猎。或在路上拉住墨丘利（Mercury，商人，旅客，盗贼，及狡猾者之保护神）和他闲谈，正如和美国西方联合电报局的送差闲谈一样。如果谈得很有趣，我们可以想象出墨丘利说："不错，好的，对不起，我要走了，要把这封电报送到七十二号街去。"希腊人并不神圣，希腊的神却具有人性。这些神跟基督教完美的上帝相较起来是多么不同啊！希腊的神不过是另一个种族的人，是一族能够永生的巨人，地球上的人却不能够。由于这个背景，便产生一切关于得墨忒耳（Demeter，司农业的女神）、珀耳塞福涅（Proserpina，地狱的女王）和俄耳甫斯（Orpheus，音乐的鼻祖）等的绝美故事。希腊人对神的信仰是理所当然的，甚至苏格拉底将饮毒酒的时候，也举杯向神祷告，求神使他快一点到另一个世界里去。这点很像孔子的态度。在那个时候，人们的态度必须是这样的；至于希腊精神如果在现代，其对于人类和上帝将取什么态度，我们不幸没有知道的机会。希腊的异教世界不是现代的，而现代的基督教世界也不是希腊的，这是很可惜的。

大体说来，希腊人承认人类是总有一死的，有时还要受残酷命运所支配。人类一接受了这种命运后，便感到十分愉快。因为希腊人酷爱这个人生和这个宇宙，他们除了专心致志，科学地去理解物质世界外，也应注意于理解人生的真美善。希腊人的思想里没有类似伊甸园式的"黄金时代"，也没有人类堕落的讽喻；希腊人自己不过是丢卡利翁（Deucalion）和他的妻皮拉（Pyrrha）在洪水后，走下平原时，从地上拾起来向后抛去的石子所变成的人类罢了。他们对疾病和忧虑是用滑稽的方法去解释；他们以为疾病和忧虑似一个少妇有一种难于压制的欲望，想打开一箱珍宝——潘多拉的箱子。希腊人的想象是美丽的。他们大都把人性就当人性看待，但是

基督徒或许会说他们是被"总有一死"的命运所支配。但总有一死的命运
是美丽的，人类在这里可以理解，人生可以让自由推究的精神去发展。有
些诡辩家认为人性本善，有些认为人性本恶，可是他们的理论总没有像托
马斯·霍布斯（Thomas Hobbes，十五世纪英国的哲学家，英国理性主义
传统的奠基人）和卢梭（十六世纪法国的哲学家）的互相矛盾。最后，柏
拉图认为人类似乎是欲望、情感和思想的混合物。理想的人生便是在理
论、智慧、真正理解的指导下，三方面和谐地在一起生活。柏拉图认为
"思想"是不朽的，不过个人的灵魂之或贱或贵，根据他们是否爱好正义、
学问、节制和美而定。在苏格拉底的心目中，灵魂也是一种独立和不朽的
存在；他在《斐德若》（Phaedo）里告诉我们："当灵魂独自存在时，由肉
体解放出来，而肉体也由灵魂解放出来的时候，那时除死亡之外还有什么
呢？"相信人类灵魂的不朽，显然是基督教徒、希腊人、道教和儒教的观
念上相同的地方。相信灵魂不朽的现代人，当然不能抓住这一点当做话
题。苏格拉底对灵魂不朽的信仰，在现代人看来，也许毫无意义，因为他
的许多理论根据，如化身转世之类，是现代人所不能接受的。

至于中国人对于人类的观念，人类是造物之主，"万物之灵"。在儒家
看来，人和天地并列成为"三灵"。如果以灵魂说为背景，讲起来世间万
物都有生命，或都有神灵依附，风和雷是神灵的本身，每一大山和河流都
有神灵统治，可说即是属于这个神灵的；每一种花都有花神，在天上管理
季节，看顾它们盛开凋谢。还有一个百花仙子，她的生辰是在二月十二
日。每棵柳树、松树、柏树，每一只狐狸或乌龟活了很长的岁月，达到了
很高的年龄，就变成精。

在这种灵魂说背景之下，人类自然也被视为神灵的具体表现。这神灵
和宇宙间的一切生物一样，是由雄性的、主动的、正的，或阳的成分，和

雌性的、被动的、负的，或阴的成分，结合而产生出来的——在事实上不过是对阴阳电原理的一种玄妙的猜测罢了。附在人身上的这种灵性叫做"魄"；离开人身随处飘荡时叫做"魂"（一个人有坚强的个性或是精神充沛时，便称之为有"魄力"）。人死后，"魂"依旧四处飘荡。魂是不常扰人的，但如果没有人埋葬或祭祀死者，神灵便会变成"无祀孤魂"来缠扰人，因此，中国人定七月十五日为"祭亡日"，以祭祀那些溺死的和客死异乡的鬼。更甚的，假使死者是被杀的或冤枉死的，那鬼魂便到处飘荡骚扰，直到雪冤，方才停止。

人既是神灵的具体表现，所以在世的时候，当然须有一些热情欲望和精神（Vital Energy or Nervous Energy），这些东西无所谓好坏，不过是一些和人类生活不能分离的天赋的性质而已。一切男女都有热烈的感情，自然的欲望，高尚的意志，以及良知；也有性欲、饥饿、愤怒，受着疾病、疼痛、苦恼和死亡的支配。文化的用处，便在于怎样使这些热情和欲望能够和谐地表现。这就是儒家的观念，依这种观念，假使我们能够和这种天赋的本性过着和谐的生活，便可以和天地相列；关于这一点，我将在第六章末再讲。然而佛教对于人类的肉体情欲的观念和中世纪基督教很相同——以为这些情欲是必须割弃的讨厌东西。太聪慧或思想过度的男女有时会接受这个观念，因而去做和尚或尼姑；但在大体上说来，儒家的健全意识并不赞成这种行为。同样，佛教的观念有点近于道教的意味，认为薄命红颜是"被谪下凡的神女"，因为她们动了凡心，或是在天上失了职，所以被贬入尘世来受这命运注定的人间痛苦。

人类的智能被认为是一种潜力。这种智能即我们所谓"精神"，这"精"字的意义和狐狸精的"精"字相同。我在前面已经说过，英语中和"精神"意义最相近的是Vitality或Nervous Energy，这种东西在人生中每天有许多不同的时候，正像潮水那样涨落不定。一个人生下来就有热情、欲望和这种精神，这些在幼年、壮年、老年和死亡各时期中循着不同的路线而流行。孔子说："少之时，血气未定，戒之在色；及其壮也，血气方刚，戒之在斗；及其老也，血气既衰，戒之在得。"反过来讲，就是少年爱色，壮年好斗，老年嗜财。

　　当着这个身体的、智能的，和道德的资产混合物，中国人对于人类本身所抱的一般态度，可以归纳到"让我们做合理近情的人"这句话里。就是一种中庸之道，不希望太多，也不太少。好像人类是介乎天地之间，介乎理想主义和现实主义之间，介乎崇高的思想和卑鄙的情欲之间。这样的介乎中间，便是人类天性的本质；渴求智识和渴求清水，喜欢一个好的思想和喜爱一盆美味的笋炒肉，吟哦一句美丽的诗词和向慕一个美丽的女人，这些都是人的常情。因之我们感到人间总是一个不完美的世界。要把这社会加以改良，机会当然是有的，但是中国人并不想得到完全的和平，也不想达到快乐的顶点。这里有个故事可做证明。有一个人从幽冥降生到人间去，他对阎王说："如果你要我回到人间，你须答应我的条件。""什么条件呢？"阎王问。那人回答："我要做宰相的儿子，状元的父亲，我的住宅四周要有一万亩地，有鱼池，有各种花果，我要有一位美丽的太太，和一些娇艳的婢妾，她们都须待我很好，我要满屋珠宝，满仓五谷，满箱金银，而我自己要做公卿，一生荣华富贵，活到一百岁。"阎王说："如果人间有这样的人可做，我自己也要去投生，不让你去了！"

　　然而合理近情的态度，就是说：我们既有了这种人类的天性，就让我们开始做人吧。何况要逃避这个命运，根本是办不到的。不管热情和本能本来是好是坏，空口争论是没有什么用处的。或者我们反而倒有被束缚的危险。这种近情合理的态度造成了一种宽恕的哲学，觉得人类的错误和谬行都是可以获得宽恕的，不论是法律上的、道德上的或政治上的，都可以认为是"一般的人类天性"或"人之常情"。至少，那批有教养的、心胸旷达的、遵循合理近情的精神而生活的学者，都抱着这种态度。中国人甚至以为天或上帝也是一个颇为合理近情的人物，他们以为你只要过着合理近情的生活，依着你的良知行事，你就不必再有所怕惧，他们认为良心的安宁是最大的福气，认为一个心地光明磊落的人，连鬼怪也不能侵犯他。所以，只要有一个合理近情的上帝来担任管理那些不合理不近情者的任务，世界便太平无事，诸事顺利了。专制者死了；卖国者自杀了；唯利是图者变卖他的财产了；有权有势，拥有古董的收藏家（他们是利欲熏心，靠权势来剥削人家的）的儿子们，把他们父亲用尽心机搜罗得来的珍宝一齐变卖，四散地藏在别人的家庭里了；杀人凶犯伏法了，遭辱的女人得到报复的机会了，难得有个被压迫者会喊着说："老天爷瞎了眼睛！"（正义不伸）。在道家和儒家两方面，最后都以为哲学的结论和它的最高理想，即必须对自然完全理解，以及必须和自然和谐；如果要用一个名词以便分类，我们可以把这种哲学称为"合理的自然主义"（Reasonable Naturalism），一个合理的自然主义者便带着兽性的满足在这世界上生活下去。目不识丁的中国妇人说："人家生我们，我们生人家，另外还有什么事可做呢？"

"人家生我们，我们生人家"，这一句话蕴藏着一种可怕的哲学。由于这种说法，人生将变成一种生物学的程序，而永生的问题便绝口不必谈了。这正和一个搀着孙儿到糖果店里去，一面在想着五年或十年后便要回到坟墓里去的中国祖父一样，他们在这世间最大的希望就是不至于生下羞辱门第的子孙来。中国人的整个人生范式就是这样一个观念组合起来的。

I. Christian, Greek and Chinese

THERE are several views of mankind, the traditional Christian theological
view, the Greek pagan view, and the Chinese Taoist-Confucianist view. (I do
not include the Buddhist view because it is too sad.) Deeper down in their
allegorical sense, these views after all do not differ so much from one another,
especially when the modern man with better biological and anthropological
knowledge gives them a broader interpretation. But these differences in their
original forms exist.

The traditional, orthodox Christian view was that man was created
perfect, innocent, foolish and happy, living naked in the Garden of Eden.
Then came knowledge and wisdom and the Fall of Man, to which the
sufferings of man are due, notably (1) work by the sweat of one's brow for
man, and (2) the pangs of labor for women. In contrast with man's original
innocence and perfection, a new element was introduced to explain his
present imperfection, and that is of course the Devil, working chiefly
through the body, while his higher nature works through the soul. When
the"soul"was invented in the history of Christian theology I am not aware,
but this"soul" became a something rather than a function, an entity rather
than a condition, and it sharply separated man from the animals, which have
no souls worth saving. Here the logic halts, for the origin of the Devil had to

be explained, and when the medieval theologians proceeded with their usual scholastic logic to deal with the problem, they got into a quandary. They could not have very well admitted that the Devil, who was Not-God, came from God himself, nor could they quite agree that in the original universe, the Devil, a Not-God, was co-eternal with God. So in desperation they agreed that the Devil must have been a fallen angel, which rather begs the question of the origin of evil (for there still must have been another Devil to tempt this fallen angel) , and which is therefore unsatisfactory, but they had to leave it at that. Nevertheless from all this followed the curious dichotomy of the spirit and the flesh, a mythical conception which is still quite prevalent and powerful today in affecting our philosophy of life and happiness.

Then came the Redemption, still borrowing from the current conception of the sacrificial lamb, which went still farther back to the idea of a God Who desired the smell of roast meat and could not forgive for nothing. From this Redemption, at one stroke a means was found by which all sins could be forgiven, and a way was found for perfection again. The most curious aspect of Christian thought is the idea of perfection. As this happened during the decay of the ancient worlds, a tendency grew up to emphasize the afterlife, and the question of salvation supplanted the question of happiness or simple living itself. The notion was how to get away from this world alive, a world which was apparently sinking into corruption and chaos and doomed. Hence the overwhelming importance attached to immortality. This represents a contradiction of the original Genesis story that God did not want man to live forever. The Genesis story of the reason why Adam and Eve were driven out of the Garden of Eden was not that they had tasted of the Tree of Knowledge, as is popularly conceived, but the fear lest they should disobey a second time and eat of the Tree of Life and live forever:

And the Lord God said, Behold, the man is become as one of us, to know good and evil: and now, lest he put forth his hand, and take also of the tree of life, and eat, and live for ever:

Therefore the Lord God sent him forth from the Garden of Eden, to till the ground from whence he was taken.

So he drove out the man; and he placed at the east of the Garden of Eden cherubims, and a flaming sword which turned every way, to keep the way of the tree of life.

The Tree of Knowledge seemed to be somewhere in the center of the garden, but the Tree of Life was near the eastern entrance, where for all we know, cherubims are still stationed to guard the approach by men.

All in all, there is still a belief in total depravity, that enjoyment of this life is sin and wickedness, that to be uncomfortable is to be virtuous, and that on the whole man cannot save himself except by a greater power outside. The doctrine of sin is still the basic assumption of Christianity as generally practiced today, and Christian missionaries trying to make converts generally start out by impressing upon the party to be converted a consciousness of sin and of the wickedness of human nature (which is, of course, the sine qua non for the need of the ready-made remedy which the missionary has up his sleeve) . All in all, you can't make a man a Christian unless you first make him believe he is a sinner. Some one has said rather cruelly,"Religion in our country has so narrowed down to the contemplation of sin that a respectable man does not any longer dare to show his face in the church."

The Greek pagan world was a different world by itself and therefore their conception of man was also quite different. What strikes me most is that the Greeks made their gods like men, while the Christians desired to make men like the gods. That Olympian company is certainly a jovial, amorous, loving, lying, quarreling and vow-breaking, petulant lot; hunt-loving, chariot-riding and javelin-throwing like the Greeks themselves—a marrying lot, too, and having unbelievably many illegitimate children. So far as the difference between gods and men is concerned,   the gods merely had divine powers of hurling thunderbolts in heaven and raising vegetation on earth, were immortal, and drank nectar instead of wine—the fruits were pretty much the same. One feels one can be intimate with this crowd, can go hunting with a knapsack on one's back with Apollo or Athene, or stop Mercury on the way and chat with him as with a Western Union messenger boy, and if the conversation gets too interesting, we can imagine Mercury saying,"Yeah. Okay. Sorry, but I'll have to run along and deliver this message at 72nd Street."The Greek men were not divine, but the Greek gods were human. How different from the perfect Christian God! And so the gods were merely another race of men, a race of giants, gifted with immortality, while men on earth were not. Out of this background came some of the most inexpressibly beautiful stories of Demeter and Proserpina and Orpheus. The belief in the gods was taken for granted, for even Socrates, when he was about to drink hemlock, proposed a libation to the gods to speed him on his journey from this world to the next. This was very much like the attitude of Confucius. It was necessarily so in that period; what attitude toward man and God the Greek spirit would take in the modern world there is unfortunately no chance of knowing. The Greek pagan world was not modern, and the modern Christian world is not Greek. That's the pity of it.

On the whole, it was accepted by the Greeks that man's was a mortal lot, subject sometimes to a cruel Fate. That once accepted, man was quite happy as he was, for the Greeks loved this life and this universe, and were interested in understanding the good, the true and the beautiful in life, besides being fully occupied in scientifically understanding the physical world. There was no mythical"Golden Period"in the sense of the Garden of Eden, and no allegory of the Fall of Man; the Hellenes themselves were but human creatures transformed from pebbles picked up and thrown over their shoulders by Deucalion and his wife Pyrrha, as they were coming down to the plain after the Great Flood. Diseases and cares were explained comically; they came through the uncontrollable desire of a young woman to open and see a box of jewels–Pandora's Box. The Greek fancy was beautiful. They took human nature largely as it was: the Christians might say they were"resigned"to the mortal lot. But it was so beautiful to be mortal: there was free room for the exercise of understanding and the free, speculative spirit. Some of the Sophists thought man's nature good, and some thought man's nature bad, but there wasn't the sharp contradiction of Hobbes and Rousseau. Finally, in Plato, man was seen to be a compound of desires, emotions, and thought, and ideal human life was the living together in harmony of these three parts of his being under the guidance of wisdom or true understanding. Plato thought"ideas"were immortal, but individual souls were either base or noble, according as they loved justice, learning, temperance and beauty or not. The soul also acquired an independent and immortal existence in Socrates; as we are told in"Phaedo,""When the soul exists in herself, and is released from the body, and the body is released from the soul, what is this but death?"Evidently the belief in immortality of the human soul is something which the Christian, Greek, Taoist and Confucianist views have in common. Of course this is nothing to be jumped at by modern believers in the immortality of the soul. Socrates' belief in immortality would probably mean nothing to a modern man, because many of his premises in support of it, like re-incarnation, cannot be accepted by the modern man.

The Chinese view of man also arrived at the idea that man is the Lord of the Creation ("Spirit of the Ten Thousand Things"), and in the Confucianist view, man ranks as the equal of heaven and earth in the"Trio of Geniuses". The background was animistic: everything was alive or inhabited by a spirit– mountains, rivers, and everything that reached a grand old age. The winds and thunder were spirits themselves; each of the great mountains and each river was ruled by a spirit who practically owned it; each kind of flower had a fairy in heaven attending to its seasons and its welfare, and there was a Queen of All Flowers whose birthday came on the twelfth day of the second moon; every willow tree, pine tree, cypress, fox or turtle that reached a grand old age, say over a few hundred years, acquired by that very fact immortality and became a"genius".

With this animistic background, it is natural that man is also considered a manifestation of the spirit. This spirit, like all life in the entire universe, is produced by the union of the male, active, positive or yang principle, and the female, passive, negative or yin principle–which is really no more than a lucky, shrewd guess at positive and negative electricity. When this spirit becomes incarnated in a human body, it is called p'o; when unattached to a body and floating about as spirit it is called hwen. (A man of forceful personality or"spirits"is spoken of as having a lot of p'oli, or p'o energy.) After death, this hwen continues to wander about. Normally it does not bother people, but if no one buries and offers sacrifices to the deceased, the spirit becomes a"wandering ghost,"for which reason an All Souls' Day is set apart on the fifteenth day of the seventh moon for a general sacrifice to those drowned in water or dead and unburied in a strange land. Also, if the deceased was murdered or died suffering a wrong, the sense of injustice in the ghost compels it to hang about and cause trouble until the wrong is avenged and the spirit is satisfied. Then all trouble is stopped.

While living, man, who is spirit taking shape in a body, necessarily has certain passions, desires, and a flow of"vital energy,"or in more easily understood English, just"nervous energy."In and for themselves, these are neither good nor bad, but just something given and inseparable from the characteristically human life. All men and women have passions, natural desires and noble ambitions, and also a conscience; they have sex, hunger, fear, anger, and are subject to sickness, pain, suffering and death. Culture consists in bringing about the expression of these passions and desires in harmony. That is the Confucianist view, which believes that by living in harmony with this human nature given us, we can become the equals of heaven and earth, as quoted at the end of Chapter VI. The Buddhists, however, regard the mortal desires of the flesh essentially as the medieval Christians did–they are a nuisance to be done away with. Men and women who are too intelligent, or inclined to think too much, sometimes accept this view and become monks and nuns; but on the whole, Confucian good sense forbids it. Then also, with a Taoistic touch, beautiful and talented girls suffering a harsh fate are regarded as"fallen fairies,"punished for having mortal thoughts or some neglect of duty in heaven and sent down to this earth to live through a predestined fate of mortal sufferings.

Man's intellect is considered as a flow of energy. Literally this intellect is"spirit of a genius" (chingshen), the word"genius" being essentially taken in the sense in which we speak of fox genii, rock genii and pine genii. The nearest English equivalent is, as I have suggested,"vitality"or"nervous energy"which ebbs and flows at different times of the day and of the person's life. Every man born into this world starts out with certain passions and desires and this vital energy, which run their course in different cycles through childhood, youth, maturity, old age and death. Confucius said,"When young, beware of fighting; when strong, beware of sex; and when old, beware of possession,"which simply means that a boy loves fighting, a young man loves women, and an old man loves money.

Faced with this compound of physical, mental and moral assets, the Chinese takes an attitude toward man himself, as toward all other problems, which may be summed up in the phrase:" Let us be reasonable."This is an attitude of expecting neither too much nor too little. Man is, as it were, sandwiched between heaven and earth, between idealism and realism, between lofty thoughts and the baser passions. Being so sandwiched is the very essence of humanity; it is human to have thirst for knowledge and thirst for water, to love a good idea and a good dish of pork with bamboo shoots, and to admire a beautiful saying and a beautiful woman. This being the case, our world is necessarily an imperfect world. Of course there is a chance of taking human society in hand and making it better, but the Chinese do not expect either perfect peace or perfect happiness. There is a story illustrating this point of view. There was a man who was in Hell and about to be re-incarnated, and he said to the King of Re-incarnation, "If you want me to return to the earth as a human being, I will go only on my own conditions. ""And what are they?"asked the King. The man replied,"I must be born the son of a cabinet minister and father of a future'Literary Wrangler'(the scholar who comes out first at the national examinations). I must have ten thousand acres of land surrounding my home and fish ponds and fruits of every kind and a beautiful wife and pretty concubines, all good and loving to me, and rooms stocked to the ceiling with gold and pearls and cellars stocked full of grain and trunks chockful of money, and I myself must be a Grand Councilor or a Duke of the First Rank and enjoy honor and prosperity and live until I am a hundred years old."And the King of Re-incarnation replied,"If there was such a lot on earth, I would go and be re-incarnated myself, and not give it to you! "

The reasonable attitude is, since we've got this human nature, let's start with it. Besides, there is no escaping from it anyway. Passions and instincts are originally good or originally bad, but there is not much use talking about them, is there? On the other hand, there is the danger of our being enslaved by them. Just stay in the middle of the road. This reasonable attitude creates such a forgiving kind of philosophy that, at least to a cultured, broadminded scholar who lives according to the spirit of reasonableness, any human error or misbehavior whatsoever, legal or moral or political, which can be labeled as"common human nature" (more literally,"man's normal passions"), is excusable. The Chinese go so far as to assume that Heaven or God Himself is quite a reasonable being, that if you live reasonably, according to your best lights, you have nothing to fear, that peace of conscience is the greatest of all gifts, and that a man with a clear conscience need not be afraid even of ghosts. With a reasonable God supervising the affairs of reasonable and some unreasonable beings, everything is quite all right in this world. Tyrants die; traitors commit suicide; the grasping fellow is seen selling his property; the sons of a powerful and rich collector of curios (about whom tales are told of grasping greed or extortion by power) are seen selling out the collection on which their father spent so much thought and trouble, and these same curios are now being dispersed among other families; murderers are found out and dead and wronged women are avenged. Sometimes, but quite seldom, an oppressed person cries out, "Heaven has no eyes!"(Justice is blind.) Eventually, both in Taoism and in Confucianism, the conclusion and highest goal of this philosophy is complete understanding of and harmony with nature, resulting in what I may call "reasonable naturalism," if we must have a term for classification. A reasonable naturalist then settles down to this life with a sort of animal satisfaction. As Chinese illiterate women put it, "Others gave birth to us and we give birth to others. What else are we to do?"

There is a terrible philosophy in this saying, "Others gave birth to us and we give birth to others." Life becomes a biological procession and the very question of immortality is sidetracked. For that is the exact feeling of a Chinese grandfather holding his grandchild by the hand and going to the shops to buy some candy, with the thought that in five or ten years he will be returning to his grave or to his ancestors. The best that we can hope for in this life is that we shall not have sons and grandsons of whom we need be ashamed. The whole pattern of Chinese life is organized according to this one idea.

二．与尘世结不解缘

　　人类如要生活，依然须生活在这个世界上。什么生活在天上啊等问题，必须抛弃。人类的心神哟！别张起翅膀，飞到天神那边去而忘掉这个尘世呀！我们不都是注定着要遭遇死亡命运的凡人吗？上天赐给了我们七十年的寿命，如果我们的心志太高傲，想要永生不死，这七十年确是很短促的，但是如果我们的心地稍为平静一点，这七十年也尽够长了。一个人在七十年可以学到很多的东西，享受到很多的幸福。要看看人类的愚蠢，要获得人类的智慧，七十年已是够长的时期了。一个有智慧的人如充分长寿，在七十年的兴衰中，也尽够去视看习俗、道德和政治的变迁。他在那人生舞台闭幕时，也应该可以心满意足地由座位立起来，说一声"这是一出好戏"而走开吧。

　　我们是属于这尘世的，而且和这尘世是一日不可离的。我们在这美丽的尘世上好像是过路的旅客，这个事实我想大家都承认的，即使这尘世是一个黑暗的地牢，但我们总得尽力使生活美满。况且我们并不是住在地牢里，而是在这个美丽的尘世上，要过着七八十年的生活，假如我们不尽力使生活美满，那就是忘恩负义了。有时我们太富于野心，看不起这个卑低的但也是宽大的尘世，可是如要获得精神的和谐，我们对于这么一个孕育万物的天地，必须有一种感情，对于这个身心的寄托处所，必须有一种依恋之感。

所以，我们必须有一种动物性的信仰和一种动物性的怀疑，就把这尘世当做尘世看。梭罗（Thoreau，美国十九世纪作家和自然主义者）觉得自己和土壤是属于同类，具有同样的忍耐功夫，在冬天时，期望着春日的来到，在百无聊赖的时候，不免要想到寻求神灵不是他的分内事，而应由神灵去寻求他；依他的说法，他的快乐也不过和土拨鼠的快乐很相似，他这种整个的大自然性也是我们所应该保持的。尘世到底是真实的，天堂终究是缥缈的，人类生在这个真实的尘世和缥缈的天堂之间是多么幸运啊！

凡是一种良好的、实用的哲学理论，必须承认我们都有这么一个身体。现在已是我们应该坦白地承认"我们是动物"的适当时机。自从达尔文进化论的真理成立以后，自从生物学，尤其是生物化学，获得极大的进展之后，这种承认是必然的。不幸我们的教师和哲学家都是属于所谓智识阶级，都对于智能有着一种特殊的、专家式的自负，致力于精神的人以精神为荣，正如皮鞋匠以皮革为荣一样。有时他们连"精神"一词也还觉得不够缥缈抽象，更拿什么"精粹""灵魂"或"观念"一类的词，冠冕堂皇地写出来，想拿它们来恐吓我们。人的身体便在这种人类学术的机器中蒸馏成精神，而这种精神进一步凝聚起来，变成一种精粹的东西。但是要晓得即使是酒精也须有一个"实体"——和淡水混合起来——才能味美适口。然而我们这些可怜的俗人须饮这种精神所凝聚的精华。这种过分着重精神的态度实是有害的。它使我们和自然的本能搏斗，它使我们对于天性无从造成一种整体完备的观念，这是我批评它的一个主要点。同时这种态度对于生物学和心理学，对于感官、情感，尤其是本能，在我们生命上所占的地位，也是极少认识的。人类是灵与肉所造成，哲学家的任务应该是使身心协调起来，过着和谐的生活。

II. Earthbound

THE situation then is this: man wants to live, but he still must live upon this earth. All questions of living in heaven must be brushed aside. Let not the spirit take wings and soar to the abode of the gods and forget the earth. Are we not mortals, condemned to die? The span of life vouchsafed us, threescore and ten, is short enough, if the spirit gets too haughty and wants to live forever, but on the other hand, it is also long enough, if the spirit is a little humble. One can learn such a lot and enjoy such a lot in seventy years, and three generations is a long, long time to see human follies and acquire human wisdom. Anyone who is wise and has lived long enough to witness the changes of fashion and morals and politics through the rise and fall of three generations should be perfectly satisfied to rise from his seat and go away saying,"It was a good show"when the curtain falls.

For we are of the earth, earth-born and earthbound. There is nothing to be unhappy about the fact that we are, as it were, delivered upon this beautiful earth as its transient guests. Even if it were a dark dungeon, we still would have to make the best of it; it would be ungrateful of us not to do so when we have, instead of a dungeon, such a beautiful earth to live on for a good part of a century. Sometimes we get too ambitious and disdain the humble and yet generous earth. Yet a sentiment for this Mother Earth, a feeling of true affection and attachment, one must have for this temporary abode of our body and spirit, if we are to have a sense of spiritual harmony.

We have to have, therefore, a kind of animal skepticism as well as animal faith, taking this earthly life largely as it is. And we have to retain the wholeness of nature that we see in Thoreau who felt himself kin to the sod and partook largely of its dull patience, in winter expecting the sun of spring, who in his cheapest moments was apt to think that it was not his business to be"seeking the spirit,"but as much the spirit's business to seek him, and whose happiness, as he described it, was a good deal like that of the woodchucks. The earth, after all is real, as the heaven is unreal: how fortunate is man that he is born between the real earth and the unreal heaven!

Any good practical philosophy must start out with the recognition of our having a body. It is high time that some among us made the straight admission that we are animals, an admission which is inevitable since the establishment of the basic truth of the Darwinian theory and the great progress of biology, especially bio-chemistry. It was very unfortunate that our teachers and philosophers belonged to the so-called intellectual class, with a characteristic professional pride of intellect. The men of the spirit were as proud of the spirit as the shoemaker is proud of leather. Sometimes even the spirit was not sufficiently remote and abstract and they had to use the words,"essence"or"soul"or"idea,"wr iting them with capital letters to frighten us. The human body was distilled in this scholastic machine into a spirit, and the spirit was further concentrated into a kind of essence, forgetting that even alcoholic drinks must have a"body"–mixed with plain water–if they are to be palatable at all. And we poor laymen were supposed to drink that concentrated quintessence of spirit. This over-emphasis on the spirit was fatal. It made us war with our natural instincts, and my chief criticism is that it made a whole and rounded view of human nature impossible. It proceeded also from an inadequate knowledge of biology and psychology, and of the place of the senses, emotions and, above all, instincts in our life. Man is made of flesh and spirit both, and it should be philosophy's business to see that the mind and body live harmoniously together, that there be a reconciliation between the two.

三. 一个生物学的观念

　　如果我们对自己身体的功能和智能的程序有了深一层的了解，我们对于人类就能具有较真切较广泛的观念，使"动物"一名词减掉一些旧有的恶味。"会了解便会宽恕"，这句俗语可以应用到我们自己的身心的程序上去。因为我们如果对身体的功能有更深切的认识，我们便绝不会轻视这些功能。这个事实看来似乎很奇怪，然而确是正确的。关于我们的消化程序，要点不在乎批评它的贵贱，而仅仅是在了解它，这样它已变得非常高贵了。这情形也适用于我们身体中各种生物学上的功能，如出汗、排泄、胰液、胆汁、内分泌腺，以及更微妙的情感程序和思想程序。我们不再蔑视肾脏，我们只想了解它；我们不再把一对坏牙齿当做身体最后腐败的象征，也不当做拯救灵魂的警告者，我们只跑去找一位牙医，检验一下，把那坏牙齿补好就完了。一个人由牙医处出来后，便不再轻视他的牙齿，反而增加对它们的尊敬——因为他对于啃嚼苹果和鸡骨等，将要感到更大的乐趣了。讲到那些以为牙齿属于魔鬼的超形而上主义者，和那些不承认人类是有牙齿的新柏拉图主义者，当我看见他们自己患了牙痛和乐观的诗人患了消化不良症，我就往往感到这是近于对他的一种讽刺而觉得痛快。他为什么不再继续去做他的哲学理论呢？他为什么要像你、我，或隔壁的嫂嫂那样，把手按在面颊上呢？患着消化不良症的诗人为什么不信世上有所谓乐观呢？他为什么不再唱歌了？但一旦内脏工作恢复而不骚扰他的时候，他便把内脏忘得一干二净，只知歌颂神灵，他真是忘恩负义啊！

　　科学使我们从身体的动作得到一种更奇妙的感觉，它教我们怎样更进一步去尊敬我们的身体。第一，关于遗传学方面，我们开始知道我们人类的生成，绝不是泥土做成的，而是站在动物谱系的最高处。对于这一点，一个神志清楚没有被自己精神麻醉的人，想必会感到相当的满足和快慰吧。我的意思并不是说"恐龙"在几百万年前由生存而灭亡，因而使我们在今日可以生着两条腿，在地球上行走。生物学没有立出这种无所谓的假设，所以不会损害一丝一毫的人类尊严，也不会对人类优于万物这个观念加上疑点。所以任何一个立意要看重人类尊严的人，对此也曾觉得十分满意。第二，我们对于身体上的神秘和美丽，愈久愈有深刻的印象。我们不能不感到我们身体内的各部动作，以及彼此间的微妙联系是在极端困难的情形下所做成的，而其结果又是那么简单，始终不变。科学在说明体内这些化学的程序时，非但不能把它们弄得简单易解些，反而把它们弄得更复杂更难解，使这些程序比无生理学智识者所想象的更为复杂和困难。须知宇宙外表的神秘和宇宙内里的神秘，在本质上是相同的。

　　生理学家越是努力分析人类生理上的生物物理和生物化学的程序，便越觉得莫名其妙起来。所以一个心胸宽大的生理学家，有时也不得不接受神秘的人生观念。关于这点，我们可以举亚利克西斯·卡莱尔博士（Dr. Alexis Carrel，生物学家，1912年获得诺贝尔奖）为例。不论我们是否赞成他在《未知的人类》（Man, the Unknown）一书中所发表的意见；我们不能不同意实有那些事实和那些事实都未曾解释过，而且是无法解释的。我们开始觉得物质本身也有智能了。

"器官是依靠器官液和神经系而互相联系的。身体上每一部分和其他部分互相适应。这种适应的方式是循着目的而实现的。如果我们跟机械学者及活力论者的意见一样，认为思维具有一种和我们人类相同的智能，那么那些生理上的程序好似是为着各自的目的而互相联系的，有机体具着始终不变性，这是无可否认的。每一部分似乎都知道整个身体的现在和将来的需要，因而依照这个目的而去工作。时间和空间在我们的纤维和我们的心智的应用上是不相同的。身体意识到近的东西也能意识到远的东西，意识到现在，也能意识到将来。"（《未知的人类》原文第一九七页）

例如我们的内脏受了损伤，它们自己会自愈，完全不需要我们的努力，这种现象是值得惊异的：

"受伤的地方，起初变为不能动弹，暂时瘫痪，使粪类不能通过腹部。同时其他部分肠管或是网膜的表面，移近到伤处，表现了腹膜的特性，自动地黏附着。在四五个钟点内，伤处便合口了。有时伤口是被外科医生用针线缝好的，但那伤处仍是由于腹膜表面的自动黏附性而痊愈的。"（《未知的人类》原文第二〇〇页）

肌肉本身既有着这种智能，我们为什么还轻视肉体呢？我们是终究有一个身体，它是一架机器，自己营养，自己管理，自己修补，自己发动，自己生产，在我们出世的时候已装置就绪，像我们祖父用过的那座精美的钟一样，一用就是七十余年，不用我们担心。这架机器装着无线电式的视觉和无线电式的听觉，又有一种比电话机或电报机更复杂的神经系和淋巴系。它有一个规模极大的神经复杂体，担任编排报告的工作效率极高，不重要的案卷放在屋顶的小阁上，较重要的案卷则放在较便利的台架上，放在小阁上的那些案卷即使经过三十年，不常拿出来用，却依然在那里，等要用的时候，又可以马上拿出来用。而且这架机器能像汽车般到处奔跑，机件灵活，有着不发声响的引擎；如果遇到了意外，譬如说玻璃破碎了或驾驶轮弄坏了，它便自动地流出或制造出一种质素去替代玻璃，并且另生出一个驾驶轮来，或者至少想法子不用那根驾驶轴已肿的一端去开车；我们必须知道当我们体内的一个肾脏被割掉时，另外的一个肾脏就膨胀起来，增加它的效能，使常量的尿可以照常排出。同时，它在平时总保持着差度只在华氏一度（17.2℃）的十分之一以内的温度，自己能制造化学物质，以便将食品变成活的纤维。

还有最紧要的一点，就是它有一种生命韵律的意识，有一种时间的意识，它不但意识到几个钟点和几天，甚至意识到几十年的时光；身体统制着自己的童年时期、青春时期和成年时期，到够大的时期，便不再长大，甚至在我们不知不觉的时候，它早把一颗智齿长出来了。我们的身体也能制造清除毒物的解毒剂，而且有着那样惊人的满意成绩；它在做这些事时绝对没有声息，绝没有那种通常工厂里必有的嘈杂声响，因之，超等的形而上学家尽可以不受骚扰，优游自在地去思索他的精神或他的精粹。

059

III. A Biological View

THE better knowledge of our own bodily functions and mental processes gives us a truer and broader view of ourselves and takes away from the word"animal" some of its old bad flavor. The old proverb that"to understand is to forgive"is applicable to our own bodily and mental processes. It may seem strange, but it is true, that the very fact that we have a better understanding of our bodily functions makes it impossible for us to look down upon them with contempt. The important thing is not to say whether our digestive process is noble or ignoble; the important thing is just to understand it, and somehow it becomes extremely noble. This is true of every biological function or process in our body, from perspiration and the elimination of waste to the functions of the pancreatic juice, the gall, the endocrine glands and the finer emotive and cogitative processes. One no longer despises the kidney, one merely tries to understand it; and one no longer looks upon a bad tooth as symbolic of the final decay of our body and a reminder to attend to the welfare of our soul, but merely goes to a dentist, has it examined, explained and properly fixed up. Somehow a man coming out from a dentist's office no longer despises his teeth, but has an increased respect for them–because he is going to gnaw apples and chicken bones with increased delight. As for the superfine metaphysician who says that the teeth belong to the devil, and the Neo-Platonists who deny that individual teeth exist, I always get a satirical delight in seeing a philosopher suffering from a toothache and an optimistic poet suffering from dyspepsia. Why doesn't he go on with his philosophic disquisitions, and why does he hold his hand against his cheek, just as you or I or the woman in the next house would do? And why does optimism seem so unconvincing to a dyspeptic poet? Why doesn't he sing any more? How ungrateful it is, of him, therefore, to forget the intestines and sing about the spirit when the intestines behave and give him no trouble!

Science, if anything, has taught us an increased respect for our body, by deepening a sense of the wonder and mystery of its workings. In the first place, genetically, we begin to understand how we came about, and see that, instead of being made out of clay, we are sitting on the top of the genealogical tree of the animal kingdom. That must be a fine sensation, sufficiently satisfying for any man who is not intoxicated with his own spirit. Not that I believe dinosaurs lived and died millions of years ago in order that we today might walk erect with our two legs upon this earth. Without such gratuitous assumptions, biology has not at all destroyed a whit of human dignity, or cast doubt upon the view that we are probably the most splendid animals ever evolved on this earth. So that is quite satisfying for any man who wants to insist on human dignity. In the second place, we are more impressed than ever with the mystery and beauty of the body. The workings of the internal parts of our body and the wonderful correlation between them compel in us a sense of the extreme difficulty with which these correlations are brought about and the extreme simplicity and finality with which they are nevertheless accomplished. Instead of simplifying these internal chemical processes by explaining them, science makes them all the more difficult to explain. These processes are incredibly more difficult than the layman without any knowledge of physiology usually imagines. The great mystery of the universe without is similar in quality to the mystery of the universe within.

The more a physiologist tries to analyze and study the bio-physical and bio-chemical processes of human physiology, the more his wonder increases. That is so to the extent that sometimes it compels a physiologist with a broad spirit to accept the mystic's view of life, as in the case of Dr. Alexis Carrel. Whether we agree with him or not, as he states his opinions in Man, the Unknown, we must agree with him that the facts are there, unexplained and unexplainable. We begin to acquire a sense of the intelligence of matter itself:

The organs are correlated by the organic fluids and the nervous system. Each element of the body adjusts itself to the others, and the others to it. This mode of adaptation is essentially teleological. If we attribute to tissues an intelligence of the same kind as ours, as mechanists and vitalists do, the physiological processes appear to associate together in view of the end to be attained. The existence of finality within the organism is undeniable. Each part seems to know the present and future needs of the whole, and acts accordingly. The significance of time and space is not the same for our tissues as for our mind. The body perceives the remote as well as the near, the future as well as the present.

And we should wonder, for instance, and be extremely amazed that our intestines heal their own wounds, entirely without our voluntary effort:

The wounded loop first becomes immobile. It is temporarily paralyzed, and fecal matter is thus prevented from running into the abdomen. At the same time, some other intestinal loop, or the surface of the omentum, approaches the wound and, owing to a known property of peritoneum, adheres to it. Within four or five hours the opening is occluded. Even if the surgeon's needle has drawn the edges of the wound together, healing is due to spontaneous adhesion of the peritoneal surfaces.

Why do we despise the body, when the flesh itself shows such intelligence? After all, we are endowed with a body, which is a self-nourishing, self-regulating, self-repairing, self-starting and self-reproducing machine, installed at birth and lasting like a good grandfather clock for three-quarters of a century, requiring very little attention. It is a machine provided with wireless vision and wireless hearing, with a more highly complicated system of nerves and lymphs than the most complicated telephone and telegraph system of the world. It has a system of filing reports done by a vast complexus of nerves, managed with such efficiency that some files, the less important ones, are kept in the attic and others are kept in a more convenient desk, but those kept in the attic, which may be thirty years old and rarely referred to, are nevertheless there and sometimes can be found with lightning speed and efficiency. Then it also manages to go about like a motor car with perfect knee-action and absolute silence of engines, and if the motor car has an accident and breaks its glass or its steering wheel, the car automatically exudes or manufactures a substance to replace the glass and does its best to grow a steering wheel, or at least manages to do the steering with a swollen end of the steering shaft; for we must remember that when one of our kidneys is cut out, the other kidney swells and increases its function to insure the passage of the normal volume of urine. Then it also keeps up a normal temperature within a tenth of a Fahrenheit degree, and manufactures its own chemicals for the processes of transforming food into living tissues.

Above all, it has a sense of the rhythm of life, and a sense of time, not only of hours and days, but also of decades; the body regulates its own childhood, puberty and maturity, stops growing when it should no longer grow, and brings forth a wisdom tooth at a time when no one of us ever thought of it. Our conscious wisdom has nothing to do with our wisdom tooth. It also manufactures specific antidotes against poison, on the whole with amazing success, and it does all these things with absolute silence, without the usual racket of a factory, so that our superfine metaphysician may not be disturbed and is free to think about his spirit or his essence.

四．诗样的人生

　　我以为从生物学的观点看起来，人生几乎是像一首诗。它有韵律和拍子，也有生长和腐蚀的内在循环。它开始是天真朴实的童年时期，嗣后是粗拙的青春时期，企图去适应成熟的社会，带着青年的热情和愚憨，理想和野心，后来达到一个活动较剧烈的成年时期，由经验上获得进步，又由社会及人类天性上获得更多的经验；到中年的时候，才稍微减轻活动的紧张，性格也圆熟了，像水果的成熟或好酒的醇熟一样，对于人生渐抱一种较宽容、较玩世、也较温和的态度；以后到了老年的时期，内分泌腺减少了它们的活动，假如我们对于老年能有一种真正的哲学观念，照这种观念调和我们的生活形式，那么这个时期在我们看来便是和平、稳定、闲逸和满足的时期；最后生命的火花闪灭，一个人便永远长眠不醒了。我们应当能够体验出这种人生的韵律之美，像欣赏大交响曲那样欣赏人生的主旨，欣赏它急缓的旋律，以及最后的决定。这些循环的动作，在正常的人体上是大致相同的，不过那音乐必须由个人自己去演奏。在某些人的灵魂中，那个不调和的音键变得日益宏大，竟把正式的曲调淹没了，如果那不调和的音键声音太响，使音乐不能继续演奏下去，那个人便开枪自戕或跳河自尽了。这是因为他缺乏良好的自我教育，弄得原来的主旋律遭了掩蔽。反之，正常的人生是会保持着一种严肃的动作和行列，朝着正常的目标前进。在我们许多人之中，有时震音或激越之音太多，因此听来甚觉刺耳；我们也许应该有一些恒河般伟大的音律和雄壮的音波，慢慢地永远地向着大海流去。

一个人有童年、壮年和老年，我想没有一个人会觉得这是不美满。一天有上午、中午、日落，一年有春、夏、秋、冬四季，这办法再好没有。人生没有什么好坏，只有"在那一季里什么东西是好的"的问题。如果我们抱着这种生物学的人生观念，循着季节去生活，那么除自大的呆子和无可救药的理想主义者之外，没有人会否认人生确是像一首诗那样过去的。莎士比亚曾在他的人生七阶段的文章里把这个观念极明显地表达出来，许多中国作家也曾说过与此相似的话。莎士比亚没有变成富于宗教观念的人，也不曾对宗教表示很大的关怀，这是很可怪的。我想这便是他所以伟大的地方；他把人生当做人生看，他不打扰世间一切事物的配置和组织，正如他不打扰他戏剧中的人物一样。莎士比亚和大自然本身相似，这是我们对一位作家或思想家最大的赞颂。他只是活在世界上，观察人生而终于离开了。

IV. Human Life a Poem

I think that, from a biological standpoint, human life almost reads like a poem. It has its own rhythm and beat, its internal cycles of growth and decay. It begins with innocent childhood, followed by awkward adolescence trying awkwardly to adapt itself to mature society, with its young passions and follies, its ideals and ambitions; then it reaches a manhood of intense activities, profiting from experience and learning more about society and human nature; at middle age, there is a slight easing of tension, a mellowing of character like the ripening of fruit or the mellowing of good wine, and the gradual acquiring of a more tolerant, more cynical and at the same time a kindlier view of life; then in the sunset of our life, the endocrine glands decrease their activity, and if we have a true philosophy of old age and have ordered our life pattern according to it, it is for us the age of peace and security and leisure and contentment; finally, life flickers out and one goes into eternal sleep, never to wake up again. One should be able to sense the beauty of this rhythm of life, to appreciate, as we do in grand symphonies, its main theme, its strains of conflict and the final resolution. The movements of these cycles are very much the same in a normal life, but the music must be provided by the individual himself. In some souls, the discordant note becomes harsher and harsher and finally overwhelms or submerges the main melody. Sometimes the discordant note gains so much power that the music can no longer go on, and the individual shoots himself with a pistol or jumps into a river. But that is because his original leitmotif has been hopelessly over-shadowed through the lack of a good self-education. Otherwise the normal human life runs to its normal end in a kind of dignified movement and procession. There are sometimes in many of us too many staccatos or impetuosos, and because the tempo is wrong, the music is not pleasing to the ear; we might have more of the grand rhythm and majestic tempo of the Ganges, flowing slowly and eternally into the sea.

No one can say that a life with childhood, manhood and old age is not a beautiful arrangement; the day has its morning, noon and sunset, and the year has its seasons, and it is good that it is so. There is no good or bad in life, except what is good according to its own season. And if we take this biological view of life and try to live according to the seasons, no one but a conceited fool or an impossible idealist can deny that human life can be lived like a poem. Sha- kespeare has expressed this idea more graphically in his passage about the seven stages of life, and a good many Chinese writers have said about the same thing. It is curious that Shakespeare was never very religious, or very much concerned with religion. I think this was his greatness; he took human life largely as it was, and intruded himself as little upon the general scheme of things as he did upon the characters of his plays. Shakespeare was like Nature herself, and that is the greatest compliment we can pay to a writer or thinker. He merely lived, observed life and went away.

Chapter Three   **Our Animal Heritage**

第三章

我们的动物性遗产

一．猴子的故事

　　如果那种生物学的观念能够帮助我们去欣赏人生的韵律美，那也证明我们能力有限。我们如将人类即是动物这一点描写得更准确清楚，就能使我们得到较正确的印象，使我们更能了解自己，以及人类文化的进步。人类的天性是以我们动物始祖为根据的，当我们对天性有了更正确更深切的认识时，我们就会产生一种较慷慨的同情，甚至产生一种宽裕的玩世态度。由委婉地提醒我们自己我们是尼安得特人或北京人（The Neanderthal or the Peking man）的子孙，再说远一点，我们是人猿的子孙，于是我们终于能够轻视我们的罪恶和缺点，同时赞叹我们的猴子式聪明，这就是所谓人类喜剧的意识。克劳伦斯·戴伊（Clarence Day，美国著名作家）在他那篇《人猿世界》（This Simian World）发人深省的论文里，就曾表现出这种美妙的思想，当我们阅读这篇论文时，我们会宽恕一切人类：检察官、宣传主任、法西斯的编辑、纳粹党的无线电报告员、国会议员、立法委员、独裁者、经济学家、国际会议代表，以及那种干涉别人的生活的好管闲事者。因为我们已开始了解他们了。

　　从这种意义上讲来，我现在愈加能够体会《西游记》这部中国伟大的猴子故事的智慧和见识。我们在这一观点上，对于人类历史的演进，便能得到更亲切的认识；人类历史的演进和那些半人类的动物到西天去参圣的行程，是多么相似啊——孙悟空好似代表人类的智能，猪八戒代表较卑下的天性，沙和尚代表常识，玄奘法师则代表智慧和圣道。玄奘法师在这些怪异的随从保护之下，由中国出发到印度去取经。人类进展的事迹，就是像一群都有缺点的动物的谒圣行程一样，为了他们具着愚笨和恶作剧，所以不断地遭逢着许多危险和好笑的情境。法师每每须纠正并责罚那恶作剧的猴子和风流自赏的猪，因为不完美的心思和卑鄙的情欲，常常陷入各种窘境。在这个由人类到神佛的参拜旅程中，人性脆弱的本能、愤怒、复仇、暴躁、肉欲、不宽恕，尤其是自大和不谦逊的本能，不断地暴露出来。人类的技巧增高时，破坏力同时加高，因为我们现在都像那只有法术的猴子一样，能腾云驾雾，在空中大翻筋斗（即飞机在空中倒飞侧飞），由我们的猴腿上拔下毫毛，使它们变成小猴，去攻击我们的敌人，敲打天门，粗野无礼地把守卫天兵推开，要求和天神同等并列。

这只猴子是聪慧的，但是很自大；他有厉害的法术，可以闯入天门，可是没有相当健全、平衡和冷静的精神在天上安静地过生活。所以他对于这个尘世的生活，资格很够，可是对于天上的那些他比不上的神仙生活，他的资格还差得远呢。他的品性上有一些粗鄙的、恶作剧的、叛逆的质素，好比黄金里有着未曾炼净的渣滓，所以在上半部《西游记》里，当他未曾参加西行取经时，有一次他跑到天上去，造成了一种可怕的局面，像一只从动物园里铁笼中逃出来的野性狮子一样，为了有一种不能悛改的恶作剧的习性，他曾破坏了西天王母娘娘款待天上神仙所开的年宴。他因为未曾被邀请参加蟠桃盛会，不禁大怒，假扮成上帝的使者，遇着赤脚大仙去赴会，即谎骗他宴会的地址已改，使他走错了地方，而自己变成赤脚大仙的样子，跑去参加盛会。上他当的神仙为数很多。他跑到宝阁，才知道他是最先光临的贵客。除了那些在右厢走廊下看管几瓮玉液琼浆的仆人外，一个宾客也没有到。他就使个神通，拔下几根毫毛，放入口中，嚼碎喷去，喝一声"变"，即刻变做几个瞌睡虫，把那些仆人全弄睡了，于是把那几瓮仙酒喝完，喝得半醉跌跌撞撞地跑进大厅，把摆在台子上的蟠桃也吃光。当那些客人来临看见宴会席上的那种杯盘狼藉的情形时，他已跑到太上老君的家里去弄把戏了，设法偷吃了太上老君的长生不老金丹。后来，他一则恐怕这把戏发生严重的后果，二则因为不曾被邀请去参加蟠桃盛会，心里很是愤愤不平，所以即偷偷地离开天上回到了他的花果山，又做起猴王来，并对小猴们说他厌恶上天。他便举起背叛上天的旗帜，在旗上写着"齐天大圣"的字样。接着，这只猴子就和上天发生猛烈的战争，他并没有败北，后来还亏太上老君在云中用了金刚圈把他打倒，总算把他捉住了。

我们永像这只猴子一样在做叛逆的行为，没有和平，也没有谦卑，一直到太上老君从天上抛下金刚圈，把我们克服了才止。我们直需等到科学把宇宙间的一切界线探索出来后，才会得到真正谦卑的教训。在那部故事里那只猴子被捉住后还在背叛不已，质问天上的玉皇大帝为什么不在神仙中给他一个更高的名位，最后还要和如来佛或者上帝打个赌，才肯降服。他说，以他的法力他能够跑到天地的尽头，如做到了，应实授他"齐天大圣"的名号，如不能他便情愿一辈子屈服。于是他跳到空中，一个筋斗，风驰电掣地不知过了多少路，等他停下来时，只见五根肉红柱子，他便以为一定是人迹罕到的尽头了。为证明他曾到过这地方，他在第三根柱子根下撒了一泡猴尿，很得意地跑回来，把他的行程告诉佛祖。佛祖于是张开那只手，叫他闻闻中指下边的气味，告诉他说，他始终不曾跑出佛祖的一只手掌。这时猴子才低头认输，被佛祖用铁链缚在石上，经过了五百年，才由玄奘法师将他释放，跟着到西天去取经。

这只猴子——就是我们的小影——尽管其自大和恶作剧，终究还是一只极其可爱的动物。所以人类尽管有许多弱点，尽管有许多缺点，我们仍必须爱人类。

I. *The Monkey Epic*

BUT if this biological view helps us to appreciate the beauty and rhythm of life, it also shows our ludicrous limitations. By presenting to us a more correct picture of what we are as animals, it enables us to better understand ourselves and the progress of human affairs. A more generous sympathy, or even tolerant cynicism, comes with a truer and deeper understanding of human nature which has its roots in our animal ancestry. Gently reminding ourselves that we are children of the Neanderthal or the Peking man, and further back still of the anthropoid apes, we eventually achieve the capacity of laughing at our sins and limitations, as well as admiring our monkey cleverness, which we call a sense of human comedy. This is a beautiful thought suggested by the enlightening essay of Clarence Day, This Simian World. Reading that essay of Day's, we can forgive all our fellowmen, the censors, publicity chiefs, Fascist editors, Nazi radio announcers, senators and lawmakers, dictators, economic experts, delegates to international conferences and all the busybodies who try to interfere with other people's lives. We can forgive them because we begin to understand them.

In this sense, I come more and more to appreciate the wisdom and insight of the great Chinese monkey epic, Hsiyuchi. The progress of human history can be better understood from this point of view; it is so similar to the pilgrimage of those imperfect, semi-human creatures to the Western Heaven—the Monkey Wuk'ung representing the human intellect, the Pig Pachieh representing our lower nature, Monk Sand representing common sense, and the Abbot Hsüantsang representing wisdom and the Holy Way. The Abbot, protected by this curious escort, was engaged upon a journey from China to India to procure sacred Buddhist books. The story of human progress is essentially like the pilgrimage of this variegated company of highly imperfect creatures, continually landing in dangers and ludicrous situations through their own folly and mischief. How often the Abbot has to correct and chastise the mischievous Monkey and the sensuous Pig, forever led by their sadly imperfect minds and their lower passions into all sorts of scrapes! The instincts of human frailty, of anger, revenge, impetuousness, sensuality, lack of forgiveness, above all self-conceit and lack of humility, forever crop up during this pilgrimage of mankind toward sainthood. The increase of destructiveness goes side by side with the increase of human skill, for like the Monkey with magical powers, we are able today to walk upon the clouds and turn somersaults in the air (which is called"looping-the-loop"in modern terms), to pull monkey hair out of our monkey legs and transform them into little monkeys to harass our enemy, to knock at the very gates of Heaven, brush the Heavenly Gate Keeper brusquely aside and demand a place in the company of the gods.

The Monkey was clever, but he was also conceited; he had enough monkey magic to push his way into Heaven, but he had not enough sanity and balance and temperance of spirit to live peacefully there. Too good perhaps for this earth and its mortal existence, he was yet not good enough for Heaven and the company of the immortals. There was something raw and mischievous and rebellious in him, some dregs unpurged in his gold, and that was why when he entered Heaven he created a terrific scare there, like a wild lion let loose from a menagerie cage in the streets of a city, in the preliminary episode before he joined the pilgrims' party. Through his inborn incorrigible mischief, he spoiled the Annual Dinner Party given by the Western Queen Mother of Heaven to all the gods, saints, and immortals of Heaven. Enraged that he was not invited to the party, he posed as a messenger of God and sent the Bare-Footed Fairy on his way to the feast in a wrong direction by telling him that the place of the party had been changed, and then transformed himself into the shape of the Bare-Footed Fairy and went to the feast himself. Quite a number of other fairies had been misled by him in this way. Then entering the courtyard, he saw he was the first arrival. Nobody was there except the servants guarding the jars of fairy wine in the corridor. He then transformed himself into a sleepingsickness insect and stung the servants into sleep and drank the jars of wine. Half intoxicated, he tumbled into the hall and ate up the celestial peaches laid out at table. When the guests arrived and saw the despoiled dinner, he was already off for some other exploits at the home of Laotse, trying to eat his pills of immortality. Finally, still in disguise, he left Heaven, partly afraid of the consequences of his drunken exploits, but chiefly disgusted because he had not been invited to the Annual Dinner. He returned to his Monkey Kingdom where he was the king and told the little monkeys so, and set up a banner of rebellion against Heaven, writing on it the words"The Great Sage, Equal to Heaven."There followed then terrific combats between this Monkey and the heavenly warriors, in which the Monkey was not captured until the Goddess of Mercy knocked him down with a gentle sprig of flowers from the clouds.

So, like the Monkey, forever we rebel and there will be no peace and humility in us until we are vanquished by the Goddess of Mercy, whose gentle flowers dropped from Heaven will knock us off our feet. And we shall not learn the lesson of true humility until science has explored the limits of the universe. For in the epic, the Monkey still rebelled even after his capture and demanded of the Jade Emperor in Heaven why he was not given a higher title among the gods, and he had to learn the lesson of humility by an ultimate bet with Buddha or God Himself. He made a bet that with his magical powers he could go as far as the end of the earth, and the stake was the title of"The Great Sage, Equal of Heaven,"or else complete submission. Then he leaped into the air, and traveled with lightning speed across the continents until he came to a mountain with five peaks, which he thought must be as far as mortal beings had ever set foot. In order to leave a record of his having reached the place, he passed some monkey urine at the foot of the middle peak, and having satisfied himself with this feat, he came back and told Buddha about his journey. Buddha then opened one hand and asked him to smell his own urine at the base of the middle finger, and told him how all this time he had never left the palm. It was only then that the Monkey acquired humility, and after being chained to a rock for five hundred years, was freed by the Abbot and joined him in his pilgrimage.

After all, this Monkey, which is an image of ourselves, is an extremely lovable creature, in spite of his conceit and his mischief. So should we, too, be able to love humanity in spite of all its weaknesses and shortcomings.

二. 猴子般的形象

因之,《圣经》上所说我们是以上帝的形象来造成的那种观念我们必须抛开,我们觉得我们是由猴子的形象而来的,同时,如把我们和那完美的上帝相比,相差之远,犹如蚂蚁和我们一样的小巫见大巫。我们是聪明的,这一点,我们十分相信。因为我们确有心智,所以对自己的聪明常常有点骄傲。可是生物学家来对我们说,这个心智,可以用言语来表示的思想而论,尚是一种晚近的发展,在那些构成道德本质的要素中,除了心智外,还有一些动物的,或可说野蛮的本能,这些动物力比心智更大,而事实上也就是这些东西使我们在团体生活中做出各个错误行为。这样我们更能了解那个自傲的人类心智的性质。第一,我们见到这个心智是一个相当智慧的心智,但也颇有缺憾。我们考据人类头颅的进化,知道它不过是一根脊椎骨长大而成的,所以它跟脊髓的功用一样,只是在意识到危险,应付外边的环境和保存生命——但不在于思想。思想的工作大多是做得极笨拙的。亚瑟·巴尔福勋爵(Lord Balfour,英国首相中的哲学家)曾说:"人类的头脑对于寻求食物,和猪鼻一样的重要。"这一句话已可使他不朽了。我以为这句话并不代表有真正的玩世态度。我以为他说这话，不过是基于他对人类的一般理解而已。

我们由创始的方面了解我们人类的不完美。不完美吗？很对，造物主就是把我们造成这个样子的。不过问题不在这里。主要的一点是：我们的远祖都像人猿泰山那样，在森林中游憩，由这个树枝荡到那个树枝，或像长尾猴那样，用一只臂膀或尾巴钩住树枝倒悬着[6]。在我的心目中，以人类的进化而论，把各个阶段分开来看，可说都是极其完美的。可是现在，我们须做一种困难万倍的调整工作。

当人类在创造自己的文化时，在生物学方面讲来，所走的路径也许会使造物主吓一大跳。以适应大自然而论，生于大自然的一切动物是极完美的，因为造物主已把那些不能适应大自然的动物都灭尽了。可是现在我们毋庸适应大自然，我们只须适应自己，适应文化。在大自然的怀抱中，一切本能都是美好的、健全的，但是在社会中，我们把一切本能都叫做野蛮。每只老鼠都偷吃东西——但它并不因这种行为而有损于道德或变成更不道德。每一只狗都吠，每一只猫晚上总不回家或是破坏物件，每只狮子都杀害其他动物，每匹马看见危险都跑开，每只乌龟都把一天宝贵的光阴在睡眠中消磨掉，每只虫儿、爬行动物、鸟儿和兽类都在大庭广众之间生产子嗣。以文明世界的语词来说，每只老鼠都是盗贼，每只狗都太会吵闹，每只猫假如不是艺术品的野蛮破坏者，便是"不忠实的丈夫"，每只

---

[6] 这就是我们在秋千上，刚要由后荡向前面之时，我们觉得脊髓的末端——从前长着尾巴的地方——有一种刺激的感觉的理由：反射作用还在那里，我们还在想用一条早已脱掉的尾巴去钩住旁边的东西。

狮子或老虎都是嗜杀者，每匹马都是懦怯者，每只乌龟都是懒鬼，最后千百种虫儿、爬行动物、鸟儿和兽类一律都是淫猥的，世间事的评价有着多么重大的变动啊！这就是使我们惊讶造物主把我们造得这样不完美的理由。

II. In the Image of the Monkey

SO then, instead of holding on to the Biblical view that we are made in the image of God, we come to realize that we are made in the image of the monkey, and that we are as far removed from the perfect God, as, let us say, the ants are removed from ourselves. We are very clever, we are quite sure of that; we are often a little cocky about our cleverness, because we have a mind. But the biologist comes in to tell us that the mind after all is a very late development, as far as articulate thinking is concerned, and that among the things which go into the make-up of our moral fiber, we have besides the mind a set of animal or savage instincts, which are much more powerful and are in fact the explanation why we misbehave individually and in our group life. We are the better able to understand the nature of that human mind of which we are so proud. We see in the first place that, besides being a comparatively clever mind, it is also an inadequate mind. The evolution of the human skull shows us that it is nothing but an enlargement of one of the spinal vertebrae and that therefore its function, like that of the spinal cord, is essentially that of sensing danger, meeting the external environment and preserving life—not thinking. Thinking is generally very poorly done. Lord Balfour ought to go down to posterity on the strength of his one saying that"the human brain is as much an organ for seeking food as the pig's snout."I do not call this real cynicism, I call it merely a generous understanding of ourselves.

We begin to understand genetically our human imperfections. Imperfect? Lord, yes, but the Lord never made us otherwise. But that is not the point. The whole point is, our remote ancestors swam and crawled and swung from one branch to another in the primeval forest in Tarzan fashion, or hung suspended from a tree like a spider monkey by an arm or a tail. At each stage, considered by itself, it was rather marvelously perfect, to my way of thinking. But now we are called upon to do an infinitely more difficult job of readjustment.

When man creates a civilization of his own, he embarks upon a course of development that biologically might terrify the Creator Himself. So far as adaptation to nature is concerned, all nature's creatures are marvelously perfect, for those that are not perfectly adapted she kills off. But now we are no longer called upon to adapt ourselves to nature; we are called upon to adapt ourselves to ourselves, to this thing called civilization. All instincts were good, were healthy in nature; in society, however, we call all instincts savage. Every mouse steals—and he is not the less moral or more immoral for stealing—every dog barks, every cat doesn't come home at night and tears everything it can lay its paws upon, every lion kills, every horse runs away from the sight of danger, every tortoise sleeps the best hours of the day away, and every insect, reptile, bird and beast reproduces its kind in public. Now in terms of civilization, every mouse is a thief, every dog makes too much noise, every cat is an unfaithful husband, when he is not a savage little vandal, every lion or tiger is a murderer, every horse a coward, every tortoise a lazy louse, and finally, every insect, reptile, bird and beast is obscene when he performs his natural vital functions. What a wholesale transformation of values! And that is the reason why we sit back and wonder how the Lord made us so imperfect.

三. 论不免一死

因为我们有这么个会死的身体，以至于遭到下面一些不可逃避的后果：第一，我们都不免一死；第二，我们都有一个肚子；第三，我们有强壮的肌肉；第四，我们都有一个喜新厌旧的心。这些事实各有它根本的特质，所以对于人类文明有很重要的影响。因为这种现象太明显了，我们反而不曾想起它。我们如果不把这些后果看清楚，便不能认识我们自己和我们的文明。

人类无论贵贱，身躯总是五六尺高，寿命总是五六十岁。我疑惑这世间的一切民主政治、诗歌和哲学是否都是以上帝所定的这个事实为出发点。大致说来，这种办法颇为妥当。我们的身子长得恰到好处，不太高，也不太低，至少我对于我这个五尺四寸之躯是很满意的。同时五六十年在我看来已是够悠长的时期，事实上五六十年便是两三个世代（Generation）了。依造物主的安排方法，当我们呱呱坠地后，一些年高的祖父即在相当时期内死掉。当我们自己做祖父的时候，我们看见又有小婴儿出世了。看起来，这办法真是再好也没有。这里的整个哲学便是依据下面的这句中国俗语——"家有千顷良田，只睡五尺高床。"即使是一个国王，他的床似乎不需超过七尺，一到晚上，他也非到那边去躺着不可，所以我是跟国王一样幸福的。无论这个人怎么样富裕，但能超过《圣经》中所说的七十年限度的就不多见，活到七十岁，在中国便称为"古稀"，因为中国有一句诗："人生七十古来稀。"

关于财富，也是如此。我们在这生命中人人有份，但没有一个人握着全部抵押权。因此我们对人生可以抱着比较轻快随便的态度：我们不是这个尘世永久的房客，而是过路的旅客。地主、佃户，都是一样的旅客。这种观念减弱了"地主"一词的意义。没有一个人能实在地说，他拥有一所房子或一片田地。一位中国诗人说得好：

> 苍田青山无限好，
> 前人耕耘后人收；
> 寄语后人且莫喜，
> 更有后人乐逍遥！

人类很少能够体念到死的平等意义。世间假如没有死，那么即使是圣赫勒拿岛（St. Helena）之于拿破仑也要觉得毫不在乎，而欧洲将不知是要变成个什么样子。世间如果真没有死，我们便没有英雄豪杰的传记，就是有，作者也一定会有一种较不宽恕、较无同情心的态度。我们宽恕世界的一切伟人，因为他们死了。他们一死，我们便觉得已和他们消除了仇恨。每个葬礼的行列都似有着一面旗帜，上边写着"人类平等"的字样。万里长城的建造者、专制暴君秦始皇焚书坑儒，制定"腹诽"处死的法律；中国人民在下面那首讲到秦始皇之死的歌谣里，表现了多么伟大的生之欢乐啊！

秦始皇奄僵[7]！

开吾民，

据吾床，

饮吾酒，

唾吾浆，

餐吾饮，

以为粮；

张吾弓，

射东墙，

前至沙丘当灭亡！

　　人类喜剧的意识，与诗歌和哲学的资料，大多是如此而产生的。能鉴到死亡的人，也能见到人类喜剧的意识，于是很迅速地变成诗人了。莎士比亚写哈姆雷特寻找亚力山大大帝的高贵残骸遗灰，"后来他发现这灰土也被人家拿去塞一个啤酒桶的漏洞"；"亚力山大死了，亚力山大葬了，亚力山大变成尘土了，我们拿尘土来做黏土，为什么不可以去塞一个啤酒桶的漏洞呢？"莎士比亚写这段文字时，已经变成了一个深刻的诗人了。莎士比亚使理查二世谈到坟墓、虫儿、墓志铭，谈到皇帝死后，虫儿在他的头颅中也玩着朝廷上的滑稽剧，又谈到"有一个购买田地的大买主，经过着法令、具结、罚金、双重证据和收回，结果他虽花了如许罚

---

[7] 在中国历史学家的心目中，这些歌谣是先知的预言，是上帝借人民的声音表现出来的预言，所以这首歌谣中的动词都是将来式，秦始皇后来的确死于沙丘。

金（Fines），但仍变成一个良好的头顶满装着精致的尘土（Fine Plate full of fine dirt）"。莎士比亚在这地方即表现着最优越的喜剧意识。欧玛尔·海亚姆（Omar Khayyam，十世纪波斯诗人）及中国的贾凫西（别名木皮子，木皮散客，明末鼓词作家），都是从死亡的意识上获得他们的诙谐心情，以及对历史的诙谐解释。他们从那些在皇帝的坟墓里住着的狐狸来借题发挥庄子的全部哲学，也是基于庄子对一个骷髅的言论。中国的哲学到庄子的时代，才第一次蕴含着深刻的理论和幽默的成分：

> 庄子之楚，见空骷髅，然有形；撽以马捶，因而问之曰："夫子贪生失理，而为此乎？将子有亡国之事，斧钺之诛，而为此乎？将子有不善之行，愧遗父母妻子之丑，而为此乎？将子有冻馁之患，而为此乎？将子之春秋故及此乎？"于是语卒，援骷髅，枕而卧……
>
> 庄子妻死，惠子吊之。庄子则方箕踞鼓盆而歌。惠子曰："与人居，长子、老、身死，不哭亦足矣；又鼓盆而歌，不亦甚乎？"
>
> 庄子曰："不然。是其始死也，我独何能无概然？察其始而本无生；非徒无生也，而本无形；非徒无形也，而本无气。杂乎芒芴之间，变而有气；气变而有形；形变而有生；今又变而之死，是相与为春秋冬夏四时行也。人且偃然寝于巨室，而我嗷嗷然随而哭之，自以为不通乎命。故止也。"

当我们承认人类不免一死的时候，当我们意识到时间消逝的时候，诗歌和哲学才会产生出来。这种时间消逝的意识是藏在中西一切诗歌的背面的——人生本是一场梦；我们正如划船在一个落日余晖返照的明朗下午，沿着河划去；花不常好，月不常圆，人类生命也随着在动植物界的行列中永久向前走，出生、长成、死亡，把空位又让给别人。等到人类看透了这尘世的空虚时，方才开始觉悟起来。庄子说，有一次做个梦，梦见自己变成蝴蝶，他也觉得能够展开翅膀来飞翔，好像一切都是真的，可是当他醒来时，他觉得他才是真实的庄子；但是后来，他陷入颇滑稽的沉思中，他不知道到底是庄子在梦做蝴蝶，还是一只蝴蝶在梦做庄子。所以人生真是一场梦，人类活像一个旅客，乘在船上，沿着永恒的时间之河驶去，在某一个地方上船，在另一个地方上岸，好让其他河边等候的旅客上船。假如我们不以为人生实是一场梦，或是过路的旅客所走的一段旅程，或是一个连演员自己也不知道是在做戏的舞台，那么，人生的诗歌连一半也不曾存在了。一个名叫刘达生的中国学者在给他朋友的信中写着：

世间极认真事，曰："做官"；极虚幻事，曰："做戏"；而弟曰愚甚。每于场上遇见歌哭笑骂，打诨插科，便确认为真实；不在所打扮古人，而在此扮古人之戏子。——俱有父母妻儿，——俱要养父母活妻儿，——俱靠歌哭笑骂，打诨插科去养父母活妻儿，此戏子乃真古人也。又每至于顶冠束带，装模作样之际，俨然自道一真官，天下亦无一人疑我为戏子者，正不知打恭看坐，欢颜笑口；与夫作色正容，凛莫敢犯之官人，实即此养父母活妻儿，歌哭笑骂打诨插科，假扮之戏子耳！乃拿定一场戏目，戏本戏腔，至五脏六腑，全为戏用，而自亦不觉为真戏子。悲夫！

III. On Being Mortal

THERE are grave consequences following upon our having this mortal body: first our being mortal, then our having a stomach, having strong muscles and having a curious mind. These facts, because of their basic character, profoundly influence the character of human civilization. Because this is so obvious, we never think about it. But we cannot understand ourselves and our civilization unless we see these consequences clearly.

I suspect that all democracy, all poetry, and all philosophy start out from this God-given fact that all of us, princes and paupers alike, are limited to a body of five or six feet and live a life of fifty or sixty years. On the whole, the arrangement is quite handy. We are neither too long nor too short. At least I am quite satisfied with five feet four. And fifty or sixty years seems to me such an awfully long time; it is, in fact, a matter of two or three generations. It is so arranged that when we are born, we see certain old grandfathers, who die in the course of time, and when we become grandfathers ourselves, we see other tiny tots being born. That seems to make it just perfect. The whole philosophy of the matter lies in the Chinese saying that"A man may own a thousand acres of land, and yet he still sleeps upon a bed of five feet"or sixty inches. It doesn't seem as if a king needed very much more than seven feet at the outside for his bed, and there he will have to go and stretch himself at night. I am therefore as good as a king. And no matter how rich a man is, few exceed the Biblical limit of threescore and ten. To live beyond seventy is to be called in Chinese"ancient-rare,"because of the Chinese line that"it is rare for man to live over seventy since the ancient times. "

And so in respect of wealth. Of this life, everybody has a share, but no one owns the mortgage. And so we are enabled to take this life more lightly; instead of being permanent tenants upon this earth, we become its transient guests, for guests we all are of this earth, the owners of the land no less than the share-croppers. It takes something out of the meaning of the word"landlord."No one really owns a house and no one really owns a field. As a Chinese poet says:

> *What pretty, golden fields against a hill!*
> *Newcomers harvest crops that others till.*
> *Rejoice not, O newcomers, at your harvest;*
> *One waits behind—a new newcomer still!*

The democracy of death is seldom appreciated. Without death, even St. Helena would have meant nothing to Napoleon, and I do not know what Europe would be like. There would be no biographies of heroes or conquerors, and even if there were, their biographers certainly would be less forgiving and sympathetic. We forgive the great of this world because they are dead. By their being dead, we feel that we have got even with them. Every funeral procession carries a banner upon which are written the words,"Equality of Mankind."What joy of life is seen in the following ballad that the oppressed people of China composed about the death of Ch'in Shihhuang, the builder of the Great Wall and the tyrant, who, while he lived, made"libellous thoughts in the belly"punishable by death, burned the Confucian books and buried hundreds of Confucian scholars alive:

Ch'in Shih-huang is going to die!
He opened my door,
And sat on my floor,
He drank my gravy,
 And wanted some more.
He sipped my wine,
And couldn't tell what for;
I'll bend my bow,
And shoot him at the wall.
When he arrives at Shach'iu,
Then he is going to fall!

From this, then, a sense of human comedy and the very stuff of human poetry and philosophy take their rise. He who perceives death perceives a sense of the human comedy, and quickly becomes a poet. Shakespeare became a deep poet, when he had Hamlet trace the noble dust of Alexander, "till he find it stopping a bunghole"; "Alexander died, Alexander was buried, Alexander returneth into dust; the dust is earth; of earth we make loam; and why of that loam, whereto he was converted, might they not stop a beerbarrel?" There is, after all, no more superb sense of comedy in Shakespeare than when he let King Richard II talk of graves, of worms and epitaphs and the antic that keeps court within the hollow crown that rounds the mortal temples of a king, or where he speaks of "a great buyer of land, with his statutes, his recognizances, his fines, his double vouchers, his recoveries," with all his fines ending in a "fine pate full of fine dirt." Omar Khayyam and his Chinese counterpart, Chia Fuhsi (alias Mup'itse, an obscure Chinese poet), derived all their comic spirit and comic interpretation of history from the sense of death itself, by pointing to the foxes making their homes in the kings' graves. And Chinese philosophy first acquired depth and humor with Chuangtse, who based his entire philosophy, too, on a comment on the sight of a skull:

Chuangtse went to Ch'u and saw an empty skull with its empty and dried outline. He struck it with a horsewhip and asked it, "Hast thou come to this because thou loved pleasures and lived inordinately? Wert thou a refugee running away from the law? Didst thou do something wrong to bring shame upon thy parents and thy family? Or wert thou starved to death? Or didst thou come to thy old age and die a natural death?" Having said this, Chuangtse took the skull and slept upon it as a pillow...

When Chuangtse's wife died, Hueitse went to express his condolence but found Chuangtse squatting on the ground and singing a song, beating time by striking an earthen basin. "Why, this woman has lived with you and borne you children. At the worst, you might refrain from weeping when her old body dies. Is it not rather too much that you should beat the basin and sing?"

And Chuangtse replied, "You are mistaken. When she first died, I could not also help feeling sad and moved, but I reflected that in the beginning she had no life, and not only no life, she had no bodily shape; and not only no bodily shape, she had no ghost. Caught in this everchanging flux of things, she became a ghost, the ghost became a body, and the body became alive. Now she has changed again and become dead, and by so doing she has joined the eternal procession of spring, summer, autumn and winter. Why should I make so much noise and wail and weep over her while her body lies quietly there in the big house? That would be a failure to understand the course of things. That is why I stopped crying. "

Thus I see both poetry and philosophy began with the recognition of our mortality and a sense of the evanescence of time. This sense of life's evanescence is back of all Chinese poetry, as well as of a good part of Western poetry—the feeling that life is essentially but a dream, while we row, row our boat down the river in the sunset of a beautiful afternoon, that flowers cannot bloom forever, the moon waxes and wanes, and human life itself joins the eternal procession of the plant and animal worlds in being born, growing to maturity and dying to make room for others. Man began to be philosophical only when he saw the vanity of this earthly existence. Chuangtse said that he once dreamed of being a butterfly, and while he was in the dream, he felt he could flutter his wings and everything was real, but that on waking up, he realized that he was Chuangtse and Chuangtse was real. Then he thought and wondered which was really real, whether he was really Chuangtse dreaming of being a butterfly, or really a butterfly dreaming of being Chuangtse. Life, then, is really a dream, and we human beings are like travelers floating down the eternal river of time, embarking at a certain point and disembarking again at another point in order to make room for others waiting below the river to come aboard. Half of the poetry of life would be gone, if we did not feel that life was either a dream, or a voyage with transient travelers, or merely a stage in which the actors seldom realized that they were playing their parts. So wrote a Chinese scholar, Liu Tasheng, to his friend:

Of all the things in the world, that in which we are most earnest is to be an official, and that which we call the most frivolous is to be an actor in a play. But I think this is all foolishness. I have often seen on the stage how the actors sing and weep and scold each other and crack jokes, believing that they are real people. But the real thing in a play is not the ancient characters thus being enacted, but rather these actors who enact them. They all have their parents, wives and children, all want to feed their parents, wives and children, and all do so by singing and weeping and laughing and scolding and cracking jokes. They are the real ancient characters that they try to portray. I have also seen how some of these actors, who wear an official cap and gown and by their own acting believe themselves to be real officials, so much so that they think no one in the world ever suspects they are acting. They do not realize that while they bow and kowtow to each other and sit and talk and look about, and even while they are the dignified officials before whom the prisoners tremble, they are only actors who by their singing and weeping and laughing and scolding and cracking jokes are trying to feed their parents, wives and children! Alas! that there are people who stick to a certain play, a certain role, a certain text and a certain accent or style of delivery, until the entire asset of their bowels and internal organs (i. e., instincts and emotions) are dominated by the play, without realizing once that they are really actors!

四. 论肚子

　　凡是动物便有这么一个叫做"胃"的无底洞。这无底洞曾影响了我们整个的文明。中国号称"美食家"的李笠翁在《闲情偶寄》卷十二《饮馔部》的序言里，对于这个无底洞颇有怨尤之言：

　　吾观人之一生，眼、耳、鼻、舌、手、足、躯骸，件件都不可少，其尽可不设而必欲赋之，遂为万古生人之累者，独是口腹二物。口腹具而生计繁矣，生计繁而诈伪奸险之事出矣。诈伪奸险之事出，而五刑不得不设。君不能施其爱育，亲不能遂其恩私，造物好生而亦不能逆行其志者，皆当日赋形不善，多此二物之累也。草木无口腹，未尝不生；山石土壤无饮食，未闻不长养；何事独异其形，而赋以口腹？即生口腹，亦当使如鱼虾之饮水，蜩螗之吸露，尽可滋生气力，而为蹩跃飞鸣。若是，则可与世无求，而生人之患熄矣。乃既生以口腹，又复多其嗜欲，使如谿壑之不可厌，多其嗜欲，又复洞其底里，使如江河之不可填，以致人之一生，竭五官百骸之力，供一物之所耗而不足者。吾反复推详，不能不于造物主是咎，亦知造物于此，未尝不自悔其非，但以制定难移，只得终遂其过。甚矣，作法慎初，不可草草定制！

　　我们既有了这个无底洞，自须填满。那真是无可奈何的事，我们有这个肚子，它的影响确已及于人类历史的过程。孔子对于人类的天性有着深切的了解，他把人生的大欲简括于营养和生育二事之下，简单地说来，就是饮食男女。许多人曾抑制了色，可是我们不曾听见过有一位圣人克制过饮食。即使是最神圣的人，总不能把饮食忘记到四五小时之上。我们每隔几小时脑海中便要浮起"是吃的时候了吧"这一句话，每天至少要想到三次，多者四五次。国际会议在讨论到政治局势的紧要关头时，也许因吃午餐而暂告停顿。国会须依吃饭的钟点去安排议程。一个需要五六小时之久而碍于午餐的加冕典礼，将立被斥为有碍公众生活。上天既然赋予了我们肚子，那么当我们聚在一起，想对祖父表示敬意的时候，最好是替他举行一次庆寿的宴会。

　　所以这是不无原因的，朋友在餐席上的相见就是和平的相见。一碗燕窝汤或一盆美味的炒面，对于激烈的争辩有缓和的效用，使双方冲突的意见会和缓下来。叫两个空着肚子的好朋友在一起，总是要发生龃龉的。一餐丰美的饮食，效力之大，不只是延长到几小时，直可以达到几星期，甚至几个月之久。如果要我们写一篇书评去骂三四个月以前曾经请我们吃过一餐丰盛晚餐的作家的作品，我们真要犹豫不能落笔。正因为如此，所以洞烛人类天性的中国人，他们不拿争论去对簿公庭，却解决于筵席之上。他们不但是在杯酒之间去解决纷争，而且可用来防止纷争。在中国，我们常设宴以联欢。事实上，也是政治上的登龙术。假使有人去做一次统计，那么他会发现一个人的宴客次数与他的升官速度是有一种绝对的关系存在的。

　　既然天生如此，我们又怎能背道而驰呢？我不相信这是东方的特殊情形。一个西洋邮务总长或部长，对于一个曾请他到家里去吃过五六次饭的朋友和私人请托，怎么能够拒绝呢？我敢说西洋人是与东方人一样有人性的。那唯一的不同点，是西洋人未曾洞察人类天性，或未曾按着这人类天性去合理地组织他们的政治生活。我猜想的西洋政治圈子中，也有与这种东方人生活方式相同的地方，因为我始终相信人类天性是大抵相同的，而同在这皮肉包裹之下，我们都是一样的，只是那习惯没有像中国那样普遍而已。我所听见的事情，只有政府官吏候选人摆了露天茶会请区内选民的眷属，拿冰激凌和苏打水给他们的小孩子吃以贿赂他们的母亲。这样请了大家一顿之后，人们自然不免相信"他是一个和气的好人"了，这句话是常常被当做歌曲唱着的。欧洲中世纪的王公贵族，在婚事或寿辰的时候，总要以丰盛的酒肉，设宴请佃户们开怀大吃一餐，这也无非是这种事情的另一表现方式而已。

　　我们基本上受这种饮食的影响非常之深，食物深刻地影响着和平、战争、爱国主义、国际交流，我们的日常生活和整个人类社会组织。法国大革命的起因是什么？是卢梭、伏尔泰和狄德罗？不，正是食物。俄国革命和苏联经济实验的起因是什么？还是食物。更有甚者，拿破仑还说过"兵马未动，粮草先行"，显示了他智慧的精华。还有诸如说着"和平，和平"时候，在横膈膜下面就没有和平吗？这在国家和个人中都存在。当民不聊生时，帝国会崩溃，最强有力的恐怖政权统治会瓦解。在饥饿的时候，人们不肯工作，主角歌女不肯唱歌，参议员不肯辩论，除了在家里图一顿饱餐这目的之外，做丈夫的为什么要整天在办公室里工作流汗呢？因此有一句俗话说，博得男人欢心最好的办法，便是从他的肚子入手。当他的肉体满足了以后，他的精神便比较平静舒适，他也比较多情服帖了。妻子们总是埋怨他的丈夫不注意她们的新衣服、新鞋子、新眉样，或新椅套，可是妻子们可曾有埋怨他的丈夫不注意一块好肉排一客好煎蛋吗？……除了爱我们幼时所爱吃的好东西之外还有什么呢？我在别处说过，对"山姆大叔"的忠诚，是对炸面圈和火腿、番薯的忠诚，德国人对祖国的忠诚，是对煎饼蛋和圣诞果子面包的忠诚。至于国际交流，我觉得我们对意大利通心粉的欣赏要远大于墨索里尼……那都是因为食物，就像死亡一样，都是人类不可缺的手足。

　　一个东方人在盛宴当前时是多么精神焕发啊！当他的肚肠填满了的时候，他是多么轻易地会喊出人生是美妙的啊！从这个填满了的肚子里透射出了一种精神上的快乐。东方人是靠着本能的，而他的本能告诉他，当肚子好着的时候，一切事物也都好了，所以我说在东方人生活是靠近于本能，以及有一种使他们更能公开承认他们的生活近于本能的哲学。我曾在别处说过，中国人对于快乐的观念是"温、饱、黑、甜"——指吃完了一顿美餐上床去睡觉的情景。所以有一个中国诗人说："肠满诚好事，余者皆奢侈。"

　　因为中国人有着这种哲学，所以对于饮食就不固执，吃时不妨吃得津津有味。当喝一口好汤时，也不妨啜唇作响。这在西方人就是无礼貌。所谓西方的礼节，是强使我们鸦雀无声地喝汤，静静地吃饭，几乎无法表达愉悦，我想这或许就是阻碍西方烹调技术发展的真原因。西方人士在吃饭的时候，为什么谈得那么有气无力，吃得那么阴森，规矩高尚呢？多数的美国人都没有那种聪明，把一根鸡腿啃个一干二净；反之，他们仍用刀叉玩弄着，感到非常苦恼，而不敢说一句话。假如鸡肉真真是烧得很好，这真是一种罪过。讲到餐桌上的礼貌，我觉得当母亲禁止小孩啜唇作响的时候，就是使他开始感觉到人生的悲哀。依照人类的心理讲，假使我们不表示我们的快乐，我们就不会再感觉到快乐，于是消化不良、忧郁、神经衰弱，以及成人生活中所特有的精神病等都接踵而来了。当堂倌儿端上一盘美味的小牛排时，我们应该跟法国人学学说一声"啊"，尝过第一口后，像动物那样哼一声"嗯！"欣赏食物不是什么可羞的事。有健康的胃口不是很好吗？不，中国人就两样。我们吃东西时礼貌虽不好，可是善于享受盛宴。

事实上，中国人之所以对动植物学一无贡献，是因为中国的学者不能冷静地观察一条鱼，而只想着鱼在口中的滋味，想吃掉它。我之所以不信任中国的外科医生，是因为我怕他们在割我的肝脏找石子的时候，也许会忘记了石子，而想把我的肝脏放到油锅里去。当中国人看见一只豪猪时，便会想出种种吃法来，只要在不中毒的原则之下吃掉它。在中国人看来，不中毒是唯一实际而重要的问题。豪猪的刺毛引不起他们的兴趣。这些刺毛怎样会竖立的？有什么功用？它们和皮怎样生连着？当它看见仇敌时，这些刺毛怎样会有竖立的能力？这些问题，在中国人看来是极其无聊的。中国人对于动植物都是这样，主要的观念是怎样欣赏它，享受它，而不是它们是什么。鸟的歌声，花的颜色，蓝的花瓣，鸡肉的肌理，才是我们所关心的东西。东方人须向西方人学习动植物的全部科学，可是西方人须向东方人学习怎样欣赏花鱼鸟兽，怎样能赏心悦目地赏识动植物各种的轮廓与姿态，因而从它们联想到各种不同的心情和感觉。

这样看来，饮食是人生中难得的乐事之一。肚子饿不像性饥渴那样受着社会的戒律和禁例，也大致不会发生什么有损于道德的问题，这是值得愉快的。人类在饮食方面比在性方面较少矫揉造作。哲学家、诗人、商贾能跟艺术家坐在一起吃饭，在众目昭彰之下，做喂饲自己的工作而毫不害羞，这真是不幸中的大幸，虽则也有些野蛮民族对于饮食尚有一些羞怯的意识，仍愿独个儿到没有旁人的地方才敢吃。关于性的问题，以后再讨论，我们在这里，至少可以看见一种本能，这本能如不受阻碍，即可减少变态及疯狂和犯罪的行为。在社会的接触中，饥饿的本能和性的本能其差异是显然的。可是事实上饥饿这种本能，前面已经讲过，不会牵涉到我们的心理生活，而实是人类的一种福利。其理由即因人类能对这个本能非常坦白，毫不讳饰。因为饮食没有拘束，所以也就没有精神病、神经官能症或各种变态。临唇之杯不免有失手之虞，可是一进唇内，就几乎没有什么意外。我们坦白地承认人类都要吃饭，可是对于性的本能，非但不如此，并加以抑制。假如食欲满足了，麻烦就少。顶多有些人患消化不良症、胃疡，或肝硬化，或有些人以牙齿自掘坟墓——现代中国少数的要人颇有几个是如此的——但即使如此，他们也并不以为可羞。

所以社会的罪恶从性欲问题产生的多，而从饮食问题产生的少。刑事条文为奸淫、离婚，和侵犯女性等案而设者为多，因饮食而违犯不合法、不道德或背信罪者就很少。顶多不过是有些丈夫去搜索冰箱里的食物，但是我们很少听见因此而遭绞杀的。假如真有这么一件案件上了法庭，法官对于被告一定也会表示同情。因为我们都愿坦白承认大家必须饮食。我们对饥民表示同情，却不曾对尼姑庵里的尼姑表示同情。

这种推论并不是无中生有的，因为我们对于饮食的问题，总比性欲问题明白得多。满族人的女孩儿在出嫁之前，必须受烹调的训练，同时受关于恋爱之术的训练，但世界上可有别处的人实行这种教育吗？饮食问题已接受知识之光，可是性的问题仍是被神仙故事、神话和迷信所包围。饮食问题可以说是见到天日了，但性的问题依然处于暗中。

在另一方面讲，我们人类没有沙囊或浮囊，真是莫大的缺憾，假如有，人类社会的过程一定会有极大的变更，可以说，我们将变为一种完全不同的人类。如有沙囊，人类一定会有最和平、最知足、最可爱的天性，和小鸡、小羊一样。我们也许会长出一个跟鸟嘴一样的嘴巴，因而改变了我们审美的观念，或者也许会生着一些啮齿类动物的牙齿。植物的种子和果实或许已足为我们的食物，也许我们会在青翠的山边吃草。大自然的产物是那样丰盛，我们不必再为食物而斗争，不必再用牙齿去咬仇敌的肉，也一定不会像我们今日这样的好斗。

食物与性情的关系，比我们所想象的更加密切。凡是蔬食动物的天性都是和平的，如羊、马、牛、象、麻雀等；凡是肉食动物都是好斗嗜杀的，像狼、狮、虎、鹰等。如果我们是属于前一类的，我们的天性就会比较像牛羊了。在无须战斗的地方，大自然并不造出好斗的天性。公鸡的搏斗，不是为食物，是为雌性，人类社会中的男人也还有着这种斗争，但今日的欧洲为了输出罐头食物的权利而斗争，其原因又有天壤之别了。

我不曾听见过猴子会吃猴子，可是我知道人会吃人。考据我们的人类学，证明确有人吃人的习俗，而且非常普遍。我们的祖先便是这种肉食的动物。所以，在几种意义上——个人的、社会的、国际的——如说我们依然在互相吞食，并不足为怪。蛮子和杀戮好像是有连带性，他们虽承认杀人是一种不合情理的事，是一种无可避免的罪恶，可是依然很干脆地把已杀死的仇敌的腰肉、肋骨和肝脏吃掉。吃人的蛮子吃掉已死了的仇敌，而文明的人类把杀死的仇敌埋葬了，并在墓上竖起十字架来，为他们的灵魂祷告。我们实在自傲和劣性之外，又加上愚蠢了。

我也以为我们是在向着完美之路前进，那就是说，我们在目前还未达到完善的境地，我们要有沙囊动物的性情时，才可以称为真文明的人类。在现代人类之间，肉食动物和蔬食动物都有之——前者就是性情可爱的，后者便是那种性情不可爱的。蔬食的人终身以管自己的事为主，而肉食的人专以管别人的事为生。十年前我曾尝试过政治生涯，但四个月后便弃绝仕途，因为我发现我不是天生的肉食动物，吃好肉排当然例外。世界上一半人是消磨时间去做事，另外一半人则强迫他人去替他们服役，或是弄到别人不得做事。肉食者的特点是喜欢格斗、操纵、欺骗、斗智，以及先下手为强，而且都出于真兴趣和全副本领，可是我得声明我对于这种手段是绝对反对的。但这完全是本能问题；天生有格斗本能的人似乎喜欢陶醉在这种举动中，同时真有创造性的才能（即能做自己事情的才能和能认清自己目标的才能）却似乎太不发达了。那些善良的、沉静的、蔬食类的教授，在和别人竞争之中，似乎全然没有越过别人的贪欲和才能，不过我是多么称赞他们啊！事实上，我敢说，全世界有创造才能的艺术家，只管他们自己的事，实比去管别人的事情好得多，因此他们都可说是属于蔬食类的。蔬食人种的繁殖率胜过肉食人种，这就是人类的真进化。可是在目前，肉食人种终究还是我们的统治者，在以强壮肌肉为信仰的现世界中其情其势必如此。

IV. On Having a Stomach

ONE of the most important consequences of our being animals is that we have got this bottomless pit called the stomach. This fact has colored our entire civilization. The Chinese epicure Li Liweng wrote a complaint about our having this bottomless pit, in the prefatory note to the section on food in his book on the general art of living.

I see that the organs of the human body, the ear, the eye, the nose, the tongue, the hands, the feet and the body, have all a necessary function, but the two organs which are totally unnecessary but with which we are nevertheless endowed are the mouth and the stomach, which cause all the worry and trouble of mankind throughout the ages. With this mouth and this stomach, the matter of getting a living becomes complicated, and when the matter of getting a living becomes complicated, we have cunning and falsehood and dishonesty in human affairs. With the coming of cunning and falsehood and dishonesty in human affairs, comes the criminal law, so that the king is not able to protect with his mercy, the parents are not able to gratify their love, and even the kind Creator is forced to go against His will. All this comes of a little lack of forethought in His design for the human body at the time of the creation, and is the consequence of our having these two organs. The plants can live without a mouth and a stomach, and the rocks and the soil have their being without any nourishment. Why, then, must we be given a mouth and a stomach and endowed with these two extra organs? And even if we were to be endowed with these organs, He could have made it possible for us to derive our nourishment as the fish and shellfish derive theirs from water, or the cricket and the cicada from the dew, who all are able to obtain their growth and energy this way and swim or fly or jump or sing. Had it been like this, we should not have to struggle in this life and the sorrows of mankind would have disappeared. On the other hand, He has given us not only these two organs, but has also endowed us with manifold appetites or desires, besides making the pit bottomless, so that it is like a valley or a sea that can never be filled. The consequence is that we labor in our life with all the energy of the other organs, in order to supply inadequately the needs of these two. I have thought over this matter over and over again, and cannot help blaming the Creator for it. I know, of course, that He must have repented of His mistake also, but simply feels that nothing can be done about it now, since the design or pattern is already fixed. How important it is for a man to be very careful at the time of the conception of a law or an institution!

There is certainly nothing to be done about it, now that we have got this bottomless pit to fill, and the fact of our having possessed a stomach has, to say the least, colored the course of human history. With a generous understanding of human nature, Confucius reduced the great desires of human beings to two: alimentation and reproduction, or in simpler terms, food and drink and women. Many men have circumvented sex, but no saint has yet circumvented food and drink. There are ascetics who have learned to live a continent life, but even the most spiritual of men cannot forget about food for more than four or five hours. The most constant refrain of our thought occurring unfailingly every few hours is, "When do I eat?" This occurs at least three times a day, and in some cases four or five times. International conferences, in the midst of discussion of the most absorbing and most critical political situations, have to break up for the noon meal. Parliaments have to adjust their schedule of sessions to meal hours. A coronation ceremony that lasts more than five or six hours or conflicts with the midday meal, will be immediately denounced as a public nuisance. And stomach-gifted that we all are, the best arrangement we can think of when we gather to render public homage to a grandfather is to give him a birthday feast.

There is a reason for it. Friends that meet at meals meet at peace. A good birds' nest soup or a delicious chow mein has the tendency to assuage the heat of our arguments and tone down the harshness of our conflicting points of view. Put two of the best friends together when they are hungry, and they will invariably end up in a quarrel. The effect of a good meal lasts not only a few hours, but for weeks and months. We rather hesitate to review unfavorably a book written by somebody who gave us a good dinner three or four months ago. It is for this reason that, with the Chinese deep insight into human nature, all quarrels and disputes are settled at dinner tables instead of at the court of justice. The pattern of Chinese life is such that we not only settle disputes at dinner, after they have arisen, but also forestall the arising of disputes by the same means. In China, we bribe our way into the good will of everybody by frequent dinners. It is, in fact, the only safe guide to success in politics. Should some one take the trouble of compiling statistical figures, he would be able to find an absolute correlation between the number of dinners a man gives to his friends and the rate or speed of his official promotion.

But, constituted as we all are, how can we react otherwise? I do not think this is peculiarly Chinese. How can an American postmaster-general or chief of department decline a private request for a personal favor from some friend at whose home he has eaten five or six good meals? I bet on the Americans being as human as the Chinese. The only difference is the Americans haven't got insight into human nature or haven't proceeded logically to organize their political life in accordance with it. I guess there is something similar to this Chinese way of life in the American political world, too, since I cannot but believe human nature is very much the same and we are all so much alike under the skin. Only I don't notice it practiced so generally as in China. The only thing I have heard of is that candidates for public office give outings for the families in the districts, bribing the mothers by feeding their children with ice cream and soda pop. The inevitable conviction of the people after such a public feeding is that"He's a jolly good fellow,"which usually bursts out in song. This is merely another form of the practice of the medieval lords and nobles in Europe who, on the occasion of a wedding or a noble's birthday, gave their tenants a generous feast with liberal meats and wine.

How a Chinese spirit glows over a good feast! How apt is he to cry out that life is beautiful when his stomach and his intestines are well-filled! From this well-filled stomach suffuses and radiates a happiness that is spiritual. The Chinese relies upon instinct and his instinct tells him that when the stomach is right, everything is right. That is why I claim for the Chinese a life closer to instinct and a philosophy that makes a more open acknowledgment of it possible. The Chinese idea of happiness is, as I have noted elsewhere, being"warm, well-filled, dark and sweet"–referring to the condition of going to bed after a good supper. It is for this reason that a Chinese poet says,"A well-filled stomach is indeed a great thing; all else is luxury. "

With this philosophy, therefore, the Chinese have no prudery about food, or about eating it with gusto. When a Chinese drinks a mouthful of good soup, he gives a hearty smack. Of course, that would be bad table manners in the West. On the other hand, I strongly suspect that Western table manners, compelling us to sip our soup noiselessly and eat our food quietly with the least expression of enjoyment, are the true reason for the arrested development of the art of cuisine. Why do the Westerners talk so softly and look so miserable and decent and respectable at their meals? Most Americans haven't got the good sense to take a chicken drumstick in their hand and chew it clean, but continue to pretend to play at it with a knife and fork, feeling utterly miserable and afraid to say a thing about it. This is criminal when the chicken is really good. As for the so–called table manners, I feel sure that the child gets his first initiation into the sorrows of this life when his mother forbids him to smack his lips. Such is human psychology that if we don't express our joy, we soon cease to feel it even, and then follow dyspepsia, melancholia, neurasthenia and all the mental ailments peculiar to the adult life. One ought to imitate the French and sigh an"Ah!"when the waiter brings a good veal cutlet, and makes a sheer animal grunt like"Um-mm!"after tasting the first mouthful. What shame is there in enjoying one's food, what shame in having a normal, healthy appetite? No, the Chinese are different. They have bad table manners, but great enjoyment of a feast.

In fact,I believe the reason why the Chinese failed to develop botany and zoology is that the Chinese scholar cannot stare coldly and unemotionally at a fish without immediately thinking of how it tastes in the mouth and wanting to eat it. The reason I don't trust Chinese surgeons is that I am afraid that when a Chinese surgeon cuts up my liver in search of a gall-stone, he may forget about the stone and put my liver in a frying pan. For I see a Chinese cannot look at a porcupine without immediately thinking of ways and means of cooking and eating its flesh without being poisoned. Not to be poisoned is for the Chinese the only practical, important aspect of it. The taste of the porcupine meat is supremely important, if it should add one more flavor known to our palate. The bristles of the porcupine don't interest us. How they arose, what is their function and how they are connected with the porcupine's skin and endowed with the power of sticking up at the sight of an enemy are questions that seem to the Chinese eminently idle. And so with all the animals and plants, the proper point of view is how we humans can enjoy them and not what they are in themselves. The song of the bird, the color of the flower, the petals of the orchid, the texture of chicken meat are the things that concern us. The East has to learn from the West the entire sciences of botany and zoology, but the West has to learn from the East how to enjoy the trees, the flowers, and the fishes, birds and animals, to get a full appreciation of the contours and gestures of different species and associate them with different moods or feelings.

Food, then, is one of the very few solid joys of human life. It is a happy fact that this instinct of hunger is less hedged about with taboos and a social code than the other instinct of sex, and that generally speaking, no question of morality arises in connection with food. There is much less prudery about food than there is about sex. It is a happy condition of affairs that philosophers, poets, merchants and artists can join together at a dinner, and without a blush perform the function of feeding themselves in open public, although certain savage tribes are known to have developed a sense of modesty about food and eat only when they are individually alone. The problem of sex will come in for consideration later, but here at least is an instinct which, because less hampered, produces fewer forms of perversion and insanity and criminal behavior. This difference between the instinct of hunger and the instinct of sex in their social implications is quite natural. But the fact remains that here is one instinct which does not complicate our psychological life, but is a pure boon to humanity. The reason is because it is the one instinct about which humanity is pretty frank. Because there is no problem of modesty here, there is no psychosis, neurosis or perversion connected with it. There is many a slip between the cup and the lip, but once food gets inside the lips, there is comparatively little sidetracking. It is freely admitted that everybody must have food, which is not the case with the sexual instinct. And being gratified, it leads to no trouble. At the worst, some people eat their way into dyspepsia, or an ulcered stomach or a hardened liver, and a few dig their graves with their own teeth—there are cases of Chinese dignitaries among my contemporaries who do this—but even then, they are not ashamed of it.

For the same reason, fewer social crimes arise from food than from sex. The criminal code has comparatively little to do with the sins of illegal, immoral and faithless eating, while it has a large section on adultery, divorce, and assault on women. At the worst, husbands may ransack the icebox, but we seldom hang a man for spiking a Frigidaire. Should such a case ever be brought up, the judge will be found to be full of compassion. The frank admission of the necessity of every man feeding himself makes this possible. Our hearts go out to people in famine, but not to the cloistered nuns.

This speculation is far from being idle because there is little public ignorance about the subject of food, as compared with public ignorance on the subject of sex, which is appalling. There are Manchu families which school their daughters in the art of love as well as in the art of cooking before their marriage, but how much of this is done elsewhere in the world? The subject of food enjoys the sunshine of knowledge, but sex is still surrounded with fairy tales, myths and superstitions. There is sunshine about the subject of food, but very little sunshine about the subject of sex.

On the other hand, it is highly unfortunate that we haven't got a gizzard or a crop or a maw. In that case, human society would be altered beyond recognition; in fact, we should have an altogether different race of men. A human race endowed with gullets or gizzards would be found to have the most peaceful, contented and sweet nature, like the chicken or the lamb. We might grow a beak, which would alter our sense of beauty, or we might have merely done with rodent teeth. Seeds and fruits might be sufficient, or we might pasture on the green hillsides, for Nature is so abundant. Because we should not have to fight for our food and dig our teeth into the flesh of our defeated enemy, we would not be the terrible warlike creatures that we are today.

There is a closer relation between food and temperament—in Nature's terms—than we thought. All herbivorous animals are peaceful by nature: the lamb, the horse, the cow, the elephant, the sparrow, etc.; all carnivorous animals are fighters: the wolf, the lion, the tiger, the hawk, etc. Had we been an herbivorous race, our nature would certainly be more elephantine. Nature does not produce a pugnacious temperament where no fighting is needed. Cocks still fight with each other, but they fight not about food, but about women. There would still be a little fighting of this sort among the males in human society, but it would be vastly different from this fighting for exported canned goods that we see in present-day Europe.

I do not know about monkeys eating monkeys, but I do know about men eating men, for certainly all evidences of anthropology point to a pretty universal practice of cannibalism. That was our carnivorous ancestry. Is it therefore any wonder that we are still eating each other in more senses than one—individually, socially and internationally? There is this much to be said for the cannibals, that they are sensible about this matter of killing. Conceding that killing is an undesirable but unavoidable evil they proceed to get something out of it by eating the delicious sirloins, ribs and livers of their dead enemies. The difference between cannibals and civilized men seems to be that cannibals kill their enemies and eat them, while civilized men kill their foes and bury them, put a cross over their bodies and offer up prayers for their souls. Thus we add stupidity to conceit and a bad temper.

I quite realize that we are on the road to perfection, which means that we are excusably imperfect at present. That, I think, is what we are. Not until we develop a gizzard temper can we call ourselves truly civilized. I see in the present generation of men both carnivorous and herbivorous animals—those who have a sweet temper and those who have not. The herbivorous men go their way through life minding their own business, while the carnivorous men make their living by minding that of others. If I abjured politics ten years ago, after having a foretaste of it during four months, it was because I early made the discovery that I was not by nature a carnivorous animal, although I enjoy a good steak. Half of the world spends its time doing things, and half of the world spends its time making others do things for them, or making it impossible for others to do anything. The characteristic of the carnivorous is a certain sheer delight in pugilism, logrolling, wire-pulling, and in double-crossing, outwitting and forestalling the enemy, all done with a genuine interest and real ability, for which, however, I confess I fail to have the slightest appreciation. But it is all a matter of instinct; men born with this pugilistic instinct seem to enjoy and revel in it, while real creative ability,

ability in doing their own jobs or knowing their own subjects, seems at the same time usually to be underdeveloped. How many good, quiet herbivorous professors are totally lacking in rapacity and the ability to get ahead in competition with others, and yet how truly I admire them! In fact I may essay the opinion that all the world's creative artists are vastly better at minding their own business than minding that of others, and are therefore of the herbivorous species. True evolution of mankind consists in the multiplication of the herbivorous homo sapiens over against the carnivorous variety. For the moment, however, the carnivorous must still be our rulers. That must be so in a world believing in strong muscles.

五．论强壮的肌肉

因为我们是动物，有一个会死的身体，所以我们也就有被杀的可能，一般的人当然是不喜欢被杀的。我们有一种追求智识和智慧的神圣欲望，可是我们一旦有了智识，便产生各人不同的见解，争论也就此产生。在长生不死的神灵世界里，争论是永不会停止的，如果有异见的双方都不肯认错，我真想不出有什么方法可以解决它。在人类的世界里，便不同了，争论者的对方便是他的眼中钉——越看越觉得看不顺眼，自己的论据也越觉合理——于是把对方干脆杀死，争端就此解决。如果甲杀死乙，甲便是对的；如果乙杀死甲，乙便是对的。毋庸讳言，这就是禽兽解决争端的老法子。所以在动物世界里，狮子始终站在对的地位。

人类的社会情形就是这样，所以我们可以根据这种现象，把人类的历史——一直到现代——做一种适当的解释。关于地球圆形说及太阳系的问题，伽利略（Galileo）曾发现了一些观念，但他不能不把他的观念改变一些，因为他有一个会死会被杀戮和被苦刑的身体。和伽利略辩论是件吃力的事，假如伽利略少了一个会死的身体的话，你休想叫他认错，这就变成讨厌的事情了。但在当时，只要有一间行刑房或一间监牢——更不必说断头台和炮烙柱——就可以叫他认错。当时的传教士和绅士们决心要和伽利略一决雌雄，后来伽利略认错了，于是传教士和绅士们更相信他们自己是对的，争端就此爽快地解决。

这种解决方法极为便当，极有效力。侵略战争、宗教战争，萨拉丁（Saladin，十二世纪埃及和叙利亚，也门等的苏丹，历史英雄）跟基督教的战争，宗教的肃清，烧死神巫的事件，以至于近代用战舰去宣传基督福音……这一切的事件——全是依据于这种人类由遗传所得的动物的逻辑……

我们都有一些高贵的狮子性格，我们都鄙视争论。我们崇敬军人，因为他能把意见不同者毫不犹豫地杀死。如果一个人要证明他自己是对的，要使对方闭口无言，最敏捷的方法是把他绞杀，当人们无力强迫人家认错时，才会用说话这方法。所以实际行动的人是少说话的，他们鄙视争论。我们说话的目的是想影响人家，如果我们知道力量足以影响人家或统治他们，那又何必多说话呢？这样看来，国际联盟在上次东三省战争和埃塞俄比亚战争时说了那么许多的话，岂不有点无聊？那是够可悲的，所以国际联盟这种特性是一个不祥的预兆……

我常以为国际联盟实是一所优良的现代语言学校，注重现代语言的翻译，起先由一个演说家用英语做了一次完美的演说，等到听众熟识了演词的要旨和内容后，又由一个翻译专家把这篇演词译成流利畅达、优雅的法语。关于发音声调之抑扬顿挫等，务必达于上乘，使听众对语言学得到一次极美满的实习，事实上比倍立兹学校更好；它是一所现代语言学兼演说学的学校。有一个朋友甚至对我说，当他在日内瓦住了六个月后，他多年发音含糊的旧习居然纠正了。但是这里也有一个令人诧异的事实，就是在这个虽然是专为交换意见的国际联盟，说话不做别用的机关里，居然也有

"大说话者"和"小说话者"之别,"大说话者"是那些有"大拳头"者,"小说话者"是那些有"小拳头"者,可见这种玩意儿根本是骗人的勾当,是十分无聊的,"小拳头"国家的口才不能像"大拳头"国家那么流利似的。我以为信服"大拳头"者的口才的固有观念,仍是上述那种动物遗传性的一部分(我在此不愿用畜生 Brute 一词,然而用在这里似乎是再适当也没有了)。

这件事的要点当然是在人类除了有斗争的本能外,也还有说话的本能。从历史的意义上说来,舌头是和拳头或粗臂膀同其久远的。人类之异于其他动物,便是能把说话跟拳脚混合应用,这就是人类特有的性格。这点似乎说明了国际联盟、美国议会或职工大会这一类的组织——只要是人类有机会说话的任何组织——会永远存在着的,我们人类似乎是注定必须要先用谈论的方法去决定正或误,这并不错,因为谈论也是天使们的一个特性。糟糕的是:当我们谈论到某一个程度时,臂膀较粗的一边便会恼羞成怒,由恼羞成怒而捏紧拳头向台一敲,揪住对方的颈项痛殴一番,然后回过头来问那些好似陪审官的观众道:"我对,还是他对?"这种解决方法只有人类会用。天使完全以说话去解决争端;禽兽完全以肌肉和爪牙解决争端;唯有人类拿拳脚和说话去解决争端。天使绝对相信公理;禽兽绝对相信强权;只有人类以为强权就是公理。两者比较起来,谈论本能或辩白是非的努力当然比较高尚一些。我们相信终会有一天人类将完全以谈论方式去解决争端,到那时候,人类才是真的得救了。在现在我们只好暂时让茶馆方法和茶馆心理去解决争端,不管是在茶馆里或国际联盟里解决;这两个地方始终一贯地同样表现着人类的特性。

这种茶馆式的解决方式,我曾见到过两次,一次是在一九三一——一九三二年,一次是在一九三六年。最有趣的是:在这两次的争论中,又夹杂了人类的第三种本能——谦让。在一九三一年那桩事件中，两边发生了争端，我们在茶馆里据说是做陪审官的。起诉的原因是一边犯了偷窃产业之罪。那个臂膀粗大的家伙起初也参加争论，做了一次替自己辩争的演说，他说他对这邻人已表示无上的忍耐——他是多么有自制力，多么大量慷慨，他是要替他邻人整顿花园，动机是多么纯洁！但有桩可笑的事情，当他一边在督促我们继续谈论下去时，一边却溜出屋外，在那偷来的产业四周筑了一道篱笆，然后回来请我们去看看他的行径是否正当。我们都去看，看见他那道新筑的篱笆还在慢慢地向西扩大开去，还继续不断地移动着呢，"好吧！我对，还是他对？"我们的判决是："你错了。"——我们说这句话确有一点轻率。那个臂膀粗大的家伙以为他在大庭广众之间遭了凌辱！他的谦让之心受了冤枉，他的荣誉遭了玷污，便提出抗议，并且又生气又骄傲地走出会场，用着带讥笑的鄙视态度把鞋上的尘埃拂去，认为我们都不够朋友。试想这样的一个家伙居然以为是受了凌辱！所以我说，谦让这第三种本能把事情弄得愈加复杂。这次之后，这家以科学方法解决私人争端为标榜的茶馆便失掉了不少威信。

后来我们在一九三六年又去评判另一个争端。另外一个臂膀粗大的家伙说，他要把这次争论的始末和盘托出，要求大家主持正义。我听得"正义"一词，不禁打了一个寒噤，我们鉴于局势的恶劣和我们陪审官的才能不足，所以早具戒心。但因为我们决心要表明我们是名副其实的公正裁判者，所以几乎全体一致地当面对他说，你的行为是错误的，是恃强凌弱的。他也以为他是受了凌辱，谦让之心受了冤枉，荣誉受了玷污，于是揪住对方的颈项，拖到外边把他杀死，然后回转来问我们："我对，还是他对？"我们齐声说："你对，你对！"一边说一边还深深地向他鞠一躬。他还是不满足，又问我们："现在我可有资格做你们的朋友吗？"我们都像茶馆里的顾客一般，嚷道："你当然有资格做我们的朋友！"杀人者是多么谦逊啊！

　　这是救世主降生后一九三六年的人类文明。我想法律和正义的演进，在最古当我们还是野蛮人的时代，一定也有着上述那种情形。由茶馆式的解决方式演进到最高法院——在那里被判罪者并不抗议他是受了凌辱——似乎已经过了一个很长时期的发展。十年前当我们创办那片茶馆时，我们以为我们是走上文明之路了，可是一个更明慧的上帝，一个认识人类和人类的主要性格的上帝，也许早就预料到中途会发生挫折的。他也许起始就知道我们一定会失败，一定会踌躇不前，我们又回复了从前的行为，像森林中的野蛮人一样，互相攻击，揪住对方的头发，咬着对方的肉……但我并不完全绝望。因为谦让或廉耻这种本能究竟是好的，谈论也是好的本能。在我看来，现在的人类完全不知道什么叫做羞耻。但我们还是应该继续假想着我们是有廉耻观念的，继续去谈论吧，让我们这样一直谈论下去，总会有一天能够达到天使那种幸福的境界。

V. On Having Strong Muscles

ANOTHER important consequence of our being animals and of our having mortal bodies is that we are susceptible to murder, and the average man doesn't like murder. True, we have a divine desire for knowledge and wisdom, but with knowledge come also differences of point of view and therefore arguments. Now in a world of immortals, arguments would last forever, for I can conceive of no way of settling a dispute, if neither of the disputing immortals is willing to admit that he is wrong. In a world of mortals, the situation is different. The disputing party generally gets so obnoxious in the eyes of his opponent—and the more obnoxious he will appear, the more embarrassingly right his arguments are—so that the latter just kills him, and that settles the argument. If"A"kills"B", "A"is right; and if"B"kills"A", "B"is right. This, we hardly need remind ourselves, is the old, old method of settling arguments among brutes. In the animal kingdom, the lion is always right.

This is basically so true of human society that it offers a good interpretation of human history, even down to the present time. After all, Galileo retracted as well as discovered certain ideas about the roundness of the earth and the solar system. He retracted because he had a mortal body, susceptible to murder or torture. It would have taken infinite trouble to have argued with Galileo, and if Galileo had had no mortal body, you could never have convinced him that he was wrong, and that would have been an eternal nuisance. As it was, however, a torture chamber or a prison cell, not to speak of the gallows or the stake, sufficed to show how wrong he was. The clergy and the gentlemen of the period were determined to have a showdown with Galileo. The fact that Galileo was convinced that he was wrong strengthened the belief of the clergy of that period that they were right. That settled the matter very neatly.

There is something convenient and handy and efficient about this method of settling quarrels. Wars of depredation, religious wars, the conflict of Saladin and the Christians, the Inquisition, the burning of witches, the more modern preaching of the Christian gospel and proselytizing of heathens by gun-boats...—all these proceed upon this animal logic to which all mankind is heir...

There is something of the noble lion in us that disdains arguments. Hence our glorification of the soldier because he makes short shrift with dissenters. The quickest way to shut up a man who believes he is right, and who shows the propensity to argue, is to hang him. Men resort to talking only when they haven't the power to enforce their convictions upon others. On the other hand, men who act and have the power to act seldom talk. They despise arguments. After all, we talk in order to influence people, and if we know we can influence people, or control them, where is the need for talking at all? In this connection, is it not somewhat disheartening that the League of Nations talked so much during the last Manchurian and Ethiopian wars? It was altogether pathetic. There is something ominous about this quality of the League of Nations...

I have always thought that the League of Nations was an excellent School for Modern Languages, specializing in translation of the modern tongues, giving the hearers excellent practice by first making an accomplished orator deliver a perfect address in English, and after the audience is thus made acquainted with the gist and content of the speech, having it rendered into fluent, flawless, classical French by a professional translator, with intonation, accent and all. In fact, it is better than the Berlitz School; it is a school of modern languages and public speaking to boot. One of my friends, in fact, reported that, after a six months' stay at Geneva, his lisping habit which had bothered him for years was cured. But the amazing fact is that even in this League of Nations, consecrated to the exchange of opinions, in an institution that conceivably has no other purpose than talking, there should be a distinction between Big Talkers and Small Talkers, the Big Talkers being those having Big Fists, and the Small Talkers being those having Small Fists, which shows the whole thing is quite silly, if not a fake. As if the nations with Small Fists couldn't talk as fluently as the others! That is to say, if we mean just talking... I cannot but think that this inherent belief in the eloquence of the Big Fist belongs to that animal heritage we have spoken of. (I shouldn't like to use the word"brute"here, and yet it would seem most appropriate in this connection.)

Of course, the gist of the matter lies in the fact that mankind is endowed with a chattering instinct as well as the fighting instinct. The tongue is, historically speaking, as old as the fist or the strong arm. The ability to talk distinguishes man from animals, and the mixture of verbiage and barrage seems to be a peculiarly human trait. This would seem to point to the permanency of institutions like the League of Nations, or the American Senate, or a tradesmen's convention–anything that affords men an opportunity to talk. It seems we humans are destined to chatter in order to find out who is right. That is all right; chattering is a characteristic of the angels. The peculiarly human trait lies in the fact that we chatter to a certain point until one of the parties of the dispute who has a stronger arm feels so embarrassed or angered–"Embarrassment leads naturally to anger, "the Chinese say–until that embarrassed and therefore angered party thinks that this chattering has gone far enough, bangs the table, takes his opponent by the neck, gives him a wallop, and then looks about and asks the audience, which is the jury, "Am I right or am I wrong?" And as we learn at every tea house, the audience invariably replies, "You are right!" Only humans ever settle a thing like that. Angels settle arguments all by chatter; brutes settle arguments all by muscle and claws; human beings alone settle them by a strange confusion of muscle and chatter. Angels believe sheerly in right; brutes believe sheerly in might; and human beings alone believe that might is right. Of the two, the chattering instinct, or the effort to find out who is right, is of course the nobler instinct. Someday we must all just chatter. That will be the salvation of mankind. At present, we must be content with the tea house method and tea house psychology. It doesn't matter whether we settle an argument in a tea house or at the League of Nations; at both places, we are consistently and characteristically human.

I have witnessed two such tea house scenes, one in 1931-1932 and one in 1936. And the most amusing thing is, there was an admixture of a third instinct, modesty, in these two squabbles. In the 1931 affair, we were at the tea house and there was one party in dispute with another and we were supposed to be the jury in the matter. The charge was some sort of a theft or stealing of property. The fellow with the strong arm at first joined in the argument, made an address to justify himself, spoke of his infinite patience with his neighbor—what restraint, what magnanimity, what unselfishness of motive in his desire to cultivate his neighbor's garden! The funny thing was, he encouraged us to go on with our chatter while he stole outside the room and completed the stealing by staking up a fence around the stolen property, and then came in to ask us to go and see for ourselves if he wasn't right. We all went, saw the new fence being steadily pushed farther and farther to the West, for even then the fence was being constantly shifted. "Now, then, am I right or wrong?" We returned the verdict of "You are wrong!"—a little impudent of us to have said that. There upon the fellow with the strong arm protested that he was publicly insulted, that his sense of modesty was injured and his honor be smirched. Angrily and proudly he walked out of the room, wiping the dust off his shoes with sneering contempt, thinking us not good company for him. Imagine a man like that feeling insulted! That is why I say the third instinct of modesty complicates the matter. Thereupon the tea house lost a good bit of its reputation as a place for scientific settling of private quarrels.

Then in 1936 we were called upon to judge another dispute. Another fellow with a strong arm said he would lay the facts of the dispute before the table and ask for justice. I heard the word"justice"with a shudder. And we believed him—not without a premonition as to the awkwardness of the situation or our questionable capacity as a jury. Determined to justify our reputation as fair-minded and competent judges, we, almost to a man, told him to his face that he was wrong, that he was nothing but a bully. He, too, felt insulted; again his sense of modesty was injured and his honor was besmirched. Well, then, he took the opponent by the neck and went outside and killed him, and then he came back and asked us,"Now am I right or wrong?"And we echoed,"You are right!"with a profound bow. Still not satisfied, he asked us,"Am I good enough company for you now?"and we shouted like a regular tea house crowd,"Of course you are!"But what modesty on the part of the killer!

That is human civilization in the year of Our Lord 1936. I think the evolution of law and justice must have passed through scenes like the above in its earliest dawn, when we were little better than savages. From that tea house scene to the Supreme Court of Justice, where the convicted does not protest that he is insulted by the conviction, seems a long, long way of development. For some ten years, while we started the tea house, we thought we were on the road to civilization, but a wiser God, knowing human beings and our essential human traits, might have predicted the setback. He might have known how we must fail and falter at the beginning, being only half civilized as we are at present. For the present, the reputation of the tea house is gone, and we are back to falling upon each other and tearing each other's hair out and digging our teeth into each other's flesh, in the true grand style of the jungle... Still I am not in total despair. That thing called modesty or shame is after all a good thing, and the chattering instinct also. The way I look at it is we are quite devoid of real shame at present. But let us continue to pretend that we have a sense of shame, and continue to chatter. By chattering we shall one day attain the blessed state of the angels.

六．论灵心

　　你也许说人类的灵心是造物主最高贵的产物。这话大多数人以为如此，尤其是指像爱因斯坦的那种灵心一般，能以一个长的数学方程式去证明弯曲的空间。或像爱迪生的灵心那样，发明留声机和活动影戏，或像其他物理学家的灵心那样，能测量出一颗行近地球或远离地球的星辰的光线，或去研究无从捉摸的原子构造，或是像彩色电影摄影机发明家的灵心一样；和猴子无目的、善变的、暗中探索的好奇心比较之下，不得不使我们承认我们确有一个高贵的、伟大的灵心，有一个能够了解这宇宙的灵心。

　　然而普通的灵心只是可爱而不是高贵的。如果人类的灵心都是高贵的，那么我们将变成完全合理的动物，没有罪恶，没有弱点，也没有错误的行为。如果真是这样，这世界将变成一个多么乏味的世界，我们一定会变成极讨厌的动物。我是一个人性主义者，所以一无罪恶的圣人引不起我的兴趣。而在我们的不合理中，自相矛盾中，戏要和假日的欢乐中、成见中、顽固中和健忘中，我觉得我们都是可爱的，如果我们都有一个十全十美的头脑，我们在每一新年里便不用做新的计划。当我们在大除夕回想到新年里所决定的计划时，我们发现我们只做到了三分之一，另外三分之一不曾发现，还有三分之一则已经忘却了。人生之美便在这里。一个计划如果可以完全实现，便不能引起我们的兴趣。一个将军如果预先知道可以绝对获胜，连双方死伤的确数也能预料得到，他对战事便会失掉兴趣，远不如把它放弃不干爽快些；下棋的人，如果知道对方的灵心——不管是比他好的、坏的，或平常的——而无错误，便不会再想下棋。如果我们看小说时，确知书中每个人物的未来灵心动作，因此而料到小说的最后结果，那么所有的小说无一读的价值了。阅读一部小说，便是在追求一个多变动的、不可测度的灵心，这个灵心是一条以许多连续发生的情势而造成的迷路，在相当的时候，实现其不可测摸的决定。如在小说中写一个严峻的、无宽恕心的父亲，假如一直没有宽容子女的时候，在我们看来便不再像是一个人，甚至是一个不忠实的丈夫。如果永远是这样，读者的兴趣不久就会失掉。你可以假想一位骄傲的作曲家，人家无论怎样规劝他，总不愿替某一位美丽的女人写一出歌剧，可是当他一听见有一位他所憎恶的作曲家想做这工作时，便会马上答应。或试想一位科学家，发誓不把他的著作刊在报纸上，可是一看见一位和他竞争的科学家弄错了一个字，他便会忘掉自己所定的规律，拿着作品去发表。这里，我们把握到人类灵心的特性了。

人类的灵心是不合理的，是固执的、偏见的，是任性的，是不可预料的，因此可爱。如果我们不承认这个真理，那么我们费去一百年在人类心理学上的研究工作，便不能算为有结果。换言之，我们的灵心仍保存着人猿智力上那种无目的、暗中摸索的性质。

试看人类灵心的演进程序。我们灵心的功用原本是一个觉察危险而保全生命的器官，而它终于能够体会逻辑和准确的数学方程式，仅是一桩偶然的事。我们的这个灵心确不是为这种功用而创造的。它的原来功用是仅想嗅嗅食物。但除了嗅嗅食物外，如也能嗅嗅一个抽象的数学公式，那固然也不坏。以我的观念，人类的头脑像章鱼或海星，长了一些触角以便摸索真理，待摸到后就把他吃掉（我对其他动物的头脑观念也是如此）。我们今日总说"摸索"（Feeling）真理，而不说"思索"（Think）真理。脑部及其他的感官就是摸索用具。头脑的触角怎样摸索真理，在物理学上有着一个很奥妙的现象，正如眼睛网膜中的紫色怎样感光一样奥妙。当头脑每次和其他有关的知觉器官脱离联系，从事所谓"抽象的思维"时，当每次离开威廉·詹姆斯（William James，美国本土第一位哲学家和心理学家）所谓知觉的现实（Perceptual Reality）而逃进意念的现实世界（The world of Conceptual Reality）时，它的活力消灭了，人性也消失了，也退了。我们都被一种错误的见解所困惑，以为灵心的真真功用便是思维，如果我们不更正对"思维"这个名词的错误观念，我们一定会在哲学上犯下很笨拙的错误。当一个哲学家走出他们的书房，观察市场上的往来群众时，这个错误的见解一定会使他感到幻灭，好像思维与我们日常的行为是很有关系似的。

已故的罗宾逊（James Harvey Robinson，美国历史学家，二十世纪初美国"新史派"的奠基人和倡导者）在《创造中的灵心》（The Mind in the Making）里，曾经想证明我们的灵心是怎样由四个基本阶段而产生，他以为人类的灵心，是由动物的灵心、野蛮人的灵心、孩童的灵心和传统的文明人的灵心渐渐产生出来，现在还在这四个基本阶段上进展着；他又更进一步说，如果现代的人类要想把文明继续发展下去，我们还须产生一颗更善于批评的心。我的思虑比较科学化的时候，颇赞同这个见解，在比较明慧的时候，却怀疑这个阶段在一般的进步上是否能办得到，或甚至是否适宜。我颇愿让我们的灵心像现在一样不合理下去，这是可爱的。我不愿见到我们在这世界上都变成十全十美合理的人类。我不相信科学的进步吗？不，我不信任圣者的境界。我反对智识吗？或许是，或许不是，我只是爱好人生，因爱好人生，所以我极端不信任智能。你可以幻想出一个完美的世界，在那里报纸上没有杀人的新闻，因为那时大家都是无所不通、无所不知，因此没有一所房屋会发生火警，没有一架飞机会失事，没有一个丈夫会遗弃他的老婆，没有一个牧师会跟歌女私奔，没有一个皇帝会因恋爱而牺牲皇位，每个人的心思都千篇一律，大家都各照着自己在十岁时所决定的计划去实行，丝毫不苟——这么一个幸福的人世还是省了吧！在这么一个世界里，人生的一切兴奋和骚动全都消灭了。世界没有文学了，因为那时已没有罪恶，没有错误的行为，没有人类的弱点，没有混乱的情欲，没有不规则的举动，最坏的是，没有令人惊异的事物，那就等于四五万观众在看他们预先已知道哪一匹马得锦标的跑马比赛一样，毫无趣味。人类易生错误的本性是人生色彩的精粹所在，正如障碍跑马比赛上的出冷门一样的有趣。试想塞缪尔·约翰逊博士（Dr. Samuel Johnson，花了九年时间独力编写《英文字典》，十八世纪英国文坛执牛耳者）如果没有他的固执偏见将成为怎样一个人？如果我们全是十全十美合理性的人，那么我们非但不能变成十全十美的智者，反而将退化而成自动机器，而人类灵心也只在记录某一些冲动，像煤气表那样机械地记录下来。这便是不人道的行为，而不人道便是不好。

读者或许疑心我在故意视罪恶为美德，竭力替人类的弱点辩护。这是不对的。如果我们一方面有了一个完全合理的灵心，而获得了合理完美的行为，另一方面，却会失去人生的欢乐和色彩。跟一个具着美德但是平凡模样的丈夫或妻子同过一生一样，是再无聊也没有的事。我相信种种极其合理的人类所造成的社会确是适于生存的，但我疑惑在这种情境之下的生存是否值得。我们固然要想尽种种方法去造成一个有秩序的社会——可是我们不要一个太过于有秩序的社会。我想世界上，也许蚂蚁是最合理的动物。它们无疑已经创立了一个十全十美的社会主义国家，在这种制度之下生活了近一百万年。单以合理的行为方面而论，我想蚂蚁应当占第一位，人类占第二位（但我还是怀疑我们是否有这个资格）。蚂蚁是一种耐劳的、健全的、好储蓄的、肯节俭的动物，它们的生活都受着社会的统制和进行自我的训练，但是我们不然。它们为了国家或社会，肯一天工作十四小时；它们只知道义务而很少想到权利；它们有恒心、有秩序、有礼貌、有毅力，尤其有着更严明的纪律。人类在纪律方面是拙劣的标本，拙劣到连做博物院里的标本也够不上。

你可以跑到名人纪念堂去看看那些陈列在甬道上的伟大人物的雕像，便能发觉在他们一生中合理行为是最缺少的东西。那个爱上克里奥佩特拉（Cleopatra）的恺撒（Julius Caesar）——高贵的恺撒，他的行为太不合理了，几乎为了一个女人而忘掉了帝国[安东尼（Anthony）却是完全忘掉帝国的]。那个摩西——在一怒之下，把他那花了四十天工夫跟上帝在西奈山上铭刻的神圣石版敲碎，以这一点而论，他并不比那些叛弃上帝去崇拜金牛犊的以色列人更有理性。那个大卫王（King David）——有时残暴，有时慷慨，有时虔诚，有时亵渎，有时敬拜上帝，有时犯罪，后来写了诗篇来表示忏悔，重新敬拜上帝。所罗门王（King Solomon）——他是智慧的象征，但对他的儿子一筹莫展……孔子——他回答一个宾客他不在家，等那客人刚走到门口时，他又在楼上唱歌，使客人知道他确是在家。耶稣——在喀西马尼流泪，在十字架上怀着疑心。莎士比亚——把"次好的床"遗赠给他的老婆。弥尔顿（Milton）——因为不能和他十七岁的妻子共同生活，写了一篇离婚的论文，后因受人攻击，便在《论出版自由》（Areopagitica）一文里替言论自由辩护。歌德（Goethe）和他的夫人在礼拜堂举行婚礼时，他们那十九岁的儿子就站在旁边哩。斯威夫特（Jonahan Swift）和史黛拉（Stella）……易卜生（Ibsen）和巴达奇（Emilie Bar-dach）（他保持着合理的行为——这对他是有益的）……

统治这世界的是热情，不是理智，这已是很明显了。所以使这些伟大人物都成为可爱者，使他们有人性者，实是他们的缺乏"理性"，而不是"合理性"。中国人为他们祖先所写的讣闻和传记，大多是无趣味的，不正确的，所以不堪一读，因为他们已把他们祖先写成变态的、完全伪善的人。——他们对于我所著的《吾国与吾民》最大的批评是：我把中国人描写得太有人性了，因为我把他们的长处和缺点都描写出来。他们（至少那些小官僚）相信如果我能把中国写成一个乐园，有儒家圣贤居住着，永远过着和平和理智的幸福生活，我就能够替祖国做更有力的宣传！官僚们的愚蠢真是没有办法。——传记之有魔力和传记值得一读，全在其表现伟大人物所具有和我们相同的人性方面的特性，传记里面每一个不合理的行为，都能显示其更有真实性。利顿·斯特拉奇（Lytton Strachey，英国著名传记作家）作品之所以成功，便是他在描画人物时能注意这一点。

英国人的健全灵心，可以做极佳的例证，英国人对于逻辑尚欠高明，但是他们的头脑有着很好的触角去察觉危险，保全生命。不过我在他们的民族行为上或他们的理性的历史里，还寻不出合于逻辑的东西。他们的大学、宪法、英格兰教会，都是杂凑成章的东西，因为它们都是在历史的发展过程中逐渐累积起来的。不列颠帝国的力量就是在于英国人的脑筋欠高明，在于他们完全不能了别人的意见，他们深信英国人的工作方法是唯一正当的方法，英国人的食品是唯一精美的食品。英国人一旦懂得了讲道理并失去了倔强的自信心时，不列颠帝国便会倾覆灭亡。一个人如果对自己尚且怀疑，便不能出去征服世界。我们全然不能了解英国人对他们国王的态度，一方面如此忠诚和真实敬爱；另一方面却剥夺了国王的言论自由，毫无顾忌地告诉他行为要谨慎，否则"滚你的蛋"……英国在伊丽莎白女皇时代需要海盗来保护帝国，便居然能够有着相当数目的海盗以应付当时的局势，因而也就崇敬他们。英国在每一时代都能在适当的时候，有着适当的同盟国，对付着相当的仇敌，从事适当的战事，可是总用着一个不适当的名称。英国人从不依逻辑去行事，而是靠着他们的触角去行事。

英国人有着红润的肤色，无疑这是伦敦的雾和玩板球运动的结果。这么一个健康的皮肤在他们思想上当然占着极重要的地位，换句话说，在他们摸索中过着一生的程序上占着重要的地位。英国人用他们健康的皮肤去思想，正如中国人用他们的伟大的肚肠去思想一样。这一回事，凡是中国人大多是承认的。我们中国人以为我们确是用肚肠去思想的；我们说一个学者有学问便称他为"满腹思想""满腹经纶""满腹诗文"。此外还有"满腹"的"牢骚""愤怒""悔恨""郁闷"或"期望"等话。中国的情人分别之后写信时总说"愁肠百结"，或在别离的当儿说"肝肠寸断"。中国学者把一篇文章或演讲词的大意想好还没写上纸时，我们便说他们已打好了"腹稿"。他们已经把他们要写的东西在肚里安排好了。这一点是绝对科学化的，可以拿得出证据来，尤其是在现代心理学家对我们思想的情感性质和构造更为明了的今日。可是中国人并不要科学上的证据，他们只要肚里有数。中国曲调的情感性质，全由唱曲者的横膈膜下发出来：如果你不懂得这个，你就无法了解中国音乐及其浓厚的情感色彩。

我们在研究自然的宇宙或和人类无关的任何东西时，不应该否认人类灵心的伟大才能。我对于科学的成就很乐观，可是对善于批评的灵心在应付人类事件时，或对于人类达到一种超过情欲支配的宁静和理解的境地时的发展，怀着较小的希望。以我个人的意思，人类也许已经达到崇高的阶段，但是从社会集团这方面说来，人类还受着原始时代的情欲所支配。因之，在进化的过程中有时不免要开倒车，野蛮的本能有时要暴露出来，疯狂的行为和集团的歇斯底里有时也要产生出来。

我们既然了解了我们人类的弱点，所以更有理由可以诅骂那许许多多的坏蛋：即利用我们的弱点来煽动我们参加二次大战的坏蛋；那个灌输仇恨心理（我们之间的仇恨已经太多了）的人；那个称颂自夸和自私（这两种东西本已不少了）的人；那个利用我们人类的顽固和种族观念的人；那个在训练青年时取消上帝第五诫的人；那个推崇残杀和战争（好像我们还不够好战似的）的人；那个煽动我们人类的情欲（好像我们还不够像禽兽似的）的人。这种坏蛋的灵心，无论怎样的机巧，怎样的聪明，终于是禽兽的灵心。智慧的优美精神被一只禽兽或一个魔鬼绊缠着，这种情形我们现在才知道也是我们的动物遗传性之一；也可说：智慧的优雅精神拿着一条破旧的皮带暂时把这个魔鬼缚住，使之驯服，不过这条皮带随时有扭断的可能，魔鬼也随时可以获得自由，在和散那（Hosanna，希伯来颂赞上帝之声或祈福之语）的颂赞声中，黑天〔Juggernaut，印度神话中毗湿奴神（Vishnu）第八化身克利希那（Krishna）的称号〕每年的纪念日，人民以巨车载其偶像游行各处（如信徒有自伏地下被车辗死得升天国），车子将毫无顾忌地在我们身上辗过去，暗示着我们是始终如何近于野蛮，和我们的文明是多么肤浅，于是世界将变成一个伟大的舞台，在舞台上，摩尔人（Moors）将杀死基督徒，基督徒将弄杀摩尔人，黑种人将攻击白种人，白种人将杀死黑种人，野鼠将由沟渠里跑出来吃人类的尸身，鸷鸟将盘旋于一个丰盛的人肉宴席上——这一切不过是要提醒我们，使我们知道动物间的关系罢了，大自然是善于做这种实验的。

　　精神分析学家在医治有精神病的病人时，常常使他们回忆过去的事情，使他们用客观的眼光去观察他们自己的生活。所以人类对他们的过去多多回忆一下，对于他们自己的驾驭力也许会有更大的进步。我们如果知道我们有动物的遗传性以及跟禽兽相差无几，或许就会晓得怎样去抑止那些禽兽般的行为。我们有了这个动物遗性，更容易在动物寓言和讥讽文章里，如《伊索寓言》(Aesop's Fables)、乔叟的《百鸟会议》(Parliament of Fowles)、斯威夫特的《格列佛游记》(Swift's Gulliver's Travels)和法郎士的《企鹅岛》(Anatole Frances Penguin Island)等里边看见我们的原形。这些动物寓言在伊索时代就很合时宜，就是在救主降生后四千年仍旧是很适合的。我们有补救的方法吗？那善于批评的灵心太浅陋、太冷酷了，要用这个灵心来思考是不能得益处的，智理也没多大用处；只有那种合理的，有理性的精神，那种温暖的、朝气的、情感的、直觉的思想，跟着同情混合起来，才不至于使我们重复退化到我们祖先的典型。只有把我们的生命发展起来，和我们的本能调和着，我们才会得救。我们为培植我们的感觉和情感，比教育我们的思想更为重要。

VI. On Having a Mind

THE human mind, you say, is probably the noblest product of the Creation. This is a proposition that most people will admit, particularly when it refers to a mind like Albert Einstein's that can prove curved space by a long mathematical equation, or Edison's that can invent the gramaphone and the motion picture, or the minds of other physicists who can measure the rays of an advancing or receding star or deal with the constitution of the unseen atoms, or that of the inventor of natural-color movie cameras. Compared with the aimless, shifting and fumbling curiosity of the monkeys, we must agree that we have a noble, a glorious intellect that can comprehend the universe in which we are born.

The average mind, however, is charming rather than noble. Had the average mind been noble, we should be completely rational beings without sins or weaknesses or misconduct, and what an insipid world that would be! We should be so much less charming as creatures. I am such a humanist that saints without sins don't interest me. But we are charming in our irrationality, our inconsistencies, our follies, our sprees and holiday gaieties, our prejudices, bigotry and forgetfulness. Had we all perfect brains, we shouldn't have to make new resolutions every New Year. The beauty of the human life consists in the fact that, as we review on New Year's Eve our last New Year resolutions, we find we have fulfilled a third of them, left unfulfilled another third, and can't remember what the other third was. A plan that is sure to be carried out down to its last detail already loses interest for me. A general who goes to battle and is completely sure of his victory beforehand, and can even predict the exact number of casualties, will lose all interest in the battle, and might just as well throw up the whole thing. No one would play chess if he knew his opponent's mind—good, bad or indifferent—was infallible. All novels would be unreadable did we know exactly how the mind of each character was going to work and were we able consequently to predict the exact outcome. The reading of a novel is but the chase of a wayward and unpredictable mind making its incalculable decisions at certain moments, through a maze of evolving circumstances. A stern, unforgiving father who does not at some moment relax ceases to impress us as human, and even a faithless husband who is forever faithless soon forfeits the reader's interest. Imagine a renowned, proud composer, whom no one could induce to compose an opera for a certain beautiful woman, but who, on hearing that a hated rival composer is thinking of doing it, immediately snatches at the job; or a scientist who in his life has consistently refused to publish his writings in newspapers, but who, on seeing a rival scientist make a slip with one single letter, forgets his own rule and rushes into print. There we have laid our finger upon the singularly human quality of the mind.

The human mind is charming in its unreasonableness, its inveterate prejudices, and its waywardness and unpredictability. If we haven't learned this truth, we have learned nothing from the century of study of human psychology. In other words, our minds still retain the aimless, fumbling quality of simian intelligence.

Consider the evolution of the human mind. Our mind was originally an organ for sensing danger and preserving life. That this mind eventually came to appreciate logic and a correct mathematical equation I consider a mere accident. Certainly it was not created for that purpose. It was created for sniffing food, and if after sniffing food, it can also sniff an abstract mathematical formula, that's all to the good. My conception of the human brain, as of all animal brains, is that it is like an octopus or a starfish with tentacles, tentacles for feeling the truth and eating it. Today we still speak of "feeling" the truth, rather than "thinking" it. The brain, together with other sensory organs, constitutes the feelers. How its tentacles feel the truth is still as great a mystery in physics as the sensitivity to light of the purple in the eye's retina. Every time the brain dissociates itself from the collaborating sensory apparatus and indulges in so-called "abstract thinking," every time it gets away from what William James calls the perceptual reality and escapes into the world of conceptual reality, it becomes devitalized, dehumanized and degenerate. We all labor under the misconception that the true function of the mind is thinking, a misconception that is bound to lead to serious mistakes in philosophy unless we revise our notion of the term "thinking" itself. It is a misconception that is apt to leave the philosopher disillusioned when he goes out of his studio and watches the crowd at the market. As if thinking had much to do with our everyday behavior!

The late James Harvey Robinson has tried to show, in The Mind in the Making, how our mind gradually evolved from, and is still operating upon, four underlying layers: the animal mind, the savage mind, the childish mind and the traditional civilized mind, and has further shown us the necessity of developing a more critical mind if the present human civilization is to continue. In my scientific moments, I am inclined to agree with him, but in my wiser moments, I doubt the feasibility, or even the desirability, of such a step of general progress. I prefer to have our mind charmingly unreasonable as it is at present. I should hate to see a world in which we are all perfectly rational beings. Do I distrust scientific progress? No, I distrust sainthood. Am I anti-intellectualistic? Perhaps yes; perhaps no. I am merely in love with life, and being in love with life, I distrust the intellect profoundly. Imagine a world in which there are no stories of murder in newspapers, every one is so omniscient that no house ever catches fire, no airplane ever has an accident, no husband deserts his wife, no pastor elopes with a choir girl, no king abdicates his throne for love, no man changes his mind and everyone proceeds to carry out with logical precision a career that he mapped out for himself at the age of ten—good-by to this happy human world! All the excitement and uncertainty of life would be gone. There would be no literature because there would be no sin, no misbehavior, no human weakness, no upsetting passion, no prejudices, no irregularities and, worst of all, no surprises. It would be like a horse race in which every one of the forty or fifty thousand spectators knew the winner. Human fallibility is the very essence of the color of life, as the upsets are the very color and interest of a steeplechase. Imagine a Doctor Johnston without his bigoted prejudices! If we were all completely rational beings, we should then, instead of growing into perfect wisdom, degenerate into automatons, the human mind serving merely to register certain impulses as unfailingly as a gas meter. That would be inhuman, and anything inhuman is bad.

My readers may suspect that I am trying a desperate defense of human frailties and making virtues of their vices, and yet it is not so. What we gained in correctness of conduct through the development of a completely rational mind, we should lose in the fun and color of life. And nothing is so uninteresting as to spend one's life with a paragon of virtue as a husband or wife. I have no doubt that a society of such perfectly rational beings would be perfectly fitted to survive, and yet I wonder whether survival on such terms is worth having. Have a society that is well-ordered, by all means—but not too well-ordered! I recall the ants, who, to my mind, are probably the most perfectly rational creatures on earth. No doubt ants have evolved such a perfect socialist state that they have been able to live on this pattern for probably the last million years. So far as complete rationality of conduct is concerned, I think we have to hand it to the ants, and let the human beings come second (I doubt very much whether they deserve that). The ants are a hard-working, sane, saving and thrifty lot. They are the socially regimented and individually disciplined beings that we are not. They don't mind working fourteen hours a day for the state or the community; they have a sense of duty and almost no sense of rights; they have persistence, order, courtesy and courage, and above all, self-discipline. We are poor specimens of self-discipline, not even good enough for museum pieces.

Run across any hall of honor, with statues of the great men of history lining the corridor, and you will perceive that rationality of conduct is probably the last thing to be recalled from their lives. This Julius Caesar, who fell in love with Cleopatra-noble Julius Caesar, who was so completely irrational that he almost forgot (as Anthony did entirely forget) an empire for a woman. That Moses, who in a fit of rage shattered the sacred stone tablets which had taken him forty days on Mount Sinai to inscribe in company with God, and in that he was no more rational than the Israelites who forsook God and took to worshiping the Golden Calf during his absence. That King David, who was alternately cruel and generous, alternately religious and impious, who worshiped God and sinned and wrote psalms of repentance and worshiped God again. King Solomon, the very image of wisdom, who couldn't do a thing about his son... Confucius, who told a visitor he was not at home and then, as the visitor was just outside the door, sang upstairs in order to let him know that he was at home... Jesus, with his tears at Gethsemane and his doubts on the cross... Shakespeare, who bequeathed his"second-best bed' to his wife... Milton, who couldn't get along with his seventeen-year-old wife and therefore wrote a treatise on divorce and, being attacked, then burst forth into a defense of the liberty of speech in Areopagitica... Goethe, who went through the Church's wedding ceremony with his wife, their nineteen-year-old son standing by their side... Jonathan Swift and Stella... Ibsen and Emilie Bardach (he kept rational–good for him) ...

Is it not plain that passion rather than reason rules the world? And that what made these great men lovable, what made them human, was not their rationality, but their lack of rationality? Chinese obituary notices and biographical sketches of men and women written by their children are so unreadable, so uninteresting and so untrue, because they make all their ancestors appear abnormally and wholly virtuous beings... The great criticism of my book on China by my countrymen is that I make the Chinese too human, that I have painted their weaknesses as well as their strength... But the very charm of biography, its very readability, depends on showing the human side of a great character which is so similar to ours. Every touch of irrational behavior in a biography is a stroke in convincing reality. On that alone, the success of Lytton Strachey's portraits depends.

An excellent illustration of a perfectly sound mind is provided by the English. The English have got bad logic, but very good tentacles in their brains for sensing danger and preserving life. I have not been able to discover anything logical in their national behavior or their rational history. Their universities, their constitution, their Anglican Church are all pieces of patchwork, being the steady accretions of a process of historical growth. The very strength of the British Empire consists in the English lack of cerebration, in their total inability to see the other man's point of view, and in their strong conviction that the English way is the only right way and English food is the only good food. The moment the Englishmen learn to reason and lose their strong confidence in themselves, the British Empire will collapse. For no one can go about conquering the world if he has doubts about himself. You can make absolutely nothing out of the English attitude toward their king, their loyalty to, and their quite genuine affection for, a king who is deprived by them of the liberty of speech and is summarily told to behave or quit the throne... When Elizabethan England needed pirates to protect the Empire, she was able to produce enough pirates to meet the situation and glorified them. In every period, England was able to fight the right war, against the right enemy, with the right ally, on the right side, at the right time, and call it by a wrong name. They didn't do it by logic, did they? They did it by their tentacles.

The English have a ruddy complexion, developed no doubt by the

London fog and by cricket. A skin that is so healthy cannot but help playing an important part in their thinking, that is, in the process of feeling their way through life. And as the English think with their healthy skin, so the Chinese think with their profound intestines. That is a pretty generally established matter in China. We Chinese know that we do think with our intestines; scholars are said to have "a bellyful of ideas," or "of scholarship," "of poetry and literature," or "a bellyful of sorrow," or "of anger," "remorse," "chagrin" or "lon ging." Chinese lovers separated from each other write letters to say that "their sorrowful intestines are tied into a hundred knots," or that at their last parting "their intestines were broken." Chinese scholars who have arranged their ideas for an essay or a speech, but have not written them down on paper, are said to have their "belly manuscript" ready. They have got their ideas all arranged down there. I'm quite sure they have. This is, of course, all strictly scientific and capable of proof, especially when modern psychologists come to understand better the emotional quality and texture of our thought. But the Chinese don't need any scientific proof. They just feel it down there. Only by appreciating the fact that the emotional quality of Chinese melodies all starts from below the diaphragm of the singers,   can one understand Chinese music with its profound emotional color.

One must never deprecate the capacity of the human mind when dealing with the natural universe or anything except human relationships. Optimistic about the conquests of science, I am less hopeful about the general development of a critical mind in dealing with human affairs, or about mankind reaching a calm and understanding far above the sway of passions. Mankind as individuals may have reached austere heights, but mankind as social groups are still subject to primitive passions, occasional back-slidings and outcroppings of the savage instincts, and occasional waves of fanaticism and mass hysteria.

Knowing then our human frailties, we have the more reason to hate the despicable wretch who in demagogue fashion makes use of our human foibles to hound us into another world war; who inculcates hatred, of which we already have too much; who glorifies self-aggrandizement and self-interest, of which there is no lack; who appeals to our animal bigotry and racial prejudice; who deletes the fifth commandment in the training of youth and encourages killing and war as noble, as if we were not already warlike enough creatures; and who whips up and stirs our mortal passions, as if we were not already very near the beast. This wretch's mind, no matter how cunning, how sagacious, how worldly-wise, is itself a manifestation of the beast. The gracious spirit of wisdom is tied down to a beast or a demon in us, which by this time we have come to understand is nothing but our animal heritage, or rather it ties this demon down by an old and worn leash and holds it but in temporary submission. At any time the leash may snap, and the demon be unleashed, and amidst hosannas the car of Juggernaut will ride roughshod over us, just to remind us once more how terribly near the savage we have been all this time, and how superficial is our civilization. Civilization will then be turned into a magnificent stage, on which Moors will kill Christians and Christians kill Moors and Negroes fall upon whites and whites stab Negroes and field mice emerge from sewers to eat human corpses and hawks circle in the air over an abundant human feast—all just to remind ourselves of the brotherhood of animals. Nature is quite capable of such experiments.

Psychoanalyists often cure mental patients by making them review their past and see their life objectively. Perhaps if mankind will think more of their past, they will also have a better mastery over themselves. The knowledge that we have an animal heritage and that we are very near the beasts might help to check our behaving like beasts. This animal heritage of ours makes it easier to see ourselves as we are in animal fables and satires, as in Aesop's Fables, Chaucer's Parliament of Fowles, Swift's Gulliver's Travels and Anatole France's Penguin Island. These animal fables were good in Aesop's day and will still be good in the year A. D. 4000. How can we remedy the situation? The critical mind is too thin and cold, thinking itself will help little and reason will be of small avail; only the spirit of reasonableness, a sort of warm, glowing, emotional and intuitive thinking, joined with compassion, will insure us against a reversion to our ancestral type. Only the development of our life to bring it into harmony with our instincts can save us. I consider the education of our senses and our emotions rather more important than the education of our ideas.

第四章

论近人情

一．论人类的尊严

　　我在前一章里已经讨论过人类的不免一死和在动物界里的地位，以及人类文明本质上因此而发生的后果。可是我们觉得这个轮廓还不完全，我们还需要一种用以造成一个关于人类天性和人类尊严二者联合而成的圆满观念。噢，有了，人类的尊严——这就是我们所找寻的名词！我们必须对这一点多加说明，我们必须了解这尊严是由什么东西所造成，否则就会有谬误，而失掉它的踪迹。尤其是在二十世纪的现代和后代，我们随时有失掉我们尊严的危险。

　　如果你一定说我们是动物，那么你以为人类是最奇妙的动物吗？这一点我很同意。只有人类发明了一种文明，这就是难能可贵的事情。世间也许有形式更优良，构造更完美的高超动物，如马就是一例；其他如狮子，有着更优美的肌肉；狗有着更灵敏的嗅觉，更驯良、更忠义的心地；鹰有着更锐利的视觉；鸽子有着更清晰的方向感觉；蚂蚁有着更节俭、更有纪律、更有劳作的能力；鸠鸽和鹿有着更温驯的脾气；牛有着更大的忍耐性和满足性；百灵有着更悦耳的歌喉；鹦鹉和孔雀有着更美丽的服饰。最后还有猴子有着一种更好的才能，使我比较喜欢猴子而不喜欢上述那些动

物。人类因有一些猴子的好奇心和聪明，所以我就宁愿做人。就算蚂蚁像
我上面所说，比我们有着更合理、更有纪律的社会，有着一个比今日西班
牙更稳固的政府，可是它们没有图书馆和博物院，是不是？假如蚂蚁或象
有一日发明一个大望远镜，或发现一颗新行星或预知月食，或海豹微积分
学，或海獭开浚巴拿马运河，我便要把锦标赠给它们，称它们为世界之主
和宇宙的主宰。现在我们觉得很可自负，不过，这使我们足以自负的究竟
是什么东西？人类尊严的精华究竟在哪里？最好都把它探索出来。

　　我在本书的卷首已约略暗示过，这种人类的尊严是由放浪者（中国文
学上所尊敬的人物）的四种特质所造成。就是：一种嬉戏的好奇心，一种
梦想的能力，一种纠正这些梦想的幽默感，一种在行为上任性的、不可测
度的质素。这些特质并合起来便是由美国的个人学说所蜕变而来的中国人
的观念。中国文学上所表现的放浪者是一幅极其生动的个人主义者的肖
像，拥戴美国个人主义最有力的文学家沃尔特·惠特曼（Walt Whitman）
之所以被人家称为"伟大的闲逸者（Magnificent Idler）"，确是有来由的。

●

I. On Human Dignity

IN the preceding chapter, we have seen man's mortal heritage, the part he shares with the animal world, and its consequences on the character of human civilization. But still we find the picture is not complete. There is still something missing for a well-rounded view of human nature and human dignity. Ah, human dignity—that is the word! There is a need of emphasizing that and there is a need of knowing what that dignity consists of, lest we confuse the issue and lose it. For there is a very evident danger of our losing that dignity in the twentieth century and especially in the present and immediate following decades.

"Don't you think a man is the most amazing of animals, if you insist that we are animals?"I quite agree. Man alone has invented a civilization, and this is not something to be lightly dismissed. There are perhaps finer animals with better forms and nobler structures, like the horse; with finer muscles, like the lion; with a finer sense of smell and greater docility and loyalty, like the dog; or better vision, like the eagle; or a better sense of direction, like the homing pigeon; with greater thrift and discipline and capacity for hard work, like the ant; with a sweeter temper like the dove or the deer; more patience and contentment like the cow; better singers, like the lark; and better-dressed beings, like the parrot and the peacock. Still there is something in a monkey that makes me prefer the monkey to all these animals, and something of the monkey curiosity and monkey cleverness in man that makes me prefer to be a man. Granted that ants are more rational and better-disciplined beings than ourselves, as I have pointed out, and granted that they have a more stable form of government than present-day Spain, still they haven't got a library or a museum, have they? Any time ants or elephants can invent a giant telescope or discover a new variable star or predict a solar eclipse, or seals can discover the science of calculus or beavers can cut the Panama Canal, I will hand them the championship as masters of the world and Lords of Creation. Yes, we can be proud of ourselves, but we had better find out what it is that we have got to be proud of, what is the essence of human dignity.

This human dignity, as I have already hinted at the beginning of this book, consists of four characteristics of the scamp, who has been glorified by Chinese literature. They are: a playful curiosity, a capacity for dreams, a sense of humor to correct those dreams, and finally a certain waywardness and incalculability of behavior. Together they represent the Chinese version of the American doctrine of the individual. It is impossible to paint a more glowing portrait of the individualist than has been done for the scamp in Chinese literature, and it is certainly no accident that Walt Whitman, the greatest literary champion of American individualism, is himself called the "Magnificent Idler. "

二. 近乎戏弄的好奇：人类文明的勃兴

　　人类的放浪者怎样开始去爬上文明的山巅？他的远大的前程，或他的发展中的智力，事先可有一些什么征兆？这问题的答案，无疑便是人类有一种嬉戏的好奇心，他开头就用他的双手去摸索，把一切东西都翻过来考验研究，像猴子在闲逸的时候把同伴的眼皮或耳朵拨开来，捉一捉虱，或竟是无目的地翻着玩玩。你到动物园去看一对猴子在彼此玩弄耳朵，便可意会到一个牛顿或一个爱因斯坦的前程。

　　人类以手去做嬉戏摸捉的活动形象，实是一个科学上的真理，而不仅是一个形象。当人类能够把身体直立起来，变成了两足动物后，他的双手已得到了解放，人类文明的根基就这样造了起来。我们甚至在猫的身上，当它们的前肢不必担负走路和支撑身体的任务时，也常可以看见这种嬉戏的好奇心。猫和猴子都有产生文明的可能，只可惜，虽然猴子的手指因为攀握树枝已经有了特异的发展，而猫掌依然还是一个掌—— 一块块的肉和软骨造成的东西。

　　让我暂时忘掉我不是一个及格的生物学家，而由这两只手的解放来考据一些人类的文明兴起，因为我有几句话要说，不过这话也许已经有人说过亦未可知。直立的姿势和双手的解放发生了极重大的后果，人类开始能够运用器具，产生了谦逊的意识，而且征服了女人；在这方面也许还有语言的发展，以及嬉戏的好奇心，和探索本能的增强。大家都知道人类文明是由发现器具而开始，而这种发现就是因为人类晓得怎样运用他的双手的结果。当那只大人猿的一部分身体由树上伸下来时（也许因为他的身体太笨重吧），它的进化有两种趋向，不是变成用四只脚走路的狒狒，就是变成学用两只后脚走路的猩猩。由此可以猜想到人类的祖先一定不是狒狒，因为狒狒的前掌太忙了。在另一方面，猩猩至少已经能够直立起来，因此两只手可以得到自由，这种自由在全世界的文明发展上，它的意义是多么重大啊！那个时候，这人猿一定已经学会用手去采摘果子，而不必再用他的大颚。不久，他就居住在高崖上的山洞里，能搬起石头，由崖上滚下来攻打他的敌人，这是人类所用的第一个器具。我们可以想象到他当时怎样地无时无刻用双手去探索操纵，漫无目的地攫取着各式各样的东西。他在漫无目标的摸索之中，偶然摸到一些尖锐的燧石或凹凸棱角的石块，而发觉比圆形的石块更适于斗争杀戮。把东西翻覆察看的简单动作，例如：看看耳后或耳前的简单动作，一定已经增加了他对一切东西的思考力，因而增加了他脑中的印象，于是他头脑的前部因受刺激而发育起来。

我相信人类对于性欲方面的羞怯观念——这在动物是完全没有的——其原因也是由于这种直立的姿势。因为这种直立的姿势（造物主在创造万物的时候也许不曾有过这个意思）使身体后面的某些部分立刻变成了身体的中心，而本来是在后边的东西现在在前边了。此外还有其他不良的部位调整与这种新姿势有关，这些不良的部位调整尤其影响到女人，使她们时常发生流产及月经方面的烦恼。依解剖学说来，我们的肌肉构造原本是根据四足动物而成。例如，当母猪怀胎的时候，它就很合理地将它的胎儿由身上的横脊骨悬挂下来，像已洗涤干净的衣服晒挂在一条绳索上一样，重量的分配非常均称。如果要怀孕的女人直立着，正如把那绳索垂直起来，而仍希望挂在上边的衣服保持着原来的地位。我们的腹膜肌肉是极端不适于这种姿势的：如果我们本来是两足动物，那么这种肌肉一定会很适宜地连在肩膀上，而一切也比较合理称职。对于人类的子宫和卵巢的构造有着相当研究的人，在看见这些东西能保持原来的位置并且发生了它们本来的作用，而并没有脱节和引起更多的月经烦扰等时，应当感觉奇怪。月经的整个神秘现象还没有人有过满意的解释，可是我非常相信纵使卵巢里的卵子有按期更换的必要，我们敢说这种机能的现时程序实是太不适当、太冗长和太痛苦，我相信现在这些缺点都是二足的姿势所造成。

于是发生了女人屈服的事情，而人类社会及其现在的特征或许也就这样发展起来。假如人类的母亲能以四足走路，我想绝不会被她的丈夫所征服。这里有着两种力量同时在进行。在一方面当时男女已经是闲逸的、好奇的、嬉戏的动物，好色的本能有了新的表现。接吻还不够快乐或完全美满，这点我们可由两只黑猩猩用坚硬的、突出的下颚互相接吻的情形看出来。有了手却产生出一些较灵巧较温柔的新动作：轻拍、抚摩、搔抓、拥抱，这些完全是因为互相捕捉身上的虱子而偶然发现的。假如我们多毛的人类祖宗身上没有虱子，我相信世界上一定不会有抒情诗之类的东西产生，所以对于这好色本能的发展一定是有着很大帮助。

在另一方面，那两足而怀孕的母亲，现在须过着长期的、微弱无力的、一筹莫展的生活。在较早的时候，当人类对于直立的姿势还没有完全适应时，我知道那怀孕的母亲如要带着她的重负到各处去走动是很困难的，尤其是在脚腿和后跟还没有完全改变成功，盘骨还没有向后突出，而使前面的负担可以平衡的时候。在最早的时期，直立的姿势尚是拙劣不便的，所以一个洪积期的母亲一定会在没有人看见的地方匍匐而行，以减轻其背脊的疼痛。人类的母亲因为有了这些不便利和其他方面的女人的困难，于是开始运用其他的手段，而以爱情为媒介，但因此失掉了她的一些独立精神。她在这种困于床褥的时候，当然需要人家的安慰和体贴啊！并且直立的姿势使婴孩难于学步，因此，总须经过一段长的婴孩时期方能走路，但小牛或小象几乎一出世就会走路，而初生的婴孩要二三年的工夫才能学会。担任养育婴孩的责任除了母亲还有更适宜的人吗[8]？

于是人类走上了一条崭新的发展之路。广义的性问题开始渲染了人类的日常生活；人类社会就是依据这一个事实而发展起来的。女人比其他的雌性动物——黑女比雌虎，伯爵夫人比雌狮——更有着女人特具的女性化，始终不变地成为女性。在文明的意义上说来，男女开始有了明显分野。从前是男人注意修饰，现在却是女人专意修饰了；她们第一步想必是把脸上和胸前的毫毛拔去。这不足为怪，纯然是生存策略之一，在动物间尤其明显。老虎有攻击的策略，乌龟有退避的策略，马匹有逃走的策略——究其目的，无非是求生存。女性的可爱和美丽，以及温柔的和狡猾的手段，在生存目的上自有其价值。男人有着较强壮的臂膀，跟他们战斗是不能取胜的；所以唯有贿赂他，谄媚他，博他的欢心，这便是现代文明的特性。女人不用抵抗和进攻的策略，而用迷惑的手段，不用武力去达到她的目的，而尽力用温柔的方法去求实现。所以总括地说，温和即文明。

---

[8] 这种父母的养护时期越来越延长。一个六七岁的野蛮小孩子，差不多可以自立了，文明社会的孩子，却需二十多年的工夫去做求生的准备，甚至在二十多年的教育之后，还得从头学习。

我认为，人类的文明是由女人开始的，而不是由男人。

同时，我认为在说话（即今日之所谓语言）的进展史上，女人也比男子占着更重要的地位。说话的本能在女人是得天独厚的，所以我深信她们对于人类语言的演进，一定有着比男人更重大的贡献。我想古代的人类一定是沉默寡欢的动物。当男性人猿离开洞穴去打猎时，邻居的女人在她们的洞穴前无所事事，一定谈论威廉是否比哈罗好，或哈罗是否比威廉好，或是哈罗昨晚多情得讨厌，他性子多么暴躁。我想人类的语言必是在这个时候开始，此外别无他途。人类以手去取食，使腭部不必再担负去拿食物的任务，结果腭部逐渐低平，逐渐变小，这对于人类语言的发展也有很大的帮助。

不过，这种姿势最重要的结果，是把两手解放了，使它们可以把东西拿起来观察研究，像猴子捉虱为乐那样，这种动作便是研究精神发展的起点。今日人类进步大抵还须归功于捕捉那些扰乱人类的虱子。一种好奇的本能也发展出来了，使人类的灵心可以很自由地用嬉戏的态度去探究各种题目和社会疾患。这智能上的活动和寻索食物并没有关系，而完全是一种人类精神上的训练。猴子捉虱的目的，不是想把它们吃掉，而是当一种游戏玩着。这便是有价值的人类学术和智识的特征，对事物本身发生兴趣，心中存着嬉戏的、闲逸的欲望想把它们了解，而并不是因为那种学问可以直接使我们的肚子不饿（如果这里有自相矛盾，那么，以中国人的立场来说，我对于自相矛盾这件事是觉得快活的）。我以为这是人性的特征，对于人类尊严有着极大的帮助。追求知识的方式不过是一种游戏：所有一切伟大的科学家和发明家，以及创立过有价值的伟大事业者，他们都是如此。从事医学研究的人觉得对细菌所引起的兴趣比对人类更大；天文学家很起劲地记录一颗距离我们有几万万里远的星星的动作，虽则这颗星星和我们人类一点没有关系。一切动物，尤其是年轻的，都有这种游戏的本能，但也只有人类把这种嬉戏的好奇心发展到值得重视的地步。

II. On Playful Curiosity: the Rise of Human Civilization

HOW did the human scamp begin his ascent to civilization? What were the first signs of promise in him, or of his developing intelligence? The answer is undoubtedly to be found in man's playful curiosity, in his first efforts to fumble about with his hands and turn everything inside out to examine it, as a monkey in his idle moments turns the eyelid or the ear-lobe of a fellow-monkey, looking for lice or for nothing at all—just turning about for turning about's sake. Go to the zoo and watch a pair of monkeys picking each other's ears, and there you have the promise of an Isaac Newton or an Albert Einstein.

This figure of the playful, fumbling activities of the exploring human hand is more than a figure. It is a scientific truth. The very basis of human civilization started with the emancipation of the hands consequent upon man's assuming an erect stature and becoming a biped. Such playful curiosity we see even in cats, the moment their front paws are relieved from the duty of walking and supporting the body. It might have been quite as possible for a civilization to be developed from cats as well as from monkeys, except for the fact that in the case of the monkeys, the fingers were already well developed through the clasping of branches, whereas the cat's paws were still paws—merely lumps of flesh and cartilage.

Let me for a moment forget that I am not a qualified biologist and speculate about the rise of human civilization from this emancipation of the hands, because I have a few things to say here, which may or may not have been observed by others. The assumption of an erect stature and the consequent emancipation of the hand had extremely far-reaching results. It brought about the use of tools, the sense of modesty, the subjection of women, and in this connection probably also the development of language, and finally a prodigious increase in playful curiosity and the instinct of exploration. It is pretty well known that human civilization began with the discovery of tools and that this came from the development of the human hands. When the big anthropoid ape descended partially from the tree, probably because his body was too heavy, he had two roads to follow, either that of a baboon, going on all fours, or that of the orang-outang, learning to walk on its hind legs. Human ancestry could not possibly have come from the baboon, a quadruped (or quadrumanum), because the baboon's front paws were too much occupied. On the other hand, with an erect posture more or less successfully acquired by the orangoutang, the hands acquired freedom, and how significant was this freedom for all civilization! By that time, the anthropoid ape certainly had learned already to pick fruit with his hands, instead of with his big jaws. It was but a simple step, when he took to living in a cave on a high cliff, to pick stones and pebbles and roll them down from the cliff on his enemies. That was the first tool man ever used. There we must picture a constant fumbling and manipulating activity of his hands, grasping at things for some purpose or for no purpose. There would be sharp flints or jagged pieces of rocks which through his aimless fumbling were accidently discovered to be more useful for killing than round pieces of stone. The mere act of turning things about, for instance, of looking at the back as well as the front of an ear-lobe, must have already increased his power for conceiving things in their totality and therefore also the number of images he carried in his brain, thus stimulating the growth of the frontal lobes of the brain.

I believe the mystery of the origin of sexual modesty in man, which is totally absent in animals, is also due to this erect posture. For by this new posture, which Father Nature in his scheme of things probably never intended, certain posterior parts of the body at one stroke came to occupy the center of the body, and what was naturally behind came in front. Allied to this terrible new situation were other maladjustments chiefly affecting women, causing frequent abortions and menstrual troubles. Anatomically, our muscles were designed and developed for the quadruped position. The mother pig, for instance, carries its litter of pig embryos logically suspended from its horizontal spine, like wash hung on a line with its weight properly distributed. Asking the human pregnant mother to stand erect is like tipping the wash line vertically and expecting the clothes to remain in position. Our peritoneal muscles are badly designed for that: if we were originally bipeds, such muscles should be nicely attached to the shoulder, and the whole thing would be a more pleasant job. Anybody with a knowledge of the anatomy of the human womb and ovaries should be surprised that they keep in position and function at all, and that there are not more dislocations and menstrual troubles. The whole mystery of menstruation has never yet been satisfactorily explained, but I am quite sure that, even granted that a periodical renewal of ova is necessary, we must admit that the function is carried out in a most inefficient, unnecessarily long and needlessly painful manner, and I have no doubt that this inefficiency is due to the biped position.

This, then, led to the subjection of women and probably also to the development of human society with its present characteristics. I do not think that if the human mother could walk on all fours, she would have been subjected by her husband at all. Two forces came into play simultaneously. On the one hand, men and women were already by that time idle, curious and playful creatures. The amorous instinct developed new expressions. Kissing was still not entirely pleasant, or wholly successful, as we can see it between two chimpanzees kissing each other with hard, clapping, protruding jaws. But the hand developed new, more sensitive and softer movements, the movements of patting, pawing, tickling and embracing, all as incidental results of chasing lice on each other's body. I have no doubt that lyrical poetry would not have developed if our hairy human ancestors had had no lice on their bodies. This, then, must have helped considerably to develop the amorous instinct.

On the other hand, the biped human pregnant mother was now for a considerably longer period subjected to a state of grievous helplessness. During the earlier period of imperfect adjustment to the erect position, I can see that it was even more difficult for the pregnant mother to carry her load and go about, especially before the legs and heels were properly modified, and the pelvis was properly projected backwards to counter-balance the burden in front. At the earliest stages, the biped position was so awkward that a Pleistocene mother must have shamefacedly gone on all fours when nobody was looking, to relieve her aching spine. What with these inconveniences and other women's troubles, the human mother began to use other tactics and play for love, thereby losing some of her spirit of independence. Good Lord, she had need of being patted and pawed during those times of confinement! The erect posture prolonged, too, the period of infancy by making it difficult for the human baby to learn to walk. While the baby calf or baby elephant can trot about practically as soon as it is born, the human baby took two or three years to learn the job, and who was the most natural person to look after him except the mother?

Man then went off into a completely new path of development. Human society developed from the single fact that sex, in the broadest sense of the word, began to color human daily life. The human female was more consciously and constantly a female than a female animal–the negress more than the tigress, and the countess more than the lioness. Specialization between men and women in the civilized sense began to develop, and the female, instead of the traditional male, began to decorate herself, probably by picking hair out of her face and her breast. It was all a matter of tactics for

survival. We see these tactics clearly in animals. The tiger attacks, the tortoise
hides, and the horse runs away—all for survival. Love or beauty and the gentle
cunning of womanhood had then a survival value. The man probably had a
stronger arm, and there was no use fighting him; why not, therefore, bribe
and flatter and please him? That is the very character of our civilization even
today. Instead of learning to repel and attack, woman learned to attract, and
instead of trying to achieve her goal by force, she tried her best to achieve it
by softer means. And after all, softness is civilization. I rather think therefore
that human civilization began with woman rather than with man.

And then I also cannot help but think that woman played a greater role
than man in the development of chattering, which we call language today.
The instinct for chatter is so deep in women that I firmly believe they must
have helped to create human language in a more important manner than
men. Early men, I imagine, were quite morose, silent creatures. I suppose
human language began when the first male anthropoids were away from their
cave dwellings hunting, and two women neighbors were discussing before
their caves whether William was a better fellow than Harold or Harold was
a better fellow than William, and how Harold was disgustingly amorous
last night, and how easily he could be offended. In some such form, human
language must have begun. It cannot be otherwise. Of course the taking
of food by hands, thus relieving the original double duty of the jaw in both
taking and eating food, eventually made it possible also for the jaw gradually
to recede and diminish in size, and thus also helped toward the development
of human language.

But, as I have suggested, the most important consequence of this new
posture was the emancipation of the hands for turning things about and
examining them inside out, as symbolized in the pastime of chasing lice by
monkeys. From this chasing of lice, the development of the spirit of free
inquiry into knowledge had its start. Today human progress still consists very
largely in chasing after some form or other of lice that is bothering human

society. An instinct for curiosity has been developed which compels the human mind to explore freely and playfully into all kinds of subjects and social diseases. This mental activity has nothing to do with seeking food; it is an exercise of the human spirit pure and simple. The monkeys do not chase after lice in order to eat them, but for the sheer fun of it. And this is the characteristic of all worthwhile human learning and human scholarship, an interest in things in themselves and a playful, idle desire to know them as they are, and not because that knowledge directly or immediately helps in feeding our stomach. (If I contradict myself here as a Chinese, I am happy as a Chinese that I contradict myself.) This I regard as characteristically human and contributing very largely to human dignity. Knowledge, or the process of seeking knowledge, is a form of play; it is certainly so with all scientists and inventors who are worth anything and who truly accomplish worthwhile results. Good medical research doctors are more interested in microbes than in human beings, and astronomers will try to record or register the movements of a distant star hundreds of millions of miles away from us, although the star cannot possibly have any direct bearing on human life on this planet. Almost all animals, especially the young, have also the play instinct, but it is in man alone that playful curiosity has been developed to an important extent.

三. 论梦想

有人说过，不知足是神圣的；我却以为不知足是人性的。猴子是第一号阴沉动物，在动物中，我只看见黑猩猩有一个真正忧郁的面孔。我往往觉得这种动物很像哲学家，因为唯有哲学家才会有忧郁和沉思的表情。牛似乎不会思想，至少似乎不在推究哲理，因为它们看起来是那么知足；象也许会怀着盛怒，可是它们那不断摆动象鼻的动作似乎代替了思想，而把胸怀中的一切不满都排除。唯有猴子能够显示出彻底讨厌生命的表情。猴子是真够伟大啊！

九九归原地说起来，哲学或许是由讨厌的感觉而开始。无论怎样，人类的特征便是怀着一种追求理想的期望，一种忧郁的、模糊的、沉思的期望。人类住在这个现实的世界里，还有梦想另一个世界的能力和倾向。人类和猴子的差异点也许是猴子仅仅觉得讨厌无聊，而人类除讨厌无聊外，还有着想象力。我们都有一种脱离旧辙的欲望，我们都希望变成另一种人物，大家都有着梦想。兵卒梦想做伍长，伍长梦想做上尉，上尉想做少校或上校。一个气魄宽宏的上校是不把上校当做一回事的，用文雅的词语说起来，他仅仅称之为服务人群的一个机会而已。在事实上讲起来，这种工作确没有什么别的意义。老实说，琼·克劳馥（Joan Crawford，好莱坞黄金时代著名女影星）对于自己并不像世人那么注意，珍妮·盖诺（Janet Gaynor，1928年获得第一届奥斯卡影后）对于自己也不像世人那么注意。世人对一切伟人说："你们不都是很伟大吗？"如果那些伟人真正是伟大的，他们总会回答："伟大又算什么呢？"所以这个世界很像一家照菜单零点的餐馆，每一个顾客总以为邻桌顾客所点的菜肴比他自己所点的更有味、更好吃。一位现代中国大学教授说过一句诙谐语："老婆别人的好，文章自己的好。"在这种意义上说来，世间没有一个人会感到绝对的满足。大家都想做另一个人，只要这另一个人不是他现在的自己。

这种特性无疑是由于我们有想象力和梦想才能。一个人的想象力越大，就越不能得到满足。所以一个富于想象力的小孩，往往比较难于教育；他常常像猴子那样阴沉忧郁，而不像牛那样感到快乐知足。同样，离婚的案件在理想主义者和富有想象力的人们当中，一定比无想象力的人们多。一个理想的终身伴侣的幻想会生出一种不可抗拒的力量，这种力量若在缺乏想象和理想的人们便永远不会感觉。笼统地说来，人类有时也被这种理想的力量引入歧途，有时则辅导上进；可是人类终是完全靠这种想象力而进步的。

我们晓得凡是人都有志向和抱负。有这种东西是可贵的，因为志向和抱负大都被视为高尚的东西。无论个人和国家，都有梦想，我们的行动多少都依照梦想而行事。有些人比一般普通人多做了一些梦，正如每个家庭里都有一个梦想较多的孩子，或是有一个梦想较少的孩子。我得承认我私下比较喜欢那个有梦想的孩子，虽则是个比较忧郁的孩子，也没有关系；他有时也会享受到更大的欢乐、兴奋和狂喜。我觉得人类的构造和无线电收音机很相像，所差者我们收来的不是播来的音乐，而是我们自己所产生的观念和思想。有些灵敏的收音机能够收到其他收音机所收不到的短波，因为这些更远更细的音乐不大容易收到，所以更觉宝贵。

而且我们幼时的那些梦想并不是没有实现性的，这些梦想常和我们终身共存着。因此，如果我自己可以自选做世界上作家之一，我颇愿做个安徒生，能够写《小人鱼》（The Little Mermaid，又名《海的女儿》）的故事，想着那《小人鱼》的思想，渴望着到了长大的时候到水面上来，那真是人类所能感到的最深沉最美妙的快乐了。所以，无论一个孩子是在屋顶的小阁上，或在谷仓里，或是躺在水边，随处都有他的梦想，而这些梦想也是真实的。爱迪生、史蒂文森、司各特（Sir Walter Scott）这三个人在幼年时都梦想过。这种奇妙的梦想，结出了最优美最瑰丽的果。较平庸的孩子也曾多少有过这些梦想，他们梦想中的幻象或许各不相同，但是他们感觉到的快乐是一样的。每个小孩子都有一颗思慕的和切望的灵魂，怀着一种热望去睡觉，希望在早晨醒转来时发现他的梦想已成为事实。他并不把这些梦想告诉大家，因为这些是他自己的，是他正在生长的自我的一部分。小孩子的梦想当中有些较为清晰，有些较模糊，清晰者产生了迫使这梦想实现的力量；而那些较不明晰的便在长成的时候逐渐消失。我们一生中总想把我们幼时的梦想说出来，但是，"有时还没有找到我们所要说的话，我们已经死了"。

讲到国家也是这样，她也有其梦想。这种梦想可以经过许多的年代和世纪依然存在着。有些梦想是高尚的，有些却是歹恶的。征服人家和那些独霸世界一类的梦想，都可说是噩梦；这种国家比那些较有和平梦想的国家不安得多。不过另外还有较好的梦想，梦想着一个更好的世界，梦想着和平，梦想着各国和睦共处，梦想着减少残酷，减少不公平，减少贫穷和痛苦。噩梦常想破坏好梦，因之，二者之间不断地搏斗苦战。人们为梦想而斗争，正如为财产而斗争一样。于是梦想即由幻象的世界走进了现实的世界，而成为我们生命中的一个真实力量。梦想无论怎样模糊，总潜伏在我们心底，使我们的心境永远得不到宁静，直到这些梦想成为事实才止；像种子在地下一样，一定要萌芽滋长，伸出地面来，寻找阳光。所以梦想是真实的。

我们有时也会有混乱的梦想和不符现实的梦想，那是很危险的。因为梦想也是逃避的方法之一。一个做梦者常常梦想要逃避这个世界，但是又不知道要逃避到哪里去。知更鸟常常引动浪漫主义者的幻想。人类有一种热烈的欲望，想把今日的我们变成另一种人，脱离现在的常轨。只要是可以促成变迁的事物，一般人便趋之若鹜。战争总是有吸引力的，因为它使城市里的事务员有机会可以穿起军服，扎起绑腿布，可以有机会免费旅行；在战壕里已经度过三四年生活的兵士，而觉得厌倦了的时候，休战也是情愿的，因为这又使他们有机会回家再穿起平民的衣服，打上一条红领带了。人类显然需要这种刺激。假如世界真要避免战争，最好各国政府行一种制度，每隔十年募集二十岁至四十五岁的人，送他们到欧洲大陆去做一次旅行，去参观博览会一类的盛会。现在英国政府正在动用五十万万英镑去重整军备，我想这笔款子尽够送每个英国人到里维埃拉（Riviera,法国东南地中海边名胜区）去旅行一次了。他们以为战争的费用是必需的，而旅行是奢侈，我不很同意！旅行是必需的，而战争才是奢侈哩。

此外还有其他的梦想，如乌托邦的梦想和长生不老的梦想。长生不老的梦想虽则也像其他的梦想一样模糊，但是十分近于人情，而且极其普遍。不过人类如果真的可以长生不死，到了那时恐怕他们也要不知所为。长生不死的欲望跟站在另一极端的自杀心理属于同类，二者都厌恶这世界，以为现在的世界还不够好。如果问为什么现在的世界还不够好呢？我们只要在春天到乡间去游览一次，就能知道这句问话不应该问而觉得惊异了。

　　乌托邦的梦想情形也是如此。理想仅是一种信仰另一世态的心境，不管它是一种什么世态，总之只要和现代人类的世态不同就是了。理想的自由主义者往往相信本国是国家中最坏的国家，他所生活的社会是最坏不过的社会。他依然是那个在餐馆里照单点菜的家伙，相信邻桌所点的菜总比自己所点的好吃……

III. On Dreams

DISCONTENT, they say, is divine; I am quite sure anyway that discontent is human. The monkey was the first morose animal, for I have never seen a truly sad face in animals except in the chimpanzee. And I have often thought such a one a philosopher, because sadness and thoughtfulness are so akin. There is something in such a face which tells me that he is thinking. Cows don't seem to think, at least they don't seem to philosophize, because they look always so contented, and while elephants may store up a terrific anger, the eternal swinging of their trunks seems to take the place of thinking and banish all brooding discontent. Only a monkey can look thoroughly bored with life. Great indeed is the monkey!

Perhaps after all philosophy began with the sense of boredom. Anyway it is characteristic of humans to have a sad, vague and wistful longing for an ideal. Living in a real world, man has yet the capacity and tendency to dream of another world. Probably the difference between man and the monkeys is that the monkeys are merely bored, while man has boredom plus imagination. All of us have the desire to get out of an old rut, and all of us wish to be something else, and all of us dream. The private dreams of being a corporal, the corporal dreams of being a captain, and the captain dreams of being a major or colonel. A colonel, if he is worth his salt, thinks nothing of being a colonel. In more graceful phraseology, he calls it merely an opportunity to serve his fellow men. And really there is very little else to it. The plain fact is, Joan Crawford thinks less of Joan Crawford and Janet Gaynor thinks less of Janet Gaynor than the world thinks of them." Aren't you remarkable?" the world says to all the great, and the great, if they are truly great, always reply, "What is remarkable?" The world is therefore pretty much like an à la carte restaurant where everybody thinks the food the next table has ordered is so much more inviting and delicious than his own. A contemporary Chinese professor has made the witticism that in the matter of desirability, "Wives are always better if they are others', while writing is always better if it is one's own." In this sense, therefore, there is no one completely satisfied in this world. Everyone wants to be somebody so long as that somebody is not himself.

This human trait is undoubtedly due to our power of imagination and our capacity for dreaming. The greater the imaginative power of a man, the more perpetually he is dissatisfied. That is why an imaginative child is always a more difficult child; he is more often sad and morose like a monkey, than happy and contented like a cow. Also divorce must necessarily be more common among the idealists and the more imaginative people than among the unimaginative. The vision of a desirable ideal life companion has an irresistible force which the less imaginative and less idealistic never feel. On the whole, humanity is as much led astray as led upwards by this capacity for idealism, but human progress without this imaginative gift is itself unthinkable.

Man, we are told, has aspirations. They are very laudable things to have, for aspirations are generally classified as noble. And why not? Whether as individuals or as nations, we all dream and act more or less in accordance with our dreams. Some dream a little more than others, as there is a child in every family who dreams more and perhaps one who dreams less. And I must confess to a secret partiality for the one who dreams. Generally he is the sadder one, but no matter; he is also capable of greater joys and thrills and heights of ecstasy. For I think we are constituted like a receiving set for ideas, as radio sets are equipped for receiving music from the air. Some sets with a finer response pick up the finer short waves which are lost to the other sets, and why, of course, that finer, more distant music is all the more precious if only because it is less easily perceivable.

And those dreams of our childhood, they are not so unreal as we might think. Somehow they stay with us throughout our life. That is why, if I had my choice of being any one author in the world, I would be Hans Christian Andersen rather than anybody else. To write the story of The Mermaid, or to be the Mermaid ourselves, thinking the Mermaid's thoughts and aspiring to be old enough to come up to the surface of the water, is to have felt one of the keenest and most beautiful delights that humanity is capable of. And so, out in an alley, up in an attic, or down in the barn or lying along the waterside, a child always dreams, and the dreams are real. So Thomas Edison dreamed. So Robert Louis Stevenson dreamed. So Sir Walter Scott dreamed. All three dreamed in their childhood. And out of the stuff of such magic dreams are woven some of the finest and most beautiful fabrics we have ever seen. But these dreams are also partaken of by lesser children. The delights they get are as great, if the visions or contents of their dreams are different. Every child has a soul which yearns, and carries a longing on his lap and goes to sleep with it, hoping to find his dream come true when he wakes up with the morn. He tells no one of these dreams, for these dreams are his own, and for that reason they are a part of his innermost growing self. Some of these children's dreams are clearer than others, and they have a force which compel their own realization; on the other hand, with growing age, those less clear dreams are forgotten, and we all live through life trying to tell those dreams of our childhood, and"sometimes we die ere we find the language. "

And so with nations, too. Nations have their dreams and the memories of such dreams persist through generations and centuries. Some of these are noble dreams, and others wicked and ignoble. The dreams of conquest and of being bigger and stronger than all the others are always bad dreams, and such nations always have more to worry about than others who have more peaceful dreams. But there are other and better dreams, dreams of a better world, dreams of peace and of nations living at peace with one another, and dreams of less cruelty, injustice, and poverty and suffering. The bad dreams tend to destroy the good dreams of humanity, and there is a struggle and a fight between these good and bad dreams. People fight for their dreams as much as they fight for their earthly possessions. And so dreams descend from the world of idle visions and enter the world of reality, and become a real force in our life. However vague they are, dreams have a way of concealing themselves and leave us no peace until they are translated into reality, like seeds germinating under ground, sure to sprout in their search for the sunlight. Dreams are very real things.

There is also a danger of our having confused dreams and dreams that do not correspond to reality. For dreams are escapes also, and a dreamer often dreams to escape from the present world, hardly knowing where. The Blue Bird always attracts the romanticist's fancy. There is such a human desire to be different from what we are, to get out of the present ruts, that anything which offers a change always has a tremendous appeal to average humanity. A war is always attractive because it offers a city clerk the chance of donning a uniform and wearing puttees and a chance for travel gratis, while an armistice or peace is always desirable after three or four years in the trenches because it offers the soldier a chance to come back home and wear civilian dress and a scarlet necktie once more. Some such excitement humanity evidently needs, and if war is to be avoided, governments may just as well recruit people between twenty and forty-five under a conscript system and send them on European tours to see some exposition or other, once every ten years. The British Government is spending five billion pounds on its Rearmament Program, a sum sufficient to send every Englishman on a trip to the Riviera.

The argument is, of course, that expenditures on war are a necessity while travel is a luxury. I feel inclined to disagree: travel is a necessity, while war is a luxury.

There are other dreams too. Dreams of Utopia and dreams of immortality. The dream of immortality is entirely human—note its universality—although it is vague like the rest, and few people know what they are going to do when they find eternity hanging on their hands. After all, the desire for immortality is very much akin to the psychology of suicide, its exact opposite. Both presume that the present world is not good enough for us. Why is the present world not good enough for us? We should be more surprised at the question than at any answer to the question if we were out on a visit to the country on a spring day.

And so with dreams of Utopia also. Idealism is merely that state of mind which believes in another world order, no matter what kind of an order, so long as it is different from the present one. The idealistic liberal is always one who thinks his own country the worst possible country and the society in which he lives the worst of all possible forms of society. He is still the fellow in the à la carte restaurant who believes that the next table's order of dishes is better than his own...

四．论幽默感

我很怀疑世人是否曾体验过幽默的重要性，或幽默对于改变我们整个文化生活的可能性——幽默在政治上，在学术上，在生活上的地位。它的机能与其说是物质上的，还不如说是化学上的。它改变了我们的思想和经验的根本组织，我们须默认它在民族生活上的重要。德皇威廉缺乏笑的能力，因此丧失了一个帝国，或者如一个美国人所说，使德国人民损失了几十万万元。威廉二世在私生活中也许会笑，可是在公共场所中，他胡须总是高翘着，给人以可怕的印象，好像他是永远在跟谁生气似的。并且他那笑的性质和他所笑的东西——因胜利而笑，因成功而笑，因高踞人上而笑——也是决定他一生命运的重要因素。德国战败是因为威廉二世不知道什么时候应该笑，或对什么东西应该笑，他的梦想是脱离笑的管束的。

据我看来最深刻的批评就是：民主国的总统会笑，而独裁者总是那么严肃——牙床凸出，下颌鼓起，下唇缩进，像煞是在做一些非可等闲的事情，好像没有他们，世界便不成为世界。罗斯福常常在公共场所中微笑，这对于他是好的，对于喜欢看他们总统微笑的美国人也是好的。可是欧洲独裁者们的微笑在哪里？他们的人民不喜欢看他们的微笑吗？他们一定要装着吃惊、庄严、愤怒，或非常严肃的样子，才能保持他们的政权吗？

现在我们讨论独裁者的微笑，并不是无聊地寻开心；当我们的统治者没有笑容时，这是非常严重的事，他们有的是枪炮啊。在另一方面只有当我们冥想这个世界，由一个嬉笑的统治者去管理时，我们才能够体味出幽默在政治上的重要性。比如说，派遣五六个世界上最优秀的幽默家，去参加一个国际会议，给予他们全权代表的权力，那么世界有救了。因为幽默一定和明达及合理的精神联系在一起，再加上心智上的一些会辨别矛盾、愚笨和坏逻辑的微妙力量，使之成为人类智能的最高形式，我们可以肯定，必须这样才能使每一个国家都有思想最健全的人物去做代表。让萧伯纳代表爱尔兰，史蒂芬·里科克（Stephen Leacock，加拿大著名作家和经济学家）代表加拿大；吉尔伯特·基思·切斯特顿（G. K. Chesterton，英国推理小说著名角色"布朗神父"创造者）已经死了，可是佩德勒·G·伍德豪斯（P. G. Wodehouse，英国幽默小说家，散文大师）或奥尔德斯·赫胥黎（Aldous Huxley，英国作家，代表作《美丽新世界》）可以代表英格兰。威尔·罗杰斯（Will Rogers，美国幽默大师及演员）可惜已经死了，不然他倒可以做一个美国代表。现在我们可以请罗伯特·本奇利（Robert Benchley，美国幽默作家）或海伍德·勃朗恩（Heywood Broun，美国著名记者）去代替他。意大利、法国、德国、俄国也有他们的幽默代表，如果派遣这些人物在大战前夕去参加一个国际会议，我想无论他们怎样拼命地努力，也不能掀起一次欧洲的大战来。你不会想象到这一批国际外交家会掀起一次战争，甚至企谋一次战争，幽默感会禁止他们这样做。当一个民族向另一民族宣战时，他们是太严肃了，他们是半疯狂的，他们深信自己是对的，上帝是站在他们这一边的。具有健全常识的幽默家是不会这么样想。你可以听见萧伯纳在大喊爱尔兰是错误的，一位柏林的讽刺画家说一切错误都是我们的，勃朗恩宣称大半的蠢事应由美国负责，可以看见里科克坐在椅子上向人类道歉，温和地提醒我们说，在愚蠢和愚戆这一点上，没有一个民族可以自誉强过其他民族。在这种情形之下，大战又何至于能引起呢？

那么是谁在掀起战争呢？是那些有野心的人、有能力的人、聪明的人、有计划的人、谨慎的人、有才智的人、傲慢的人、太爱国的人，那些有"服务"人类欲望的人，那些想创造一些事业给世人一个"印象"的人，那些希望在什么场地里造一个骑马的铜像来睥睨古今的人。很奇怪，那些有能力的人、聪明的人、有野心的人、傲慢的人，也就是最懦弱而糊涂的人，缺乏幽默家的勇气、深刻和机巧。他们永远在处理琐碎的事情，他们并不知那些心思较旷达的幽默家更能应付伟大的事情。如果一个外交家不低声下气地讲话，装得战战兢兢、胆怯、拘束、谨慎的样子，便不成其为外交家。——事实上，我们并不一定需要一个国际幽默家的会议来拯救这世界。我们大家都充分潜藏着这种所谓幽默感的东西。当欧洲大战的爆发，真在一发千钧的当儿，那些最劣等的外交家，那些最"有经验"和自信的，那些最有野心的，那些最善于低声下气讲话的，那些最会装得战战兢兢、拘束、谨慎模样的，甚至那些最切望于"服务"人类的外交家，在他们被派遣到会议席上去时，只稍在每次上午及下午的开会议程中，拨出十分钟的时间来放映米老鼠影片，令全体外交家必须参加，那么任何战争仍旧是可以避免的。

我以为这就是幽默的化学作用：改变我们思想的特质。这作用直透到文化的根底，并且替未来的人类对于合理时代的来临，开辟了一条道路。在人道方面我觉得没有再比合理时代更合崇高的理想。因为一个新人种的兴起，一个浸染着丰富的合理精神，丰富的健全常识，简朴的思想，宽和的性情，及有教养眼光的人种的兴起，终究是唯一的重要事情。人类的理想世界不会是一个合理的世界，在任何意义上说来，也不是一个十全十美的世界，而是一个缺陷会随时被看出，纷争也会合理地解决的世界。对于人类，这是我们所希冀的最好的东西，也是我们能够合理地冀望它实现的最崇高的梦想。这似乎是包含着几样东西：思想的简朴性，哲学的轻逸性，微妙的常识，才能使这种合理的文化创造成功。而微妙的常识、哲学的轻逸性和思想的简朴性，恰巧也正是幽默的特性，而且非由幽默不能产生。

这样的一个新世界是很难想象的，因为它跟我们现在的世界是那么不同。一般讲起来，我们的生活是过于复杂了，我们的学问是太严肃了，我们的哲学是太消沉了，我们的思想是太纷乱了。这种种严肃和纷乱的复杂性，使现在的世界成为这么一个凄惨的世界。

我们现在必须承认：生活及思想的简朴性是文明与文化的最崇高最健全的理想，同时必须承认当一种文明失掉了它的简朴性，而浸染习俗，熟悉世故的人们不再回到天真淳朴的境地时，文明就会到处充满困扰，日益退化下去。于是人类变成在他自己所产生的观念、思想、志向和社会制度之下的奴隶，担荷这个思想、志向和社会制度的重担，而似乎无法摆脱它。幸而人类的心智尚有一种力量，能够超脱这一切观念、思想、志向而付之一笑，这种力量就是幽默家的微妙处。幽默家运用思想和观念，就像高尔夫球或弹子戏的冠军运用他们的球，或牧童冠军运用他们的缰绳一样。他们的手法，有一种因熟练而产生的从容，有着把握和轻快的技巧。总之，只有那个能轻快运用他的观念的人，才是他的观念主宰，只有那个能做他的观念主宰的人，才不被观念奴役。严肃终究不过是努力的标记，而努力又只是不熟练的标志。一个严肃的作者在观念的领域里是呆笨而局促的，正如一个暴发户在社交场中那样呆笨而不自然一样。他很严肃，因为他和他的观念相处还不曾达于自然。

说起来有点矛盾，简朴也就是思想深刻的标志和象征。在我看来，在研究学问和写作上，简朴是最难实现的东西。欲求思想明澈已经是一桩困难的事情，然而简朴必须从明澈中产生出来。当一个作家在役使一个观念时，我们也可说那观念在役使他。这里有一桩普通的事实可以证明：一个刚以优异的成绩毕业出来的大学助教，他的讲辞总是深奥繁杂，极难于理解，而只有资格较老的教授们才能把他的思想单纯地用着简明易解的字句表达出来。如果一个年轻的助教不用他自矜博学多才的语句来讲解，他确有出类拔萃而远大的前途的。由技术到简朴，由专家到思想家，其间的过程，根本是一种知识的消化过程，我认为是和新陈代谢的作用完全一样的。一个博学的学者，须把那专门的知识消化了，并且和他的人生观察联系起来，才能够用平易简明的语句把这专门知识贡献出来。在他刻苦追求知识的时间中[我们就假定说是威廉·詹姆斯（William James）的心理学知识吧]，我觉得一定有许多次"心神清爽的休息"，好像一个人在疲乏的长途旅行中停下来喝一杯清凉的饮料一样。在那休息的时间中，那些真正的人类专家，会自己反省一下，"我们到底在做什么？"简朴必须先消化和成熟，当我们渐渐长大成人的时候，思想会变得更明澈，无关紧要的一点或虚假的一面，将尽被剔去，不再骚扰我们。等到观念有了较明确的形态后，一大串的思想便渐渐变成一个简括的公式，突然有一天在一个明朗的清晨跑进了我们的脑子，于是我们的知识达到了真正光辉的境界。嗣后便再不用努力了，真理已变得简单易解，读者也将觉得真理本身是简易的，公式的形成是自然的，因此获得很大的快乐。这种思想上和风格上的自然性——中国的诗人和批评家那么羡慕着——常常被视为一种逐渐成熟的发展过程。当我们讲到苏东坡的散文逐渐成熟时，我们便说他"渐近自然"——这种风格已经把青年人的爱好华丽、夸炫、审美技艺和文艺夸张等心理一概消除。

幽默感滋养着这种思维的简朴性，这是很自然的事。一般地说，幽默家比较接近事实，而理论家比较注重观念，当一个人跟观念本身发生关系时，他的思想会变得非常复杂。在另一方面，幽默家沉浸于突然触发的常识或智机，它们以闪电般的速度显示我们的观念与现实的矛盾。这样使许多问题变得简单。不断地和现实相接触，给了幽默家不少的活力、轻快和机巧。一切装腔作势、虚伪，学识上的胡诌、学术上的愚蠢和社交上的欺诈，将完全扫除净尽。因为人类变得有机巧有机智了，所以也显见得更有智慧。一切都是简单清楚。所以我相信只有当幽默的思维方式普遍盛行时，那种以生活和思维的简朴为特性的健全而合理的精神才会实现。

IV. On the Sense of Humor

I doubt whether the importance of humor has been fully appreciated, or the possibility of its use in changing the quality and character of our entire cultural life—the place of humor in politics, humor in scholarship, and humor in life. Because its function is chemical, rather than physical, it alters the basic texture of our thought and experience. Its importance in national life we can take for granted. The inability to laugh cost the former Kaiser Wilhelm an empire, or as an American might say, cost the German people billions of dollars. Wilhelm Hohenzollern probably could laugh in his private life, but he always looked so terribly impressive with his upturned mustache in public, as if he was always angry with somebody. And then the quality of his laughter and the things he laughed at—laughter at victory, at success, at getting on top of others—were just as important factors in determining his life fortune. Germany lost the war because Wilhelm Hohenzollern did not know when to laugh, or what to laugh at. His dreams were not restrained by laughter.

It seems to me the worst comment on dictatorships is that presidents of democracies can laugh, while dictators always look so serious—with a protruding jaw, a determined chin, and a pouched lower lip, as if they were doing something terribly important and the world could not be saved, except by them. Franklin D. Roosevelt often smiles in public—good for him, and good for the American people who like to see their president smile. But where are the smiles of the European dictators? Or don't their people want to see them smile? Or must they indeed look either frightened, or dignified, or angry, or in any case look frightfully serious in order to keep themselves in the saddle?...

We are not indulging in idle fooling now, discussing the smiles of dictators; it is terribly serious when our rulers do not smile, because they have got all the guns. On the other hand, the tremendous importance of humor in politics can be realized only when we picture for ourselves (by that faculty for dreaming known as"D") a world of joking rulers. Send, for instance, five or six of the world's best humorists to an international conference, and give them the plenipotentiary powers of autocrats, and the world will be saved. As humor necessarily goes with good sense and the reasonable spirit, plus some exceptionally subtle powers of the mind in detecting inconsistencies and follies and bad logic, and as this is the highest form of human intelligence, we may be sure that each nation will thus be represented at the conference by its sanest and soundest mind. Let Shaw represent Ireland, Stephen Leacock represent Canada; G. K. Chesterton is dead, but P. G. Wodehouse or Aldous Huxley may represent England. Will Rogers is dead, otherwise he would make a fine diplomat representing the U. S. ; we can have in his stead Robert Benchley or Heywood Broun. There will be others from Italy and France and Germany and Russia. Send these people to a conference on the eve of a great war, and see if they can start a European war, no matter how hard they try. Can you imagine this bunch of international diplomats starting a war or even plotting for one? The sense of humor forbids it. All people are too serious and half-insane when they declare a war against another people. They are so sure that they are right and that God is on their side. The humorists, gifted with better horse-sense, don't think so. You will find George Bernard Shaw shouting that Ireland is wrong, and a Berlin cartoonist protesting that the mistake is all theirs, and Heywood Broun claiming the largest share of bungling for America, while Stephen Leacock in the chair makes a general apology for mankind, gently reminding us that in the matter of stupidity and sheer foolishness no nation can claim itself to be the superior of others. How in the name of humor are we going to start a war under these conditions?

For who have started wars for us? The ambitious, the able, the clever, the scheming, the cautious, the sagacious, the haughty, the over-patriotic, the people   inspired with the desire to"serve"mankind, people who have a"career"to carve and an"impression"to make on the world, who expect and hope to look down the ages from the eyes of a bronze figure sitting on a bronze horse in some square. Curiously, the able, the clever, and the ambitious and haughty are at the same time the most cowardly and muddleheaded, lacking in the courage and depth and subtlety of the humorists. They are forever dealing with trivialities, while the humorists with their greater sweep of mind can envisage larger things. As it is, a diplomat who does not whisper in a low voice and look properly scared and intimidated and correct and cautious is no diplomat at all... But we don't even have to have a conference of international humorists to save the world. There is a sufficient stock of this desirable commodity called a sense of humor in all of us. When Europe seems to be on the brink of a catastrophic war, we may still send to the conferences our worst diplomats, the most"experienced" and self-assured, the most ambitious, the most whispering, most intimidated and correct and properly scared, even the most anxious to"serve" mankind. If it be required that, at the opening of every morning and afternoon session, ten minutes be devoted to the showing of a Mickey Mouse picture, at which all the diplomats are compelled to be present, any war can still be averted.

This I conceive to be the chemical function of humor: to change the character of our thought. I rather think that it goes to the very root of culture, and opens a way to the coming of the Reasonable Age in the future human world. For humanity I can visualize no greater ideal than that of the Reasonable Age. For that after all is the only important thing, the arrival of a race of men imbued with a greater reasonable spirit, with greater prevalence of good sense, simple thinking, a peaceable temper and a cultured outlook. The ideal world for mankind will not be a rational world, nor a perfect world in any sense, but a world in which imperfections are readily perceived and quarrels reasonably settled. For mankind, that is frankly the best we can hope for and the noblest dream that we can reasonably expect to come true. This seems to imply several things: a simplicity of thinking, a gaiety in philosophy and a subtle common sense, which will make this reasonable culture possible. Now it happens that subtle common sense, gaiety of philosophy and simplicity of thinking are characteristic of humor and must arise from it.

It is difficult to imagine this kind of a new world because our present world is so different. On the whole, our life is too complex, our scholarship too serious, our philosophy too somber, and our thoughts too involved. This seriousness and this involved complexity of our thought and scholarship make the present world such an unhappy one today.

Now it must be taken for granted that simplicity of life and thought is the highest and sanest ideal for civilization and culture, that when a civilization loses simplicity and the sophisticated do not return to unsophistication, civilization becomes increasingly full of troubles and degenerates. Man then becomes the slave of the ideas, thoughts, ambitions and social systems that are his own product. Mankind, overburdened with this load of ideas and ambitions and social systems, seems unable to rise above them. Luckily, however, there is a power of the human mind which can transcend all these ideas, thoughts and ambitions and treat them with a smile, and this power is the subtlety of the humorist. Humorists handle thoughts and ideas as golf or billiard champions handle their balls, or as cowboy champions handle their lariats. There is an ease, a sureness, a lightness of touch, that comes from mastery. After all, only he who handles his ideas lightly is master of his ideas, and only he who is master of his ideas is not enslaved by them. Seriousness, after all, is only a sign of effort, and effort is a sign of imperfect mastery. A serious writer is awkward and ill at ease in the realm of ideas as a nouveau riche is awkward, ill at ease and self-conscious in society. He is serious because he has not come to feel at home with his ideas.

Simplicity, then, paradoxically is the outward sign and symbol of depth of thought. It seems to me simplicity is about the most difficult thing to achieve in scholarship and writing. How difficult is clarity of thought, and yet it is only as thought becomes clear that simplicity is possible. When we see a writer belaboring an idea we may be sure that the idea is belaboring him. This is proved by the general fact that the lectures of a young college assistant instructor, freshly graduated with high honors, are generally abstruse and involved, and true simplicity of thought and ease of expression are to be found only in the words of the older professors. When a young professor does not talk in pedantic language, he is then positively brilliant, and much may be expected of him. What is involved in the progress from technicality to simplicity, from the specialist to the thinker, is essentially a process of digestion of knowledge, a process that I compare strictly to metabolism. No learned scholar can present to us his specialized knowledge in simple human terms until he has digested that knowledge himself and brought it into relation with his observations of life. Between the hours of his arduous pursuit of knowledge (let us say the psychological knowledge of William James), I feel there is many a "pause that refreshes," like a cool drink after a long fatiguing journey. In that pause many a truly human specialist will ask himself the all important question, "What on earth am I talking about?" Simplicity presupposes digestion and also maturity: as we grow older, our thoughts become clearer, insignificant and perhaps false aspects of a question are lopped off and cease to disturb us, ideas take on more definite shapes and long trains of thought gradually shape themselves into a convenient formula which suggests itself to us one fine morning, and we arrive at that true luminosity of knowledge which is called wisdom. There is no longer a sense of effort, and truth becomes simple to understand because it becomes clear, and the reader gets that supreme pleasure of feeling that truth itself is simple and its formulation natural. This naturalness of thought and style, which is so much admired by Chinese poets and critics, is often spoken of as a process of gradually maturing development. As we speak of the growing maturity of Su Tungpo's prose, we say that he has "gradually approached naturalness"—a style that has shed off its youthful love of pomposity, pedantry, virtuosity and literary showmanship.

Now it is natural that the sense of humor nourishes this simplicity of thinking. Generally, a humorist keeps closer touch with facts, while a theorist dwells more on ideas, and it is only when one is dealing with ideas in themselves that his thoughts get incredibly complex. The humorist, on the other hand, indulges in flashes of common sense or wit, which show up the contradictions of our ideas with reality with lightning speed, thus greatly simplifying matters. Constant contact with reality gives the humorist a bounce, and also a lightness and subtlety. All forms of pose, sham, learned nonsense, academic stupidity and social humbug are politely but effectively shown the door. Man becomes wise because man becomes subtle and witty. All is simple. All is clear. It is for this reason that I believe a sane and reasonable spirit, characterized by simplicity of living and thinking, can be achieved only when there is a very much greater prevalence of humorous thinking.

五．论任性与不可捉摸

看起来现在的军人代替了放浪者而成为人类的最高理想人物了。我们不要那种任性的、无从捉摸的、难于测度的自由人，而要合理化的、有纪律的、受统制的、穿制服的、有爱国心的工人，要在有效的管理和组织之下，五六千万人所结成的一个民族共同信仰同一种主义，皈依同一种思想，喜欢同一样的食物。关于人类的尊严，我们有两种相反的见解：一种以放浪者为理想；另一种以军人为理想；前者认为保持其自由和个性的人，是最崇高的典型，后者认为丧失了独立的判断力，将私人意见完全受制于统治者和国家，那才是最优越最崇高的人类。两种见解都可加以辩护，前者以常识为辩护，后者以逻辑为辩护。用逻辑去替爱国的自动机式理想做辩护，是不很困难的。爱国的自动机式模范公民，可以当做达到另一个外在目标的工具，这就是国家的力量，而这种力量又是为了另一个目标而存在，这个目标就是去克服另外的国家。这一切都可用逻辑很容易证明出来——又简单又坦白，所有的呆子都会死心塌地地相信。欧洲许多"文明的"和"开化的"的国家，在过去和现在都抱着这种见解，这实在使人好似难以相信。理想的公民是那种以为被遣到埃塞俄比亚首都去，结果却是在西班牙登岸的军人。这种公民又可分为"甲""乙"二等。那"甲"等的是那些在统治者所认为较好的公民，这种人晓得了他们被运到西班牙去时，仍是非常温顺、愉快，自己祷告，或由军中的牧师代为祷告，感谢上帝派遣他们到枪林弹雨中去为国牺牲。那"乙"等的都是些未曾充分开化的人，那些知道了被运错了地方而心中觉得愤恨的人。在我看来，那种内心的愤恨反抗情绪，是人类尊严的唯一标志，是那幅阴森惨淡图画中仅有的希望之火花，是人类在未来世界中恢复原位的唯一希望。

所以，不管它是什么逻辑，我自然还是拥护放浪者。我绝对拥护放浪者或流浪者，而口中或者也并不如此。我们这种矛盾心理就是我们的文明唯一的希望，我的理由很简单：我们是猴子的嫡系而不是牛的嫡系，因为我们有矛盾的心理，所以已经变成更优越、更高尚的猴子。我的自私使我愿意让牛有一种温驯而满足的脾性，在人类命令下，无论是被领到草地上，或是屠宰场里，都能保持同样旷达高尚的心思，一心一意地去为主人而牺牲。也因为十分钟爱人类，所以我不希望我们自己也变成牛。等到牛能开始反抗，心中生出反抗的情绪，或等到它们现出任性的样子，现出较不服从的样子时，我就要把它们称做有人性的动物了。我以为一切独裁制度都是不对的，这是一种生物学上的理由，独裁者可以跟牛和睦相处，却不能跟猴子和睦相处。

老实说，我从一九二〇年后，对于西洋的文明已减少了尊敬。我过去对中国的文明总感到惭愧，因为我觉得我们还没有创造出一个宪法和公权的观念，这是中国文明上的一个缺点：我始终相信建立一个共和或君主的立宪政府，是人类文化上的一种进步。可是现在在西洋文明的发祥地，我居然也看到人权、个人自由，甚至个人的信仰自由权（这自由权在中国过去和现在都享有着）都可以被蹂躏，看到西洋人不再视立宪政府为最高的政府，看见欧里庇得斯悲剧（Euripidean）的奴隶在中欧比在封建时代的中国还要多，看到一些西方国家比我们中国只有更多的逻辑而缺少常识，这真使人暗中觉得欣慰，觉得中国是足以自傲的。现在我除了将中国人观念中人类最高文化的理想表现出来，把那个中国人理想中听天由命、逍遥自在的放浪者、流浪者和漂泊者表现出来之外，我还有什么更便当的制胜良策呢？西方可也有这么一个势均力敌的良策吗？可也有什么东西足以证明它的个人自由和公权学说是一种严肃的、健全的信仰或本能吗？我拭目以待。

欧洲个人自由的传统怎么会消失，摆子在今日为什么会摆到错误的方向去，这是很容易明白的。这里有两个原因：第一是由于现代集体经济运动的结果，第二是由于维多利亚时代中叶的机械观念的遗传。在今日的各种集团主义——社会的、经济的、政治的——方兴未艾的时候，人类似乎自然地放弃了他的反抗权利，忘掉了他的个人尊严。当经济问题和经济思想占了优势，遮蔽了其他一切人类思想的时候，我们对于那种较有人性的知识和哲学，尤其是关于个人生活问题的哲学，便完全不加理会，而淡然置之了，这是极自然的。一种患胃溃疡的人时时在想到他的胃疾，一个社会有着经济弊病时，永远是被经济的思想纠缠着，结果把我们自己完全忘记了，几乎记不起还有个人……

可是我们能知道人类这种机械观是怎样在机械科学完成工业、征服自然的当中创造出来的。人类偷窃了这种科学，把这种机械式的逻辑拿来应用于人类社会，于是研究人事的人们便竭力利用"自然律"这个严肃的名词。因之我们就有"环境比人类伟大"，及"人类个性可以化成方程式"这一类的流行理论。这也许是精湛的经济学，但总是拙劣的生物学。良好的生物学，承认一个人的反应力量跟物质环境在生命的发展上两者是同样重要的因素，正如一位良医承认病人的性情和身体的反应在抗拒疾病时是同样重要的因素。现代的医生已能确定每一个人都有一种不能测算的因素。有很多病人如依逻辑和前例诊断起来，实在是应该死的，结果却会不死，反而复原起来，使医生也觉得惊奇。医生开着一式一样的药方给两个患同样疾病的人去吃，而不问他们的反应如何，我们真可以把他当做危害社会的人。社会哲学家如果忘掉个人，忘掉每个人都有不同的反应，忘掉他一般任性的、不可捉摸的行为，那么社会哲学家也是危害社会的人了。

我也许是不了解经济学，可是经济学也不见得会了解我。今日的经济学还是在失败中，还不敢昂头来置身于科学之列。经济学如果只谈商品而不更向前谈到人类的动机，它当然不是科学；即使能谈到人类的动机而要想以统计的平均数去研究，也不是科学，充其量不过是拟科学。这是经济学的悲哀。经济学甚至还不曾创造出可以检查人类心智的技术。如果它以数学方法和统计的平均律去研究人类的活动，那更有着暗中摸索的危险。所以每当一个重要的经济政策要决定的时候，总有两派经济专家和权威者站在绝对相反的地位。经济学终究和人类心智上的特癖是有关的，然而专家们对这些特癖一点也没有认识。一位专家相信如果英国放弃"金本位"，就会发生大变乱，但另一位专家坚决地相信如果英国要得救，唯有放弃金本位。人们什么时候要买什么时候要卖，就是最优异的经济专家也无法预测。证券交易之所以会变成投机事业，完全是这个缘故。纵使证券交易所能搜集到世界各国最可靠的经济资料，还是不能像天文台预测天气那样，正确地预测金银或商品市价的涨落。原因是经济学上掺有人类的要素，当很多人想卖出的时候便有一些人想买进，当很多的人想买进的时候，便有一些人想卖出，这里就有着人类的弹力和不可捉摸的要素。当然卖出的人总当那个买进的人是傻子，而那买进的人也以为卖出的人是傻子，到底谁是傻子，只有事实来证明。这种情形不但在商业交易上如此，在人类心理创造历史的过程中也是如此，在人类对于道德、风俗和社会改革的一切反应上，也都是如此。

V. On Being Wayward and Incalculable

IT seems that today the scamp is being displaced by the soldier as the highest ideal of a human being. Instead of wayward, incalculable, unpredictable free individuals, we are going to have rationalized, disciplined, regimented and uniformed, patriotic coolies, so efficiently controlled and organized that a nation of fifty or sixty millions can believe in the same creed, think the same thoughts and like the same food. Clearly two opposite views of human dignity are possible: the one regarding the scamp, and the other regarding the soldier, as the ideal; the one believing that a person who retains his freedom and individuality is the noblest type, and the other believing that a person who has completely lost independent judgment and surrendered all rights to private beliefs and opinions to the ruler or the state is the best and noblest being. Both views are defensible, one by common sense, and the other by logic. It should not be difficult to defend by logic the ideal of the patriotic automaton as a model citizen, useful as a means to serve another external goal, which is the strength of the state, which exists again for another goal, the crushing of other states. All that can be easily demonstrated by logic-a logic so simple and naive that all idiots fall for it. Incredible as it may seem, such a view has been upheld and is still being upheld in many"civili zed"and"enlightened"European countries. The ideal citizen is the soldier who thought he was being transported to Ethiopia and found himself in Guadalajara. Among such ideal citizens two classes, "A" and "B", may again be distinguished. The "A" class, consisting of the better citizens from the point of view of the state or its ruler, are those who, on discovering that they have been landed in Spain, are still extremely sweet and amiable and offer up thanks to God directly or through the army chaplain for sending them, by a kind of providential miracle, to the thick of the battle to die for the state. Class "B" would be those insufficiently civilized beings who feel an inner resentment at the discovery. Now for myself, that inner resentment, that human recalcitrancy, is the only sign of human dignity, the only spark of hope illuminating for me the otherwise somber and dismal picture, the only hope for a restoration of human decency in some future more civilized world.

It is clear then that, in spite of all logic, I am still for the scamp I am all for the scamp, or the tramp, and for "Mary, Mary, quite contrary." Our contrary-mindedness is our only hope for civilization. My reason is simple: that we are descended from the monkeys and not from the cows, and that therefore we are better monkeys, nobler monkeys, for being contrary-minded. I am selfish enough as a human being to desire a sweet and contented temper for the cows, who can be led to the pasture or to the slaughter-house at human behest with equal magnanimity and nobility of mind, motivated by the sole desire to sacrifice themselves for their master. At the same time, I am such a lover of humanity as not to desire that we become cows ourselves. The moment cows rebel and feel our recalcitrancy, or begin to act waywardly and less mechanically, I call them human. The reason I think all dictatorships are wrong is a biological reason. Dictators and cows go well together, but dictators and monkeys don't.

In fact, my respect for Western civilization has been considerably lowered since the nineteen-twenties. I had been ashamed of Chinese civilization, and I had honored the West, for I regarded it as a stain upon Chinese civilization that we had not developed a constitution and the idea of civil rights, and I decidedly thought that a constitutional government, republican or monarchical, was an advance in human culture. Now in the very home of Western civilization, I have the pleasure and satisfaction of seeing that human rights and individual liberty, even the common sense rights of individual freedom of belief that we in China enjoy and have always enjoyed, can be trampled upon, that a constitutional government is no longer thought of as the highest form of government, that there are more Euripidean slaves in central Europe than in feudalistic China, and that some Western nations have more logic and less common sense than we Chinese. What more easy than for me to play the trump card from my sleeve by producing the Chinese ideal of the happy-go-lucky, carefree scamp, tramp and vagabond, which is the highest cultural ideal of a human being according to the Chinese conception? Has the West a trump card to match, something to show that its doctrine of individual liberty and civil rights is a serious and deep-rooted belief or instinct... ? I am waiting to see it.

It is easy to see how the European tradition of individual liberty and freedom has been forgotten, and why the pendulum is swinging in the wrong direction today. The reasons are two: first, the consequences of the present economic movement toward collectivism, and second, a heritage from the mechanistic outlook of mid-Victorian times. It seems that in the present age of rising collectivism of all sorts-social, economic and political-mankind is naturally forgetting and forfeiting its right to human recalcitrancy and losing sight of the dignity of the individual. With the predominance of economic problems and economic thinking, which is overshadowing all other forms of human thinking, we remain completely ignorant of, and indifferent to, a more humanized knowledge and a more humanized philosophy, a philosophy that deals with the problems of the individual life. This is natural. As a man who has an ulcered stomach spends all his thought on his stomach, so a society with a sick and aching economics is forever preoccupied with thoughts of economics. Nevertheless, the result is that we remain totally indifferent to the individual and have almost forgotten that he exists...

But we can understand how this mechanistic view of man came about at a time when mechanistic science was proud of its achievements and its conquests over nature. This science was pilfered, its mechanistic logic transferred to apply to human society, and the always imposing name of"natural laws"was very much sought after by the students of human affairs. Hence the prevalent theory that the surroundings are greater than the man and that human personalities can be almost reduced to equations. That may be good economics, but bad biology. Good biology recognizes the individual's power of reaction as just as important a factor in the development of life as the physical environment, as any wise doctor will admit that the patient's temperament and individual reactions are an all-important factor in the fight against a disease. Medical doctors today recognize more and more the incalculable factor of the individual. Many patients, who by all logic and precedents ought to die, simply refuse to do so and shock the doctor by their recovery. A doctor who prescribes an identical treatment for an identical disease in two individuals and expects an identical development may be properly classified as a social menace. No less a social menace are the social philosophers who forget the individual, his capacity for reacting in a different manner from others, and his generally wayward and incalculable behavior.

Perhaps I don't understand economics, but economics does not understand me, either. That is why economics is still floundering today and hardly dares pop up its head as a science. The sad thing about economics is that it is no science, if it stops at commodities and does not go beyond to human motives, and if it does go beyond to human motives, it is still no science, or at best a pseudo-science, if it tries to reach human motives by statistical averages. It hasn't developed even a technique suitable to the examination of the human mind, and if it carries over to the realm of human activities its mathematical approach and its love of drawing statistical averages, it stands in still greater danger of floundering in ignorance. That is why every time an important economic measure is about to be adopted, two economic experts or authorities will come out exactly on opposite sides. Economics after all goes back to the idiosyncrasies of the human mind, and of these idiosyncrasies the experts have no ghost of an idea. One believed that, should England go off the gold standard, there would be a catastrophe, while another believed, with equal cocksureness, that England's going off the gold standard would be the only salvation. When people begin to buy and when people begin to sell are problems that the best experts cannot reasonably foretell. It is entirely due to this fact that speculations on the stock exchange are possible. It remains true that the stock exchange cannot, with the best assemblage of world economic data, scientifically predict the rise and fall of gold or silver or commodities, as the weather bureau can forecast the weather. The reason clearly lies in the fact that there is a human element in it, that when too many people are selling out, some will start buying in, and when too many people are buying in, a few people will start selling out. Thus is introduced the element of human resilience and human uncertainty. It is to be presumed, of course, that every person who is selling out regards as a fool the other person who is buying in what he is selling out, and vice versa. Who are the fools only future events can prove. This is merely an illustration of the incalculableness and waywardness of human behavior, which is true not only in the hard and matter-of-fact dealings of business, but also in the shaping of the course of history by human psychology, and in all human reactions toward morals, customs, and social reforms.

六．个人主义

　　哲学以个人为开端，亦以个人为依归。个人便是人生的最后事实。他自己本身即是目的，而绝不是人类心智创造的工具。世界最伟大的不列颠帝国，存在的目的便是使一个住在苏赛克斯（Sussex）的人民可以过着幸福合理的生活；但是谬误的哲学理论会说那个人是为了大不列颠帝国而生活的。社会哲学的最高目标，也无非是希望每个人都可以过着幸福的生活。如果有一种社会哲学不认为个人的生活幸福是文明的最后目标，那么这种哲学理论是一个病态的、不平衡的心智的产物。

　　要批判一种文化的价值，我以为应该以这种文化能产生何等的男女为标准。惠特曼这位最有智慧、最有远见的美国人在他的《民主远景》（Democratic Vistas，也译《民主的展望》）一文里就是基于这种意义去阐明个人原则之为一切文化的最终目的：

　　　　我们应该想一想，文明本身所根据的是什么——文明跟它的宗教、艺术、学校等，除要达到一个丰富美丽而多变的个人主义外，还有什么目的呢？一切事物都是向着这个目标而进展；民主主义本身就是因为要实现这个目的，才仿着大自然的规模把广漠无垠的人类荒田开垦起来，播了种子，给大家以均等的机会，所以它的地位仍在他种主义之前。一国的文学、诗歌、美学等之所以被重视，乃是由于它们能把个性的材料和暗示供给该国的男女，并以种种有效的方法去增强他们的力量。

惠特曼讲到个性的人生的最后事实时，说：

当一个人神志在最清明的时候，他有一种意识，一种独立的思想，解脱一切而高升起来，像星辰那么沉静永恒不灭。这就是和同思想——不管你是哪一种人，自己的思想终是属于自己的，我为我，你为你，各不相混。这确是奇迹中的奇迹，是人世间最神奇最模糊的梦想，但也是最明确的基本事业，是进向一切事实的大门。在那种虔诚的一瞬间，在意义深远的宇宙奇途中，信条和惯例在这个简单观念之下显觉不足轻重了。在真正的幻象之光的照射下，它是唯一有内容、有价值的东西。像寓言中的黑影矮人一旦被解放了，就能扩展到整个大地天上。

对于这位美国哲学家推崇个人的言论，我本想多介绍几段出来；可是为节省篇幅起见我就用下面几句话做一个结束：

……我们可用一处简单的观念来做最后的结论（不然整个事物的体系将成为无目标的、欺人的）：最后的和最好的方法，是依赖人类本身，及其自己天生的、常态的、充分发展的质素，而绝对不掺杂迷信的成分。

在种种变迁，在不断的嘲笑、抗辩和表面的失败中，民主主义的目的，就是要冒着任何危险去证明一个学说或原理，就是：在那最健全崇高的自由下训练出的人，他本身就是他自己的一种法律。

　　我们所应关心的是我们对于环境的反应而不是环境的本身。法德英美都生活在一样的机械文明中，不过生活的形式和趣味都各自不同；用着各不相同的方法去解决政治上的问题。当我们知道人生有许多变化的可能时，当我们看见两个汽车夫同坐在一辆货车上，听了同样的笑话而有着不同的反应时，我们即不应假定人类都须很服帖地受着机械式的统制。一个父亲对于他的两个儿子可以给予同样的教育、同样的生活基础，可是到后来，他们会渐渐依照各自内在的个性去创造自己的前程。纵使两个人都做银行的行长，有着完全相等的资本，然而在各项重要的事务，与一切造成快乐和幸福的事物中，他们完全是两样的，他们的处世态度、腔调和性情无不两样；他们和属下职员间的关系也有相异之处，职员们或许怕他们，也许爱他们，他们也许是好吹毛求疵的，也许是和蔼而宽大的；他们储蓄和用钱的方法也不同；他们的私人生活，他们的癖好、朋友、俱乐部、书籍和妻子也都是两样的。在同样环境下生活的人居然有那么大的差异，所以我们看见报纸上的许多讣告时，我们也不禁有些奇怪，生于同代，死在一天的人，两者的生活竟是那么不同，有的安居乐业，专心致志地努力着，在工作中获到乐趣；有的没有固定的职业，到处浮沉着；有的成了发明家，有的从事探险；也有些人喜欢说笑话，有的却终日沉默寡欢；有的正在飞黄腾达名利双收之时，而结果无声无息地在角落里死掉了；有的做着卖冰卖煤的买卖，在他们的地下室里被人刺死，身后遗下黄金两万元。是的，虽然在这工业时代，人生依然是奇妙的，只要人类还是人类，变化总还是人生的滋味。

　　不管是政治的或社会的革命，在人事上是没有宿命论这回事的。人性的因素使那些新原理和新制度创造者的计算完全失败，也击败了法律、制度和社会改革政策的创造者，不管所创造的是奥内达团体制度（Oneida, 奥内达人系指居住纽约奥尼湖附近的美国印第安人）或美国劳工联盟，或法官林赛（Judge Lindsay）所定的伴侣婚姻制度。新娘和新郎的性格比婚姻和离婚的惯例更为重要，那些执行法律或维护法律的人也比法律本身更重要。

　　讲到个人之重要，不单是因为个人生活是一切文明的最终目标，并且因为我们的社会生活、政治生活和国际关系的进步都是由许多个人（个人造成国家）的集体行动和集体脾性而产生，所以也完全以个人的脾气和性格为基础。国家的政治和国家每一时期的进化，决定的因素完全是人民的脾性。因为在工业发展的原则之外，一个民族做事和解决问题的方法是比较重要的因素。卢梭不会预料到法国革命的演变和拿破仑的突然出现，法国革命的演变绝不是由一般"自由、平等、博爱"的口号所决定的……

　　所以一个国家内的政事进展以及社会和政治的发展趋势，都是以各种个人的内在观念为根据。这种民族的脾性，这种称为"人民的天才"的抽象的东西，终究是许多个组成国家的个人的集合表现，也就是当一个国家在应付某项问题或危机时动作中所表现的性格。一部分人有着一种谬误的观念，以为这种"天才"的本质是像中世纪神学中的"灵魂"那样一个神话的实物，而不仅是一个比喻。实则国家的天才，不过是它行为的一种性质和做事的方法罢了。对于这种天才，我们又有错误的观念，以为它和国家的"命运"般同样是一个独立存在的抽象本质，这是不对的。这种天才是只能在动作中看得出来，只不过是一种选择的问题，即取舍和倾向，在危急时的特殊局势下，决定着国家的最后行动途径。旧式历史学家往往跟黑格尔（Hegel）一样，认为一个国家的历史仅是观念在机械的必然性下的一种发展，然而较微妙而现实的历史观念以为这大概是机缘的问题。每当一个危急的时期，国家即做一次选择，在选择的时候我们看见相反的势力和互相冲突的欲望在战斗着，情感的多寡即决定天平的倾侧。国家在危机中所表现的所谓天才，即是那个国家对于一件事情所做的取舍决定。每个国家总依着自己的意思去选择他们所喜欢的而排斥所不能容忍的东西，这种选择基于思想的潮流、一些道德观念和社会的成见。

最近在英国宪政的危机中（结果迫使一个皇帝的逊位），我们清楚地看到这种所谓的民族性格在发生作用，有时在所赞成和所反对的事物中表现出来，有时在变动的情感狂潮中表现出来，有时在自以为正常的动机冲突中表现出来。这类动机即是对一位深孚众望的君王的私人爱戴，英格兰教会对一个离婚者的偏见，英国人对于国王的传统观念。国王的私事是否真是私事，是否可以以私事视之，国王是否应该做超越傀儡式的人物，他是否应该同情工党，在这些冲突的情感中，只要任何一种情感稍微多一些，便可生出不同的结果。

在现代历史中……不管德国天主教会和基督新教会在纳粹政权的高压下能不能保持它们的完整（这须看德国人民有着多少人类的弹力），不管英国会不会真正变成工党的国家，不管美国的共产党社会大众对其好感或增或减：这些都须取决于有关国家内的个人思想、个人情感和个人性格。在一切人类历史的活动中，我只看见人类自身任性的不可捉摸、难于测度的选择所决定的波动和变迁。

根据这种意义，儒家就把世界和平问题和我们私人生活的培养联系起来，宋朝以后的儒家学者认为每一个人都该懂得这个道理，所以在儿童入学时，所读的第一课包括下列一段话：

古之欲明明德于天下者，先治其国；欲治其国者，先齐其家；欲齐其家者，先修其身；欲修其身者，先正其心；欲正其心者，先诚其意；欲诚其意者，先致其知。致知在格物。物格而后知至，知至而后意诚，意诚而后心正，心正而后身修，身修而后家齐，家齐而后国治，国治而后天下平。自天子以至于庶人，壹是皆以修身为本。其本乱而末治者，否矣，其所厚者薄，而其所薄者厚，未之有也。

物有本末，事有终始，知所先后，则近道矣。

VI. *The Doctrine of the Individual*

PHILOSOPHY not only begins with the individual, but also ends with the individual. For an individual is the final fact of life. He is an end in himself, and not a means to other creations of the human mind. The greatest empire of the world, like the British Empire, exists in order that an Englishman in Sussex may live a fairly happy and reasonable life; a false philosophy would assume that the Englishman in Sussex lives in order that there may be the great British Empire. The best social philosophies do not claim any greater objective than that the individual human beings living under such a regime shall have happy individual lives...

As far as culture is concerned, I am inclined to think that the final judgment of any particular type of culture is what type of men and women it turns out. It is in this sense that Walt Whitman, one of the wisest and most far-seeing of Americans, struggles in his essay Democratic Vistas to bring forth the principle of individuality or"personalism,"as the end of all civilization:

And, if we think of it, what does civilization itself rest upon—and what object has it, with its religions, arts, schools, etc., but rich, luxuriant, varied personalism? To that all bends; and it is because toward such result democracy alone, on anything like Nature's scale, breaks up the limitless fallows of humankind, and plants the seed, and gives fair play, that its claims now precede the rest. The literature, songs, esthetics, etc., of a country are of importance principally because they furnish the materials and suggestions of personality for the women and men of that country, and enforce them in a thousand effective ways.

Speaking of the individuality as a final fact, Whitman says:

There is, in sanest hours, a consciousness, a thought that rises, independent, lifted out from all else, calm, like the stars, shining eternal. This is the thought of identity— yours for you, whoever you are, as mine for me. Miracle of miracles, beyond statement, most spiritual and vaguest of earth's dreams, yet hardest basic fact, and only entrance to all facts. In such devout hours, in the midst of the significant wonders of heaven and earth, (significant only because of the Me in the centre), creeds, conventions, fall away and become of no account before this simple idea. Under the luminousness of real vision, it alone takes possession, takes value. Like the shadowy dwarf in the fable, once liberated and look'd upon, it expands over the whole earth, and spreads to the roof of heaven.

The temptation is strong to quote more from this typically American philosopher's most eloquent glorification of the individual, summed up in the following manner;

....and, as an eventual conclusion and summing up (or else the entire scheme of things is aimless, a cheat, a crash), the simple idea that the last, best dependence is to be upon humanity itself, and its own inherent, normal, full-grown qualities, without any superstitious support whatever.

The purpose of democracy... is, through many transmigrations, and amid endless ridicules, arguments and ostensible failures, to illustrate, at all hazards, this doctrine or theory that man, properly train'd in sanest, highest freedom, may and must become a law, and series of laws, unto himself...

After all, it is not our surroundings, but our reactions toward them that count. France, Germany, England and America are all living in the same machine civilization, yet their patterns and flavors of life are all different, and all solve their political problems in different ways. It is foolish to assume that man must be swamped by the machine in a uniform, helpless manner, when we realize there is such room for variety of life, when we see that two drivers on the same truck will take a joke differently. A father of two sons who gives them the same education and the same start in life, will see how they gradually shape their lives according to the inner laws of their own being. Even if both turn out to be presidents of banks with exactly the same capitalization, yet in all things that matter, in all things that make for happiness, they are different, different in their address, accent and temperament; in their policies and ways of handling problems; in the way they get on with their staff, whether they are feared or loved, harsh and exacting or pleasant and easygoing; in the way they save and spend their money; and different in their personal lives as colored by their hobbies, their friends, their clubs, their reading and their wives. Such is the rich variety possible in identical surroundings that no one can take up the obituary page of a newspaper, without wondering how persons living in the same generation and dying or the same day have led entirely different lives, how some plodded on in a chosen vocation with a singular devotion and found happiness in it, how others had a checkered and varied career, how some invented, some explored, some cracked jokes, some were morose and without a sense of humor, some skyrocketed to fame and wealth and died in the cold, dark cinders of the rocket, and some sold ice and coal and were stabbed to death in their cellar homes with a hoard of twenty thousand dollars in gold. Yes, human life is wondrous strange still, even in an industrial age. So long as man is man, variety will still be the flavor of life.

There is no such thing as determinism in human affairs, whether politics or social revolution. The human factor is what upsets the calculations of the propounders of new theories and systems, and what defeats the originators of laws, institutions and social panaceas, whether it be the Oneida Community, or the American Federation of Labor, or Judge Lindsay's companionate marriage. The quality of the bride and bridegroom is more important than the conventions of marriage and divorce, and the men administering or upholding the laws are more important than the laws themselves.

But the importance of the individual comes not only from the fact that the individual life is the end of all civilization, but also from the fact that the improvement of our social and political life and international relationships comes from the aggregate action and temper of the individuals which compose a nation and is eventually based on the temper and quality of the individual. In national politics and the evolution of a country from one stage to another, the determining factor is the temperament of the people. For above the laws of industrial development, there is the more important factor of a nation's way of doing things and solving problems. Rousseau little foresaw the course of the French Revolution and the appearance of Napoleon. The course of the French Revolution was not determined by the slogan of Liberty, Equality and Fraternity, but by certain traits in human nature in general and in the French temperament in particular...

And so the conduct of a nation's affairs and the course of its social and political development are eventually based on the ideas which govern the individuals. This racial temperament, the thing we abstractly call"the genius of the people,"is after all an aggregate of the individuals who comprise that nation, for it is nothing but the character of a nation in action, as it faces certain problems or crises. There is nothing more false than the notion that this"genius" is a mythological entity like the"soul" in medieval theology, as if it were something more than a figure of speech. The genius of a nation is nothing but the character of its conduct and its way of doing things. So far from being an abstract entity with an independent existence of its own, as we sometimes think of the"destiny"of a nation, this genius can be seen only in action; it is a matter of choice, of certain selections and rejections, preferences and prejudices, which determine the nation's final course of action in a given crisis or situation. Historians of the old school would like to think with Hegel that the history of a nation is but the development of an idea proceeding by a kind of mechanical necessity, whereas a more subtle and realistic view of history is that it was very largely a matter of chance. At every critical period, the nation made a choice, and in the choice we see a struggle of opposing forces and conflicting passions, and a little more of this kind of sentiment or a little less of that kind might have tipped the scale in the other direction. The so-called genius of a nation, as expressed in such a given crisis, was a decision by the nation that they would like to have a little more of one thing, or had had enough of it. For after all every nation went ahead with what it liked, or what appealed to its sentiments, and rejected what it would not tolerate. Such a choice was based on a current of ideas and a set of moral sentiments and social prejudices.

In the last constitutional crisis of England, eventually compelling the abdication of a king, we see most clearly this thing called the character of a people in action, revealed by its approvals and disapprovals, its tide of changing emotions, in a conflict among many working motives of assumed validity. Such motives were personal loyalty to a popular prince, the Church of England's prejudice against a divorcée, the Englishman's traditional conception of a king, the question whether a king's private affair was or could be a private affair, and whether a king should be more than a figurehead, or whether he should have definite Laborite sympathies. A little more of any one of these conflicting sentiments might have brought about a different solution of the crisis.

And so throughout current history,... whether the German Catholic and Protestant churches might or might not hold their integrity in their resistance against the Nazi regime (that is, how much human resilience there is in Germany), whether England might turn truly Laborite, and whether the American Communist party might grow or lose in public favor, are things which eventually are determined by the ideas, sentiments and character of the individual members of the states concerned. In all this moving panorama of human history, I see only flux and change, determined by man's own wayward and incalculable and unpredictable choice.

In this sense, Confucianism connected the question of world peace with the cultivation of our personal lives. The very first lesson that Confucian scholars since the Sung Dynasty have decided should be learned by the child at school contains this passage:

The ancient people who desired to have a clear moral harmony in the world would first order their national life; those who desired to order their national life would first regulate their home life; those who desired to regulate their home life would first cultivate their personal lives; those who desired to cultivate their personal lives would first set their hearts right; those who desired to set their hearts right, would first make their wills sincere; those who desired to make their wills sincere would first arrive at understanding; understanding comes from the exploration of knowledge of things. When the knowledge of things is gained, then understanding is reached; when understanding is reached, then the will is sincere; when the will is sincere, then the heart is set right; when the heart is set right, then the personal life is cultivated; when the personal life is cultivated, then the home life is regulated; when the home life is regulated, then the national life is orderly; and when the national life is orderly; then the world is at peace. From the Emperor down to the common man, the cultivation of the personal life is the foundation for all. It is impossible that when the foundation is disorderly, the superstructure can be orderly. There has never been a tree whose trunk is slender and whose top branches are heavy and strong. There is a cause and a sequence in things, and a beginning and end in human affairs. To know the order of precedence is to have the beginning of wisdom.

/ Chapter Five　**Who Can Best Enjoy Life** /

第五章

谁最会享受人生

一. 发现自己: 庄子

　　在现代生活中，如果真有哲学家的话，那么"哲学家"这名词已变成一个仅是社交上恭维人家的名称了。哲学家差不多是世界上最受人尊崇，也最不受人注意的人物。只要是一个神秘暧昧深奥不易了解的人物即可称之为"哲学家"，一个对现状漠不关心的人也被称为"哲学家"。然而，后者的这种意义中还有着相当的真理。在莎士比亚的《皆大欢喜》(As You Like It) 一剧中，丑角"试金石"(Touchstone) 所说的"牧羊人，你也懂得一些哲学吧？"这句话就是包含后者这种意义的。从这一种意义说来，哲学仅是对事物和人生的一种普通而粗浅的观念，而且这种观念每个人多少总有一点。如果某一个人否认现实的表面价值或不肯尽信报纸上所说的话，他就有哲学家的意味，他是一个不愿被骗的人。

　　哲学总带着一种如梦初醒的意味。哲学家观察人生，正如艺术家观察风景一样——是隔着一层薄纱或一层烟雾的。这种看法使生硬的人生琐事软化，容易使我们看出其中的意义。至少中国的艺术家或哲学家是如此思想的。所以，哲学家和彻底的现实主义者的观念完全相反；后者熙来攘往忙碌终日，以为他的成败盈亏完全是绝对的、真实的。这种人真是无药可救，他连一些怀疑的念头也没有，所以不能得到一个起点。孔子说："不曰如之何如之何者，吾末如之何也已矣！"——在孔子少数而有意的诙谐语句中，这句实得我心。

我想在这章中介绍一些中国哲学家对生活图案的观念。他们之间的意见越是参差，越是一致以为人类必须有智慧和过着幸福生活的勇气。孟子的那种比较积极的人生观念和老子的那种比较圆滑和顺的观念，协调起来成为一种中庸的哲学，这种中庸的哲学可说已成了一般中国人的宗教。动和静的冲突，结果却产生了一种妥洽的观念，使人们对于这个不得完美的地上天堂也感到了满足，这种智慧而愉快的人生哲学就此产生。陶渊明——在我的心目中他是中国最伟大的诗人，有着最和谐的性格——就是这种生活的一种典型。

一切中国哲学家在不知不觉中所认为最重要的问题就是：我们要怎样去享受人生？谁最会享受人生？我们不去追求完美的理想，不去寻找那势不可得的事物，不去穷究那些不可得知的东西；我们认识的只是些不完美的、曾死的人类的本性；最重要的问题是怎样去调整我们的人生，使我们得以和平地工作，旷达地忍耐，幸福地生活。

我们是谁？这是第一个问题。这个问题似乎是不能解答的。不过我们都已承认，我们日常忙碌生活中的自我并不是完全真正的自我，在生活的追求中我们已经丧失一些东西。例如：我们看见一个人在田野里东张西望地在寻找东西。聪明的人可以提出一个难题来让那些旁观者去猜猜：那个人究竟失掉了什么东西？有的猜一只表，有的猜一只钻石别针；各人有各人的猜测。聪明人其实也不知道那人失掉了些什么，可是当大家猜不着时，他可以说："我告诉你们吧，他失掉魂儿了。"我想没有人会说他这句话不对。我们往往在生活的追求中忘记了真正的自我，正如庄子在一个美妙的譬喻里所讲的那只鸟一样，为了要吃一只螳螂而忘记自身的危险，而那只螳螂又为了要捕捉一只蝉也忘了自身的危险。

庄周游于雕陵之樊，睹一异鹊自南方来者。翼广七尺，目大运寸。感周之颡，而集于栗林。

庄周曰："此何鸟哉？翼殷不逝，目大不睹？"

蹇裳躩步，执弹而留之。睹一蝉，方得美荫而忘其身；螳螂执翳而搏之，见得而忘其形；异鹊从而利之，见利而忘

其真。

庄周怵然曰："噫！物固相累，二类相召也。"捐弹而反走，虞人逐而谇之。

庄周反入，三月不庭。蔺且从而问之："夫人何为顷间甚不庭乎？"

庄周曰："吾守形而忘身。观于浊水而迷于清渊。且吾闻诸夫子曰：'入其俗，从其俗。'今吾游于雕陵而忘吾身。异鹊感吾颡。游于栗林而忘真，栗林虞人以吾为戮。吾所以不庭也。"

庄子乃是老子的门生，正如孟子是孔子的门生一样，二人都富于口才，二人的生存年月都和他们老师距离约一百年。庄子和孟子生在同时，老子和孔子大约也在同时。可是孟子很赞成庄子人性已有所亡，而哲学之任务就是去发现并去取回那些失掉了的东西这句话。据孟子的见解，以为失掉的便是"赤子之心"。他说："大人者，不失其赤子之心者也。"孟子认为，文明的人为生活，其影响及于人类赤子之心，有如山上的树木被斧斤伐去一样。

牛山之木尝美矣。以其郊于大国也，斧斤伐之，可以为美乎？是其日夜之所息，雨露之所润，非无萌蘖之生焉，牛羊又从而牧之。是以若彼濯濯也；人见其濯濯也，以为未尝有材焉。此岂山之性也哉？虽存乎人者，岂无仁义之心哉？其所以放其良心者，亦犹斧斤之于木也；旦旦而伐之，可以为美乎？其日夜之所息，平旦之气，其好恶与人相近也者几希。则其旦昼之所为，有梏亡之矣。梏之反覆，则其夜气不足以存；夜气不足以存，则其违禽兽不远矣。人见其禽兽也，而以为未尝有才焉者。是岂人之情也哉？

I. Find Thyself: Chuangtse

IN modern life, a philosopher is about the most honored and most unnoticed person in the world, if indeed such a person exists. "Philosopher" has become merely a term of social compliment. Anyone who is abstruse and unintelligible is called "a philosopher." Anyone who is unconcerned with the present is also called "a philosopher." And yet there is some truth in the latter meaning. When Shakespeare made Touchstone say in As You Like It, " Hast any philosophy in thee, shepherd?" he was using it in the second meaning. In this sense, philosophy is but a common, rough and ready outlook on things or on life in general, and every person has more or less of it. Anyone who refuses to take the entire panorama of reality on its surface value, or refuses to believe every word that appears in a newspaper, is more or less a philosopher. He is the fellow who refuses to be taken in.

There is always a flavor of disenchantment about philosophy. The philosopher looks at life as an artist looks at a landscape—through a veil or a haze. The raw details of reality are softened a little to permit us to see its meaning. At least that is what a Chinese artist or a Chinese philosopher thinks. The philosopher is therefore the direct opposite of the complete realist who, busily occupied in his daily business, believes that his successes and failures, his losses and gains, are absolute and real. There is nothing to be done about such a person because he does not even doubt and there is nothing in him to start with. Confucius said:"If a person does not say to himself 'What to do? What to do?'indeed I do not know what to do with such a person!"–one of the few conscious witticisms I have found in Confucius.

I hope to present in this chapter some opinions of Chinese philosophers on a design for living. The more these philosophers differ, the more they agree–that man must be wise and unafraid to live a happy life. The more positive Mencian outlook and the more roguishly pacifist Laotsean outlook merge together in the Philosophy of the Half-and-Half, which I may describe as the average Chinaman's religion. The conflict between action and inaction ends in a compromise, or contentment with a very imperfect heaven on earth. This gives rise to a wise and merry philosophy of living, eventually typified in the life of T'ao Yüanming–in my opinion China's greatest poet and most harmonious personality.

The only problem unconsciously assumed by all Chinese philosophers to be of any importance is: how shall we enjoy life, and who can best enjoy life? No perfectionism, no straining after the unattainable, no postulating of the unknowable; but taking poor, mortal human nature as it is, how shall we organize our life that we can work peacefully, endure nobly and live happily?

Who are we? That is the first question. It is a question almost impossible to answer. But we all agree that the busy self occupied in our daily activities is not quite the real self. We are quite sure we have lost something in the mere pursuit of living. When we watch a person running about looking for something in a field, the wise man can set a puzzle for all the spectators to solve: what has that person lost? Some one thinks it is a watch; another thinks it is a diamond brooch; and others will essay other guesses. After all these guesses have failed, the wise man who really doesn't know what the person is seeking after tells the company:"I'll tell you. He has lost some breath."And no one can deny that he is right. So we often forget our true self in the pursuit of living, like a bird forgetting its own danger in pursuit of a mantis, which again forgets its own danger in pursuit of another prey, as is so beautifully expressed in a parable by Chuangtse:

When Chuangtse was wandering in the park at Tiao-ling, he saw a strange bird which came from the south. Its wings were seven feet across. Its eyes were an inch in circumference. And it flew close past Chuangtse's head to alight in a chestnut grove.

"What manner of bird is this?"cried Chuangtse. "With strong wings it does not fly away. With large eyes it does not see."

So he picked up his skirts and strode towards it with his crossbow, anxious to get a shot. Just then he saw a cicada enjoying itself in the shade, forgetful of all else. And he saw a mantis spring and seize it, forgetting in the act its own body, which the strange bird immediately pounced upon and made its prey. And this it was which had caused the bird to forget its own nature.

"Alas!"cried Chuangtse with a sigh, "how creatures injure one another. Loss follows the pursuit of gain."So he laid aside his bow and went home, driven away by the park-keeper who wanted to know what business he had there.

For three months after this, Chuangtse did not leave the house; and at length Lin Chü asked him, saying, "Master, how is it that you have not been out for so long?"

"While keeping my physical frame,"replied Chuangtse, "I lost sight of my real self. Gazing at muddy water, I lost sight of the clear abyss. Besides, I have learnt from the Master as follows:—'When you go into the world, follow its customs.'Now when I strolled into the park at Tiao-ling, I forgot my real self. That strange bird which flew close past me to the chestnut grove, forgot its nature. The keeper of the chestnut grove took me for a thief. Consequently I have not been out. "

Chuangtse was the eloquent follower of Laotse, as Mencius was the eloquent follower of Confucius, both separated from their masters by about a century. Chuangtse was a contemporary of Mencius, as Laotse was probably a contemporary of Confucius. But Mencius agreed with Chuangtse that we have lost something and the business of philosophy is to discover and recover that which is lost–in this case,"a child's heart,"according to Mencius."A great man is he who has not lost the heart of a child,"says this philosopher. Mencius regards the effect of the artificial life of civilization upon the youthful heart born in man as similar to the deforestation of our hills:

There was once a time when the forests of the Niu Mountain were beautiful. But can the mountain any longer be regarded as beautiful, since being situated near a big city, the woodsmen have hewed the trees down? The days and nights gave it rest, and the rains and the dew continued to nourish it, and a new life was continually springing up from the soil, but then the cattle and the sheep began to pasture upon it. That is why the Niu Mountain looks so bald, and when people see its baldness, they imagine that there was never any timber on the mountain. Is this the true nature of the mountain? And is there not a heart of love and righteousness in man, too? But how can that nature remain beautiful when it is hacked down every day, as the woodsman chops down the trees with his ax? To be sure, the nights and days do the healing and there is the nourishing air of the early dawn, which tends to keep him sound and normal, but this morning air is thin and is soon destroyed by what he does in the day. With this continuous hacking of the human spirit, the rest and recuperation obtained during the night are not sufficient to maintain its level, and when the night's recuperation does not suffice to maintain its level, then the man degrades himself to a state not far from the beast's. People see that he acts like a beast and imagine that there was never any true character in him. But is this the true nature of man?

二．情智勇：孟子

　　最合于享受人生的理想人物，就是一个热诚的、悠闲的、无恐惧的人。孟子列述"大人"的三种"成熟的美德"是"仁、智、勇"。我以为把"仁"字改为"情"字更为确当，而以"情、智、勇"为大人物的特质。在英语中幸亏尚有Passion这个字，其用法和华语中的"情"字差不多。这两个字起首都含有"情欲"的那种狭义，但现在都有了更广大的意义。张潮说："多情者必好色，而好色者未必尽属多情。"又说："情之一字，所以维持世界；才之一字，所以粉饰乾坤。"如果我们没有"情"，我们便没有人生的出发点。情是生命的灵魂，星辰的光辉，音乐和诗歌的韵律，花草的欢欣，飞禽的羽毛，女人的艳色，学问的生命。没有情的灵魂是不可能的，正如音乐不能不有表情一样。这种东西给我们以内心的温暖和活力，使我们能快乐地去对付人生。

　　我把中国作家笔下所用的"情"字译作Passion也许不很对，或者我可用Sentiment一字（代表一种较温柔的情感，较少激越的热情所生的冲动性质）去译它？"情"这一字或许也含着早期浪漫主义者所谓Sensibility一字的意义，即属于一个有温情的大量艺术化的人的质素。在西洋的哲学家中，除了爱默生（Emerson）、亨·费·埃米尔（Amiel，十九世纪瑞士哲学家）、约瑟夫·儒贝尔（Joubert，十九世纪法国著名诗人）和伏尔泰（Voltair）外，很少对于热情能说些好话的人，这是奇怪的。也许我们所用的词语虽不同，而我们所指的实是同一样东西。但是，假如说"热情"（Passion）异于"情感"（Sentiment），两者意义不同，而前者只是专指一种暴躁的冲动的情感，那么在中国字中找不到一个相应的字可以代表它，我们只好依然用"情"这个字了。我很疑惑这是否就是种族脾性不同的表征？这是否就是中国民族缺乏那种侵蚀灵魂去造成那种西洋文学里悲剧材料的伟大热情的表征？这可就是中国文学中没有产生过希腊意义上的悲剧的原因？这可就是中国悲剧角色在危急之时饮泣吞声，让敌人带去了他们的情人，或如楚霸王那样，先杀死情人，然后自刎的原因？这种结局是不会使西洋观众满意的，可是中国人的生活是这样的，所以在文学上当然也就是这样的了。一个人跟命运挣扎，放弃了争斗，事过之后，随之在悲剧回忆中，发生了一阵徒然的后悔和想望。正如唐明皇的悲剧那样，他谕令他的爱妃自杀以满足叛军的要求，过后，便神魂颠倒地成天思念她。这种悲剧的情感是在那出戏剧结束后，在一阵悲哀中才表现出来的。当他在出狩生活中旅行时，在雨中听见隔山相应的铃声，便作了那首《雨霖铃》曲以纪念她；所能看到或扪触到的事事物物，无论是一条余香未尽的小领巾，或是她的一个老婢，都使他想起他的爱妃，这悲剧的结束便是由一位道士替他在仙境里寻觅她的芳魂。如此我们就在这里看到一种浪漫的敏感性，如不能称之为热情，不过这热情已变成一种圆熟而温和的了。所以，中国哲学家有着一种特点，他们虽鄙视人类的"情欲"（即"七情"的意思），却不鄙视热情或情感本身，反使之成为正常人类的生活基础，甚至于视夫妇之情为人伦之本。

我们的热情或情感是随生命而同来，无可选择，正如我们不能择拣父
母一样，我们不幸天生就有一种冷静或热烈的天性，这是事实。在另一方
面，没有一个小孩生来就是冷心的；当我们渐次失掉那种少年心时，我们
才会逐渐失掉我们内在的热情。在我们生活的某一时期中，我们热情的天
性被一种邪恶的环境摧残压制，挫折或剥削，其所以如此，大概是由于我
们没有留意使之继续生长，或者是我们不能从这种环境里解脱出来。我们
在获取"世事经验"的过程中对我们的天性曾多方摧残，我们学会了硬心
肠，学会了虚伪矫饰，学会了冷酷残忍，因此在一个人自夸他已获得了很
多的人世经验时，他的神经显然已变成不敏锐而麻木迟钝——此种现象尤
其是在政界为最多。结果世界上多了一个伟大的"进取者"（Go-getter），
把别人挤在一旁，自己爬到顶上，世界上从此多了一个意志刚强、心志坚
定的人，不过感情——他称之为愚笨的理想主义或多情的东西——在他胸
怀中的最后一些灰烬，也渐渐地熄灭了。我很看不起这种人，这世界上冷
酷心肠的人实在太多了。如果国家有一天要施行消灭那些不适于生存者的
生殖机能，第一步，应该把那些无道德感念的人、艺术观念陈腐的人、铁
石心肠的人、残酷而成功的人、意志坚决一无情义的人，以及那一切失掉
生之欢乐的人，一起把他们的生殖机能割掉——而不必亟亟于那些疯狂的
人和患肺痨的人。因为在我看来，一个热情而有情感的人或许会做出一些
愚蠢和鲁莽的事情，可是一个无热情也无情感的人好像是一个笑话或一
幅讽刺画了。他跟都德（Dandet）和萨福（Sappho）两者比较起来，只好
算一条虫、一架机器、一座自动机、尘世上的一点污点而已。有许多妓女
的一生比大腹便便的商人来得高洁。萨福虽然犯罪，但也懂得爱；我们对
于那些会显示深爱的人，应该给予较大的宽恕，无论怎样，她从一个冷酷
的商业环境中走出来的时候，总比我们周遭那些百万富翁怀着更热烈的心

情。对抹大拉的马利亚（Mary Magdalene，被耶稣拯救的妓女）崇拜是对的。热情和情感有时免不了使我们做错事，因而受罪是应该的。但是有许多宽容的母亲因为过于纵容子女，往往因爱子之心而失掉了理智的判断，不过她们到了老年的时候，她们一定会回忆到从前那种融融洽洽的家庭生活，以为比那些苛刻严峻的人的家庭生活来得快乐。有一个朋友曾告诉我一个故事。他说有一个年纪已七十八岁的老妇人对他说："回溯过去的七十八年中，每想到我所做的错事时，我还是觉得快乐的；不过又想到我的愚蠢，我甚至到今天还不能饶恕我自己。"

可是人生是残酷的，一个有着热烈的、慷慨的、天性多情的人，也许容易受他比较聪明的同伴之愚。那些天性慷慨的人，常常因慷慨而错了主意，常常因对付仇敌过于宽大，或对于朋友过于信任，而走了失着。慷慨的人有时会感到幻灭，因而跑回家中写出一首悲苦的诗。在中国有许多的诗人和学者就是这样，例如喝茶大家张岱，很慷慨地替亲友出力帮忙，甚至把家产也因此花完，结果吃了他最亲密的亲友的亏；后来他把这遭遇写成十二首诗，那诗要算是我所曾读到过的最辛酸最悲苦的了。可是我很相信直到他老死还是那么慷慨大量的，即使是在他很穷困的时候，有几次几乎穷得要饿死，也必仍然如此。我相信那些悲哀的情绪不久就会烟消雾散，而他依旧会快乐的。

虽说如此，但这种慷慨热烈的心情须有一种哲学加以保护，人生是严酷的，热烈的心性不足以应付环境，热情必须和智勇联结起来，方能避免环境的摧残。我觉得智和勇是同样的东西，勇乃是了解人生之后的产物；一而二，二而一，一个完全了解人生的人始能有勇。如果智不能生勇，智便无价值。智抑制了我们愚蠢的野心，使我们从这个世界的骗子（Humbug）——无论是思想上的或人生上的——手中解放出来而生出勇气。

在我们这个世界里，骗子真是不胜其多，不过中国佛教已经把许多小骗子归纳于两个大骗子之中：就是名和利。据说乾隆皇帝游江南的时候，有一次在一座山上眺望景色，望见中国海上帆船往来如梭，便问他身旁的大臣那几百只帆船是干什么的，他的大臣回答，只看见两只船，一只叫做"名"，一只叫做"利"。有修养的人士只能避免利的诱惑，只有最伟大的人物才能够逃避名的诱惑。有一次，一位僧人跟他的弟子谈到这两种俗念的根源时说："绝利易，绝名心难。即退隐之学者僧人仍冀得名。彼乐与大众讲经说法，而不愿隐处小庵与弟子作日常谈。"那个弟子道："然则师傅可为世上唯一绝名心之人矣。"师傅微笑而不言。

据我的人生观察，佛教徒的那种分类是不完全的。人生的大骗子不止两个，而实有三个：即名、利、权。在美国惯用的字中，可以拿"成功"（Success）这名词把这三个骗子概括起来。但是有许多智者以为成功和名利的欲望实是失败、贫穷和庸俗无闻的恐惧之一种讳称，而这些东西是支配着我们的生活的。有许多人已经名利双全，可是他们还在费尽心机去统治别人，他们就是竭一生心力为祖国服役的人，这代价常是巨大的。如果你去请一个真真的智者来，要选他做总统，要他随时向一群民众脱帽招呼，一天中要演说七次，这种总统他一定不要做的。詹姆斯·布莱斯以为美国民主政府现行的制度不能招致国中最优秀的人才去入政界服役。我觉得单是竞选的吃力情形足已吓退美国的智者了。从政的人顶了竭毕生心力为人群服役的名义，一星期须参加六次的宴会。他为什么不坐在家里，自己吃一顿简单的晚餐，随后穿上睡衣，舒舒服服地上床去睡呢？一个人在名誉或权力的迷惑下，不久也会变成其他骗子的奴隶，越陷越深，永无止日。他不久便开始想改革社会，想提高人们的道德，想维护教会，想消弭罪恶，做一些计划给人家去实行，推翻别人所定的计划，在大会中读一篇他的下属替他预备好的统计报告，在委员会的席上研究展览会的蓝纸图样，甚至想创设一间疯人院（真厚脸皮啊）——总之一句话，想干涉人家的生活。但是不久，这些自告奋勇而负起的责任，什么改造人家、实施计划、破坏竞争者的计划等问题一股脑儿抛在脑后，或甚至不曾跑进过他的脑筋呢。一个在总统竞选中失败了的候选人，两星期过后，对于劳工、失业关税等诸大问题都忘得一干二净！他是什么人，干吗要改造人家，提高人们的道德，送人家进疯人院去呢？可是他如果成功了，那些大骗子和小骗子会使他踌躇满志地奔忙着，而使他想象着以为他的确是在做一些事情，确是一个"重要的人物"。

然而，世间还有一个次等的社会骗子，和上述的骗子有同样的魅力，一样普遍，就是时尚（Fashion）。人类原来的自我本性很少有表现出来的勇气。希腊哲学家德谟克利特（Democritus）以为已把人类从畏惧上帝和死亡这两个大恐怖的压迫下解放出来，是一种对人类的伟大贡献，虽然如此，可是他还不曾把我们从另一个普遍的恐惧——畏惧周遭的人——中解放出来。人们虽由畏惧上帝和畏惧死亡的压迫中解放了出来，但还有许多人仍不能解除畏惧人们的心理，不管我们是有意或无意，在这尘世中一律都是演员，在一些观众面前，演着他们所认可的戏剧。

这种演戏的才能加上模仿的才能（其实即演戏才能的一部分），是我们猴子的遗传中最出色的质素。这种表演才能无疑可以得到实在利益，最显而易见的就是博得观众的喝彩。但是喝彩声越高，台后的心绪也越加紧张。这才能也帮助一个人去谋生，所以我们不能怪谁迎合观众心理去扮演他的角色。

唯一不合之处就是那演员或许会篡夺了那个人的位置，而完全占有了他；在这世上享盛名居高位的人，能够保存本性的真少而又少，也只有这一种人自知是在做戏，他们不会被权势、名号、资产、财富等人造的幻象所欺蒙。当这些东西跑来时，他们只用宽容的微笑去接受，他们并不相信他们如此便变成特殊，便和常人不同。这一类的人物是精神上的伟大，也只有这些人的个人生活始终是简朴的。因为他们永不重视这些幻象，所以简朴才永远是真真伟大人物的标志。小官员幻想着自己的伟大；交际场中的暴发户夸耀他的珠宝；幼稚的作家幻想自己跃登作家之林，马上变成较不简朴、较不自然的人，这些都足以表示心智之狭小。

我们的演戏本能是根深蒂固的，以至于我们常常忘记离开舞台，忘记还有一些真正的生活可过。因此，我们一生辛辛苦苦地工作，并不依照自己的本性，为自己而生活，而只是为社会人士的喝彩而生活，如中国俗语所说老处女"为他人做嫁衣裳"。

II. Passion, Wisdom and Courage: Mencius

THE ideal character best able to enjoy life is a warm, carefree and unafraid soul. Mencius enumerated the three"mature virtues"of his"great man"as"wisdom, compassion and courage."I should like to lop off one syllable and regard as the qualities of a great soul passion, wisdom and courage. Luckily, we have in the English language the word"passion". which in its usage very nearly corresponds to the Chinese word ch'ing. Both words start out with the narrower meaning of sexual passion, but both have a much wider significance. As Chang Ch'ao says,"A passionate nature always loves women, but one who loves women is not necessarily a passionate nature."And again,"Passion holds up the bottom of the world, while genius paints its roof."For unless we have passion, we have nothing to start out in life with at all. It is passion that is the soul of life, the light in the stars, the lilt in music and song, the joy in flowers, the plumage in birds, the charm in woman, and the life in scholarship. It is as impossible to speak of a soul without passion as to speak of music without expression. It is that which gives us inward warmth and the rich vitality which enables us to face life cheerily.

Or perhaps I am wrong in choosing the word"passion"when I speak of what the Chinese writers refer to as ch'ing. Should I translate it by the word"sentiment,"which is gentler and suggests less of the tumultuous qualities of stormy passion? Or perhaps we mean by it something very similar to what the early Romanticists call "sensibility," which we find in a warm, generous and artistic soul. It is strange that among the Western philosophers so few, except Emerson, Amiel, Joubert and Voltaire, have a good word to say for passion. Perhaps we are arguing about words merely, while we mean the same thing. But then, if passion is different from sentiment and means something tumultuous and upsetting, then we haven't got a Chinese word for it, and we still have to go back to the old word ch'ing. Is this an index of a difference in racial temperament, of the absence among the Chinese people of grand and compelling passions, which eat up one's soul and form the stuff of tragedy in Western literature? Is this the reason why Chinese literature has not developed tragedy in the Greek sense; why Chinese tragic characters at the critical moment weep, give up their sweethearts to their enemy, or as in the case of Ch'u Pawang, stab their sweethearts and then plunge the knife into their own breasts? It is a sort of ending that will be found unsatisfying to a Western audience, but as Chinese life is, so is Chinese literature. Man struggles with fate, gives up the battle, and the tragedy comes in the aftermath, in a flood of reminiscences, of vain regret and longing, such as we see in the tragedy of Emperor T'ang Minghuang, who after granting the suicide of his beloved queen to placate a rebellious army, lives in a dream world in memory of her. The tragic sense is shown in the remaining part of the Chinese play long after the dénouement, in a swelling crescendo of sorrow. As he travels in his exile, he hears the distant music of cowbells in the hills on a rainy day and he composes the "Song of Rain on Cowbells" in her honor, everything he sees or touches, a little perfumed scarf that still retains its old scent, or an old maid servant of hers, reminds him of his beloved queen, and the play ends with him searching for her soul with the help of Taoist priests in the abode of the Immortals. So then, we have here a romantic sensibility, if we are not allowed to speak of it as passion. But it is passion mellowed down to a gentle glow. So it is characteristic of Chinese philosophers that while they disparage the human"desires"(in the sense of the" seven passions"), they have never disparaged passion or sentiment itself, but made it the very basis of a normal human life, so much so that they regard"the passion between husband and wife as the very foundation of all normal human life. "

It is unfortunately true that this matter of passion, or still better, sentiment, is something born in us, and that as we cannot choose our parents, we are born with a given cold or warm nature. On the other hand, no child is born with a really cold heart, and it is only in proportion as we lose that youthful heart that we lose the inner warmth in ourselves. Somewhere in our adult life, our sentimental nature is killed, strangled, chilled or atrophied by an unkind surrounding, largely through our own fault in neglecting to keep it alive, or our failure to keep clear of such surroundings. In the process of learning "world experience,"there is many a violence done to our original nature, when we learn to harden ourselves, to be artificial, and often to be cold-hearted and cruel, so that as one prides oneself upon gaining more and more worldly experience, his nerves become more and more insensitive and benumbed—especially in the world of politics and commerce. As a result, we get the great"go-getter"pushing himself forward to the top and brushing everybody aside; we get the man of iron will and strong determination, with the last embers of sentiment, which he calls foolish idealism or sentimentality, gradually dying out in his breast. It is that sort of person who is beneath my contempt. The world has too many cold-hearted people. If sterilization of the unfit should be carried out as a state policy, it should begin with sterilizing the morally insensible, the artistically stale, the heavy of heart, the ruthlessly successful, the cold-heartedly determined and all those people who have lost the sense of fun in life—rather than the insane and the victims of tuberculosis. For it seems to me that while a man with passion and sentiment may do many foolish and precipitate things, a man without passion or sentiment is a joke and a caricature. Compared with Daudet's Sappho, he is a worm, a machine, an automaton, a blot upon this earth. Many a prostitute lives a nobler life than a successful business man. What if Sappho sinned? For although she sinned, she also loved, and to those who love much, much will be forgiven. Anyway she emerged out of an equally harsh business environment with more of the youthful heart than many of our millionaires. The worship of Mary Magdalene is right. It is unavoidable that passion and sentiment should lead us into mistakes for which we are duly punished, yet there is many an indulgent mother who by her indulgence often let her love get the better of her judgment, and yet who, we feel sure, in her old age felt that she had had a more happy life with her family than many rigorous and austere souls. A friend told me the story of an old lady of seventy-eight who said to him, "As I look back upon my seventy-eight years, it still makes me happy to think of when I sinned; but when I think I was stupid, I cannot forgive myself even at this late day. "

But life is harsh, and a man with a warm, generous and sentimental nature may be easily taken in by his cleverer fellowman. The generous in nature often make mistakes by their generosity, by their too generous regard of their enemies and faith in their friends. Sometimes the generous man comes home disillusioned to write a poem of bitterness. That is the case of many a poet and scholar in China, as for instance that great tea-drinker, Chang Tai, who generously squandered his fortune, was betrayed by his own closest friends and relatives, and set down in twelve poems some of the bitterest verses I have ever read. But I have a suspicion that he kept on being generous to the end of his days, even when he was quite poor and destitute, being many times on the verge of starvation, and I have no doubt that those bitter sentiments passed away like a cloud and he was still quite happy.

Nevertheless this warm generosity of soul has to be protected against life by a philosophy, because life is harsh, warmth of soul is not enough, and passion must be joined to wisdom and courage. To me wisdom and courage are the same thing, for courage is born of an understanding of life; he who completely understands life is always brave. Anyway that type of wisdom which does not give us courage is not worth having at all. Wisdom leads to courage by exercising a veto against our foolish ambitions and emancipating us from the fashionable humbug of this world, whether humbug of thought or humbug of life.

There is a wealth of humbug in this life, but the multitudinous little humbugs have been classified by Chinese Buddhists under two big humbugs: fame and wealth. There is a story that Emperor Ch'ienlung once went up a hill overlooking the sea during his trip to South China and saw a great number of sailing ships busily plying the China Sea to and fro. He asked his minister what the people in those hundreds of ships were doing, and his minister replied that he saw only two ships, and their names were"Fame"and"Wealth."Many cultured persons were able to escape the lure of wealth, but only the very greatest could escape the lure of fame. Once a monk was discoursing with his pupil on these two sources of worldly cares, and said:"It is easier to get rid of the desire for money than to get rid of the desire for fame. Even retired scholars and monks still want to be distinguished and well-known among their company. They want to give public discourses to a large audience, and not retire to a small monastery talking to one pupil, like you and me now." The pupil replied:"Indeed, Master, you are the only man in the world who has conquered the desire for fame!"And the Master smiled.

From my own observation of life, this Buddhist classification of life's humbugs is not complete, and the great humbugs of life are three, instead of two: Fame, Wealth and Power. There is a convenient American word which again combines these three humbugs into the One Great Humbug: Success. But many wise men know that the desires for success, fame and wealth are euphemistic names for the fears of failure, poverty and obscurity, and that these fears dominate our lives. There are many people who have already attained both fame and wealth, but who still insist on ruling others. They are the people who have consecrated their lives to the service of their country. The price is often very heavy. Ask a wise man to wave his silk hat to a crowd and make seven speeches a day and give him a presidency, and he will refuse to serve his country. James Bryce thinks the system of democratic government in America is such that it is hardly calculated to attract the best men of the country into politics. I think the strenuousness of a presidential campaign alone is enough to frighten off all the wise souls of America. A public office often demands that a man attend six dinners a week in the name of consecrating his life to the service of mankind. Why does he not consecrate himself to a simple supper at home and to his bed and his pyjamas? Under the spell of that humbug of fame or power, a man is soon prey to other incidental humbugs. There will be no end to it. He soon begins to want to reform society, to uplift others' morality, to defend the church, to crush vice, to map programs for others to carry out, to block programs that other people have mapped out, to read before a convention a statistical report of what other people have done for him under his administration, to sit on committees examining blueprints of an exposition, even to open an insane asylum (what cheek!)–in general, to interfere in other people's lives. He soon forgets that these gratuitous assumptions of responsibility, these problems of reforming people and doing this and preventing one's rivals from doing that, never existed for him before, perhaps had not even entered his mind. How completely the great problems of labor, unemployment and tariffs leave the mind of a defeated presidential candidate even two weeks after an election! Who is he that he should want to reform other people and uplift their morals and send other people into an insane asylum? But these primary and secondary humbugs keep him happily busy, if he is successful, and give him the illusion that he is really doing something and is therefore "somebody."

Yet there is a secondary social humbug, quite as powerful and universal, the humbug of fashion. The courage to be one's own natural self is quite a rare thing. The Greek philosopher Democritus thought he was doing a great service to mankind by liberating it from the oppression of two great fears: the fear of God and the fear of death. But even that does not liberate us from another equally universal fear; the fear of one's neighbors. Few men who have liberated themselves from the fear of God and the fear of death are yet able to liberate themselves from the fear of man. Consciously or unconsciously, we are all actors in this life playing to the audience in a part and style approved by them.

This histrionic talent, together with the related talent for imitation, which is a part of it, are the most outstanding traits of our simian inheritance. There are undoubted advantages to be derived from this showmanship, the most obvious being the plaudits of the audience. But then the greater the plaudits, the greater also are the flutterings of heart backstage. And it also helps one to make a living, so that no one is quite to blame for playing his part in a fashion approved by the gallery.

The only objection is that the actor may replace the man and take entire possession of him. There are a few select souls who can wear their reputation and a high position with a smile and remain their natural selves; they are the ones who know they are acting when they are acting, who do not share the artificial illusions of rank, title, property and wealth, and who accept these things with a tolerant smile when they come their way, but refuse to believe that they themselves are thereby different from ordinary human beings. It is this class of men, the truly great in spirit, who remain essentially simple in their personal lives. It is because they do not entertain these illusions that simplicity is always the mark of the truly great. Nothing shows more conclusively a small mind than a little government bureaucrat suffering from illusions of his own grandeur, or a social upstart displaying her jewels, or a halfbaked writer imagining himself to belong to the company of the immortals and immediately becoming a less simple and less natural human being.

So deep is our histrionic instinct that we often forget that we have real lives to live off stage. And so we sweat and labor and go through life, living not for ourselves in accordance with our true instincts, but for the approval of society, like"old spinsters working with their needles to make wedding dresses for other women,"as the Chinese saying goes.

三. 玩世、愚钝、潜隐: 老子

　　老子刁慈的"老猾"哲学却产生了和平、容忍、简朴和知足的崇高理想，这看来似乎是矛盾的。这类教训包括愚笨者的智慧，隐逸者的长处，柔弱者的力量和熟悉世故者的简朴。中国艺术的本身，和它那诗意的幻象以及对于樵夫渔夫的简朴生活之赞颂，都不能脱离这种哲学而存在。中国和平主义的根源，就是能忍受暂时的失败，静待时机，相信在天地万物的体系中，在大自然动力和反动力的规律运行之下，没有一个人能永远占着便宜，也没有一个人永远做"傻子"。

　　　大巧若拙，
　　　大辩若讷。
　　　躁胜寒，
　　　静胜热。
　　　清静为天下正。

　　（老子《道德经》，下同）

我们既知道大自然的运行中，没有一个人能永远占着便宜或是做着傻子，所以其结论竞争是徒劳的。老子曰："圣人夫唯不争，故天下莫能与之争。"又曰："强梁者不得其死，吾将以为教父。"当今的作家也可加上一句："世间的独裁者如能不要密探来卫护，我愿做他的党徒。"

因此，老子曰："天下有道，却走马以粪；天下无道，戎马生于郊。"

善为士者不武；
善战者不怒。
善胜敌者不与；
善用人者为之下。
是谓不争之德，
是谓用人之力，
是谓配天古之极。

有了动力与反动力的规律，便产生了暴力对付暴力的局势：

以道佐人主者，
不以兵强天下；
其事好还。
师之所处，荆棘生焉。
大军之后，必有凶年。
善者果而已；
不敢以取强。
果而勿矜；

果而勿伐；

果而勿骄；

果而不得已。

果而勿强；

物壮则老。

是谓不道，

不道早已。

凡尔赛会议如果请老子去做主席，我想今日一定不会有这么一个希特勒。希特勒自以为他在政治上当权之速，证明他得到"上帝的庇佑"。但我以为事情还要简单，他是得到克里孟梭（Clemenceau，"一战"法国总理）神魂的庇佑。中国的和平主义不是那种人道的和平主义——不以博爱为本，而以一种近情的微妙的智慧为本。

将欲歙之，

必固张之。

将欲弱之，

必固强之。

将欲废之，

必固兴之。

是谓微明。

柔弱胜强。

鱼不可脱于渊；

邦利器不可以示人。

关于柔弱者的力量，爱好和平者之总能得到胜利，以及隐逸者的长处这一类训诲，没有一个人再能比老子讲得更有力量。在老子看来，水便是柔弱者的力量的象征——轻轻地滴下来，能在石头上穿一个洞；水有道家最伟大的智慧，向最低下的地方去求它的水平线：

> 江海所以能为百谷王者，
> 以其善下之，故能为百谷王。

"谷"是空洞象征，代表世间万物的子宫和母亲，代表"阴"或"牝"。

> 谷神不死，
> 是谓元牝。
> 元牝之门，
> 是谓天地之根。
> 绵绵呵！其若存！
> 用之不堇。

以"牝"来代表东方文化，而以"牡"来代表西方文化，这不会是牵强附会之谈吧。无论如何，在中国的消极力量里，有些东西很像子宫或山谷，老子说："……为天下谷；于天下谷，常德乃足。"

恺撒要做乡村中第一个人，而老子反之，他的忠告是："不敢为天下先。"讲到出名是一桩危险的事，庄子曾写过一篇讽刺的文章去反对孔子夸耀知识的行为。庄子著作里，有许多诽议孔子的文章，好在庄子写文章时，孔子已死，当时中国又没有关于毁坏名誉的法律。

孔子围于陈蔡之间，七日不火食。

大公任往吊之，曰："子几死乎？"

曰："然。"

"子恶死乎？"

曰："然。"

任曰："予尝言不死之道。

"东海有鸟焉，其名曰'意怠'。其为鸟也，翂翂翐翐，而似无能。引援而飞；迫胁而栖。进不敢为前；退不敢为后。食不敢先尝；必取其绪。是故其行列不斥，而外人卒不得害，是以免于患。

"直木先伐。甘井先竭。于其意者饰知以惊愚；修身以明污。昭昭乎若揭日月而行，故不免也。……"

孔子曰："善哉！"辞其交游，去其弟子，逃于大泽，衣裘褐，食杼栗。入兽不乱群，入鸟不乱行。鸟兽不恶，而况人乎！

我曾写过一首诗概括道家思想：

愚者有智慧，

缓者有雅致，

钝者有机巧，

隐者有益处。

在信仰基督教的读者们看来，这几句话或者很像耶稣的"山上训言"，而也许同样地对他们不生效力。老子说，愚者得福，因他们是世上最快乐的人，这句话好似替"山上训言"加了一些诙谐的成分。庄子继老子"大巧若拙，大辩若讷"的名句而说"弃智"。八世纪时的柳宗元把他比邻的山叫做"愚山"，附近的水叫做"愚溪"。十八世纪时的郑板桥说了一句名言："聪明难，糊涂亦难，由聪明转入糊涂更难。"中国文学上有诸如此类不少赞颂愚钝的话。美国有一句俚语是"不要太精明"（Don't be too smart），也可看出抱这种态度者的智慧。大智是常常如愚的。

所以，在中国文化上我们看见一种稀奇的现象，就是一个大智对自己发生怀疑，因而产生（据我所知）唯一的愚者的福音和潜隐的理论，认为是人生斗争的最佳武器。由庄子的创说"弃智"，到尊崇愚者的观念，其中只是一个短短的过程；在中国的绘画中和文章中，有着不少的乞丐，不朽的隐逸者、癫僧，或如屠隆的《冥寥子游》中的奇隐士等，在那上面，我们都可以看出这种尊崇愚者观念的反映。当这个可怜的褴褛癫道变成了我们心目中最高智慧和崇高性格的象征时，智人即从人生的迷恋中清醒过来，接受一些浪漫的或宗教润色，而进入诗意的幻想境界。

傻子的受人欢迎是一桩实事。我相信无论在东方或西方，人们总是憎恶那个过于精明的同伴。袁中郎曾写过一篇文字，说明他和他的兄弟为什么要用那四个极愚笨但是忠心的仆人。任何人只要把他所有的朋友同伴细细想一想，就可以发现我们究竟喜欢怎样的人。我们喜欢愚笨的仆人是因为他比较老实可靠。和他在一起过日子，我们尽可以写写意意，不必处处提心吊胆。智慧的男人多数要不太精明的妻子，而智慧的女子也多数愿嫁不太精明的丈夫。

中国历史上那些著名的傻子，都是因为他们的真癫或假癫而讨人欢喜，受人敬爱。例如宋朝著名画家米芾号"米颠"（即癫），有一次穿了礼服去拜一块岩石，要那块岩石做他的"丈人"，因此得了"米颠"的名号。他和元朝的著名画家倪云林都有好洁之癖。又有一个著名的疯诗人赤了足，往来于各大寺院，在厨房里打杂，吃人家的残羹冷饭，不朽的诗便写在庙寺里厨房的墙壁上。最受中国人民爱戴的，要算是伟大的疯和尚颠僧了，他名叫济公，是一部通俗演义的主人公；这部演义越演越长，篇幅比《堂吉诃德》（Don Quixote）还长三倍，但好像还没有完结。他生活于一个魔术、能医、恶作剧和醉酒的世界里，他有一种神力，能在相距几百英里的不同城市里同时出现，纪念他的庙宇至今还屹立于杭州西子湖边的虎跑。十六世纪和十七世纪的伟大浪漫天才，如徐文长、李卓吾、金圣叹（他自号"圣叹"，据他说，当他出世时，孔庙里曾发出一阵神秘的叹息）虽然和我们一样是人，可是他们在外表和举动上多少违背着传统的习惯，所以给人一种疯狂的印象。

III. *Cynicism, Folly and Camouflage: Laotse*

PARADOXICALLY, Laotse's most wicked philosophy of"the old rogue"has been responsible for the highest ideal of peace, tolerance, simplicity and contentment. Such teachings include the wisdom of the foolish, the advantage of camouflage, the strength of weakness, and the simplicity of the truly sophisticated. Chinese art itself, with its poetic illusion and its glorification of the simple life of the woodcutter and the fisherman, cannot exist apart from this philosophy. And at the bottom of Chinese pacificism is the willingness to put up with temporary losses and bide one's time, the belief that, in the scheme of things, with nature operating by the law of action and reaction, no one has a permanent advantage over the others and no one is a"damn fool"all the time.

> *The greatest wisdom seems like stupidity.*
> *The greatest eloquence like stuttering.*
> *Movement overcomes cold,*
> *But staying overcomes heat.*
> *So he by his limpid calm*
> *Puts everything right under heaven.*

Knowing then that in Nature's ways no man has a permanent advantage over others and no man is a damn fool all the time, the natural conclusion is that there is no use for contention. In Laotse's words, the wise man"does not contend, and for that very reason no one under Heaven can contend with him."Again he says,"Show me a man of violence that came to a good end, and I will take him for my teacher. "A modern writer might add,"Show me a dictator that can dispense with the services of a secret police, and I will be his follower. "

For this reason, Laotse says,"When the Tao prevails not, horses are trained for battle; when the Tao prevails, horses are trained to pull dungcarts. "

> The best charioteers do not rush ahead;
> The best fighters do not make displays of wrath.
> The greatest conqueror wins without joining issue;
> The best user of men acts as though he were their inferior.
> This is called the power that comes of not contending,
> Is called the capacity to use men,
> The secret of being mated to heaven, to what was of old.

The law of action and reaction brings about violence rebounding to violence:

> He who by Tao purposes to help a ruler of men
> Will oppose all conquest by force of arms;
> For such things are wont to rebound.
> Where armies are, thorns and brambles grow.
> The raising of a great host
> Is followed by a year of dearth.
> Therefore a good general effects his purpose and then stops;
> he does not take further advantage of his victory.

Fulfills his purpose and does not glory in what he has done;
Fulfills his purpose and does not boast of what he has done;
Fulfills his purpose, but takes no pride in what he has done;
Fulfills his purpose, but only as a step that could not be avoided.
Fulfills his purpose, but without violence;
For what has a time of vigor also has a time of decay.
This is against Tao,
And what is against Tao will soon perish.

My feeling is that, if Laotse had been invited to take the chair at the Versailles Conference, there would not be a Hitler today. Hitler claims that he and his work must have been "blessed by God," on the evidence of his miraculous rise to power. I am inclined to think that the matter is simpler than that, that he was blessed by the spirit of Clemenceau. Chinese pacifism is not that of the humanitarian, but that of the old rogue—based not upon universal love, but upon a convincing type of subtle wisdom.

What is in the end to be shrunk
Must first be stretched.
Whatever is to be weakened
Must begin by being made strong.
What is to be overthrown
Must begin by being set up.
He who would be a taker
Must begin as a giver.
This is called "dimming" one's light.
It is thus that the soft overcomes the hard
And the weak, the strong.
"It is best to leave the fish down in his pool;
Best to leave the State's sharpest weapons

where none can see them."

There has never been a more effective sermon more effectively preached on the strength of weakness, the victory of the peaceloving and the advantage of lying low than by Laotse. For water remains for Laotse forever as the symbol of the strength of the weak—water that gently drips and makes a hole in a rock, and water which has the great Taoistic wisdom of seeking the lowest level:

How did the great rivers and seas get their kingship over the hundred lesser streams?
Through the merit of being lower than they; that was how they got their kingship.

An equally common symbol is that of" the Valley,"representing the hollow, the womb and mother of all things, the yin or the Female.

The Valley Spirit never dies.
It is named the Mysterious Female.
And the Doorway of the Mysterious Female
Is the base from which Heaven and Earth sprang.
It is there within us all the while;
Draw upon it as you will, it never runs dry.

It would not be at all far-fetched to say that Oriental civilization represents the female principle, while the Occidental civilization represents the male principle. Anyway, there is something terribly resembling the womb or valley in China's passive strength that, in Laotsean language,"receives into it all things under heaven, and being a valley, has all the time a power that suffices. "

Against the desire of Julius Caesar to be the first man in a village, Laotse

gives the opposite counsel of" Never be the first in the world."This thought
of the danger of being eminent is expressed by Chuangtse in the form of a
satire against Confucius and his display of knowledge. There were many such
libels against Confucius in the books of Chuangtse, for Confucius was dead
when Chuangtse wrote, and there was no libel law in China.

*When Confucius was hemmed in between Ch'en and Ts'ai, he passed seven days
without food.*

*The minister Jen went to condole with him, and said,"You were near, Sir, to
death. "*

"I was indeed,"replied Confucius.

"Do you fear death, Sir?"inquired Jen.

"I do,"said Confucius.

"Then I will try to teach you,"said Jen,"the way not to die. "

*"In the eastern sea there are certain birds, called the i-erh. They behave themselves
in a modest and unassuming manner, as though unpossessed of ability. They fly
simultaneously; they roost in a body. In advancing, none strives to be first; in retreating,
none venture to be last. In eating, none will be the first to begin; it is considered proper
to take the leavings of others. Therefore, in their own ranks they are at peace, and the
outside world is unable to harm them. And thus they escape trouble.*

*"Straight trees are the first felled. Sweet wells are soonest exhausted. And you, you
make a show of your knowledge in order to startle fools. You cultivate yourself in contrast
to the degradation of others. And you blaze along as though the sun and moon were
under your arms; consequently, you cannot avoid trouble... "*

*"Good indeed!"replied Confucius; and forthwith he took leave of his friends and
dismissed his disciples and retired to the wilds, where he dressed himself in skins and serge
and fed on acorns and chestnuts. He passed among the beasts and birds and they took no
heed of him.*

I have made a poem which sums up for me the message of Taoistic thought:

> *There is the wisdom of the foolish,*
> *The gracefulness of the slow,*
> *The subtlety of stupidity,*
> *The advantage of lying low.*

This must sound to Christian readers like the Sermon on the Mount, and perhaps seem equally ineffective to them. Laotse gave the Beatitudes a cunning touch when he added:" Blessed are the idiots, for they are the happiest people on earth."Following Laotse's famous dictum that"The greatest wisdom is like stupidity; the greatest eloquence like stuttering," Chuangtse says:"Spit forth intelligence." Liu Chungyüan in the eighth century called his neighborhood hill"the Stupid Hill" and the nearby river"the Stupid River." Cheng Panch'iao in the eighteenth century made the famous remark:"It is difficult to be muddle-headed. It is difficult to be clever, but still more difficult to graduate from cleverness into muddle-headedness."The praise of folly has never been interrupted in Chinese literature. The wisdom of this attitude can at once be understood through the American slang expression:"Don't be too smart."The wisest man is often one who pretends to be a"damn fool. "

In the Chinese culture, therefore, we see the curious phenomenon of a high intellect growing suspicious of itself and developing, so far as I know, the only gospel of ignorance and the earliest theory of camouflage as the best weapon in the battle of life. From Chuangtse's advice to"spit forth intelligence,"it is but a short step to the glorification of the idiot, which we see constantly reflected in Chinese paintings and literary sketches of the beggar, or the disguised immortal, or the crazy monk, or the extraordinary recluse, as seen in The Travels of Mingliaotse (Chapter XI). The wise disenchantment with life receives a romantic or religious touch and enters the realm of poetic fantasy, when the poor, ragged and half-crazy monk becomes for us the symbol of highest wisdom and nobility of character.

The popularity of fools is an undeniable fact. I have no doubt that, East or West, the world hates a man who is too smart in his dealings with his fellowmen. Yüan Chunglang wrote an essay showing why he and his brothers chose to keep four extremely stupid and extremely loyal servants. Anyone can run over the names of his friends and associates in his mind and verify this fact for himself, that those we like are not those we respect for distinguished ability and those we respect for distinguished ability are not those we like, and that we like a stupid servant because he is more reliable, and because in his company we can better relax and do not have to set up a condition of defense against his presence. Most wise men choose to marry a not too smart wife, and most wise girls choose a not too smart husband as a life companion.

There have been a number of famous fools in Chinese history, all extremely popular and beloved for their real or affected craziness. Among these, for instance, is the famous Sung painter, Mi Fu, styled"Mi Tien"(or"Mi the Crazy One"), who got this title because he once appeared in a ceremonial robe to worship a piece of jagged rock that he called his"father-in-law."Both Mi Fu and the famous Yüan painter, Ni Yunlin, had a mild form of dirt-phobia or fastidious cleanliness. There was the famous crazy poet-monk Hanshan, who went about with disheveled hair and bare feet, doing odd kitchen jobs at different monasteries, eating the leftovers, and scribbling immortal poetry on the temple and kitchen walls. The greatest crazy monk who has captured the imagination of the Chinese people is undoubtedly Chi Tien ("Chi the Crazy One"), or Chi Kung ("Master Chi"), who is the hero of a popular romance steadily being lengthened and added to until it is about three times the size of Don Quixote, and still seems endless. For he lives in a world of magic, medicine, roguery, and drunkenness, and possesses the gift of appearing at different cities several hundred miles apart on the same day. The temple to his honor still stands at Hupao near the West Lake of Hangchow today. To a lesser degree, the great romantic geniuses of the sixteenth

and seventeenth centuries, while decidedly as normal as we, tended through their
unconventionalities of appearance and conduct to give people an impression of
being crazy, such as Hsü Wench'ang, Li Chowu and Chin Shengt'an (literally, "the
Sigh of the Sage," a name he gave himself because he said that when he was born,
a mysterious sigh was heard in the village temple of Confucius).

四.“中庸哲学”：子思

我相信主张无忧虑和心地坦白的人生哲学，一定要叫我们摆脱过于繁忙的生活和太重大的责任，因而使人们渐渐减少实际行动的欲望。在另一方面，生于现代的人，大都需要这种玩世主义之熏陶，因为这对他是很有益的。那种引颈前瞻徒然使人类在无效果和浪费的行动中过生活的哲学，它的遗毒或许比古今哲学中的全部玩世思想为害更大。每个人都有许多生理上的工作行动，随时能把这种哲学的力量抵消；这种放浪者的伟大哲学虽到处受欢迎，可是中国人至今还是世界上最勤勉的民族，大多数人都未成为玩世者，因为大多数人都不是哲学家。

所以这样说来，玩世主义很少会有变成大众所崇拜的流行思想的危险，这一点可以不必担忧。中国道家哲学虽已获得了中国人心胸中的感应，已经存在了几千年，在每首诗歌和每幅山水画里都可看得出来，但是大多数中国人依旧过着熙来攘往的生活，依旧相信财富、名誉、权力，肯为他们的国家服役。如若不这样，人类生活便不能维持下去。所以中国并没有人人都服从玩世主义，他们只在失败后才做玩世者和诗人，我们的多数同胞依旧还是出力的演员。道家玩世主义的影响，仅在于减低紧张生活，同时在天灾人祸的时候，引导人民去信仰自然律的动作和反动作，信仰正义必能因此而得伸张。

　　然而，在中国的思想上还有一种相反的势力，它和这种无忧无虑的哲学、自然放浪者的哲学，站在对立的地位。自然绅士哲学的对面有社会绅士的哲学，道家哲学的对面有儒家哲学。如道家哲学和儒家哲学的含义，一个代表消极的人生观，一个代表积极的人生观，那么，我相信这两种哲学不仅是中国人有之，也是人类天性所固有的东西。我们大家都是生就一半道家主义，一半儒家主义。一个彻底的道家主义者理应隐居山中，去竭力模仿樵夫和渔夫的生活，无忧无虑，简单朴实如樵夫一般去做青山之王，如渔夫一般去做绿水之王。道家主义者的隐士隐现于山上的白云中，一面俯视樵夫和渔夫在相对闲谈；一面默念着青山、流水，全然不理会这里还有着两个渺小的谈话者，他在这种凝想中获得一种彻底的和平感觉。不过要叫我们完全逃避人类社会的那种哲学，终究是拙劣的。

　　此外还有一种比这自然主义更伟大的哲学，就是人性主义的哲学。所以，中国最崇高的理想，就是一个人不必逃避人类社会和人生，而本性仍能保持原有快乐。如果一个人离开城市，到山中去过着幽寂的生活，那么他也不过是第二流隐士，还是环境的奴隶。"城中隐士实是最伟大的隐士"，因为他对自己具有充分的节制力，不受环境的支配。如果一个僧人回到社会上去喝酒、吃肉、交女人，同时并不腐蚀他的灵魂，那么他便是一个"高僧"了。因此，这两种哲学有互通性，颇有合并的可能。儒教和道家的对比是相对的，而不是绝对的；这两种学说只是代表了两个极端的理论，而在这两个极端的理论之间，还有着许多中间的理论。

　　我以为半玩世者是最优越的玩世者。生活的最高典型终究应属子思所倡导的中庸生活，他即是《中庸》作者，孔子的孙儿。与人类生活问题有关的古今哲学，还不曾发现过一个比这种学说更深奥的真理。这种学说，就是指介于两个极端之间的那一种有条不紊的生活——酌乎其中学说。这种中庸精神，在动作和静止之间找到了一种完全的均衡，所以理想人物，应属一半有名，一半无名；懒惰中带用功，在用功中偷懒；穷不至于穷到付不出房租，富也不至于富到可以完全不做工，或是可以称心如意地资助朋友；钢琴也会弹弹，可是不十分高明，只可弹给知己的朋友听听，而最大的用处还是给自己消遣；古玩也收藏一点，可是只够摆满屋里的壁炉架；书也读读，可是不很用功；学识颇广博，可是不成为任何专家；文章也写写，可是寄给《泰晤士报》的稿件一半被录用一半退回——总而言之，我相信这种中等阶级生活，是中国人所发现最健全的理想生活。李密庵（清代诗人）在他的《半半歌》里把这种生活理想很美妙地表达出来：

看破浮生过半，
半之受用无边。
半中岁月尽幽闲，
半里乾坤宽展。
半郭半乡村舍，
半山半水田园；
半耕半读半经廛，
半士半姻民眷；
半雅半粗器具，
半华半实庭轩；
衾裳半素半轻鲜，
肴馔半丰半俭；
童仆半能半拙，

妻儿半朴半贤；

心情半佛半神仙；

姓字半藏半显。

一半还之天地；

让将一半人间。

半思后代与沧田，

半想阎罗怎见。

饮酒半酣正好，

花开半吐偏妍；

帆张半扇免翻颠，

马放半缰稳便。

半少却饶滋味，

半多反厌纠缠。

百年苦乐半相参，

会占便宜只半。

所以，我们如把道家的现世主义和儒家的积极观念配合起来，便成中庸的哲学。因为人类是生于真实的世界和虚幻的天堂之间，所以我相信这种理论在一个抱前瞻观念的西洋人看来，一瞬间也许很不满意，但这总是最优越的哲学，因为这种哲学是最近人情的。总而言之，半个查尔斯·奥古斯都·林白（Charles Augustus Lindbergh，又译林德伯格，美国飞行员，首个进行单人不着陆的跨大西洋飞行的人）比一个整的林白更好，因为半个能比较快乐。如果林白只飞了大西洋的半程，我相信他一定会更快乐。我们承认世间非有几个超人——改变历史进化的探险家、征服者、大发明家、大总统、英雄——不可，但是最快乐的人还是那个中等阶级者，所赚的钱足以维持独立的生活，曾替人群做过一点点事情，可是不多；在社会上稍具名誉，可是不太显著。只有在这种环境之下，名字半隐半显，经济适度宽裕，生活逍遥自在，而不完全无忧无虑的那个时候，人类的精神才是最为快乐的，才是最成功的。我们必须在这尘世上活下去，所以我们须把哲学由天堂带到地上来。

IV. "Philosophy of Half-and-Half": Tsesse

I have no doubt that a philosophy which enjoins the carefree and conscience-
free life has a strong tendency to warn us away from a too busy life and from
too great responsibilities, and therefore tends to decrease the desire for action.
On the other hand, the modern man needs this refreshing wind of cynicism
which cannot but do him some good. Probably more harm is done by a
forward-looking philosophy delivering man over to a life of futile, wasteful
activities than is ever done by all the cynicism of the ancient and modern
philosophies combined. There are too many physiological impulses for
action in every man, ready to counteract this philosophy, and in spite of the
popularity of this great Philosophy of the Scamp, the Chinese people are still
one of the most industrious on earth. The majority of men cannot be cynics,
simply because the majority of men are not philosophers.

As far as I can see, therefore, there is very little danger of cynicism being
transformed into a general vogue followed by the herd. Even in China, where
the Taoist philosophy finds an instinctive response in the Chinese breast, and
where that philosophy has been at work for several thousand years, staring at
us from every poem and every scroll of landscape painting, life still goes on
merrily with lots of people believing in wealth and fame and power, quite
determined and anxious to serve their country. Were it not so, human life

would not be able to get along at all. No, the Chinese are cynics and poets only when they have failed; most of my countrymen are still very good showmen. The effect of Taoistic cynicism has been only to slow down the tempo of life, and in the case of natural calamities and human misrule, to promote trust in the law of action and reaction, which brings about justice in the end.

And yet there is an opposite influence in Chinese thought in general which counteracts this carefree philosophy, the philosophy of the natural vagabond. Opposed to the philosophy of nature's gentlemen, there is the philosophy of society's gentlemen; opposed to Taoism, there is Confucianism. Insofar as Taoism and Confucianism mean merely the negative and positive outlooks on life, I do not think they are Chinese, but are inherent in all human nature. We are all born half Taoists and half Confucianists. The logical conclusion of a thorough-going Taoist would be to go to the mountains and live as a hermit or a recluse, to imitate as far as possible the simple carefree life of the woodcutter and the fisherman, the woodcutter who is lord of the green hills and the fisherman who is the owner of the blue waters. The Taoist recluse, half-hidden in the clouds on top of the mountain, looks down at the woodcutter and the fisherman holding an idle conversation, remarking that the hills go on being green and the waters go on flowing just to please themselves, entirely oblivious of the two tiny conversationalists. From this reflection, he gets a sense of perfect peace. And yet it is poor philosophy that teaches us to escape from human society altogether.

There is still a greater philosophy than this naturalism, namely, the philosophy of humanism. The highest ideal of Chinese thought is therefore a man who does not have to escape from human society and human life in order to preserve his original, happy nature. He is only a second-rate recluse, still slave to his environment, who has to escape from the cities and live away in the mountains in solitude. "The Great Recluse is the city recluse," because he has sufficient mastery over himself not to be afraid of his surroundings.

He is therefore the Great Monk (the kaoseng) who returns to human society and eats pork and drinks wine and mixes with women, without detriment to his own soul. There is, therefore, the possibility of the merging of the two philosophies. The contrast between Confucianism and Taoism is relative, the two doctrines setting forth only two great extremes, and between them there are many intermediate stages.

Those are the best cynics who are half-cynics. The highest type of life after all is the life of sweet reasonableness as taught by Confucius' grandson, Tsesse, author of The Golden Mean. No philosophy, ancient or modern, dealing with the problems of human life has yet discovered a more profound truth than this doctrine of a well-ordered life lying somewhere between the two extremes—the Doctrine of the Half-and-Half. It is that spirit of sweet reasonableness, arriving at a perfect balance between action and inaction, shown in the ideal of a man living in half-fame and semi-obscurity: half-lazily active and half-actively lazy; not so poor that he cannot pay his rent, and not so rich that he doesn't have to work a little or couldn't wish to have slightly more to help his friends; who plays the piano, but only well enough for his most intimate friends to hear, and chiefly to please himself; who collects, but just enough to load his mantelpiece; who reads, but not too hard; learns a lot but does not become a specialist; writes, but has his correspondence to the Times half of the time rejected and half of the time published—in short, it is that ideal of middle-class life which I believe to be the sanest ideal of life ever discovered by the Chinese. This is the ideal so well expressed in Li Mian's "The Half-and-Half Song":

> By far the greater half have I seen through
> This floating life—Ah, there's a magic word—
> This "half"—so rich in implications.
> It bids us taste the joy of more than we
> Can ever own. Halfway in life is man's

Best state, when slackened pace allows him ease;
A wide world lies halfway, twixt heaven and earth;
To live halfway between the town and land,
Have farms halfway between the streams and hills;
Be half-a-scholar, and half-a-squire, and half
In business; half as gentry live,
And half related to the common folk;
And have a house that's half genteel, half plain,
Half elegantly furnished and half bare;
Dresses and gowns that are half old, half new,
And food half epicure's, half simple fare;.
Have servants not too clever, not too dull;
A wife who's not too simple, nor too smart—
So then, at heart, I feel I'm half a Buddha,
And almost half a Taoist fairy blest.
One half myself to Father Heaven I
Return; the other half to children leave—
Half thinking how for my posterity
To plan and provide, and yet half minding how
To answer God when the body's laid at rest.
He is most wisely drunk who is half drunk;
And flowers in half-bloom look their prettiest;
As boats at half-sail sail the steadiest,
And horses held at half-slack reins trot best.
Who half too much has, adds anxiety,
But half too little, adds possession's zest.
Since life's of sweet and bitter compounded,
Who tastes but half is wise and cleverest.

We have here, then, a compounding of Taoistic cynicism with the

Confucian positive outlook into a philosophy of the half-and-half. And because man is born between the real earth and the unreal heaven, I believe that, however unsatisfactory it may seem on the first look to a Westerner, with his incredibly forward-looking point of view, it is still the best philosophy, because it is the most human. After all, a half Lindbergh would be better, because more happy, than a complete Lindbergh. I am quite sure Lindbergh would be much happier if he had flown only halfway across the Atlantic. After all allowances are made for the necessity of having a few supermen in our midst—explorers, conquerors, great inventors, great presidents, heroes who change the course of history—the happiest man is still the man of the middle-class who has earned a slight means of economic independence, who has done a little, but just a little, for mankind, and who is slightly distinguished in his community, but not too distinguished. It is only in this milieu of well-known obscurity and financial competence with a pinch, when life is fairly carefree and yet not altogether carefree, that the human spirit is happiest and succeeds best. After all, we have to get on in this life, and so we must bring philosophy down from heaven to earth.

五．爱好人生者：陶渊明

　　所以我们已经知晓，如果把积极的人生观念和消极的人生观念适度地配合起来，我们便能得到一种和谐的中庸哲学，介于动作和静止之间，介于尘世的徒然匆忙和完全逃避现实人生之间；世界上所有的哲学中，这一种可说是人类生活上最健全最完美的理想了。还有一种结果更加重要，就是这两种不同观念相混合后，和谐的人格也随之产生；这种和谐的人格也就是那一切文化和教育所欲达到的目的，我们即从这种和谐的人格中看见人生的欢乐和爱好。这是值得注意的。

　　要描写这种爱好人生的性质是极困难的；如用譬喻，或叙述一位爱好人生者的真事实物，那就比较容易。在这里，陶渊明这位中国最伟大的诗人和中国文化上最和谐的产物，不期然地浮上我的心头。陶渊明也是整个中国文学传统上最和谐最完美的人物，我想没有一个中国人会反对我的话吧。他没有做过大官，很少权力，也没有什么勋绩，除了本薄薄的诗集和三四篇零星的散文外，在文学遗产上也不曾留下什么了不得的著作，但至今还是照彻古今的炬火，在那些较渺小的诗人和作家心目中，他永远是最高人格的象征。他的生活和风格是简朴的，令人自然敬畏，会使那些较聪明与熟识的人自惭形秽。他是今日真正爱好人生者的模范，因为他心中虽有反抗尘世的欲望，但并不沦于彻底逃避人世，而反使他和七情生活洽调起来。文学的浪漫主义和道家闲散生活的崇尚以及对儒家教义的反抗，在那时的中国已活动了两百多年，这种种和前世纪的儒家哲学配合起来，就产生了这么一种和谐的人格。以陶渊明为例，我们看见积极人生观已经丧失了愚蠢的自满心，玩世哲学已经丧失了尖锐的叛逆性，在梭罗身上还可找出这种特质——这是一个不成熟的标志，而人类的智慧第一次在宽容和嘲弄的精神中达到成熟的时期。

在我看来，陶渊明代表一种中国文化的奇怪特质，即一种耽于肉欲和灵的妄尊的奇怪混合，是一种不流于制欲的精神生活和耽于肉欲的物质生活的奇怪混合，在这奇怪混合中，七情和心灵始终是和谐的。所谓理想的哲学家即是一个能领会女人的妩媚而不流于粗鄙，能爱好人生而不过度，能够察觉到尘世间成功和失败的空虚，能够生活于超越人生和脱离人生的境地而不仇视人生的人。陶渊明的心灵已经发展到真正和谐的境地，所以我们看不见他内心有一丝一毫的冲突，因之，他的生活也像他的诗一般那么自然而冲和。

陶渊明生于第四世纪的末叶，是一位著名学者兼贵官的曾孙。这位学者在家无事，常于早上搬运一百个甓到斋外，至薄暮又搬运回斋内。陶渊明幼时，因家贫亲老，任为州祭酒，但不久即辞了官职去过他的耕种生活，因此得了一种疾病。有一天，他对亲朋说："聊欲弦歌以为三径之资，可乎？"有一个朋友听了这句话，便荐他去做彭泽令。他因为喜欢喝酒，所以命令县里都种秫谷，可是他的妻子不以为善，固请种粳，才使一顷五十亩种秫，五十亩种粳。后因郡里的督邮将到，县吏说他应该束带相见，陶渊明叹曰："吾不能为五斗米折腰。"于是官也不愿做了，写了《归去来兮辞》这首名赋。此后，他就过着农夫的生活，好几次有人请他做官，他一概拒绝。他家里本穷，故和穷人一起生活，在给他儿子的一封信里，曾慨叹他们的衣服褴褛，做着贱工。有一次他送一个农家的孩子到他的儿子那里去帮做挑水取柴等事，在给他儿子的信里说："此亦人子也，可善遇之。"

他唯一的弱点就是喜欢喝酒。他平常过着孤独的生活，很少和宾客接触，可是一看见酒，纵使他不认识主人，也会坐下来和大家一起喝酒。有时他做主人的时候，在席上喝酒先醉，便对客人说："我醉欲眠卿且去。"（语出李白《山中与幽人对酌》）他有一张无弦的琴，这种古代的乐器只能在心情很平静的时候，慢慢地弹起来才有意思。他和朋友喝酒时，或是有兴致想玩玩音乐时，便抚抚这张无弦的琴。他说："但识琴中趣，何劳弦上声？"

他心地谦逊，生活简朴，且极自负，交友尤慎。江州刺史王弘很钦仰他，想和他交朋友，可是无从谋面。他曾很自然地说："我性不狎世，因疾守闲，幸非洁志慕声。"王弘只好和一个朋友用计骗他，由这个朋友去邀他喝酒，走到半路停下来，在一个凉亭里歇脚，那朋友便把酒拿出来。陶渊明真的欣欣然就坐下来喝酒，那时王弘早已隐身在附近的地方，这时候便走出来和他相见，他非常高兴，于是欢宴终日，连朋友的地方也忘记去了。王弘见陶渊明无履，就叫他的左右为他造履。当请他量履的时候，陶渊明便把脚伸出来。此后，凡是王弘要和他见面时，总是在林泽间等候他。有一次，他的朋友们在煮酒，就把他头戴的葛巾来漉酒，用过了还他，他又把葛巾戴在头上了。

他那时的住处，位于庐山之麓，当时庐山有一个闻名的禅宗白莲社，是由一位大学者所主持。这位学者想邀他入社，有一天便请他赴宴，请他加入。他提出的条件是在席上可以喝酒，本来这种行为是违犯佛门戒条的，可是主人答应他。当他正要签名入社时，却又"攒眉而去"。另外一个大诗人谢灵运很想加入这个白莲社，可是不得其门而入。后来那位方丈想跟陶渊明做个朋友，所以请了另一位道人和他一起喝酒。他们三个人，那个方丈代表佛教，陶渊明代表儒教，那个朋友代表道家。那位方丈曾立誓说终生不再走过某一座桥，可是有一天，当他和他的朋友送陶渊明回家时，他们谈得非常高兴，大家都不知不觉地走过了那桥。当三人明白过来时，不禁大笑。这三位大笑的老人，后来便成为中国绘画上常用的题材，这个故事象征着三位无忧无虑的智者的欢乐，象征着三个宗教的代表人物在幽默感中团结一致的欢乐。

他就是这样过他的一生，做一个无忧无虑的、心地坦白的、谦逊简朴的乡间诗人，一个智慧而快乐的老人。在他那本关于喝酒和田园生活的小诗集，三四篇偶然冲动而写出来的文章，一封给他儿子的信，三篇祭文（一篇是自祭文）和遗留给子孙的一些话里，我们看出一种造成那和谐生活的情感和天才，这种和谐的生活已达到了炉火纯青的境地，没有一个人能比他更卓越。他在《归去来兮辞》那首赋里所表现的就是这种爱好人生的情感。这篇名作是在公元四〇五年十一月，就是在决定辞去那县令的时候写的。

归去来兮辞

归去来兮，田园将芜胡不归！既自以心为形役，奚惆怅而独悲？悟已往之不谏，知来者之可追；实迷途其未远，觉今是而昨非。舟遥遥以轻飏，风飘飘而吹衣。问征夫以前路，恨晨光之熹微。乃瞻衡宇，载欣载奔，僮仆欢迎，稚子候门。三径就荒，松菊犹存；携幼入室，有酒盈樽。引壶觞以自酌，眄庭柯以怡颜；倚南窗以寄傲，审容膝之易安。园日涉以成趣，门虽设而常关；策扶老以流憩，时矫首而遐观。云无心以出岫，鸟倦飞而知还；景翳翳以将入，抚孤松而盘桓。

归去来兮，请息交以绝游。世与我而相违，复驾言兮焉求！悦亲戚之情话，乐琴书以消忧。农人告余以春及，将有事于西畴。或命巾车，或棹孤舟；既窈窕以寻壑，亦崎岖而经丘。木欣欣以向荣，泉涓涓而始流；羡万物之得时，感吾生之行休。已矣乎，寓形宇内复几时，曷不委心任去留。胡为乎遑遑欲何之？富贵非吾愿，帝乡不可期。怀良辰以孤往，或植杖而耘耔。登东皋以舒啸，临清流而赋诗。聊乘化以归尽，乐夫天命复奚疑？

也许有人以为陶渊明是"逃避主义者"，但事实上他绝对不是。他要逃避的仅是政治，而不是生活的本身。如果他是逻辑家，他或许早已出家做和尚，彻底地逃避人生了。可是陶渊明不愿完全逃避人生，他是爱好人生的。在他的眼中，他的妻儿太真实了，他的花园，那伸到他庭院里的枝丫，他所抚摸的孤松，这许多太可爱了。他仅是一个近情近理的人，他不是逻辑家，所以他要周旋于周遭的景物之间。他就是这样爱好人生，由种种积极的、合理的人生态度，去获得他所特有的能产生和谐的那种感觉。这种生之和谐便产生了中国最伟大的诗歌。他为尘世所生，而又属于尘世，所以他的结论不是逃避人生，而是"怀良辰以孤往，或植杖而耘耔"。陶渊明仅是回到他的田园和他的家庭里去。所以，结果是和谐，不是叛逆。

V. A Lover of Life : T'ao Yüanming

IT has been shown, therefore, that with the proper merging of the positive and the negative outlooks on life, it is possible to achieve a harmonious philosophy of the"half-and-half"lying somewhere between action and inaction, between being led by the nose into a world of futile busyness and complete flight from a life of responsibilities, and that so far as we can discover with the help of all the philosophies of the world, this is the sanest and happiest ideal for man's life on earth. What is still more important, the mixture of these two different outlooks makes a harmonious personality possible, that harmonious personality which is the acknowleged aim of all culture and education. And significantly, out of this harmonious personality, we see a joy and love of life.

It is difficult for me to describe the qualities of this love of life; it is easier to speak in a parable or tell the story of a true lover of life, as he really lived. And the picture of T'ao Yüanming, the greatest poet and most harmonious product of Chinese culture, inevitably comes to my mind. There will be no one in China to object when I say that T'ao represents to us the most perfectly harmonious and well-rounded character in the entire Chinese literary tradition. Without leading an illustrious official career, without power and outward achievements and without leaving us a greater literary heritage than a thin volume of poems and three or four essays in prose, he remains today a beacon shining through the ages, forever a symbol to lesser poets and writers of what the highest human character should be. There is a simplicity in his life, as well as in his style, which is awe-inspiring and a constant reproach to more brilliant and more sophisticated natures. And he stands, today, as a perfect example of the true lover of life, because in him the rebellion against worldly desires did not lead him to attempt a total escape, but has reached a harmony with the life of the senses. About two centuries of literary romanticism and the Taoistic cult of the idle life and rebellion against Confucianism had been working in China and joined forces with the Confucian philosophy of the previous centuries to make the emergence of this harmonious personality possible. In T'ao we find the positive outlook had lost its foolish complacency and the cynic philosophy had lost its bitter rebelliousness (a trait we see still in Thoreau—a sign of immaturity), and human wisdom first reaching full maturity in a spirit of tolerant irony.

T'ao represents to me that strange characteristic of Chinese culture, a curious combination of devotion to the flesh and arrogance of the spirit, of spirituality without asceticism and materialism without sensuality, in which the senses and the spirit have come to live together in harmony. For the ideal philosopher is one who understands the charm of women without being coarse, who loves life heartily but loves it with restraint, and who sees the unreality of the successes and failures of the active world, and stands somewhat aloof and detached, without being hostile to it. Because T'ao reached that true harmony of spiritual development, we see a total absence of inner conflict and his life was as natural and effortless as his poetry.

T'ao was born toward the end of the fourth century of our era, the great grandson of a distinguished scholar and official, who in order to keep himself from being idle, moved a pile of bricks from one place to another in the morning, and moved them back in the afternoon. In his youth he accepted a minor official job in order to support his old parents, but soon resigned and returned to the farm, tilling the field himself as a farmer, from which he developed a kind of bodily ailment. One day he asked his relatives and friends,"Would it be all right for me to go out as a minstrel singer in order to pay for the upkeep of my garden?"On hearing this, certain of his friends got him a position as a magistrate of P'engcheh, near Kiukiang. Being very fond of wine, he commanded that all the fields belonging to the local government should be planted with glutinous rice, from which wine could be made, and only on the protestations of his wife did he allow one-sixth to be planted with another kind of rice. When a government delegate came and his secretary told him that he should receive the little fellow with his gown properly girdled, T'ao sighed and said,"I cannot bend and bow for the sake of five bushels of rice."And he immediately resigned and wrote that famous poem,"Ah, Homeward Bound I Go!"From then on, he lived the life of a farmer and repeatedly refused later calls to office. Poor himself, he lived in communion with the poor, and he expressed a certain paternal regret in a letter to his sons that they should be so poorly clad and do the work of a common laborer. But when he managed to send a peasant boy to his sons when he was away, to help them do the work of carrying water and gathering fuel, he said to them,"Treat him well, for he is also some one's son. "

His only weakness was a fondness for wine. Living very much to himself, he seldom saw guests, but whenever there was wine, he would sit down with the company, even though he might not be acquainted with the host. At other times, when he was the host himself and got drunk first, he would say to his guests,"I am drunk and thinking of sleep; you can all go."He had a stringed instrument, the ch'in, without any strings left. It was an ancient instrument that could be played only in an extremely slow manner and only in a state of perfect mental calm. After a feast, or when feeling in a musical mood, he would express his musical feelings by fondling and fingering this stringless instrument."I appreciate the flavor of music; what need have I for the sounds from the strings?"

Humble and simple and independent, he was extremely chary of company. A magistrate, one Wang, who was his great admirer, wanted to cultivate his friendship, but found it very difficult to see him. With his perfect naturalness he said,"I'm keeping to myself because by nature I'm not made for the life of society, and I am staying in the house because of an ailment. Far be it from me to act in this manner in order to acquire a reputation for being high and aloof."Wang therefore had to plot with a friend in order to see him; this friend had to induce him to leave his home, by inviting him to a feast and an accidental meeting. When he was halfway and stopped at a pavilion, wine was presented. T'ao's eyes brightened and he gladly sat down to drink, when Wang, who had been hiding nearby, came out to meet him. And he was so happy that he remained there talking with him the whole afternoon, and forgot to go on to his friend's place. Wang saw that he had no shoes on his feet and ordered his subordinates to make a pair for him. When these minor officials asked for the measurements, he stretched forth his feet and asked them to take the measure. And thereafter, whenever Wang wanted to see him, he had to wait in the forest or around the lake, so that perchance he might meet the poet. Once when his friends were brewing wine, they took his linen turban to use it as a strainer, and after the wine had been strained, he wound the turban again around his head.

There was then in the great Lushan Mountains, at whose foot he lived, a great society of illustrious Zen Buddhists, and the leader, a great scholar, tried to get him to join the Lotus Society. One day he was invited to come to a Party, and his condition was that he should be allowed to drink. This breaking of the Buddhist rule was granted him and he went. But when it came to putting his name down as a member, he"knitted his brows and stole away."This was a society that so great a poet as Hsieh Lingyün had been very anxious to join, but could not get in. But still the abbot courted his friendship and one day he invited him to drink, together with another great Taoist friend. They were then a company of three; the abbot, representing

Buddhism, T'ao representing Confucianism, and the other friend representing Taoism. It had been the abbot's life vow never to go beyond a certain bridge in his daily walks, but one day when he and the other friend were sending T'ao home, they were so pleasurably occupied in their conversation that the abbot went past the bridge without knowing it. When it was pointed out to him, the company of three laughed. This incident of the three laughing old men became the subject of popular Chinese paintings, because it symbolized the happiness and gaiety of three carefree, wise souls, representing three religions united by the sense of humor.

And so he lived and died, a carefree and conscience-free, humble peasant-poet, and a wise and merry old man. But something in his small volume of poems on drinking and the pastoral life, his three or four casual essays, one letter to his sons, three sacrificial prayers (including one to himself), and some of his remarks handed down to posterity shows a sentiment and a genius for harmonious living that reached perfect naturalness and never has yet been surpassed. It was this great love of life that was expressed in the poem which he wrote one day in November, A. D. 405, when he decided to lay down the burdens of the magistrate's office.

Ah, homeward bound I go! why not go home, seeing that my field and garden with weeds are overgrown? Myself have made my soul serf to my body: why have vain regrets and mourn alone?

Fret not over bygones and the forward journey take. Only a short distance have I gone astray, and I know today I am right, if yesterday was a complete mistake.

Lightly floats and drifts the boat, and gently flows and flaps my gown. I inquire the road of a wayfarer, and sulk at the dimness of the dawn.

Then when I catch sight of my old roofs, joy will my steps quicken. Servants will be there to bid me welcome, and waiting at the door are the greeting children.

Gone to seed, perhaps, are my garden paths, but there will still be the chrysanthemums and the pine! I shall lead the youngest boy in by the hand, and on the table there stands a cup full of wine!

Holding the pot and cup I give myself a drink, happy to see in the courtyard the hanging bough. I lean upon the southern window with an immense satisfaction, and note that the little place is cosy enough to walk around.

The garden grows more familiar and interesting with the daily walks. What if no one ever knocks at the always closed door! Carrying a cane I wander at peace, and now and then look aloft to gaze at the blue above.

There the clouds idle away from their mountain recesses without any intent or purpose, and birds, when tired of their wandering flights, will think of home. Darkly then fall the shadows and, ready to come home, I yet fondle the lonely pines and loiter around.

Ah, homeward bound I go! Let me from now on learn to live alone! The world and I are not made for one another, and why drive round like one looking for what he has not found?

Content shall I be with conversations with my own kin, and there will be music and books to while away the hours. The farmers will come and tell me that spring is here and there will be work to do at the western farm.

Some order covered wagons; some row in small boats. Sometimes we explore quiet, unknown ponds, and sometimes we climb over steep, rugged mounds.

There the trees, happy of heart, grow marvelously green, and spring water gushes forth with a gurgling sound. I admire how things grow and prosper according to their seasons, and feel that thus, too, shall my life go its round.

Enough! How long yet shall I this mortal shape keep? Why not take life as it comes, and why hustle and bustle like one on an errand bound?

Wealth and power are not my ambitions, and unattainable is the abode of the gods! I would go forth alone on a bright morning, or perhaps, planting my cane, begin to pluck the weeds and till the ground.

Or I would compose a poem beside a clear stream, or perhaps go up Tungkao and make a long-drawn call on the top of the hill. So would I be content to live and die, and without questionings of the heart, gladly accept Heaven's will.

T'ao might be taken as an "escapist," and yet it was not so. What he tried to escape from was politics and not life itself. If he had been a logician, he might have decided to escape from life altogether by becoming a Buddhist monk. With T'ao's great love of life, he could not have been willing to escape from it altogether. His wife and children were too real for him, his garden and the bough stretching across his courtyard and the lonely pines which he fondled were altogether too attractive, and being a reasonable man, instead of a logician, he stuck to them. That was his love of life and his jealousy over it, and it was from this positive, but reasonable, attitude toward life that he arrived at the feeling of harmony with life which was characteristic of his culture. From that harmony with life welled forth the greatest poetry in China. Of the earth and earth-born, his conclusion was not to escape from it, but "to go forth alone on a bright morning, or perhaps, planting his cane, begin to pluck the -weeds and till the ground." T'ao merely returned to the farm and to his family. The end was harmony and not rebellion.

第六章

生命的享受

Chapter Six　　**The Feast of Life**

一．快乐问题

　　生之享受包括许多东西：我们本身的享受、家庭生活的享受，树木、花朵、云霞、溪流、瀑布，以及大自然的形形色色，都足以称为享受；此外有诗歌、艺术、沉思、友情、谈天、读书等的享受，后者这些都是心灵交流的不同表现。这许多享受中，有些享受是易见的，如食物的享受，社交宴会或家庭团聚的欢乐，风和日暖时春天的野游；另外一些较不明显的，则为诗歌、艺术和沉思等享受。我觉得这些享受，不能把它分为物质的或精神的两类，一来因为我不以为应有这种区别，二来因为我在把它们分类时每每不知适从。当我看见一群男女老少在享受一个欢乐的野宴时，我怎能说得出哪一部分是属于物质，哪一部分是属于精神？当我看到一个孩子在草地上跳跃，还有一个孩子用雏菊在编造一个小花环，母亲的手里拿着一块夹肉面包，叔父在咬一只甜美的红苹果，父亲仰卧在草地上凝望着天上的白云，祖父口中含着烟斗；也许还有人在开留声机，远远地传来了音乐的声音；或是波涛的吼声。这些欢乐之中，哪一种是属于物质，哪一种是属于精神的呢？享受一块夹肉面包和享受四周的景色（后者就是我们所谓诗歌），其区别是否很容易地可以分出来呢？听音乐，我们称之为艺术的享受；吸烟斗，我们称之为物质的享受；可是我们能说前者的享受比后者更高尚吗？所以在我看来，物质上的欢乐和精神上的欢乐，它的分别是紊乱的、不易分辨的、不真确的。我疑心这种分类是根据于一种错误的哲学理论，把美和肉严加分别，而并没有将我们真正的欢乐直接严密研究一下子以为证明。

我拿人生这个未决定的正当目的问题来做论据，我这一假定是否太过分？我总以为生活的目的即是生活的真享受，其间没有是非之争，我用"目的"这个名词时有点不敢下笔。因为这种包含真正享受它的目的，大抵不是发自有意的，而是一种人生的自然态度。"目的"这个名词便含有一种企图和努力的意义。人生世上，他的问题不是拿什么做目的，或怎样去实现这目的，而是怎样去应付此生，怎样消遣这五六十年天赋给他的光阴。他应该把生活加以调整，在生活中获得最大的快乐，这个问题跟如何去享受周末那一天的快乐一样实际，而不是形而上的问题，如果人们生在这宇宙中另有什么神秘的目的，那么只可以做抽象的渺茫的答案了。

在另一面讲，我觉得哲学家们在企图判明这个人生目的问题时，他们心中大概假定人生必有一种目的。西方思想所以把这个问题看得那样重要，就是因为受了神学的影响。我总以为我们对于这计划和目的这些东西假定得太过分了。人们想解答这个问题并为这个问题争论，甚至于弄得迷惑不解，显见都是徒然的、非必要的。如果人生真有目的或计划，那么这种目的或计划不应该这样令人困惑，那么渺茫而难于发现。

这个问题可以化为两个：第一个是关于神灵的，即上帝替人类所决定的目的；第二个是关于人类的，即人类自己所决定的目的。第一个问题我不想多加讨论，因为我们心中所存什么上帝的意志，事实上都是我们人类自己心中的思想；只是在我们想象中，上帝心中有这么一种思想而已。然而要用人类的智能来猜测神灵的智能，那是办不到的。我们的这种理论，其结果就是把上帝当做我们军中保卫旗帜的军曹，以为他和我们同样具着爱国狂；我们自欺欺人地以为上帝对世界或欧洲是不会有什么"神灵目的"或"定数"的，只有对我们的祖国才有之。我相信德国纳粹党的人物心中，上帝一定也带着卐字的臂章。每个人都认为这个上帝始终是在自己这边的，绝不会是在对方那边。其实世界上的民族，抱着这种观念的也不仅日耳曼人而已。

至于第二个问题，那争论点不是人生的目的"是什么"，而是人生的目的"应该是什么"。所以这是一个实际的而不是形而上的问题，对于人生的目的"应该是什么"这个问题，每个人都有他自己的观念和评价。我们为这个问题而争论，就是因为我们每个人的评价都不相同的缘故。以我自己而论，我的观念比较实际而少抽象，我以为人生不一定要有目的或意义。惠特曼说："我这样做一个人，已够满意了。"所以我也以为我现在活着——也许还可以再活几十年——人类的生命存在着，那就已经够了。这样看法，这个问题便变为极简单，而不容有两个答语，就是人生的目的除了去享受人生外，还有什么呢？

这个快乐问题是世界上一切非宗教哲学家所注意的重要问题，可是基督教的思想家完全置之不理，这是奇怪的事。神学家把人类快乐这问题抛开，而所焦虑的重大问题是人类的"拯救"——"拯救"听来真是一个悲惨的名词，觉得怪刺耳的。因为我在中国天天总是听人家谈"救国"。大家都想要"救"中国。这种言论，使人油然而生一种好像是在快要沉没的船上的感觉，一种万事全休，大家只在想逃生方法的感觉。基督教——有人称它为"两个没落世界（希腊和罗马）的最后叹息"——在今日还保存着这种特质，它还是被拯救问题所烦扰。人们为了离尘世和得救问题而烦扰，结果反忘掉了生活问题。人类如果没有趋近灭亡的感觉，何必去为了得救的问题担忧呢？神学家总是注意拯救问题，而没想到快乐问题，因之他们对于将来，只能渺茫地说有一个天堂。假如我们问道：在那边我们要做些什么呢？在天堂我们要得到怎样的快乐呢？他们的回答只能给我们一些渺茫的观念，如唱诗穿白衣裳之类。穆罕默德至少还用醇酒、甜美的水果和有着黑发大眼多情的少女，替我们画了一幅未来的快乐景象，这是我们这些俗人能够见得到的。如果神学家不把天堂里的景象弄得更生动逼真，更近情合理，那么我们真不想离开这个尘世而到天堂里去。有人说："明日有一只鸡，不如今日有一只蛋。"即在我们计划怎样去消遣暑假的时候，我们至少也要花些时间在探听我们所要到的地方。如果去问旅行社，而所回答的是模糊影响之辞，我是不想去的，我在原来的地方过假期好了。在天堂里也须奋斗吗？努力吗（我敢说那些希望和相信努力的人一时是这样的假定）？可是一旦我们已经十全十美了，我们还要努力些什么呢？进步到哪一层呢？或者在天堂里可以过着游手好闲，无忧无虑的日子；如果真是这样，我们尽可在这尘世上先学过游手好闲的生活，以备将来惯于永生生活，那岂不更好吗？

我们如果必须要有一个宇宙观，就让我们把自己忘掉，不要把那宇宙观限制于人类生活的范围之内。我们须把宇宙观扩展开去，把整个世界——石、树和动物——的目的都包括进去。宇宙间有一个计划（"计划"这名词，和"目的"一名词一样，都是我所不喜欢的名词）——我的意思是说，万物创造中有一个图案。我们对于这整个宇宙，须先有一个观念——虽然这个观念并不是最后的固定不移的观念——然后可以在这个宇宙中确定我们应站的地位。这种关于大自然的观念，关于我们在大自然中所占地位的观念，必须出于自然，因为我们生时是大自然的一个重要部分，而死后又是回到大自然去的。天文学、地质学、生物学以及历史，只要我们不做冒昧下断语的尝试，都能给我们以材料，协助我们得到一个相当准确的观念。如果在宇宙的目的这个广大的观念中，人类退居了次要的地位，那也不要紧的。他有着一个地位已经够了，只要他能和周遭环境和谐相处，对于人生本身便能产生一个实用而合理的观念。

I. *The Problem of Happiness*

THE enjoyment of life covers many things: the enjoyment of ourselves, of home life, of trees, flowers, clouds, winding rivers and falling cataracts and the myriad things in Nature, and then the enjoyment of poetry, art, contemplation, friendship, conversation, and reading, which are all some form or other of the communion of spirits. There are obvious things like the enjoyment of food, a gay party or family reunion, an outing on a beautiful spring day; and less obvious things like the enjoyment of poetry, art and contemplation. I have found it impossible to call these two classes of enjoyment material and spiritual, first because I do not believe in this distinction, and secondly because I am puzzled whenever I proceed to make this classification. How can I say, when I see a gay picnic party of men and women and old people and children, what part of their pleasures is material and what part spiritual? I see a child romping about on the grass plot, another child making daisy chains, their mother holding a piece of sandwich, the uncle of the family biting a juicy, red apple, the father sprawling on the ground looking at the sailing clouds, and the grandfather holding a pipe in his mouth. Probably somebody is playing a gramophone, and from the distance there come the sound of music and the distant roar of the waves. Which of these pleasures is material and which spiritual? Is it so easy to draw a distinction between the enjoyment of a sandwich and the enjoyment of the surrounding landscape, which we call poetry? Is it possible to regard the enjoyment of music which we call art, as decidedly a higher type of pleasure than the smoking of a pipe, which we call material? This classification between material and spiritual pleasures is therefore confusing, unintelligible and untrue for me. It proceeds, I suspect, from a false philosophy, sharply dividing the spirit from the flesh, and not supported by a closer direct scrutiny of our real pleasures.

Or have I perhaps assumed too much and begged the question of the proper end of human life? I have always assumed that the end of living is the true enjoyment of it. It is so simply because it is so. I rather hesitate at the word"end" or"purpose."Such an end or purpose of life, consisting in its true enjoyment, is not so much a conscious purpose, as a natural attitude toward human life. The word"purpose" suggests too much contriving and endeavor. The question that faces every man born into this world is not what should be his purpose, which he should set about to achieve, but just what to do with life, a life which is given him for a period of on the average fifty or sixty years? The answer that he should order his life so that he can find the greatest happiness in it is more a practical question, similar to that of how a man should spend his weekend, than a metaphysical proposition as to what is the mystic purpose of his life in the scheme of the universe.

On the contrary, I rather think that philosophers who start out to solve the problem of the purpose of life beg the question by assuming that life must have a purpose. This question, so much pushed to the fore among Western thinkers, is undoubtedly given that importance through the influence of theology. I think we assume too much design and purpose altogether. And the very fact that people try to answer this question and quarrel over it and are puzzled by it serves to show it up as quite vain and uncalled for. Had there been a purpose or design in life, it should not have been so puzzling and vague and difficult to find out.

The question may be divided into two: either that of a divine purpose, which God has set for humanity, or that of a human purpose, a purpose that mankind should set for itself. As far as the first is concerned, I do not propose to enter into the question, because everything that we think God has in mind necessarily proceeds from our own mind; it is what we imagine to be in God's mind, and it is really difficult for human intelligence to guess at a divine intelligence. What we usually end up with by this sort of reasoning is to make God the color-sergeant of our army and to make Him as chauvinistic as ourselves; He cannot, so we conceive, possibly have a"divine purpose"and"destiny"for the world, or for Europe, but only for our beloved Fatherland. I am quite sure the Nazis can't conceive of God without a swastika arm-band. This Gott is always mit uns and cannot possibly be mit ihnen. But the Germans are not the only people who think this way.

As far as the second question is concerned, the point of dispute is not what is, but what should be, the purpose of human life, and it is therefore a practical, and not a metaphysical question. Into this question of what should be the purpose of human life, every man projects his own conceptions and his own scale of values. It is for this reason that we quarrel over the question, because our scales of values differ from one another. For myself, I am content to be less philosophical and more practical. I should not presume that there must be necessarily a purpose, a meaning of human existence. As Walt Whitman says,"I am sufficient as I am."It is sufficient that I live–and am probably going to live for another few decades–and that human life exists. Viewed that way, the problem becomes amazingly simple and admits of no two answers. What can be the end of human life except the enjoyment of it?

It is strange that this problem of happiness, which is the great question occupying the minds of all pagan philosophers, has been entirely neglected by Christian thinkers. The great question that bothers theological minds is not human happiness, but human"salvation"–a tragic word. The word has a bad flavor for me, because in China I hear everyday some one talking about

our"national salvation."Everybody is trying to"save"China. It suggests the
feeling of people on a sinking ship, a feeling of ultimate doom and the best
method of getting away alive. Christianity, which has been described as"the
last sigh of two expiring worlds"(Greek and Roman), still retains something
of that characteristic today in its preoccupation with the question of salvation.
The question of living is forgotten in the question of getting away alive from
this world. Why should man bother himself so much about salvation, unless
he has a feeling of being doomed? Theological minds are so much occupied
with salvation, and so little with happiness, that all they can tell us about
the future is that there will be a vague heaven, and when questioned about
what we are going to do there and how we are going to be happy in heaven,
they have only ideas of the vaguest sort, such as singing hymns and wearing
white robes. Mohammed at least painted a picture of future happiness with
rich wine and juicy fruits and black-haired, big-eyed, passionate maidens
that we laymen can understand. Unless heaven is made much more vivid
and convincing for us, there is no reason why one should strive to go there,
at the cost of neglecting this earthly existence. As some one says,"An egg
today is better than a hen tomorrow."At least, when we're planning a summer
vacation, we take the trouble to find out some details about the place we
are going to. If the tourist bureau is entirely vague on the question, I am
not interested; I remain where I am. Are we going to strive and endeavor in
heaven, as I am quite sure the believers in progress and endeavor must assume?
But how can we strive and make progress when we are already perfect? Or are
we going merely to loaf and do nothing and not worry? In that case, would it
not be better for us to learn to loaf while on this earth as a preparation for our
eternal life?

    If we must have a view of the universe, let us forget ourselves and
not confine it to human life. Let us stretch it a little and include in our
view the purpose of the entire creation—the rocks, the trees and the
animals. There is a scheme of things (although"scheme"is another word,

like"end"and"purpose,"which I strongly distrust)—I mean there is a pattern of things in the creation, and we can arrive at some sort of opinion, however lacking in finality, about this entire universe, and then take our place in it. This view of nature and our place in it must be natural, since we are a vital part of it in our life and go back to it when we die. Astronomy, geology, biology and history all provide pretty good material to help us form a fairly good view if we don't attempt too much and jump at conclusions. It doesn't matter if, in this bigger view of the purpose of the creation, man's place recedes a little in importance. It is enough that he has a place, and by living in harmony with nature around him, he will be able to form a workable and reasonable outlook on human life itself.

二．人类的快乐属于感觉

　　人类一切快乐都发自生物性的快乐。这观念是绝对科学化的。这一点我必须加以说明，以免被人误解，人类的一切快乐都属于感觉的快乐。我相信精神主义者一定会误解我的意思：精神主义者之所以和唯物主义者永远会有误解，就是因为他们的语气不同，或对同一句话抱着不同的见解。但是我们在这个获取快乐的问题上，难道也要被精神主义者所欺蒙而去跟着承认精神上的快乐才是真正的快乐吗？让我们马上承认并加以限制，说精神上的舒适是有赖于内分泌腺的正常动作。在我看来，快乐问题大半是消化问题。我很想直说快乐问题大抵即是大便问题，为保持我的人格和颜面起见，我得用一位美国大学校长来做我的护身符。这位大学校长过去对每年的新生演说时，总是要讲那句极有智慧的话："我要你们记住两件事情：读《圣经》和使大便通畅。"他能说出这种话来，也可想见他是一个多么贤明，多么和蔼的老人家啊！一个人大便通畅，就觉快乐，否则就会感到不快乐，事情不过如此而已。

　　谈到我们的快乐，不要陷入抽象的议论中去，我们应该注意事实，把自己分析一下，看看我们一生中在什么时候得到真正快乐。这个世界中，快乐往往须从反面看出来，无忧愁、不受欺凌、无病无痛便是快乐。但也可成为正面感觉，那就是我们所说的欢乐，我所认为真快乐的时候，例如在睡过一夜之后，清晨起身，吸着新鲜空气，肺部觉得十分宽畅，做了一会儿深呼吸，胸部的肌肤便有一种舒服的动作感觉，感到有新的活力而适宜于工作；或是手中拿了烟斗，双腿搁在椅上，让烟草慢慢均匀地烧着；或是夏月远行，口渴喉干，看见一泓清泉，潺潺的流水声已经使我觉得清凉快乐，于是脱去鞋袜，拿两脚浸在凉爽的清水里；或一顿丰盛餐饭之后，坐在安乐椅上，面前没有讨厌的人，大家海阔天空地谈笑着，觉得精神上和身体上都与世无争；或在一个夏天的下午，天边涌起乌云，知道一阵七月的骤雨就要在一刻钟内落下来，可是雨天出门不带伞，怕给人家看见难为情，连忙趁雨未降下的时候，先跑了出去；半途遇雨，淋得全身湿透，告诉人家，我中途遇雨。

当我听着我孩子说话的声音，或是看着他们肥胖的腿儿，我说不出在物质上爱他们或是精神上爱他们；我也完全不能把心灵与肉体的欢乐分别开来。世上可有什么人对于女人只在精神上爱她，而不在肉体爱她，一个人要分析和分别他所爱的女人的媚态——如大笑、微笑、摇头的姿态、对事物的态度等——是件容易的事情吗？女子在衣饰清洁整齐的时候，都会觉得快乐，口红和胭脂使人有一种精神焕发的感觉，衣饰整齐使人感到宁静与舒泰，这在女子方面看来是真实而明确的，然而精神主义者对此就会觉得莫名其妙。我们的肉体总有一日会死去的，所以我们的肉体和精神之间只有极薄的隔膜，同时，在精神的世界里，要欣赏它最优美的情感与精神之美，只有用我们的感官才能胜任愉快。触觉、听觉和视觉各方面，是无所谓道德或不道德的。我们大都会失掉享受人生正面欢乐的能力。原因是我们感官的敏感性减退，和我们不尽量去运用这些感官。

我们用不着为这问题辩护，让我们拿出一些实在的事实：从东西洋许多酷爱人生的伟大人物里面，试举几个例证出来，看看他们是什么时候最感到快乐，这快乐和他们的听觉、嗅觉及视觉有怎样的密切关系。在某一节文章里，梭罗[9]对于蟋蟀的鸣声所生的崇高美感说：

先察蟋蟀所住的孔穴。在石头中间，穴隙到处都有。一只蟋蟀的单独歌儿更使我感到趣味。它暗示"出世已迟"，但也只有当我们认识时间和永恒的意义时，"迟延"才感觉得到。其实它什么也不迟，只是赶不上世间一切琐碎而匆忙的活动罢了。它表现着成熟的智慧，超越一切俗世的思想，它就这样在春的希望和夏的炎热中间具着秋的冷静和成熟的智慧。它们对小鸟儿说："啊！你们真像孩子，随着感情说话；大自然就是借着你们而说话的；我们却两样儿了，季节不为我们而旋转；我们反唱着它们的催眠曲。"它们就这样永恒地在草根脚下唱着。它们的住处便是天堂，不论是在五月或十一月，永远是这样。它们的歌儿具有宁静的智慧，有着散文的平稳，它们不饮酒，只吃露水。当孵卵期过后，它们的宁静无声并不是恋爱心境受了阻抑，而是归荣耀于上帝，与对上帝的永恒享受。它们处于季节转变之外。它们的歌儿像真理那样永垂不朽。人类只有在精神比较

---

[9] 梭罗对于人生的整个观念，在一切的美国作家中，可说最富于中国人的色彩。因我是中国人，所以在精神上觉得很接近他。就在几个月前才发现他，至今还觉得高兴。如果我把梭罗的文章译成中文，说是一位中国诗人写的，一定不会有人疑心的。

健全的时候，才能听见蟋蟀的鸣声。

再看惠特曼的嗅觉、视觉和听觉，它们怎样促进他的精神生活，而他又怎样认为这些东西是非常重要的：

早晨大雪，至晚未停。我在雪花纷飞中，踯躅于树林里和道路上，约莫有两个钟头。微风拂过松树发出音乐般的低鸣，清晰奇妙，犹如瀑布，时而静止，时而奔流。此时视觉、听觉、嗅觉，一切的感觉，都得了微妙的满足。每一雪片都飘飘地降在常青树、冬青树、桂树的上面，静静地躺着，所有的枝叶都穿起一件臃肿的白外套，在边缘上还缀着绿宝石——这是那茂盛的、挺直的、有着红铜色的松树——还有那一阵阵轻微的树脂和雪水混合的香味（一切东西都有气味，雪也有气味，只有你辨别得出来——这种气味无论在哪一地、哪一时都不完全相同。正午的气味和半夜的气味，冬天的气味和夏天的气味，多风的气味和无风的气味都是不同的）。

我们可有人能辨别正午和半夜的气味，冬和夏的气味，或多风和无风的气味？如果人们觉得住在城市里比住在乡下较不快乐，那就是因为一律灰色墙壁和一律的水门汀（混凝土）行人道太过于单调，人们生活在这种环境中，一切视觉和听觉都引不起感应，总因麻木而消失了。

讲到快乐时刻的界限，以及它的度量和性质，中国人和美国人的观念是相同的。在我要举出一位中国学者的三十三快乐时刻之前，我另引一段惠特曼的话来做一个比较，证明我们之间感觉的相同：

是一个天气晴朗的日子——空气干燥，有微风，充满氧气。在我的四周，有着足以使我沉醉的奇迹，那些健全沉静而又美丽的树木、流水、花草、阳光和早霜——但最吸引我盯着的还是天空。它今天是那么澄清细致，那秋天特有的蓝色，又有那透明的蓝幕上浮着朵朵白云，或大或小，在伟大的苍穹中表现它们静穆的神灵动作。在上午（由七时至十一时）这天空始终保持着美丽洁净的蓝色。近正午时，渐渐转淡了，两三个钟头后，已变成了灰色——再淡下去，一直到日落的时候——我凝望一丛大树圆顶上的落日，在缝隙中闪烁着火红、淡黄、肝褐、赤红，千种颜色的华丽展览，万条灿烂的金光斜映水面那种透明的阴影、线条、闪烁生动的颜色，是图画上所从来没有看到过的。

　　我不解其然，但我只觉得这次秋天所以使我得到许多心满意足的时刻，完全是这个天空（我一生中虽天天见到天空，但事实上过去我并没有真正看见过它）。我读过拜伦的事迹，有一段说他在逝世时，对一个朋友说，他一生中仅仅有过三个快乐的时刻。另外又有一个关于国王的钟的古代日耳曼传说，也讲到同一的感觉。当我在那树林里看那美丽的落日时，我想到了拜伦的故事和那个钟的故事，心中始悟到我在这时也正在享受一个快乐的时刻呢（我也许不曾把那最快乐的时刻记下来；因为当这种时候来临时，我不愿为着要记录而打断了它。我只是任性流连悠然自在，沉醉在宁静的出神中）。

　　快乐到底是什么呢？这就是一个快乐的时刻吗？或是像一个快乐的时刻吗？——快乐的时刻是那么难于理解——是像一个呼吸，或像一点易消失的彩色吗？我不知道——还是让我怀疑下去吧。清澄的天啊，在你蔚蓝的空中，你可有灵药来医治我的病症吗（啊，我三年来损坏的身体和骚乱的精神哟）？你现在可是把这种灵药微妙地，神秘地，经过空气隐隐地撒在我的身上？

II. Human Happiness Is Sensuous

ALL human happiness is biological happiness. That is strictly scientific. At the risk of being misunderstood, I must make it clearer: all human happiness is sensuous happiness. The spiritualists will misunderstand me, I am sure; the spiritualists and materialists must forever misunderstand each other, because they don't talk the same language, or mean by the same word different things. Are we, too, in this problem of securing happiness to be deluded by the spiritualists, and admit that true happiness is only happiness of the spirit? Let us admit it at once and immediately proceed to qualify it by saying that the spirit is a condition of the perfect functioning of the endocrine glands. Happiness for me is largely a matter of digestion. I have to take cover under an American college president to insure my reputation and respectability when I say that happiness is largely a matter of the movement of the bowels. The American college president in question used to say with great wisdom in his address to each class of freshmen, "There are only two things I want you to keep in mind: read the Bible and keep your bowels open." What a wise, genial old soul he was to have said that! If one's bowels move, one is happy, and if they don't move, one is unhappy. That is all there is to it.

Let us not lose ourselves in the abstract when we talk of happiness, but get down to facts and analyze for ourselves what are the truly happy moments of our life. In this world of ours, happiness is very often negative, the complete absence of sorrow or mortification or bodily ailment. But happiness can also be positive, and then we call it joy. To me, for instance, the truly happy moments are: when I get up in the morning after a night of perfect sleep and sniff the morning air and there is an expansiveness in the lungs, when I feel inclined to inhale deeply and there is a fine sensation of movement around the skin and muscles of the chest, and when therefore, I am fit for work; or when I hold a pipe in my hand and rest my legs on a chair, and the tobacco burns slowly and evenly; or when I am traveling on a summer day, my throat parched with thirst, and I see a beautiful clear spring, whose very sound makes me happy, and I take off my socks and shoes and dip my feet in the delightful, cool water; or when after a perfect dinner I lounge in an armchair, when there is no one I hate to look at in the company and conversation rambles off at a light pace to an unknown destination, and I am spiritually and physically at peace with the world; or when on a summer afternoon I see black clouds gathering on the horizon and know for certain a July shower is coming in a quarter of an hour, but being ashamed to be seen going out into the rain without an umbrella, I hastily set out to meet the shower halfway across the fields and come home drenched through and through and tell my family that I was simply caught by the rain.

Just as it is impossible for me to say whether I love my children physically or spiritually when I hear their chattering voices or when I see their plump legs, so I am totally unable to distinguish between the joys of the mind and the joys of the flesh. Does anybody ever love a woman spiritually without loving her physically? And is it so easy a matter for a man to analyze and separate the charms of the woman he loves—things like laughter, smiles, a way of tossing one's head, a certain attitude toward things? And after all every girl feels happier when she is well-dressed. There is a soul-uplifting quality about lipstick and rouge and a spiritual calm and poise that comes from the knowledge of being well-dressed, which is real and definite for the girl herself and of which the spiritualist has no inkling of an idea. Being made of this mortal flesh, the partition separating our flesh from our spirit is extremely thin, and the world of spirit, with its finest emotions and greatest appreciations of spiritual beauty, cannot be reached except with our senses. There is no such thing as morality and immorality in the sense of touch, of bearing and vision. There is a great probability that our loss of capacity for enjoying the positive joys of life is largely due to the decreased sensibility of our senses and our lack of full use of them.

Why argue about it? Let us take concrete instances and cull examples from all the great lovers of life, Eastern and Western, and see what they describe as their own happy moments, and how intimately they are connected with the very senses of hearing and smelling and seeing. Here is a description of the high aesthetic pleasure that Thoreau got from hearing the sound of crickets:

First observe the creak of crickets. It is quite general amid these rocks. The song of only one is more interesting to me. It suggests lateness, but only as we come to a knowledge of eternity after some acquaintance with time. It is only late for all trivial and hurried pursuits. It suggests a wisdom mature, never late, being above all temporal considerations, which possesses the coolness and maturity of autumn amidst the aspiration of spring and the heats of summer. To the birds they say:"Ah! you speak like children from impulse; Nature speaks through you; but with us it is ripe knowledge. The seasons do not revolve for us; we sing their lullaby."So they chant, eternal, at the roots of the grass. It is heaven where they are, and their dwelling need not be heaved up. Forever the same, in May and in November (?). Serenely wise, their song has the security of prose. They have drunk no wine but the dew. It is no transient love-strain hushed when the incubating season is past, but a glorifying of God and enjoying of him forever. They sit aside from the revolution of the seasons. Their strain is unvaried as Truth. Only in their saner moments do men hear the crickets.

And see how Walt Whitman's senses of smell and sight and sound contribute to his spirituality and what great importance he places upon them:

A snowstorm in the morning, and continuing most of the day. But I took a walk over two hours, the same woods and paths, amid the falling flakes. No wind, yet the musical low murmur through the pines, quite pronounced, curious, like waterfalls, now still'd, now pouring again. All the senses, sight, sound, smell, delicately gratified. Every snowflake lay where it fell on the evergreens, hollytrees, laurels, etc., the multitudinous leaves and branches piled, bulging-white, defined by edge-lines of emerald—the tall straight columns of the plentiful bronze-topt pines—a slight resinous odor blending with that of the snow. (For there is a scent to everything, even the snow, if you can only detect it—no two places, hardly any two hours, anywhere, exactly alike. How different the odor of noon from midnight, or winter from summer, or a windy spell from a still one!)

How many of us are able to distinguish between the odors of noon and midnight, or of winter and summer, or of a windy spell and a still one? If man is so generally less happy in the cities than in the country, it is because all these variations and nuances of sight and smell and sound are less clearly marked and lost in the general monotony of gray walls and cement pavements.

The Chinese and the Americans are alike when it comes to the true limits and capacities and qualities of the happy moments. Before I translate the thirty-three happy moments given by a Chinese scholar, I want to quote by way of comparison another passage from Whitman, which will show the identity of our senses:

A clear, crispy day—dry and breezy air, full of oxygen. Out of the sane, silent, beauteous miracles that envelop and fuse me—trees, water, grass, sunlight, and early frost—the one I am looking at most today is the sky. It has that delicate, transparent blue, peculiar to autumn, and the only clouds are little or larger white ones, giving their still and spiritual motion to the great concave. All through the earlier day (say from 7 to 11) it keeps a pure, yet vivid blue. *But as noon approaches the color gets lighter, quite gray for two or three hours—then still paler for a spell, till sun-down—which last I watch dazzling through the interstices of a knoll of big trees—darts of fire and a gorgeous show of light-yellow, liver-color and red, with a vast silver glaze askant on the water—the transparent shadows, shafts, sparkle, and vivid colors beyond all the paintings ever made.*

I don't know what or how, but it seems to me mostly owing to these skies, (every now and then I think, while I have of course seen them every day of my life, I never really saw the skies before,) I have had this autumn some wondrously contented hours—may I not say perfectly happy ones? As I've read, Byron just before his death told a friend that he had known but three happy hours during his whole existence. Then there is the old German legend of the king's bell, to the same point. While I was out there by the wood, that beautiful sunset through the trees, I thought of Byron's and the bell story, and the notion started in me that I was having a happy hour. (Though perhaps my best moments I never jot down; when they come I cannot afford to break the charm by inditing memoranda. I just abandon myself to the mood, and let it float on, carrying me in its placid ecstasy).

What is happiness, anyhow? Is this one of its hours, or the like of it?—so impalpable—a mere breath, an evanescent tinge? I am not sure—so let me give myself the benefit of the doubt. Hast Thou, pellucid, in Thy azure depths, medicine for case like mine? (Ah, the physical shatter and troubled spirit of me the last three years.) And dost Thou subtly mystically now drip it through the air invisibly upon me?

三．金圣叹之不亦快哉三十三则

现在让我们来观察欣赏一位中国学者自述的快乐时刻，十七世纪印象派大批评家金圣叹在《西厢记》的批语中，曾写下他觉得最快乐的时刻，这是他和他的朋友在十日的阴雨连绵中，在一所庙宇里计算出来的。下面便是他自己认为是人生真快乐的时刻，在这种时刻中，精神是和感官错综地联系着的：

其一：夏七月，赤日停天，亦无风，亦无云；前后庭赫然如洪炉，无一鸟敢来飞。汗出遍身，纵横成渠。置饭于前，不可得吃。呼簟欲卧地上，则地湿如膏，苍蝇又来缘颈附鼻，驱之不去。正莫可如何，忽然大黑车轴，疾澍澎湃之声，如数百万金鼓。檐溜浩于瀑布。身汗顿收，地燥如扫，苍蝇尽去，饭便得吃。不亦快哉！

其一：十年别友，抵暮忽至。开门一揖毕，不及问其船来陆来，并不及命其坐床坐榻，便自疾趋入内，卑辞叩内子；"君岂有斗酒如东坡妇乎？"内子欣然拔金簪相付。计之可作三日供也。不亦快哉！

其一：空斋独坐，正思夜来床头鼠耗可恼，不知其戞戞者是损我何器，嘤嘤者是裂我何书。中心回惑，其理莫措，忽见一狻猫，注目摇尾，似有所睹。歙声屏息，少复待之，则疾趋如风，唧然一声。而此物竟去矣。不亦快哉！

其一：于书斋前，拔去垂丝海棠紫荆等树，多种芭蕉一二十本。不亦快哉！

其一：春夜与诸豪士快饮，至半醉，住本难住，进则难进。旁一解意童子，忽送大纸炮可十余枚，便自起身出席，取火放之。硫黄之香，自鼻入脑，通身怡然。不亦快哉！

其一：街行见两措大执争一理，既皆目裂颈赤，如不戴天，而又高拱手，低曲腰，满口仍用者也之乎等字。其语刺刺，势将连年不休。忽有壮夫掉臂行来，振威从中一喝而解。不亦快哉！

其一：子弟背诵书烂熟，如瓶中泻水。不亦快哉！

其一：饭后无事，入市闲行，见有小物，戏复买之，买亦已成矣，所差者甚少，而市儿苦争，必不相饶，便掏袖下一件，其轻重与前直相上下者，掷而与之。市儿改笑容，拱手连称不敢。不亦快哉！

其一：饭后无事，翻倒敝箧。则见新旧逋欠文契不下数十百通，其人或存或亡，总之无有还理。背人取火拉杂烧净，仰看高天，萧然无云。不亦快哉！

其一：夏月科头赤足，自持凉伞遮日，看壮夫唱吴歌，踏桔槔。水一时溪涌而上，譬如翻银滚雪。不亦快哉！

其一：朝眠初觉，似闻家人叹息之声，言某人夜来已死。急呼而讯之，正是——城中第一绝有心计人。不亦快哉！

其一：夏月早起，看人于松棚下，锯大竹作筒用。不亦快哉！

其一：重阴匝月，如醉如病，朝眠不起，忽闻众鸟毕作弄晴之声，急引手褰帷，推窗视之，日光晶荧，林木如洗。不亦快哉！

其一：夜来似闻某人素心，明日试往看之。入其门，窥其闺，见所谓某人，方据案面南看一文书。顾客入来，默然一揖。便拉袖命坐，曰："君既来，可亦试看此书。"相与欢笑，日影尽去。既已自饥，徐问客曰："君亦饥耶？"不亦快哉！

其一：本不欲造屋，偶得闲钱，试造一屋。自此日为始，需木，需石，需瓦，需砖，需灰，需钉，无晨无夕，不来聒于两耳。乃至罗雀掘鼠，无非为屋校计，而又都不得屋住，既已安之如命矣。忽然一日屋竟落成。刷墙扫地，糊窗挂画。一切匠作出门毕去，同人乃来分榻列坐。不亦快哉！

其一：冬夜饮酒，转复寒甚，推窗试看，雪大如手，已积三四寸矣。不亦快哉！

其一：夏日于朱红盘中，自拔快刀，切绿沉西瓜。不亦
快哉！

其一：久欲为比丘，苦不得公然吃肉。若许为比丘，又得公然吃肉，则夏月以热汤快刀，净割头发。不亦快哉！

其一：存得三四癞疮于私处，时呼热汤关门澡之。不亦
快哉！

其一：箧中无意忽检得故人手迹。不亦快哉！

其一：寒士来借银，谓不可启齿，于是唯唯亦说他事。我窥见其苦意，拉向无人处，问所需多少。急趋入内，如数给与，然后问其必当速归料理是事耶？为尚得少留共饮酒耶？不亦
快哉！

其一：坐小船，遇利风，苦不得张帆，一快其心。忽逢舟艒艒，疾行

如风。试伸挽钩，聊复挽之。不意挽之便着，因取缆，缆向其尾，口中高吟老杜"青惜峰峦过，黄知橘柚来"之句，极大笑乐。不亦快哉！

其一：久欲觅别居与友人共住，而苦无善地。忽一人传来云有屋不多，可十余间，而门临大河，嘉树葱然。便与此人共吃饭毕，试走看之，都未知屋如何。入门先见空地一片，大可六七亩许，异日瓜菜不足复虑。不亦快哉！

其一：久客得归，望见郭门，两岸童妇，皆作故乡之声。不亦快哉！

其一：佳磁既损，必无完理。翻覆多看，徒乱人意。因宣付厨人作杂器充用，永不更令到眼。不亦快哉！

其一：身非圣人，安能无过。夜来不觉私作一事，早起怦怦，实不自安。忽然想到佛家有布萨之法，不自覆藏，便成忏悔，因明对生熟众客，快然自陈其失。不亦快哉！

其一：看人作擘窠大书，不亦快哉！

其一：推纸窗放蜂出去，不亦快哉！

其一：做县官，每日打鼓退堂时，不亦快哉！

其一：看人风筝断，不亦快哉！

其一：看野烧，不亦快哉！

其一：还债毕，不亦快哉！

其一：读《虬髯客传》，不亦快哉！

可怜的拜伦，他一生中只有三个快乐的时候！如果他不是一个病态而又心地不平衡的人，他一定是被那个时代的流行忧郁症所影响了。如果忧郁的感觉不那么时髦，我相信他至少有三十个快乐时刻。这样说来，世界岂不是一席人生的宴会，摆起来让我们去享受——只是由感官去享受；同时由那种文化承认这些感官的欢乐的存在，而使我们可坦白地承认这些感官欢乐的存在；这岂不显而易见吗？我疑心我们所以装做看不见这个充满着感觉的美妙世界，乃是由于那些精神主义者弄得我们畏惧这些东西的缘故，如果我们现在有一个较高尚的哲学，我们必须重新信任这个"身体"的优美收受器官，我们把轻视感觉和畏惧情感的心理一律摒除。如果那些哲学家不能使物质升华，不能把我们的身体变成一个没有神经，没有味觉，没有嗅觉，没有色觉，没有动觉，没有触觉的灵魂，而我们也不能彻底模仿印度禁欲主义者的行为，那么我们必须勇敢地面对着这个现实的人生！唯有承认现实人生的那种哲学才能够使我们获得真正快乐，也惟有这种哲学才是合理的、健全的。

III. Chin's Thirty-Three Happy Moments

WE are now better prepared to examine and appreciate the happy moments of a Chinese, as he describes them. Chin Shengt'an, that great impressionistic critic of the seventeenth century, has given us, between his commentaries on the play Western Chamber, an enumeration of the happy moments which he once counted together with his friend, when they were shut up in a temple for ten days on account of rainy weather. These then are what he considers the truly happy moments of human life, moments in which the spirit is inextricably tied up with the senses:

I: It is a hot day in June when the sun hangs still in the sky and there is not a whiff of wind or air, nor a trace of clouds; the front and back yards are hot like an oven and not a single bird dares to fly about. Perspiration flows down my whole body in little rivulets. There is the noon-day meal before me, but I cannot take it for the sheer heat. I ask for a mat to spread on the ground and lie down, but the mat is wet with moisture and flies swarm about to rest on my nose and refuse to be driven away. Just at this moment when I am completely helpless, suddenly there is a rumbling of thunder and big sheets of black clouds overcast the sky and come majestically on like a great army advancing to battle. Rain water begins to pour down from the eaves like a cataract. The perspiration stops. The clamminess of the ground is gone. All flies disappear to hide themselves and I can eat my rice. Ah, is this not happiness?

I: A friend, one I have not seen for ten years, suddenly arrives at sunset. I open the door to receive him, and without asking whether he came by boat or by land, and without bidding him to sit down on the bed or the couch, I go to the inner chamber and humbly ask my wife: "Have you got a gallon of wine like Su Tungp'o's wife?" My wife gladly takes out her gold hairpin to sell it. I calculate it will last us three days. Ah, is this not happiness?

I: I am sitting alone in an empty room and I am just getting annoyed at a mouse at the head of my bed, and wondering what that little rustling sound signifies—what article of mine he is biting or what volume of my books he is eating up. While I am in this state of mind, and don't know what to do, I suddenly see a ferocious-looking cat, wagging its tail and staring with its wide open eyes, as if it were looking at something. I hold my breath and wait a moment, keeping perfectly still, and suddenly with a little sound the mouse disappears like a whiff of wind. Ah, is this not happiness?

I: I have pulled out the hait'ang and chihching in front of my studio, and have just planted ten or twenty green banana trees there. Ah, is this not happiness?

I: I am drinking with some romantic friends on a spring night and am just half intoxicated, finding it difficult to stop drinking and equally difficult to go on. An understanding boy servant at the side suddenly brings in a package of big fire-crackers, about a dozen in number, and I rise from the table and go and fire them off. The smell of sulphur assails my nostrils and enters my brain and I feel comfortable all over my body. Ah, is this not happiness?

I: I am walking in the street and see two poor rascals engaged in a hot argument of words with their faces flushed and their eyes staring with anger as if they were mortal enemies, and yet they still pretend to be ceremonious to each other, raising their arms and bending their waists in salute, and still using the most polished language of thou and thee and wherefore and is it not so? The flow of words is interminable. Suddenly there appears a big husky fellow swinging his arms and coming up to them, and with a shout tells them to disperse. Ah, is this not happiness?

I: To hear our children recite the classics so fluently, like the sound of pouring water *from a vase. Ah, is this not happiness?*

I: Having nothing to do after a meal I go to the shops and take a fancy to a little thing. After bargaining for some time, we still haggle about a small difference, but the shopboy still refuses to sell it. Then I take out a little thing from my sleeve, which is worth about the same thing as the difference and throw it at the boy. The boy suddenly smiles and bows courteously saying, "Oh, you are too generous!"Ah, is this not happiness?

I: I have nothing to do after a meal and try to go through the things in some old trunks. I see there are dozens or hundreds of I.O.U.'s from people who owe my family money. Some of them are dead and some still living, but in any case there is no hope of their returning the money. Behind people's backs I put them together in a pile and make a bonfire of them, and I look up to the sky and see the last trace of smoke disappear. Ah, is this not happiness?

I: It is a summer day. I go bareheaded and barefooted, holding a parasol to watch young people singing Soochow folk songs while treading the water wheel. The water comes up over the wheel in a gushing torrent like molten silver or melting snow. Ah, is this not happiness?

I: I wake up in the morning and seem to hear some one in the house sighing and saying that last night some one died. I immediately ask to find out who it is, and learn that it is the sharpest, most calculating fellow in town. Ah, is this not happiness?

I: I get up early on a summer morning and see people sawing a large bamboo pole under a mat-shed, to be used as a water pipe. Ah, is this not happiness?

I: It has been raining for a whole month and I lie in bed in the morning like one drunk or ill, refusing to get up. Suddenly I hear a chorus of birds announcing a clear day. Quickly I pull aside the curtain, push open the window and see the beautiful sun shining and glistening and the forest looks like having a bath. Ah, is this not happiness?

I: At night I seem to hear some one thinking of me in the distance. The next day I go to call on him. I enter his door and look about his room and see that this person is sitting at his desk, facing south, reading a document. He sees me, nods quietly and pulls me by the sleeve to make me sit down, saying"Since you are here, come and look at

*this."And we laugh and enjoy ourselves until the shadows on the walls have disappeared.
He is feeling hungry himself and slowly asks me"Are you hungry, too?"Ah, is this not
happiness?*

*I: Without any serious intention to build a house of my own, I happened,
nevertheless, to start building one because a little sum had unexpectedly come my way.
From that day on, every morning and every night I was told that I needed to buy timber
and stone and tiles and bricks and mortar and nails. And I explored and exhausted
every avenue of getting some money, all on account of this house, without, however,
being able to live in it all this time, until I got sort of resigned to this state of things. One
day, finally, the house is completed, the walls have been whitewashed and the floors
swept clean; the paper windows have been pasted and scrolls of paintings are hung up on
the walls. All the workmen have left, and my friends have arrived, sitting on different
couches in order. Ah, is this not happiness?*

*I: I am drinking on a winter's night, and suddenly note that the night has turned
extremely cold. I push open the window and see that snowflakes come down the size of
a palm and there are already three or four inches of snow on the ground. Ah, is this not
happiness?*

*I: To cut with a sharp knife a bright green watermelon on a big scarlet plate of a
summer afternoon. Ah, is this not happiness?*

*I: I have long wanted to become a monk, but was worried because I would not be
permitted to eat meat. If then I could be permitted to become a monk and yet eat meat
publicly, why then I would heat a basin of hot water, and with the help of a sharp razor
shave my head clean in a summer month! Ah, is this not happiness?*

*I: To keep three or four spots of eczema in a private part of my body and now and
then to scald or bathe it with hot water behind closed doors. Ah, is this not happiness?*

*I: To find accidently a handwritten letter of some old friend in a trunk. Ah, is this
not happiness?*

*I: A poor scholar comes to borrow money from me, but is shy about mentioning
the topic, and so he allows the conversation to drift along on other topics. I see his
uncomfortable situation, pull him aside to a place where we are alone and ask him how*

much he needs. Then I go inside and give him the sum and after having done this, I ask him:"Must you go immediately to settle this matter or can you stay a while and have a drink with me?"Ah, is this not happiness?

I: I am sitting in a small boat. There is a beautiful wind in our favor, but our boat has no sails. Suddenly there appears a big lorcha, coming along as fast as the wind. I try to hook on to the lorcha in the hope of catching on to it, and unexpectedly the hook does catch. Then I throw over a rope and we are towed along and I begin to sing the lines of Tu Fu:"The green makes me feel tender toward the peaks, and the red tells me there are oranges."And we break out in joyous laughter. Ah, is this not happiness?

I: I have been long looking for a house to share with a friend but have not been able to find a suitable one. Suddenly some one brings the news that there is a house somewhere, not too big, but with only about a dozen rooms, and that it faces a big river with beautiful green trees around. I ask this man to stay for supper, and after the supper we go over together to have a look, having no idea what the house is like. Entering the gate, I see that there is a large vacant lot about six or seven mow, and I say to myself,"I shall not have to worry about the supply of vegetables and melons henceforth."Ah, is this not happiness?

I: A traveller returns home after a long journey, and he sees the old city gate and hears the women and children on both banks of the river talking his own dialect. Ah, is this not happiness?

I: When a good piece of old porcelain is broken, you know there is no hope of repairing it. The more you turn it about and look at it, the more you are exasperated. I then hand it to the cook, asking him to use it as any old vessel, and give orders that he shall never let that broken porcelain bowl come within my sight again. Ah, is this not happiness?

I: I am not a saint, and am therefore not without sin. In the night I did something wrong and I get up in the morning and feel extremely ill at ease about it. Suddenly I remember what is taught by Buddhism, that not to cover one's sins is the same as repentance. So then I begin to tell my sin to the entire company around, whether they are strangers or my old friends. Ah, is this not happiness?

I: *To watch some one writing big characters a foot high. Ah, is this not happiness?*

I: *To open the window and let a wasp out of the room. Ah, is this not happiness?*

I: *A magistrate orders the beating of the drum and calls it a day. Ah, is this not happiness?*

I: *To see some one's kite line broken. Ah, is this not happiness?*

I: *To see a wild prairie fire. Ah, is this not happiness?*

I: *To have just finished repaying all one's debts. Ah, is this not happiness?*

I: *To read the Story of Curly-Beard. Ah, is this not happiness?*

Poor Byron, who had only three happy hours in his life! He was either of a morbid and enormously unbalanced spirit, or else he was affecting merely the fashionable Weltschmerz of his decade. Were the feeling of Weltschmerz not so fashionable, I feel bound to suspect that he must have confessed to at least thirty happy hours instead of three. Is it not plain from the above that the world is truly a feast of life spread out for us to enjoy—merely through the senses, and a type of culture which recognizes these sensual pleasures therefore makes it possible for us frankly to admit them? My suspicion is, the reason why we shut our eyes willfully to this gorgeous world, vibrating with its own sensuality, is that the spiritualists have made us plain scared of them. A nobler type of philosophy should re-establish our confidence in this fine receptive organ of ours, which we call the body, and drive away first the contempt and then the fear of our senses. Unless these philosophers can actually sublimate matter and etherealize our body into a soul without nerves, without taste, without smell, and without sense of color and motion and touch, and unless we are ready to go the whole way with the Hindu mortifiers of the flesh, let us face ourselves bravely as we are. For only a philosophy that recognizes reality can lead us into true happiness, and only that kind of philosophy is sound and healthy.

四．对唯物主义的误解

当我们读到金圣叹三十三则"不亦快哉"时，一定会觉得现实的人生中，精神欢乐和身体的快乐是不可分离的。精神的欢乐也须由身体上感觉到才能成为真实的欢乐。我甚至于认为道德的欢乐也是同样的。宣传任何学说，也须准备接受人们的误解，像伊壁鸠鲁派（Epicureans）和斯多噶派（Stoics）那样受人们的误解。许多人都不能了解马可·奥勒留（Marcus Aurelius，罗马皇帝，著有《沉思录》）那种斯多噶派所不可少的仁厚精神；同时快乐主义者的智慧和约束学说也常被人们误解为追求欢乐的学说。人们对于这种唯物主义的观念会毫不犹豫地加以攻击，说它的意识是自私的，完全缺乏社会责任的，说它造成每个人都只为自己的享受而着想。这一类的辩论完全由于愚昧和无知，说这话的人是自己也不知所云的。他们不晓得玩世主义者的仁厚，也不知道这个爱好人生者的温顺。爱人类不应该成为一种学说，或是一个信条，或是一个智能上的坚信问题，或是一个能发生辩论的题目。对人类的爱如果需要一些理由来做根基，那便不是真正的爱。这爱必须是绝对自然的，对于人类，应该像鸟鼓翼那样自然。这爱必须是一种直觉，由一个健全的接近大自然的灵魂产生出来。

一个真爱树木的人，绝不会虐待任何动物。在十分健全的精神当中，当一个人，对人生与同类都具有一种信念时，当他们对大自然具有深切的认识时，仁爱也就是自然的产物了。这一种人用不着任何哲学或任何宗教去告诉他要有仁爱。因为他自己的心灵已经从他的感官上获得适当的营养；他的心灵已经从造作的生活和人类社会的人为学问解放出来，他已能保持一种智能和道德的健全。所以，当我们挖开泥土，使这个仁爱泉源的洞口扩大时，人家不能责难我们，说我们在宣布大公无私的观念。

唯物主义是被人们误解了，而且误解得很严重。关于这一点我应该让桑塔耶纳（George Santayana，西班牙著名自然主义哲学家、美学家，美国美学的开创者）来说话。他说他自己是"一个唯物主义者——也许是仅存的唯物主义者"，可是，我们都知道在现代他或许是一个最可爱的人物。他说我们对唯物主义观念的偏见乃是一种外表观察者的偏见。人们对于某些缺点，只在拿来和自己的信条比较时，才会觉得惊异。但只有当我们的精神生活在那个新世界中的时候，我们方才能够真正了解异族的、信仰宗教或国家。所谓"唯物主义"是含有一种喜悦、一种欢乐、一种健全的情感，这是我们平日不曾仔细看到的。桑塔耶纳说："真正的唯物主义者是跟德谟克利特这一类的笑的哲学家一样的，我们都是'不情愿的唯物主义者'，希冀着精神主义，可是事实上过着自私自利的物质生活，我们只是畸形地向着智力方面去发展，而不能发笑。"

一个彻底的唯物主义者，一个生就的而不是半路出家的唯物主义者，应像那个智慧的德谟克利特那样是一个笑的哲学家。他对于那些能够表现各种美妙形状的机构，那些能鼓动极大兴奋的情感，一定会感到欣喜。在自然科学博物院里的参观者，见了那些放在匣中的数千百种蝴蝶、火烈鸟、贝类动物、毛象、大猩猩，一定会感到欣喜；这两种快乐情绪一定含有智能上同样的质素。这世界中的无量数的生命里，当然也有它们的苦痛，不过这些苦痛是马上会消失的，然而当时的行列是何等瑰丽伟大，那些普遍的交互作用也何等引人入胜，而那些专制的小情感偏又何等愚蠢，又无法避免。有活力的灵心里所产生的物质主义，大抵就是下列这种情感吧：活跃的、欢乐的、大公无私的、蔑视私人幻觉的。

唯物主义者的伦理学，对于生理的痛苦也有其感觉。它和另外的慈悲体系一样，对于痛苦也感到一些寒栗，并且想用制欲的那种克制方法把意志收束，不使意志遭遇挫折。绝对乐观的"黑天"巨车上那些颂赞上帝的驾驭者，才不会不顾念到人类的悲哀。可是那些完全虚荣和自欺所生的罪恶，那些自以为人类是宇宙最高目标，对这些笑是适当的防御方法。笑的里面有一种微妙的长处，人们可以一面笑一面仍含着一些同情和友爱，人们对于堂吉诃德所做的荒谬行为和所遭遇的灾难，虽也觉得好笑，但是并不讥笑他的意志。他的热心虽可佩，但他必须去认识世界，然后能合宜地改造世界，并且须在理智当中才能得到快乐[10]。

那么，这种值得我们那么夸耀，甚至胜于情欲生活的智能生活或精神生活，究竟是什么东西呢？可叹现代生物学有一种趋势，想把精神回溯到它的根源，想发觉它就是那么一些纤维、液体和神经组合而成的。我疑惑

---

[10] 节录自史密斯编桑塔耶纳小品集（"Little Essays of Santayana" Edited by Logan Pearsall Smith）《唯物主义者的情感》（Emotions of the materialist）一文。

乐观就是一种液体，或是由某一种循环液体而促成的一种神经状态。可是，我要问问智能生活是从哪一部分产生的呢？智能生活又从何处得到它的生命和滋养呢？哲学家早就告诉我们人类所有的知识都是由于感官之经验而产生。如果我们没有视觉、触觉、嗅觉等感官，便不能获得知识，好比照相机一样，没有了凹凸镜和感光片，便不能拍摄景物。聪明和愚笨的分别就是前者的透镜和感受器更精细更完美，因之摄取的影像更清晰，而能保持得更长久。从书本上所得的知识进展到人生的知识，只靠想象或认识是不足为用的；他必须不停地摸索前进——去感觉各色各样事物的实情，对于人生和人类天性中的万般事物，都去获得一种正确的而不是杂乱的整个印象。对于去感觉人生和觅取经验，我们所有的感官是互相合作的；只有感官的合作和心脑的合作，我们才能得到智能上的热情。智能上的热情是我们所必须的，它是生命的标志，其重要犹如植物的绿色。我们可由热情的存亡去辨认某个人的思想中的生命，犹如从叶和纤维质的水分和结构观察一株半枯的树，可以发现这株树在遇火之后枯，还是有它的生命的。

IV. Misunderstandings of Materialism

CHIN'S description of the happy moments of his life must have already convinced us that in real human life, the mental and the physical pleasures are inextricably tied up together. Mental pleasures are real only when they are felt through the body. I would include even the moral pleasures, too. He who preaches any kind of doctrine must be prepared to be misunderstood, as the Epicureans and Stoics were. How often people fail to see the essential kindness of spirit of a Stoic, like Marcus Aurelius, and how often the Epicurean doctrine of wisdom and restraint has been popularly construed as the doctrine of the man of pleasure! It will at once be brought up against this somewhat materialistic view of things, that it is selfish, that it lacks totally a sense of social responsibility, that it teaches one to enjoy one's self merely. This type of argument proceeds from ignorance; those who use it know not what they are talking about. They know not the kindness of the cynic, not the gentleness of temper of such a lover of life. Love of one's fellowmen should not be a doctrine, an article of faith, a matter of intellectual conviction, or a thesis supported by arguments. The love of mankind which requires reasons is no true love. This love should be perfectly natural, as natural for man as for the birds to flap their wings. It should be a direct feeling, springing naturally from a healthy soul, living in touch with Nature. No man who loves the trees truly can be cruel to animals or to his fellowmen. In a perfectly healthy spirit, gaining a vision of life and of one's fellowmen and a true and deep knowledge of Nature, kindness is the natural thing. That soul does not require any philosophy or man-made religion to tell him to be kind. It is because his spirit has been properly nourished through his senses, somewhat detached from the artificial life and the still more artificial learning of human society, that he is able to retain a true mental and moral health. We cannot, therefore, be accused of teaching unselfishness when we are scratching off the earth and enlarging the opening from which this spring of kindness will naturally flow.

Materialism has been misunderstood, grievously misunderstood. In this matter I must let George Santayana speak for us, who describes himself as "a materialist—perhaps the only one living,"and who, nevertheless, as we all know, is probably one of the sweetest spirits of the present generation. He tells us that our prejudice against the materialistic philosophy is a prejudice of one looking at it from the outside. One gets a feeling of shock from certain deficiencies which are only apparent by comparison with one's old creed. But one can truly understand any foreign creed or religion or country only when one enters to live in spirit in that new world. There is a bounce and a joy, a wholesomeness of feeling in this so-called "materialism" which we usually fail to see entirely. As Santayana tells us, the true materialist is always like Democritus, the laughing philosopher. It is we, the "unwilling materialists,"who aspire to spiritualism but nevertheless live a selfish materialistic life, "that have generally been awkwardly intellectual and incapable of laughter. "

But a thorough materialist, one born to the faith and not half plunged into it by an unexpected christening in cold water, will be like the superb Democritus, a laughing philosopher. His delight in a mechanism that can fall into so many marvellous and beautiful shapes, and can generate so many exciting passions, should be of the same intellectual quality as that which the visitor feels in a museum of natural history, where he views the myriad butterflies in their cases, the flamingoes and shellfish, the mammoths and gorillas. Doubtless there were pangs in that incalculable life, but they were soon over; and how splendid meantime was the pageant, how infinitely interesting the universal interplay, and how foolish and inevitable those absolute little passions. Somewhat of that sort might be the sentiment that materialism would arouse in a vigorous mind, active, joyful, impersonal, and in respect to private illusions not without a touch of scorn.

To the genuine sufferings of living creatures the ethics that accompanies materialism has never been insensible; on the contrary, like other merciful systems, it has trembled too much at pain and tended to withdraw the will ascetically, lest the will should be defeated. Contempt for mortal sorrows is reserved for those who drive with hosannas the Juggernaut car of absolute optimism. But against evils born of pure vanity and self-deception, against the verbiage by which man persuades himself that he is the goal and acme of the universe, laughter is the proper defence. Laughter also has this subtle advantage, that it need not remain without an overtone of sympathy and brotherly understanding; as the laughter that greets Don Quixote's absurdities and misadventures does not mock the hero's intent. His ardour was admirable, but the world must be known before it can be reformed pertinently, and happiness, to be attained, must be placed in reason.

What then is this mental life, or this spiritual life, of which we have been always so proud, and which we always place above the life of the senses? Unfortunately modern biology has a tendency to track the spirit down to its lair, finding it to be a set of fibers, liquids and nerves. I almost believe that optimism is a fluid, or at least it is a condition of the nerves made possible by certain circulating fluids. Whence does the mental life arise, and from what does it take its being and derive its nourishment? Philosophers have long pointed out that all human knowledge comes from sensuous experience. We can no more attain knowledge of any kind without the senses of vision and touch and smell than a camera can take pictures without a lens and a sensitive plate. The difference between a clever man and a dull fellow is that the former has a set of finer lenses and perceiving apparatus by which he gets a sharper image of things and retains it longer. And to proceed from the knowledge of books to the knowledge of life, mere thinking or cogitation is not enough; one has to feel one's way about—to sense things as they are and to get a correct impression of the myriad things in human life and human nature not as unrelated parts, but as a whole. In this matter of feeling about life and of gaining experience, all our senses cooperate, and it is through the cooperation of the senses, and of the heart with the head, that we can have intellectual warmth. Intellectual warmth, after all, is the thing, for it is the sign of life, like the color of green in a plant. We detect life in one's thought by its presence or absence of warmth, as we detect life in a half dried-up tree struggling after some unfortunate accident, by noting the greenness of its leaves and the moisture and healthy texture of its fiber.

五．心灵的欢乐怎样

　　这里让我们来讨论这种所谓心灵和精神的高等欢乐，究竟它们和我们的情感（不是智能）有什么关系，它们的关系达到何种程度。谈到这件事，我们就不期然生出以下的问题，这些有别于高等欢乐的下等情感欢乐，究竟是什么东西？它们可是同样东西的一部分，生于情感而又回到情感？它们是否和情感是难于分解的？当我们研究到这些较高的心智欢乐时——文学、艺术、宗教、哲学——我们发现，智能比之情感和感觉实占着较为无关重要的地位。一幅美丽的图画，它的功用，只是使我们回想到一片真的风景或者是一个美丽可喜的面貌，因而生出一种情欲的欢乐，此外可还有什么作用？文学也只是重作一幅人生的图画，表现它的环境和色彩，表现草地的香味和都市中沟渠的臭味，此外，可还有什么作用呢？我们大抵都有一个观念，认为一部小说必须要描写出真正的角色和真实的情感，才近于真正文学的水准。如果一本书的描写脱离了人生，或只把人生做了一个平淡的解剖，那便不是真正的文学；一本书越有真实的人性，也便越是好文学。如果一本小说只淡淡地分析一下，而不把人生的甜酸苦辣描写出来，怎能引得起读者的兴趣呢？

关于其他的东西，例如诗歌，那不过是渲染着情感的真理；音乐，是无字的情感；宗教，是由幻象中表现的智慧。诗歌之基于音韵及真理的情感，正如绘画之基于色觉及视觉一样。音乐全然是情感，绝用不着那种运用智能所必需的语言。音乐不但能表现牛铃、繁闹的鱼市场以及战场上的声响，并且能表现花朵的美妙、波浪的澎湃起伏、月光的幽丽恬静；但如果要越出感觉的界限，而想表达一个哲学的观念时，我们可说它是没落的，它是一个没落世界的产物。

那么宗教的衰落可也就是由于理智的本身而开始？桑塔耶纳曾说，宗教衰落是由于推理过多，"不幸，这种宗教历来已不是在幻象中所表现出来的智慧，而只变成了推理过多的迷信"。宗教的衰落，就是由于迂腐太过，以及由于信条、公式、学说和谢罪文的树立所致。如果要使我们的信仰变成越加正当合理的东西，一定以为我们是对，那么我们将越加变得不敬虔了。各种宗教相信只有它自己所发现的才是唯一的真理，都成为偏狭的宗派也就是这个道理。我们越是信仰我们是合理的，便越发变得偏狭，这就是目下一切宗教派别的同一现象。因此宗教慢慢地和私人生活中最可憎的偏执仄狭、自私的心理发生了关系。这种宗教造成了个人的自私，不但鄙视其他的宗教，而且使宗教的信仰变成了他自己和上帝的私人契约，在这契约之下，乙方颂赞着甲方，终日在唱着圣诗，祷祝甲方的名字，而甲方为报答起见，也将要降给较旁人更多的福给乙方，较给别家更多的降福给乙方的家庭。因此我们所看见的那些按时上礼拜堂最"虔诚"的老太太，都是自私自利的。结果，那种自以为正常的意识，那种自以为发现了唯一的真理，便代替了产生宗教的更微妙的情感了。

我觉得艺术、诗歌和宗教的存在，其目的是辅助我们恢复新鲜的视觉、富于感情的吸引力和一种更健全的人生意识。我们正需要它们，因为当我们上了年纪的时候，我们的感觉将逐渐麻木，对于痛苦、冤屈和残酷的情感将变为冷淡，我们的人生想象，也因过于注意冷酷和琐碎的现实生活而变成歪曲了。现在幸亏还有几个大诗人和艺术家，他们的那种敏锐的感觉，那种美妙的情感反应和那种新奇的想象还没失掉，还可以行使他们的天职来维持我们道德上的良知，好比拿一面镜子来照我们已经迟钝了的想象，使枯竭的神经兴奋起来。这样说来，艺术应该是一种讽刺文学，对我们麻木了的情感、死气沉沉的思想和不自然的生活的一种警告。它教我们在矫饰的世界里保持着朴实真挚。它应该可以使我们回复到健康幸福的生活，使我们从过分智能活动所产生的昏热中恢复过来。它应该可以使我们的感觉变敏锐，重使我们的理性和本有的天性发生联系，由恢复原有的本性，把那脱离生活中已毁坏的部分收集起来，重变成一个整体。如果我们在世界里有了知识而不能了解，有了批评而不能欣赏，有了美而没有爱，有了真理而缺少热情，有了公义而缺乏慈悲，有了礼貌而一无温暖的心，这种世界将成为一个多么可怜的世界啊！

讲到哲学这种运用着卓越的精神的东西，其危险比我们失去生命本身的感觉更大。我晓得这种智能上的乐趣包括写一个很长的数学方程式，或是去发现宇宙间的一个大体系这类事情。这种发现或许是一切智能欢乐中的最单纯的欢乐，但是在我看来，反不如去吃一顿丰盛的餐食来得开心。第一，这种意念本身可说就是一个畸形产物，即是我们心智活动的副产物；它确是令人愉快，因为它是不费钱的，但无论如何它对我们总好像在生活上不大需要。这种智能上的喜悦，充其量也只是和猜着了纵横字谜游戏（Crossword Puzzle）的喜悦一样。第二，哲学家在这时大都会欺瞒自己，和这个抽象的完美发生爱情，幻想这世界上有一样比现实本身所能证明者更为伟大合理的完美。这好比是我们把星画成五个尖角一样讹误——我们把一切东西都化成公式的、矫揉造作的、太简单化的东西了。只要我们不太过分，这种对于完美的东西所生的喜悦倒也是好的，不过我们也要晓得许许多多没有发现这个简单一式的图样的人们，他们照常是快乐的。我们没有这种东西也能生活。所以我情愿同一个黑种的女佣谈话，而不愿和一位数学大家谈话；她的言语比较具体，笑也笑得较有生气；和她谈话至少对于人类天性可以增长一些知识。我是唯物主义者，所以无论什么时候总是喜欢猪肉而不喜欢诗歌，宁愿放弃一宗哲学，而获得一片拌着好酱汁的椒黄松脆的精肉。

我们只有摆脱思想而生活，才能脱离这种哲学的酷热和恶浊的空气，进而重获一些孩子的新鲜自然的真见识。真正的哲学家对于一个孩子甚至是一只关在笼里的小狮子，应该会觉得汗颜的。试看大自然所赋予那只小狮子的掌爪、肌肉、美丽的皮毛、竖直的耳朵、光亮的眼睛、敏捷的动作，和嬉戏的感觉，这些是多么完美啊！自然完美的东西有时被硬弄成不完美的东西，真正的哲学家对此应该觉得惭愧；好好一个人要去戴着眼镜，胃口不好，常常感到身心不安，一无人生的乐趣，他们对这些也应该觉得惭愧。我们不能从他们那里得到什么好处，因为他所说的话大多是无关我们痛痒的。只有那种和诗歌相应的哲学，只有那种使我们对大自然和人类天性更有真切见识的哲学，于我们才有用处。

无论哪一种人生哲学，它必须以我们天赋本能的和谐为基础。太过于理想主义的哲学家，不久之后，大自然本身也将证明他的错误。依据中国儒学的观念，对于人类尊严的最高理想，是顺其自然而生活，结果达到德参造化之境。这便是孔子之孙子思在《中庸》（The Golden Mean）一书里所倡导的学说。

天命之谓性；率性之谓道[11]；修道之谓教……喜怒哀乐之未发，谓之中；发而皆中节，谓之和。中也者，天下之大本也；和也者，天下之达道也。致中和，天地位焉，万物育焉。

……

自诚明，谓之性；自明诚，谓之教；诚则明矣，明则诚矣。唯天下至诚，为能尽其性；能尽其性，则能尽人之性；能尽人之性，则能尽物之性；能尽物之性，则可以赞天地之化育；可以赞天地之化育，则可以与天地参矣。

---

[11] 儒家的意识里也有浓厚的道家质素，或许是受了道家思想的影响，但平常人是不太注意到的。无论怎样，这一段确是儒家《四书》中的文字，此外《论语》中也可以援引同类的文字。

V. How about Mental Pleasures

LET us take the supposedly higher pleasures of the mind and the spirit, and see to what extent they are vitally connected with our senses, rather than with our intellect. What are those higher spiritual pleasures that we distinguish from those of the lower senses? Are they not parts of the same thing, taking root and ending up in the senses, and inseparable from them? As we go over these higher pleasures of the mind—literature, art, music, religion and philosophy—we see what a minor role the intellect plays in comparison with the senses and feelings. What does a painting do except to give us a landscape or a portrait and recall in us the sensuous pleasures of seeing a real landscape or a beautiful face? And what does literature do except to recreate a picture of life, to give us the atmosphere and color, the fragrant smell of the pastures or the stench of city gutters? We all say that a novel approaches the standard of true literature in proportion as it gives us real people and real emotions. The book which takes us away from this human life, or merely coldly dissects it, is not literature and the more humanly true a book is, the better literature we consider it. What novel ever appeals to a reader if it contains only a cold analysis, if it fails to give us the salt and tang and flavor of life?

As for the other things, poetry is but truth colored with emotion, music is sentiment without words, and religion is but wisdom expressed in fancy.

As painting is based on the sense of color and vision, so poetry is based on the sense of sound and tone and rhythm, in addition to its emotional truth. Music is pure sentiment itself, dispensing entirely with the language of words with which alone the intellect can operate. Music can portray for us the sounds of cowbells and fishmarkets and the battlefield; it can portray for us even the delicacy of the flowers, the undulating motion of the waves, or the sweet serenity of the moonlight; but the moment it steps outside the limit of the senses and tries to portray for us a philosophic idea, it must be considered decadent and the product of a decadent world.

And did not the degeneration of religion begin with reason itself? As Santayana says, the process of degeneration of religion was due to too much reasoning:"This religion unhappily long ago ceased to be wisdom expressed in fancy in order to become superstition overlaid with reasoning."The decay of religion is due to the pedantic spirit, in the invention of creeds, formulas, articles of faith, doctrines and apologies. We become increasingly less pious as we increasingly justify and rationalize our beliefs and become so sure that we are right. That is why every religion becomes a narrow sect, which believes itself to have discovered the only truth. The consequence is that the more we justify our beliefs, the more narrow-minded we become, as is evident in all religious sects. This has made it possible for religion to be associated with the worst forms of bigotry, narrow-mindedness and even pure selfishness in personal life. Such a religion nourishes a man's selfishness not only by making it impossible for him to be broad-minded toward other sects, but also by turning the practice of religion into a private bargain between God and himself, in which the party of the first part is glorified by the party of the second part, singing hymns and calling upon His name on every conceivable occasion, and in return the party of the first part is to bless the party of the second part, bless particularly himself more than any other person and his own family more than any other family. That is why we find selfishness of nature goes so well with some of the most"religious"and regularly church-

going old women. In the end, the sense of self-justification, of having
discovered the only truth, displaces all the finer emotions from which religion
took its rise.

I can see no other reason for the existence of art and poetry and religion
except as they tend to restore in us a freshness of vision and a more emotional
glamour and more vital sense of life. For as we grow older in life, our senses
become gradually benumbed, our emotions become more callous to suffering
and injustice and cruelty, and our vision of life is warped by too much
preoccupation with cold, trivial realities. Fortunately, we have a few poets
and artists who have not lost that sharpened sensibility, that fine emotional
response and that freshness of vision, and whose duties are therefore to be our
moral conscience, to hold up a mirror to our blunted vision, to tone up our
withered nerves. Art should be a satire and a warning against our paralyzed
emotions, our devitalized thinking and our denaturalized living. It teaches us
unsophistication in a sophisticated world. It should restore to us health and
sanity of living and enable us to recover from the fever and delirium caused
by too much mental activity. It should sharpen our senses, re-establish the
connection between our reason and our human nature, and assemble the
ruined parts of a dislocated life again into a whole, by restoring our original
nature. Miserable indeed is a world in which we have knowledge without
understanding, criticism without appreciation, beauty without love, truth
without passion, righteousness without mercy, and courtesy without a warm
heart!

As for philosophy, which is the exercise of the spirit par excellence; the
danger is even greater that we lose the feeling of life itself. I can understand
that such mental delights include the solution of a long mathematical
equation, or the perception of a grand order in the universe. This perception
of order is probably the purest of all our mental pleasures and yet I would
exchange it for a well prepared meal. In the first place, it is in itself almost
a freak, a byproduct of our mental occupations, enjoyable because it is

gratuitous, but not in any case as imperative for us as other vital processes. That intellectual delight is, after all, similar to the delight of solving a crossword puzzle successfully. In the second place, the philosopher at this moment more often than not is likely to cheat himself, to fall in love with this abstract perfection, and to conceive a greater logical perfection in the world than is really warranted by reality itself. It is as much a false picture of things as when we paint a star with five points—a reduction to formula, an artificial stylizing, an over-simplification. So long as we do not overdo it, this delight in perfection is good, but let us remind ourselves that millions of people can be happy without discovering this simple unity of design. We really can afford to live without it. I prefer talking with a colored maid to talking with a mathematician; her words are more concrete, her laughter is more energetic, and I generally gain more in knowledge of human nature by talking with her. I am such a materialist that at any time I would prefer pork to poetry, and would waive a piece of philosophy for a piece of filet, brown and crisp and garnished with good sauce.

Only by placing living above thinking can we get away from this heat and the re-breathed air of philosophy and recapture some of the freshness and naturalness of true insight of the child. Any true philosopher ought to be ashamed of himself when he sees a child, or even a lion cub in a cage. How perfectly nature has fashioned him with his paws, his muscles, his beautiful coat of fur, his pricking ears, his bright round eyes, his agility and his sense of fun! The philosopher ought to be ashamed that God-made perfection has sometimes become man-made imperfection, ashamed that he wears spectacles, has no appetite, is often distressed in mind and heart, and is entirely unconscious of the fun in life. From this type of philosopher nothing is to be gained, for nothing that he says can be of importance to us. That philosophy alone can be of use to us which joins hands merrily with poetry and establishes for us a truer vision, first of nature and then of human nature.

Any adequate philosophy of life must be based on the harmony of our given instincts. The philosopher who is too idealistic is soon tripped up by nature herself. The highest conception of human dignity, according to the Chinese Confucianists, is when man reaches ultimately his greatest height, an equal of heaven and earth, by living in accordance with nature. This is the doctrine given in The Golden Mean, written by the grandson of Confucius.

What is God-given is called nature; to follow nature is called Tao (the Way); to cultivate the Way is called culture. Before joy, anger, sadness and happiness are expressed, they are called the inner self; when they are expressed to the proper degree, they are called harmony. The inner self is the correct foundation of the world, and harmony is the illustrious Way. When a man has achieved the inner self and harmony, the heaven and earth are orderly and the myriad things are nourished and grow thereby.

To arrive at understanding from being one's true self is called nature, and to arrive at being one's true self from understanding is called culture; he who is his true self has thereby understanding, and he who has understanding finds thereby his true self. Only those who are their absolute selves in the world can fulfil their own nature; only those who fulfil their own nature can fulfil the nature of others; only those who fulfil the nature of others can fulfil the nature of things; those who fulfil the nature of things are worthy to help Mother Nature in growing and sustaining life; and those who are worthy to help Mother Nature in growing and sustaining life are the equals of heaven and earth.

/ Chapter Seven   **The Importance of Loafing** /

第七章

悠闲的重要

一．人类是唯一在工作的动物

现在当着我们面前的是人生的盛宴，唯一成为问题的是我们的胃口如何，胃口比筵席更为实在。讲到人，最最难于了解的是他对工作所抱的观念，以及他自己要做的工作或社会需要他做的工作。世间万物尽在过悠闲的日子，只有人类为着生活而工作。他因为不能不去工作，于是在文明日益进步中的生活变为愈加复杂，随时随地是义务、责任、恐惧、障碍和野心，这些并不是生而有之，而是由人类社会所产生。譬如当我坐在书桌边时，我看见一只鸽子在那远处的一座礼拜堂的尖塔旁回翔，它绝不忧虑午餐要吃些什么。但是我的午餐就比那鸽子复杂得多，拿到我面前的食物，已经过了千万人的工作，已经过了种种极复杂的种植、贸易、运输、递送和烹饪，正因如此，人类要获得食物比动物困难万倍。如果一只森林里的野兽跑进人类的都市里来，看到人类为生活如此匆忙，这只野兽一定会对这个人类社会发生很大的疑惑和惊奇。

我想那森林中的野兽，它的第一个思想一定是说人类是唯一工作的动物，因为在世间除了一些驮马和磨坊里的水牛之外，所有的动物甚至家畜等都不必工作的。警犬很少去执行职务；看门的狗总是玩耍的时候多，并且在阳光温暖的时候，总要舒舒服服地在地上睡一下；那贵族化的猫更用不着为生活而工作，它有一个天赋的敏捷身体，可以随时跳过邻居的篱笆，它甚至不以为自己是一个俘囚——想到什么地方去就去。这样看来，世间只有人类辛苦地工作着，驯服地关在笼子里，为了食物，被这个文明和复杂的社会强迫着去工作，为了自己的供养而烦虑。我虽然知道人类也有人类的长处——知识的愉快、谈话的欢乐和幻想的喜悦，例如在看一出舞台戏的时候，更能表现出来，可是在这里我们不能忘掉根本的事情，就是人类的生活太复杂了，只是一个供养自己的问题，已经要费去我们十分之九以上的活动力。所以文明大约是寻觅食物的问题，而进步便是使食物难于得到的一种发展。文明如果不使人类难于得到食物，人类就绝对不用这样劳苦地工作。人类的危机是在社会太文明，是在获取食物的工作太辛苦，因而在那获取食物的劳苦中，吃东西的胃口也失掉了——我们现在已经到这个境地。由森林中的野兽或是由哲学家看来，这好像是没有多大意义的。

当我每次看到那摩天大厦或一望无际相连的房顶时，总有些心惊胆战。这种景象确是令人惊奇的。两三座水塔，两三座钉有广告牌的钢架，一两座高入云霄的尖塔，鳞次栉比的沥青屋顶，形成了一些四方形的、垂直矗立的轮廓，全没有组织或次序，只是点缀着一些泥土，褪了色的烟囱，以及几根晒着衣服的绳索和许多交叉在天空的无线电天线。俯视街道，所见的是一排灰色或已褪色的红砖墙，墙壁上开着成列的、千篇一律的阴暗小窗，窗门半开着，一半掩着阴影，有的窗槛上有一瓶牛乳，其余的窗槛上放着几盆纤弱病态的花儿。每天早晨，有一个女孩子带着她的狗儿跑到屋顶上来，坐在屋顶的楼梯边晒太阳。当我再仰起头来极目远望时，我看见一排一排的屋顶连绵数英里，形成了一些难看的四方形的轮廓，一直到极远的地方。此外不过仍是一些水塔和一些砖屋。人类在这里，他们怎样居住？每家就住在这种阴暗的窗户里面吗？他们怎样生活呢？说来令人咋舌。在那两三个窗户的后面，就住着一对夫妻，每天到了晚上就像鸽子那样回到那鸽子笼式的房子里去睡觉。早晨起来后，喝了些咖啡，丈夫出去到某个地方为家人寻求面包，妻子便在家里不断地、拼命地把尘埃扫出去，使那一块小小的地方干净一些。下午四五点钟，她跑到门边和邻居们谈谈天，吸了一些新鲜空气。到了晚上，他们又拖着疲乏的身体睡上床去。他们就是这样生活下去的。

其他家道较小康的人家便住在较好的公寓里。他们有着较"装作爱好艺术"（arty）的房间和灯罩，房间里布置得较干净！房中稍有空处，也仅是一些些而已。

租上七个房间的已算是奢侈生活，更不用说自己拥有一套七个房间的公寓了！但是住在公寓里，也不一定会有更大的快乐，只不过是少受一些经济和债务的烦扰。情感上的纠纷、离婚案件、晚上不回家的丈夫或夫妻各自在晚上出去游乐放荡等类事件，反而较多了。他们需要的是娱乐。真是天晓得，他们要离开这些单调的墙壁和发光的地板去另找刺激！于是他们去看裸体女人。因此患神经衰弱症啦，吃阿司匹林药片啦，患贵族病啦，结肠炎啦，消化不良啦，脑部软化啦，肝脏变硬啦，患十二指肠溃烂症啦，患肠部撕裂症啦，胃动作过度和肾脏负担过重啦，患膀胱炎啦，患肝脏损坏症啦，心脏胀大啦，神经错乱啦，患胸部平坦和血压过高啦，还有什么糖尿病、肾脏炎、风湿麻痹、失眠症、动脉硬化症、痔疮、瘘管、慢性痢疾、慢性便秘、食欲减退和生之厌倦等，真是比比皆是。这样还不够，还得多养几只狗和几个孩子。快乐的成分完全须看这些住在高雅公寓里的男女的性质和脾气而定。有些人确是过着欢乐的生活，可是其他的人并不见得欢乐。普遍地说来，他们甚至还比不上那些劳苦工作的人，他们只觉得无聊和厌倦。不过他们有一辆汽车，也许还有一座造在乡间的住宅。啊！乡村住宅，这便是他们的救星。人们在乡村中劳苦工作，希望能够到都市去，在都市里赚足了钱，可以再回到乡村中去隐居。

如果你在都市街上散步，你可以在大街上看见美容院、鲜花店和运输公司。在后面的一条街上可以看见药店、食品杂货店、铁器铺、理发店、洗衣店、小餐馆以及报摊。如果那都市很大，就是闲荡了一个钟头，还是在那都市里，只不过多看见一些街道，多看见一些药店、食品杂货店、铁器铺、理发店、洗衣店、小餐馆和报摊。这些人都怎样过生活？他们都来此干什么？问题很简单，就是洗衣服的去洗理发匠和餐馆堂倌的衣服，餐馆里的堂倌去侍候洗衣匠的饭食，而理发匠替洗衣匠和堂倌剃头，那便是文明。这不是太令人惊奇了吗？我敢说，有些洗衣匠和理发匠或堂倌一生不曾到过十条街以外的地方。总算还好，他们还有电影可看，可以看见鸟儿在唱歌，树木在滋长、在摇摆。也可以看见世界之大、土耳其、埃及、喜马拉雅山、安第斯山（Andes）暴风雨、船舶沉没、加冕典礼、蚂蚁、毛虫、麝鼠、蜥蜴跟蝎子的搏斗、山丘、波浪、沙土、云霞，甚至月亮——一切的一切统统在银幕上而已。

啊！聪明智慧的人类！我颂赞你。人们为了生活而任劳任怨地工作，为了要活下去而烦虑到头发发白，甚至忘掉游戏，真是不可思议的文明！

I. Man the Only Working Animal

THE feast of life is, therefore, before us, and the only question is what appetite we have for it. The appetite is the thing, and not the feast. After all, the most bewildering thing about man is his idea of work and the amount of work he imposes upon himself, or civilization has imposed upon him. All nature loafs, while man alone works for a living. He works because he has to, because with the progress of civilization life gets incredibly more complex, with duties, responsibilities, fears, inhibitions and ambitions, born not of nature, but of human society. While I am sitting here before my desk, a pigeon is flying about a church steeple before my window, not worrying what it is going to have for lunch. I know that my lunch is a more complicated affair than the pigeon's, and that the few articles of food I take involve thousands of people at work and a highly complicated system of cultivation, merchandising, transportation, delivery and preparation. That is why it is harder for man to get food than for animals. Nevertheless, if a jungle beast were let loose in a city and gained some apprehension of what busy human life was all about, he would feel a good deal of skepticism and bewilderment about this human society.

The first thought that the jungle beast would have is that man is the only working animal. With the exception of a few draught-horses or buffalos made to work a mill, even domestic pets don't have to work. Police dogs are but rarely called upon to do their duty; a house dog supposed to watch a house plays most of the time, and takes a good nap in the morning whenever there is good, warm sunshine; the aristocratic cat certainly never works for a living, and gifted with a bodily agility which enables it to disregard a neighbor's fence, it is even unconscious of its captivity–it just goes wherever it likes to go. So, then, we have this toiling humanity alone, caged and domesticated, but not fed, forced by this civilization and complex society to work and worry about the matter of feeding itself. Humanity has its own advantages, I am quite aware–the delights of knowledge, the pleasures of conversation and the joys of the imagination, as for instance in watching a stage play. But the essential fact remains that human life has got too complicated and the matter of merely feeding ourselves, directly or indirectly, is occupying well over ninety per cent of our human activities. Civilization is largely a matter of seeking food, while progress is that development which makes food more and more difficult to get. If it had not been made so difficult for man to obtain his food, there would be absolutely no reason why humanity should work so hard. The danger is that we get over-civilized and that we come to a point, as indeed we have already done, when the work of getting food is so strenuous that we lose our appetite for food in the process of getting it. This doesn't seem to make very much sense, from the point of view either of the jungle beast or the philosopher.

Every time I see a city skyline or look over a stretch of roofs, I get frightened. It is positively amazing. Two or three water towers, the backs of two or three steel frames for billboards, perhaps a spire or two, and a stretch of asphalt roofing material and bricks going up in square, sharp, vertical outlines without any form or order, sprinkled with some dirty, discolored chimneys and a few washlines and crisscross lines of radio aerials. And looking down into a street, I see again a stretch of gray or discolored red brick walls, with tiny, dark, uniform windows in uniform rows, half open and half hidden by shades, with perhaps a bottle of milk standing on a windowsill and a few pots of tiny, sickly flowers on some others. A child comes up to the roof with her dog and sits on the roof-stairs every morning to get a bit of sunshine. And as I lift my eyes again, I see rows upon rows of roofs, miles of them, stretching in ugly square outlines to the distance. More water towers, more brick houses. And humanity live here. How do they live, each family behind one or two of these dark windows? What do they do for a living? It is staggering. Behind every two or three windows, a couple go to bed every night like pigeons returning to their pigeonholes; then they wake up and have their morning coffee and the husband emerges into the street, going somewhere to find bread for the family, while the wife tries persistently and desperately to drive out the dust and keep the little place clean. By four or five o'clock they come out on their doorsteps to chat with and look at their neighbors and get a sniff of fresh air. Then night falls, they are dead tired and go to sleep again. And so they live!

There are others, more well-to-do people, living in better apartments. More "arty" rooms and lampshades. Still more orderly and more clean! They have a little more space, but only a little more.

To rent a seven-room flat, not to speak of owning it, is considered a luxury! But it does not imply more happiness. Less financial worry and fewer debts to think about, it is true. But also more emotional complications, more divorce, more cat-husbands that don't come home at night, or the couple go prowling together at night, seeking some form of dissipation. Diversion is the word. Good Lord, they need to be diverted from these monotonous, uniform brick walls and shining wooden floors! Of course they go to look at naked women. Consequently more neurasthenia, more aspirin, more expensive illnesses, more colitis, appendicitis and dyspepsia, more softened brains and hardened livers, more ulcerated duodenums and lacerated intestines, overworked stomachs and overtaxed kidneys, inflamed bladders and outraged spleens, dilated hearts and shattered nerves, more flat chests and high blood pressure, more diabetes, Bright's disease, beri-beri, rheumatism, insomnia, arterio-sclerosis, piles, fistulas, chronic dysentry, chronic constipation, loss of appetite and weariness of life. To make the picture perfect, more dogs and fewer children. The matter of happiness depends entirely upon the quality and temper of the men and women living in these elegant apartments. Some indeed have a jolly life, others simply don't. But on the whole, perhaps they are less happy than the hard-working people; they have more ennui and more boredom. But they have a car, and perhaps a country home. Ah, the country home, that is their salvation! So then, people work hard in the country so that they can come to the city so that they can earn sufficient money and go back to the country again.

And as you take a stroll through the city, you see that back of the main avenue with beauty parlors and flower shops and shipping firms is another street with drug stores, grocery stores, hardware shops, barber shops, laundries, cheap eating places, newsstands. You wander along for an hour, and if it is a big city, you are still there; you see only more streets, more drug stores, grocery stores, hardware shops, barber shops, laundries, cheap eating places and newsstands. How do these people make their living? And why do they come here? Very simple. The laundrymen wash the clothes of the barbers and restaurant waiters, the restaurant waiters wait upon the laundrymen and barbers while they eat, and the barbers cut the hair of the laundrymen and waiters. That is civilization. Isn't it amazing? I bet some of the laundrymen, barbers and waiters never wander beyond ten blocks from their place of work in their entire life. Thank God they have at least the movies, where they can see birds singing on the screen, trees growing and swaying, Turkey, Egypt, the Himalayas, the Andes, storms, shipwrecks, coronation ceremonies, ants, caterpillars, muskrats, a fight between lizards and scorpions, hills, waves, sands, clouds, and even a moon–all on the screen!

O wise humanity, terribly wise humanity! Of thee I sing. How inscrutable is the civilization where men toil and work and worry their hair gray to get a living and forget to play!

二. 中国的悠闲理论

　　美国人是闻名的伟大的劳碌者，中国人是闻名的伟大的悠闲者。因为相反者必是互相钦佩的，所以我想美国劳碌者之钦佩中国悠闲者，是跟中国悠闲者之钦佩美国劳碌者一样的，这就是所谓民族性格上的优点。我不晓得将来东西文明是否会沟通起来，可是在事实上，现在的东西文明已经联系起来了。如将来交通更进步，现代的文明更能远布时，它们间的关系将更加密切。现在至少我们可以这样说，机械的文明中国不反对，目前的问题是怎样把这二种文化加以融合——即中国古代的物质文明——使它们成为一种普遍可行的人生哲学。至于东方哲学能否侵入西洋生活中去的这一个问题，无人敢去预言。

　　机械的文明终于使我们很快地趋近于悠闲的时代，环境也将使我们必须少做工作而多过游玩的生活。这虽然是环境问题，当人类觉得有很多的闲暇工夫时，他不得不去想出一些消磨空闲的聪明方法。这种空闲是飞快进步的结果，不管他愿意不愿意，他必须接受。一个人终不能预测下一代的事物，三十年后的生活怎样，只有大胆的人们才敢去拟想。对于这世界不断的进步，人类总有一天会感到厌倦，而去清查他对于物质方面的成就。当物质环境渐渐改善了，疾病灭绝了，穷困减少了，人寿延长了，食物加多了，到那时候，人类绝不会像现在一样匆忙，而且我相信这种环境或许会产生一种较懒惰的性格。

此外，主观的因素常是和客观的因素同样重要的。哲学不但变换了人类的观念，也改变了人类的性格。人类对于机械文明的反应，是视人类本性而异的。在生物学上讲到下列一类的情形，如对刺激的敏感性，反应的缓急，以及各种动物在同样的环境之下所做的不同行为，有些动物的反应比较迟缓。就是在机械文明里（美英法德俄等包括在内）我们看见各民族的不同气质，对于这个机械时代产生不同的反应，同时，在个人方面，在同样的环境中会产生不同的反应。我认为中国未来的机械文明所创造的生活方式一定近于现代的法国生活方式，因为中国人和法国人的气质是极相近的。

今日的美国是机械文明的先导者，大家都以为世界在未来的机械控制下，一定倾向于美国那种生活形态，这种理论我却抱着怀疑，谁也不会知道未来的美国人又将是怎样的一种气质，勃鲁克（Van Wyck Brook）在新著中所描写的新的英格兰文化时代也许会重现于今日，我以为这是可能的。没有人敢说新英格兰文化的产物不是典型的美国文化，也没有人敢说惠特曼在他的《民主主义远景》里所预测的理想——自由人类和完美母亲的产生——不是民主主义进步中的理想。假如美国能有短期的休息，我相信它或许会产生新的惠特曼、新的梭罗与新的罗伯特·罗威尔（Lowells，美国现代诗人，"自白派"的鼻祖）。到那时候，那种被"淘金热"弄糟了的美国旧文化，也许会再开花结果。这样说来，美国将来的气质，不是又要跟今日的两样了吗？不是将接近于爱默生和梭罗的气质吗？我认为文化本来就是空闲的产物，所以文化的艺术就是悠闲的艺术。在中国人心目中，凡是用他的智慧来享受悠闲的人，也是受教化最深的人。在哲学的观点上看来，劳碌和智慧似乎是根本相左的。智慧的人绝不劳碌，过于劳碌的人绝不是智慧的，善于优游岁月的人才是真正有智慧的。在此我不想讲些中国人的悠闲过活技巧和分类，只是想说明那种养成他们喜闲散、优游岁月、乐天知命的性情——常常也就是诗人的性情——的哲学背景。中国人那种对成就和成功的发生怀疑和对这种生活本身如此深爱的脾性研究是怎样生出来的呢？

中国人的悠闲哲学，可以在十八世纪的一个不大出名的女词人舒白香所说的话里看出来。她以为时间之所以宝贵，乃在时间之不被利用："闲暇之时间如室中之空隙。"做女工的女人租了小小的一个房间住着，房里满是东西，一无旋转的余地，因而感到不舒服；一旦薪水略为增加，她便要搬到一间较宽敞的房子里，在那里除了放置床桌和煤气炉子外，还有一些回旋的地方，这就使她感到舒适。同样理由，我们有了闲暇，才能感到生活的兴趣。我曾听说纽约公园大道（Park Avenue）有一位富婆，她把住宅旁边的无用地皮都买了下来，原因是防止有人在她的住宅旁造摩天大厦。她仅仅是为了一些弃置不用的空地，不惜花费大量金钱，但我以为她花的钱，再没有比花在这种地方更精明的了。

关于这点，我可以报告一些我个人的经验。原先我看不出纽约市中摩天大厦的美点，后来到了芝加哥，才觉得只要在摩天大厦的前边有相当的地面，而四周又有半里多的空地，倒可成为庄严美丽的。芝加哥在这方面比较幸运，空地较纽约曼哈顿市区多一些。如果那些大建筑物间的距离比较宽阔，在远处看起来，就似乎没有什么东西阻碍了视线。这样比较起来，我们的生活太狭仄了，使我们对精神生活的美点不能有一个自由的视野，我们精神上的"屋前空地"太缺乏了。

II. The Chinese Theory of Leisure

THE American is known as a great hustler, as the Chinese is known as a great loafer. And as all opposites admire each other, I suspect that the American hustler admires the Chinese loafer as much as the Chinese loafer admires the American hustler. Such things are called the charms of national traits. I do not know if eventually the West and the East will meet; the plain fact is that they are meeting now, and are going to meet more and more closely as modern civilization spreads, with the increase of communication facilities. At least, in China, we are not going to defy this machine civilization, and there the problem will have to be worked out as to how we are going to merge these two cultures, the ancient Chinese philosophy of life and the modern technological civilization, and integrate them into a sort of working way of life. The question is very much more problematical as to Occidental life ever being invaded by Oriental philosophy, although no one would dare to prophesy.

After all, the machine culture is rapidly bringing us nearer to the age of leisure, and man will be compelled to play more and work less. It is all a matter of environment, and when man finds leisure hanging on his hand, he will be forced to think more about the ways and means of wisely enjoying his leisure, conferred upon him, against his will, by rapidly improving methods of quick production. After all, no one can predict anything about the next century. He would be a brave man who dared even to predict about life thirty years from now. The constant rush for progress must certainly one day reach a point when man will be pretty tired of it all, and will begin to take stock of his conquests in the material world. I cannot believe that, with the coming of better material conditions of life, when diseases are eliminated, poverty is decreased and man's expectation of life is prolonged and food is plentiful, man will care to be as busy as he is today. I'm not so sure that a more lazy temperament will not arise as a result of this new environment.

Apart from all this, the subjective factor is always as important as the objective. Philosophy comes in as a way of changing man's outlook and also changing his character. How man is going to react toward this machine civilization depends on what kind of a man he is. In the realm of biology, there are such things as sensibility to stimulus, slowness or quickness of reaction, and different behaviors of different animals in the same medium or environment. Some animals react more slowly than others. Even in this machine civilization, which I understand includes the United States, England, France, Germany, Italy and Russia, we see that different reactions toward the mechanical age arise from different racial temperaments. The chances of peculiar individual reactions to the same environment are not eliminated. For China, I feel the type of life resulting from it will be very much like that in modern France, because the Chinese and the French temperaments are so akin.

America today is most advanced in machine civilization, and it has always been assumed that the future of a world dominated by the machine will tend toward the present American type and pattern of life. I feel inclined to dispute this thesis, because no one knows yet what the American temperament is going to be. At best we can only describe it as a changing temperament. I do not think it at all impossible that there may be a revival of that period of New England culture so well described in Van Wyck Brooks' new book. No one can say that that flowering of New England culture was not typically American culture, and certainly no one can say that that ideal Walt Whitman envisaged in his Democratic Vistas, pointing to the development of free men and perfect mothers, is not the ideal of democratic progress. America needs only to be given a little respite, and there may be—I am quite sure there will be—new Whitmans, new Thoreaus and new Lowells, when that old American culture, cut short literally and figuratively by the gold rush, may blossom forth again. Will not, then, American temperament be something quite different from that of the present day, and very near to the temperament of Emerson and Thoreau? Culture, as I understand it, is essentially a product of leisure. The art of culture is therefore essentially the art of loafing. From the Chinese point of view, the man who is wisely idle is the most cultured man. For there seems to be a philosophic contradiction between being busy and being wise. Those who are wise won't be busy, and those who are too busy can't be wise. The wisest man is therefore he who loafs most gracefully. Here I shall try to explain, not the technique and varieties of loafing as practised in China but rather the philosophy which nourishes this divine desire for loafing in China and gives rise to that carefree, idle, happy-go-lucky—and often poetic—temperament in the Chinese scholars, and to a lesser extent, in the Chinese people in general. How did that Chinese temperament—that distrust of achievement and success and that intense love of living as such—arise?

In the first place, the Chinese theory of leisure, as expressed by a comparatively unknown author of the eighteenth century, Shu Paihsiang, who happily achieved oblivion, is as follows: time is useful because it is not being used."Leisure in time is like unoccupied floor space in a room."Every working girl who rents a small room where every inch of space is fully utilized feels highly uncomfortable because she has no room to move about, and the moment she gets a raise in salary, she moves into a bigger room where there is a little more unused floor space, besides those strictly useful spaces occupied by her single bed, her dressing table and her two-burner gas range. It is that unoccupied space which makes a room habitable, as it is our leisure hours which make life endurable. I understand there is a rich woman living on Park Avenue, who bought up a neighboring lot to prevent anybody from erecting a skyscraper next to her house. She is paying a big sum of money in order to have space fully and perfectly made useless, and it seems to me she never spent her money more wisely.

In this connection, I might mention a personal experience. I could never see the beauty of skyscrapers in New York, and it was not until I went to Chicago that I realized that a skyscraper could be very imposing and very beautiful to look at, if it had a good frontage and at least half a mile of unused space around it. Chicago is fortunate in this respect, because it has more space than Manhattan. The tall buildings are better spaced, and there is the possibility of obtaining an unobstructed view of them from a long distance. Figuratively speaking, we, too, are so cramped in our life that we cannot enjoy a free perspective of the beauties of our spiritual life. We lack spiritual frontage.

三. 悠闲生活的崇尚

中国人之爱悠闲，有很多交织着的原因。中国人的性情，是经过了文学的熏陶和哲学的认可的。这种爱悠闲的性情由于酷爱人生而产生，并受了历代浪漫文学潜流的激荡，最后又由一种人生哲学——大体上可称它为"道家哲学"——承认为合理近情的态度。中国人能囫囵吞枣地接受这种道家的人生观，可见他们的血液中原有着道家哲学的种子。

有一点我们须先行加以澄清，这种消闲的浪漫崇尚（我们已说过它是空闲的产物），绝不是我们一般想象中的那些有产阶级者的享受，那种观念是错误的。我们要明了，这种悠闲生活是穷愁潦倒的文士所崇尚的，他们中有的是生性喜爱悠闲的生活，有的是不得不如此。当我读中国的文学杰作时，或当我想到那些穷教师拿了称颂悠闲生活的诗文去教穷弟子时，我不禁要想他们一定在这些著作中获得很大的满足和精神上的安慰。所谓"盛名多累，隐逸多适"，那些应试落第的人对这种话是很听得进去的；还有什么"晚食可以当肉"这一类的俗语，养不起家的人即可以解嘲。中国无产阶级的青年作家们指责苏东坡和陶渊明等为罪恶的有闲阶级的智识分子，这可说是文学批评史上的最大错误了。苏东坡的诗中不过写了一些"江上清风"及"山间明月"，陶渊明的诗中不过说了一些"夕露沾我衣"及"鸡鸣桑树颠"。难道江上清风、山间明月和桑树颠的鸡鸣只有资产阶级者才能占有吗？这些古代的名人并不是空口白话地谈论着农村的情形，他们是事必躬亲过着穷苦的农夫生活，在农村生活中得到了和平与和谐的。

这样说来，这种消闲的浪漫崇尚，我以为根本是平民化的。我们只要想象英国小说大家劳伦斯·斯特恩（Laurence Sterne，著有《项狄传》）在他有感触的旅程上的情景，或是想象英国大诗人华兹华斯（Wordsworth）和柯勒律治（Coleridge）徒步游欧洲，心胸蕴着伟大的美的观念，而袋里不名一文。我们想象到这些，对这些个浪漫主义就比较了解了。一个人不一定要有钱才可以旅行，就是在今日，旅行也不一定是富家的奢侈生活。总之，享受悠闲生活当然比享受奢侈生活便宜得多。要享受悠闲的生活，只要有一种艺术家的性情，在一种全然悠闲的情绪中，去消遣一个闲暇无事的下午。正如梭罗在《瓦尔登湖》（Walden）里所说的，享受悠闲的生活，所费是不多的。

笼统说来，中国的浪漫主义者都具有锐敏的感觉和爱好漂泊的天性，虽然在物质生活上露着穷苦的样子，但情感很丰富，他们深切爱好人生，所以宁愿辞官弃禄，不愿心为形役。在中国，消闲生活并不是富有者、有权势者和成功者独有的权利（美国的成功者更显匆忙了），而是那种高尚自负心情的产物，这种高尚自负的心情极像那种西方流浪者的尊严的观念，这种流浪者骄傲自负到不肯去请教人家，自立到不愿意工作，聪明到不把周遭的世界看得太认真。这样子的心情是一种超脱俗世的意识而产生，并和这种意识自然联系着的，也可说是由那种看透人生的野心、愚蠢和名利的诱惑而产生出来的。那个把他的人格看得比事业的成就来得重大，把他的灵魂看得比名利更紧要的高尚自负的学者，大家认为他是中国文学上最崇高的理想。他显然是一个极简朴地去过生活，而且鄙视世欲功名的人。

这一类的大文学家——陶渊明、苏东坡、白居易、袁中郎、袁子才，都曾度过短期的官场生活，政绩都很优良，但都厌倦了那种磕头的勾当，要求辞职，以便回家去过自由自在的生活。

另一位诗人白玉蟾，为他的书斋所写《慵庵》，对悠闲的生活竭尽称赞的能事：

丹经慵读，道不在书；
藏教慵览，道之皮肤。
至道之要，贵乎清虚，
何谓清虚？终日如愚。
有诗慵吟，句外肠枯；
有琴慵弹，弦外韵孤；
有酒慵饮，醉外江湖；
有棋慵奕，意外干戈；
慵观溪山，内有画图；
慵对风月，内有蓬壶；
慵陪世事，内有田庐；
慵问寒暑，内有神都。
松枯石烂，我常如如。
谓之慵庵，不亦可乎？

从上面的题赞看来，这种悠闲的生活也必须有一个恬静的心地和乐天旷达的观念，以及一个能尽情玩赏大自然的胸怀方能享受。诗人及学者常常自题了一些稀奇古怪的别号，如江湖客（杜甫）、东坡居士（苏东坡）、烟湖散人（Carefree Man of a Misty Lake）、襟霞阁老人（The Old Man of the Haze-Girdled Tower）等。

没有金钱也能享受悠闲的生活。有钱的人不一定能真真领略悠闲生活的乐趣，那些轻视钱财的人才真正懂得此中的乐趣，他须有丰富的心灵，有简朴生活的爱好，对于生财之道不大在心，这样的人才有资格享受悠闲的生活。如果一个人真的要享受人生，人生是尽够他享受的。一般人不能领略这个尘世生活的乐趣，那是因为他们不深爱人生，把生活弄得平凡、刻板，而且无聊。有人说老子是嫉恶人生的，这话绝对不正确，我认为老子所以要鄙弃俗世生活，正因为他太爱人生，不愿使生活变成"为生活而生活"。

有爱必有妒。一个热爱人生的人，对于他应享受的那些快乐的时光，一定爱惜非常。然而同时须保持流浪汉特有的那种尊严和傲慢，甚至他的垂钓时间也和他的办公时间一样神圣不可侵犯，而成为一种教规，好像英国人把游戏当做教规郑重其事一样。他对于他在高尔夫球总会中同他人谈论股票的市况，一定会像一个科学家在实验室中受到人家骚扰那样觉得厌恶。他一定时常计算着再有几个春天就要消逝了，为了不曾做几次遨游而心中感到悲哀和懊丧，像一个市侩懊恼今天少卖出不少货物一样。

III. The Cult of the Idle Life

THE Chinese love of leisure arises from a combination of causes. It came from a temperament, was erected into a literary cult, and found its justification in a philosophy. It grew out of an intense love of life, was actively sustained by an underlying current of literary romanticism throughout the dynasties, and was eventually pronounced right and sensible by a philosophy of life, which we may, in the main, describe as Taoistic. The rather general acceptance of this Taoistic view of life is only proof that there is Taoistic blood in the Chinese temperament.

And here we must first clarify one point. The romantic cult of the idle life, which we have defined as a product of leisure, was decidedly not for the wealthy class, as we usually understand it to be. That would be an unmitigated error in the approach to the problem. It was a cult for the poor and unsuccessful and humble scholar who either had chosen the idle life or had idleness enforced upon him. As I read Chinese literary masterpieces, and as I imagine the poor schoolmaster teaching the poor scholars these poems and essays glorifying the simple and idle life, I cannot help thinking that they must have derived an immense personal satisfaction and spiritual consolation from them. Disquisitions on the handicaps of fame and advantages of obscurity sounded pleasing to those who had failed in the civil examinations, and such sayings as"Eating late (with appetite whetted) is eating meat"tended to make the bad provider less apologetic to his family. No greater misjudgment of literary history is made than when the young Chinese proletarian writers accuse the poets Su Tungp'o and T'ao Yüanming and others of belonging to the hated leisure-class intelligentsia–Su who sang about"the clear breeze over the stream and bright moon over the hills,"and T'ao who sang about"the dew making wet his skirt"and"a hen roosting on the top of a mulberry tree."As if the river breeze and the moon over the hills and the hen roosting on a mulberry tree were owned only by the capitalist class! These great men of the past went beyond the stage of talking about peasant conditions, and lived the life of the poor peasant themselves and found peace and harmony in it.

In this sense I regard this romantic cult of the idle life as essentially democratic. We can better understand this romantic cult when we picture for ourselves Laurence Sterne on his Sentimental journey, or Wordsworth and Coleridge hiking through Europe on foot with a great sense of beauty in their breast, but very little money in their purse. There was a time when one didn't have to be rich in order to travel, and even today travel doesn't have to be a luxury of the rich. On the whole, the enjoyment of leisure is something which decidedly costs less than the enjoyment of luxury. All it requires is an artistic temperament which is bent on seeking a perfectly useless afternoon spent in a perfectly useless manner. The idle life really costs so very little, as Thoreau took the trouble to point out in Walden.

The Chinese romanticists were, on the whole, men gifted with a high sensibility and a vagabond nature, poor in their worldly possessions, but rich in sentiment. They had an intense love of life which showed itself in their abhorence of all official life and a stern refusal to make the soul serf to the body. The idle life, so far from being the prerogative of the rich and powerful and successful (how busy the successful American men are!) was in China an achievement of highmindedness, a highmindedness very near to the Western conception of the dignity of the tramp who is too proud to ask favors, too independent to go to work, and too wise to take the world's successes too seriously. This highmindedness came from, and was inevitably associated with, a certain sense of detachment toward the drama of life; it came from the quality of being able to see through life's ambitions and follies and the temptations of fame and wealth. Somehow the highminded scholar who valued his character more than his achievements, his soul more than fame or wealth, became by common consent the highest ideal of Chinese literature. Inevitably he was a man with great simplicity of living and a proud contempt for worldly success as the world understands it.

Great men of letters of this class—T'ao Yüanming,Su Tungp'o, Po Chüyi, Yüan Chunglang, Yüan Tsets'ai—were generally enticed into a short term of official life, did a wonderful job of it, and then got exasperated with its eternal kowtowing and receiving and sending off of fellow officials, and gladly laying down the burdens of an official life, returned wisely to the life of retirement.

A rather extravagant example of the praise of idleness is found in the inscription of another poet, Po Yüchien, written for his studio, which he called"The Hall of Idleness":

I'm too lazy to read the Taoist classics, for Tao doesn't reside in the books;
Too lazy to look over the sutras, for they go no deeper in Tao than its looks.
The essence of Tao consists in a void, clear, and cool,
But what is this void except being the whole day like a fool?
Too lazy am I to read poetry, for when I stop, the poetry will be gone;
Too lazy to play on the ch'in, for music dies on the string where it's born;
Too lazy to drink wine, for beyond the drunkard's dream there are rivers and lakes;
Too lazy to play chess, for besides the pawns there are other stakes;
Too lazy to look at the hills and streams, for there is a painting within my heart's portals;

Too lazy to face the wind and the moon, for within me is the Isle of the Immortals;

Too lazy to attend to worldly affairs, for inside me are my hut and my possessions;

Too lazy to watch the changing of the seasons, for within me are heavenly processions.

Pine trees may decay and rocks may rot; but I shall always remain what I am.

Is it not fitting that I call this the Hall of Idleness?

This cult of idleness was therefore always bound up with a life of inner calm, a sense of carefree irresponsibility and an intense wholehearted enjoyment of the life of nature. Poets and scholars have always given themselves quaint names like"The Guest of Rivers and Lakes"(Tu Fu);"The Recluse of the Eastern Hillside"(Su Tungp'o); the"Carefree Man of a Misty Lake"; and"The Old Man of the Haze-Girdled Tower,"etc.

No, the enjoyment of an idle life doesn't cost any money. The capacity for true enjoyment of idleness is lost in the moneyed class and can be found only among people who have a supreme contempt for wealth. It must come from an inner richness of the soul in a man who loves the simple ways of life and who is somewhat impatient with the business of making money. There is always plenty of life to enjoy for a man who is determined to enjoy it. If men fail to enjoy this earthly existence we have, it is because they do not love life sufficiently and allow it to be turned into a humdrum routine existence. Laotse has been wrongly accused of being hostile to life; on the other hand, I think he taught the renunciation of the life of the world exactly because he loved life all too tenderly, to allow the art of living to degenerate into a mere business of living.

For where there is love, there is jealousy; a man who loves life intensely must be always jealous of the few exquisite moments of leisure that he has. And he must retain the dignity and pride always characteristic of a vagabond. His hours of fishing must be as sacred as his hours of business, erected into a kind of religion as the English have done with sport. He must be as impatient at having people talk to him about the stock market at the golf club, as the scientist is at having anybody disturb him in his laboratory. And he must count the days of departing spring with a sense of sad regret for not having made more trips or excursions, as a business man feels when he has not sold so many wares in the day.

四．尘世是唯一的天堂

　　我们的生命总有一日会灭绝的，这种省悟，使那些深爱人生的人，在感觉上增添了悲哀的诗意情调，这种悲感却反使中国的学者更热切深刻地要去领略人生的乐趣。这看来是很奇怪的。我们的尘世人生因为只有一个，所以我们必须趁人生还未消逝的时候，尽情地享受它。如果我们有了一种永生的渺茫希望，那么对于这尘世生活的乐趣便不能尽情地领略了。阿瑟·凯兹爵士（Sir Arthur Keith，苏格兰著名人类进行化学家）曾说过一句和中国人的感想不谋而合的话："如果人们的信念跟我的一样，认尘世是唯一的天堂，那么他们必将更竭尽全力把这个世界造成天堂。"苏东坡的诗中有"事如春梦了无痕"之句，因为如此，所以他那么深刻坚决地爱好人生。在中国的文学作品中，常常可以看到这种"人生不再"的感觉。中国的诗人和学者在欢娱宴乐的时候，常被这种"人生不再""生命易逝"的悲哀感觉烦扰，在花前月下，常有"花不常好，月不常圆"的伤悼。李白在《春夜宴桃李园序》里，有着两句名言："浮生若梦，为欢几何？"王羲之在和他的一些朋友欢宴的时候，曾写下《〈兰亭集〉序》这篇不朽的文章，把"人生不再"的感觉表现得最为典型：

　　永和九年，岁在癸丑，暮春之初，会于会稽山阴之兰亭，修禊事也。群贤毕至，少长咸集。此地有崇山峻岭，茂林修竹，又有清流激湍，映带左右，引以为流觞曲水，列坐其次，虽无丝竹管弦之盛，一觞一咏，亦足以畅叙幽情。是日也，天朗气清，惠风和畅，仰观宇宙之大，俯察品类之盛，所以游目骋怀，足以极视听之娱，信可乐也。

　　夫人之相与，俯仰一世，或取诸怀抱，悟言一室之内；或因寄所托，放浪形骸之外；虽趣舍万殊，静躁不同，当其欣于所遇，暂得于己，快然自足，不知老之将至；及其所之既倦，情随事迁，感慨系之矣！向之所欣，俯仰之间，已为陈迹，犹不能不以之兴怀；况修短随化，终期于尽。古人云："死生亦大矣"，岂不痛哉！

　　每览昔人兴感之由，若合一契，未尝不临文嗟悼，不能喻之于怀。固知一死生为虚诞，齐彭殇为妄作，后之视今，亦犹今之视昔，悲夫！故列叙时人，录其所述，虽世殊事异，所以兴怀，其致一也。后之览者，亦将有感于斯文。

　　我们都相信人总是要死的，相信生命像一支烛光，总有一日要熄灭的，我认为这种感觉是好的。它使我们清醒，使我们悲哀，它也使某些人感到一种诗意。此外还有一层最为重要：它使我们能够坚定意志，去想法过一种合理的、真实的生活，随时使我们感悟到自己的缺点。它也使我们心中平安。因一个人的心中有了那种接受恶劣遭遇的准备，才能够获得真平安。由心理学的观点看来，它是一种发泄身上储力的程序。

　　中国的诗人与平民，即使是在享受人生的乐趣时，下意识里也有一种好景不常的感觉，例如在中国人欢聚完毕时，常常说："千里搭长棚，没有不散的宴席。"所以人生的宴会便是尼布甲尼撒（Nebuchadnezzar，古代新古巴比伦王，以强猛、骄傲、奢侈著称，建造空中花园）的宴会。这种感觉使那些不信宗教的人也有一种神灵的意识。他观看人生，好比是宋代的山水画家观看山景，是被一层神秘的薄雾包围着，或者是空气中有着过多的水蒸气似的。

我们消除了永生观念，生活上的问题就变得很简单了。问题就是这样的：人类的寿命有限，很少能活到七十岁以上，因此我们必须调整生活，在现实的环境之下尽量过着快乐的生活。这种观念就是儒家的观念，它含着浓厚的尘世气息。人类的活动依着一种固执的常识而行，他的精神就是桑塔耶纳所说把人生当做人生看的"动物信念"。这个基于动物信念，人类和动物的根本关系，不必靠达尔文的帮助，我们也能做一个明慧的猜测，这个动物的信念使我们依恋人生——本能和情感的人生——因为我们相信：既然大家都是动物，所以只有在正常的本能上获得正常的满足，我们才能够获得真正的快乐，包括生活各方面的享受。

这样说起来，我们不是变成唯物主义者了吗？但是这个问题，中国人是几乎不知道怎样回答的。因为中国人的精神哲理根本是建筑在物质上的，他们对于尘世的人生，分不出精神或是肉体。无疑，他爱物质上的享受，但这种享受就是属于情感方面的。人类只有靠理智才能分得出精神和肉体的区别，但是上面已经说过，精神和肉体享受必须通过我们的感官。音乐无疑是各种艺术中最属于心灵的，它能够把人们高举到精神的境界里去，可是音乐必须基于我们的听觉，所以对于食物的味觉享受为什么不如声音的交响曲崇高纯洁这一问题，中国人实在有些不明白。我们只有在这种实际的感觉上，才能意识到我们所爱的女人，要分开女人的灵魂和肉体是不可能的。因为我们爱一个女人，不单是爱她外表的曲线美，也爱她的举止、她的仪态、她的眼波和她的微笑。那么，这些是属于肉体的呢，还是精神的呢？我想没有人能回答出来吧。

　　这种人生现实性和人生精神性的感觉，中国的人性主义是赞成的，或者可以说它是得到中国人全部思想方法和生活方法的赞成的。简单讲来，中国的哲学，可说是注重人生的知识而不注重真理的知识。中国哲学家把一切的抽象理论撇开不谈，认为和生活问题不发生关系，以为这些东西是我们理智上所产生的浅薄感想。他们只把握人生，提出一个最简单的问题："我们怎样生活？"西洋哲学在中国人看来是很无聊的。西洋哲学以论理或逻辑为基点，着重研究知识方法的获得，以认识论为基点，提出知识可能性的问题，但最后关于生活本身的知识忘记了，那真是愚蠢琐碎的事，像一个人，只谈谈恋爱求求婚，而并不结婚生子；又像操练甚勤的军队不开到战场上去正式打仗。法国的哲学家要算最无谓，他们追求真理，如追求爱人那样地热烈，但不想和她结婚。

IV. This Earth the Only Heaven

A sad, poetic touch is added to this intense love of life by the realization that this life we have is essentially mortal. Strange to say, this sad awareness of our mortality makes the Chinese scholar's enjoyment of life all the more keen and intense. For if this earthly existence is all we have, we must try the harder to enjoy it while it lasts. A vague hope of immortality detracts from our wholehearted enjoyment of this earthly existence. As Sir Arthur Keith puts it with a typically Chinese feeling, "For if men believe, as I do, that this present earth is the only heaven, they will strive all the more to make heaven of it." Su Tungp'o says, "Life passes like a spring dream without a trace," and that is why he clung to it so fondly and tenaciously. It is this sentiment of our mortal existence that we run across again and again in Chinese literature. It is this feeling of the impermanence of existence and the evanescence of life, this touch of sadness, which overtakes the Chinese poet and scholar always at the moment of his greatest feasting and merrymaking, a sadness that is expressed in the regret that "the moon cannot always be so round and the flowers cannot forever look so fair" when we are watching the full moon in the company of beautiful flowers. It was in that poem commemorating a gorgeous feast on "A Spring Night amidst Peach Blossoms" that Li Po penned the favorite line: "Our floating life is like a dream; how many times can one enjoy one's self?"And it was in the midst of a gay reunion of his happy and illustrious friends that Wang Hsichih wrote that immortal little essay,"Preface to The Orchid Pavilion Collection", which gives, better than anything else, this typical feeling about the evanescence of life:

In the ninth year of the reign Yungho [A. D. 353] in the beginning of late spring we met at the Orchid Pavilion in Shanyin of Kweich'i for the Water Festival, to wash away the evil spirits.

Here are gathered all the illustrious persons and assembled both the old and the young. Here are tall mountains and majestic peaks, trees with thick foliage and tall bamboos. Here are also clear streams and gurgling rapids, catching one's eye from the right and left. We group ourselves in order, sitting by the waterside, and drink in succession from a cup floating down the curving stream; and although there is no music from string and wood-wind instruments, yet with alternate singing and drinking, we are well disposed to thoroughly enjoy a quiet intimate conversation. Today the sky is clear, the air is fresh and the kind breeze is mild. Truly enjoyable it is to watch the immense universe above and the myriad things below, travelling over the entire landscape with our eyes and allowing our sentiments to roam about at will, thus exhausting the pleasures of the eye and the ear.

Now when people gather together to surmise life itself, some sit and talk and unburden their thoughts in the intimacy of a room, and some, overcome by a sentiment, soar forth into a world beyond bodily realities. Although we select our pleasures according to our inclinations—some noisy and rowdy, and others quiet and sedate—yet when we have found that which pleases us, we are all happy and contented, to the extent of forgetting that we are growing old. And then, when satiety follows satisfaction, and with the change of circumstances, change also our whims and desires, there then arises a feeling of poignant regret. In the twinkling of an eye, the objects of our former pleasures have become things of the past, still compelling in us moods of regretful memory. Furthermore, although our lives may be long or short, eventually we all end in nothingness. "Great indeed are life and death"said the ancients. Ah ! what sadness!

I often study the joys and regrets of the ancient people, and as I lean over their writings and see that they were moved exactly as ourselves, I am often overcome by a feeling of sadness and compassion, and would like to make those things clear to myself. Well I know it is a lie to say that life and death are the same thing, and that longevity and early death make no difference! Alas! as we of the present look upon those of the past, so will posterity look upon our present selves. Therefore, have I put down a sketch of these contemporaries and their sayings at this feast, and although time and circumstances may change, the way they will evoke our moods of happiness and regret will remain the same. What will future readers feel when they cast their eyes upon this writing!

Belief in our mortality, the sense that we are eventually going to crack up and be extinguished like the flame of a candle, I say, is a gloriously fine thing. It makes us sober; it makes us a little sad; and many of us it makes poetic. But above all, it makes it possible for us to make up our mind and arrange to live sensibly, truthfully and always with a sense of our own limitations. It gives peace also, because true peace of mind comes from accepting the worst. Psychologically, I think, it means a release of energy.

When Chinese poets and common people enjoy themselves, there is always a subconscious feeling that the joy is not going to last forever, as the Chinese often say at the end of a happy reunion, "Even the most gorgeous fair, with mat-sheds stretching over a thousand miles, must sooner or later come to an end." The feast of life is the feast of Nebuchadnezzar. This feeling of the dreamlike quality of our existence invests the pagan with a kind of spirituality. He sees life essentially as a Sung landscape artist sees mountain scenery, enveloped in a haze of mystery, sometimes with the air dripping with moisture.

Deprived of immortality, the proposition of living becomes a simple proposition. It is this: that we human beings have a limited span of life to live on this earth, rarely more than seventy years, and that therefore we have to arrange our lives so that we may live as happily as we can under a given set of circumstances. Here we are on Confucian ground. There is something mundane, something terribly earth-bound about it, and man proceeds to work with a dogged commonsense, very much in the spirit of what George Santayana calls"animal faith."With this animal faith, taking life as it is, we made a shrewd guess, without Darwin's aid as to our essential kinship with animals. It made us therefore, cling to life–the life of the instinct and the life of the senses–on the belief that, as we are all animals, we can be truly happy only when all our normal instincts are satisfied normally. This applies to the enjoyment of life in all its aspects.

Are we therefore materialistic? A Chinese would hardly know how to answer this question. For with his spirituality based on a kind of material, earth-bound existence, he fails to see the distinction between the spirit and the flesh. Undoubtedly he loves creature comforts, but then creature comforts are matters of the senses. It is only through the intellect that man attains the distinction between the spirit and the flesh, while our senses provide the portals to both, as we have already seen in the preceding chapter. Music, undoubtedly the most spiritual of our arts, lifting man to a world of spirit, is based on the sense of hearing. And the Chinese fails to see why a sympathy of tastes in the enjoyment of food is less spiritual than a symphony of sounds. Only in this realistic sense, can we feel about the woman we love. A distinction between her soul and her body is impossible. For if we love a woman, we do not love her geometrical precision of features, but rather her ways and gestures in motion, her looks and smiles. But are a woman's looks and smiles physical or spiritual? No one can say.

This feeling of the reality and spirituality of life is helped by Chinese humanism, in fact by the whole Chinese way of thinking and living. Chinese philosophy may be briefly defined as a preoccupation with the knowledge of life rather than the knowledge of truth. Brushing aside all metaphysical speculations as irrelevant to the business of living, and as pale reflections engendered in our intellect, the Chinese philosophers clutch at life itself and ask themselves the one and only eternal question:"How are we to live?"Philosophy in the Western sense seems to the Chinese eminently idle. In its preoccupation with logic, which concerns itself with the method of arrival at knowledge, and epistemology, which poses the question of the possibility of knowledge, it has forgotten to deal with the knowledge of life itself. That is so much tomfoolery and a kind of frivolity, like wooing and courtship without coming to marriage and the producing of children, which is as bad as having redcoated regiments marching in military parades without going to battle. The German philosophers are the most frivolous of all; they court truth like ardent lovers, but seldom propose to marry her.

五. 运气是什么

　　道家不信幸运和命运的这种思想，对中国人好悠闲的性格的形成有着很重要的影响。道家的重要思想是戒过度，性格胜于事业，静胜于动。一个人能不受祸福的扰动，才能获得内心的宁静。道教哲学家淮南子曾写过一篇很有名的寓言，名叫《塞翁失马》。

　　近塞上之人，有善术者，马无敌亡而入胡。人皆吊之，其父曰："此何遽不为福乎？"居数月，其马将胡骏马而归。人皆贺之，其父曰："此何遽不能为祸乎？"家富良马，其子好骑，堕而折其髀。人皆吊之，其父曰："此何遽不为福乎？"居一年，胡人大入塞，丁壮者引弦而战。近塞之人，死者十九。此独以跛之故，父子相保。

　　显而易见，这种哲学，使人能够忍受一些折磨而不烦恼，他相信祸福是相连的，正如古钱必有正反面一样。这种哲学能使人得到宁静，不喜忙劳，淡于名利。这种哲学是说："你以为不要紧，便什么都不要紧了。"成功的欲望和失败的恐惧，两者是差不多的东西，有了这个聪明的意念，成功的欲望就不会太热切了。一个人的事业越是成功也越怕失败。不可捉摸的功名报酬及不上隐晦所得的利益。在道家看来，有识之士在成功时是不以为自己成功的，在失败时也不以为自己是失败。只有一知半解的人才把外表的成功和失败当做绝对真实的。

佛道二家的区别在于佛家的意念是要一个人与世无争，道家的意念却相反，要一个人根本不为世人所求。世上最快乐的人，也就是不被世人所求的无忧无虑的人。道家最有名最有才智的哲学家庄子时常告诫我们，不要太著名，也不可太有用。太肥的猪要被人杀死，去供神；羽毛太美丽的飞禽，易遭猎户的注意。他又说了一个譬喻：说两个人协同去掘坟，偷窃死人所穿戴的衣物，为了要得到死人口中所含着的珍珠，竟连死人的头颅连同颊骨和下腭都用铁锤敲碎了。

为什么不去过悠闲的生活呢？这是这些哲学理论的必然结论。

V. *What Is Luck*

THE peculiar contribution of Taoism to the creation of the idle temperament lies in the recognition that there are no such things as luck and adversity. The great Taoist teaching is the emphasis on being over doing, character over achievement, and calm over action. But inner calm is possible only when man is not disturbed by the vicissitudes of fortune. The great Taoist philosopher Liehtse gave the famous parable of the Old Man At the Fort:

An Old Man was living with his Son at an abandoned fort on the top of a hill, and one day he lost a horse. The neighbors came to express their sympathy for this misfortune, and the Old Man asked, "How do you know this is bad luck?" A few days afterwards, his horse returned with a number of wild horses, and his neighbors came again to congratulate him on this stroke of fortune, and the Old Man replied, "How do you know this is good luck ?" With so many horses around, his son began to take to riding, and one day he broke his leg. Again the neighbors came around to express their sympathy, and the Old Man replied, "How do you know this is bad luck?" The next year, there was a war, and because the Old Man's son was crippled, he did not have to go to the front.

Evidently this kind of philosophy enables a man to stand a few hard knocks in life in the belief that there are no such things as hard knocks without advantages. Like medals, they always have a reverse side. The possibility of calm, the distaste for mere action and bustle, and the running away from success and achievement are possible with this kind of philosophy, a philosophy which says: Nothing matters to a man who says nothing matters. The desire for success is killed by the shrewd hunch that the desire for success means very much the same thing as the fear of failure. The greater success a man has made, the more he fears a climb down. The illusive rewards of fame are pitched against the tremendous advantages of obscurity. From the Taoist point of view, an educated man is one who believes he has not succeeded when he has, but is not so sure he has failed when he fails, while the mark of the half-educated man is his assumptions that his outward successes and failures are absolute and real.

Hence, the distinction between Buddhism and Taoism is this: the goal of the Buddhist is that he shall not want anything, while the goal of the Taoist is that he shall not be wanted at all. Only he who is not wanted by the public can be a carefree individual, and only he who is a carefree individual can be a happy human being. In this spirit Chuangtse, the greatest and most gifted among the Taoist philosophers, continually warns us against being too prominent, too useful and too serviceable. Pigs are killed and offered on the sacrificial altar when they become too fat, and beautiful birds are the first to be shot by the hunter for their beautiful plumage. In this sense, he told the parable of two men going to desecrate a tomb and robbing the corpse. They hammer the corpse's forehead, break his cheekbones and smash his jaws, all because the dead man was foolish enough to be buried with a pearl in his mouth.

The inevitable conclusion of all this philosophizing is: why not loaf?

六. 美国三大恶习

　　"一个人以为不要紧，就什么都不要紧了。"这种中国人所特有的美妙的观念，同美国人的观念形成了奇特的对比。人生真的是要麻烦到"心为形役"的境地吗？这种观念被悠闲哲学的崇高精神排斥。在一家工程公司的广告上，我曾看到一条大字标题："差不多正确是还不够的。"这是我们所见到的最特殊的一张广告。求全的欲望已近于淫。美国人的烦恼也就是一定要把已经近乎正确的东西弄得更正确些；而中国人以为近乎正确已经是够好的了。

　　讲求效率，讲求准时，及希望事业成功，似乎是美国的三大恶习。美国人所以那么不快乐，那么神经过敏，是因为这三种东西在作祟。于是享受悠闲生活的天赋权利被剥夺了，许多闲逸的、美丽的、可爱的下午被他们错过了。一个人第一步应相信世界上并无灾难，也应相信"把事情放着不做"比"把事情做好"更要高尚。大体上说，一个人在接信后马上写回信，结果是好坏各居其半。如果不写回信，虽然也许会错过几次良好的约会，但也会避免几次不欢而散的约会。假如把搁置在抽屉里已三个月的信件拆开来看一下，觉得多数的信是无须答复的；三个月后再拿起来看，那么竟或觉得全无答复的必要了，答复只是把光阴浪费掉。写信实也可以变成一种罪恶，它使写信者变成推销货品的优等掮客，能使大学教授变成有效率的商业经理。在此种意义上，对那些时常上邮局的美国人抱轻视心理的梭罗，使我颇能了解他。

讲求效率能够把事情做完，而且做得甚是良好，这是毋庸争论的。我老是不喜欢用中国的自来水龙头而喜欢美国制造的，那也是一种安慰，因为美国所制的自来水龙头不漏水。可是我们对大家"必须有用，必须有效率，必须做官，必须掌握大权"的这个旧观念，我们回答："世界上自有许多傻子，他们愿意做有用的人，不怕烦恼，劳碌终日，喜欢掌握大权，而自会将一切事业都办好的。"紧要的问题却是：是谁比较聪明——悠闲者，还是劳碌者呢？我们不赞成讲求效率是因为讲求效率太费功夫，为了想把事情做得十全十美，连享受悠闲的乐趣也失掉，连神经也跟着损坏了。美国有一个杂志编辑，为了要严密校正错字，就连头发也校得灰白。中国的编辑便聪明得多，他把几个没校出来的错字留下，以便增加读者发现错误的乐趣，增加读者细心观察的能力。不但这样，中国杂志上都是按期连载一篇小说，登了几期之后，便突然失踪，而读者和编者也就淡忘了，这在美国，那编辑或许因此会大受攻击，但中国的编辑是没有关系的，仅仅是因为没有关系而已。美国工程师在建设桥梁时，核算准确，两端的接榫点，一寸的十分之一也不会相差。要是两个中国工人在山的两面分掘山洞，结果会掘成两个进口，两个出口。只要山洞掘得出，中国人就觉得是没有关系的，有两个山洞反而可以筑双轨铁道了。并不匆忙的话，两个和一个是没有关系的，山洞总是山洞，掘也算掘了，工作也算完毕了，要是火车能够行走如常，那也就算不错了。中国人也极守时，不过须给予他们充足的工作时间。只要这规定的时间够长，那么他们总能把一份工作按照规定时间做完。

在现代工业生活的速度下，我们没法享受这种伟大的悠闲生活。何况，现在拿钟来计时，使每个人的脑中对于时间这件物事印下一种特异的观念，以致连我们聪明的人类也变成了钟。这种情形自然会传到中国。譬如一家雇用两万个工人的工厂，如若全数的工人都依着各人兴趣随随便便依着自己的时刻进厂做工，这情形岂不要变得非常可怕？于是这种按时按刻的上工规则出来，造成了生活之所以那样困苦，那样紧张。一个人如要在下午五时准时到达某地，结果连五时以前所有的时候都会因此牺牲在预备这件事上。在美国，几乎每个成人都参照小学生上课的方式去决定他自己的工作时刻——三时做这件事，五时做另一件事，六时三十分换衬衣，六时五十分上汽车，七时到达旅馆。这样一来，生活险乎失掉了它的重要价值了。

美国人过于注意安排时间，已使这件事渐臻于凄惨之境。他们不但把明天的工作时刻预先排定，不但把下星期的工作时刻完全排定，并且连下一个月的工作时刻也完全排好，甚至三星期后的一个约会时刻也会预先排定，这似乎太过分了一些。一个中国人接到他朋友一张请帖时，不必答复他的朋友到或不到，如在请客名单上写一个"到"字，即表示要来，不来呢，即写上一个"谢"字，这样就算了事，可是另有多数被邀者都直截了当地写上一个"知"字，意思即是已经知道，来不来不一定。一个即将离开上海的美国人或欧洲人，他会很有把握地告诉我说，他将在一九三八年四月十九日下半天三时正，在法国巴黎参加一个委员会议，之后，又将在五月二十一日乘早班七时的火车直达奥地利维也纳。假如我们要将一个人下午判处死刑，难道一定将行刑期宣布得这样早吗？一个人既然做了自己的主人翁，难道不能随着他的趣味旅行，任着自己的意思来去吗？但是美国人之所以不懂悠闲，还有一个更重要的原因：他们做事如上所述情趣太高，把工作看得高于生存，比生存来得紧要。世界上一切出名的艺术，大家都一定要求要有一个名副其实的特性，我们的生活同样该要求他具有一种特性。但特性这种奇妙的东西是跟酒的醇熟一样的，它必须要静止着不动，还需要经过一个相当长的时间，并不是马上就可以制造出来的。在东方人的心目中，一概都觉得美国男女老少十分好笑，因为他们渴望工作，用尽方法来获得宝贵的自尊心，使年轻一代尊敬。其实老年人做工作，正如在教堂上装设播音机播送爵士音乐的节目罢了。老人家做了一辈子还不够吗？难道他们一定要永远做工作吗？壮年不悠闲已经是很糟糕的了，若到了老年再不优游岁月享享清福，这真是人类天性上的一种罪恶。

特性常和那些古旧的事物，那些依靠时间去生长的事物保持着密切的联系，特性在形成中的标识很多，人到中年时，面孔上一些美丽的线条，就是这标识的表现。但特性在每个人都把旧型汽车去贴换新型式汽车的那种生活方式中，是很难找到的。我们对于自身的好坏正和我们对所造的物事一般，随着时间而变换。在一九三七年，我们男女都是一九三七年式样，到了第二年，每个人又都具有第二年的式样了。古教堂、旧式家具、版子很老的字典以及古版的书籍，我们是喜欢的，但大多数的人都忘却了老年人的美。这种美值得我们欣赏，在生活上十分需要。我以为古老的东西，圆满的东西，饱经世变的东西才是美的。

有一些时候，我会发生一种先知式的幻觉，幻想在一千年之后，纽约曼哈顿市区的住户都变成了行动缓慢者，美国的"进取者"（Go-getter）都成了东方式的悠闲人。美国的绅士们或许都披上了长袍，着上了拖鞋，要是学不会像中国人的模样将两手缩在袖中呢，就将两手插在裤袋内，在百老汇大街上踱方步。十字路口的警察同踱方步的人搭讪，车水马龙的马路中，开车者相遇，大家来寒暄一番，互问他们祖母的健康。有人在他店门口刷牙，却一边叨叨地和他邻人谈笑，偶然还有个自称满腹经纶的学者踉踉跄跄地走路，袖子里塞着一本连角都卷起来的烂书。速食餐厅的柜台拆除了，自动饮食店里低矮而有弹力的安乐椅子增多了，以供来宾的休息。有一些人则会到咖啡店坐上一个下午，半个钟头才喝完杯橘汁，喝酒也不再是一口气地灌上一大杯，而是沾唇细酌，品味谈天，体会其中无穷的乐趣。病人挂号的办法取消了，"急症室"也废除掉，病人同医生可以讨论他们的人生哲学。救火车变得像蜗牛那样地笨，慢慢地爬着，这时救火队员停下来看空中飞雁，为了它们的数目而争执。这种快乐的时代可惜在纽约曼哈顿市区没有实现的希望。一旦能实现，人们一定可以尽情享受更完美的悠闲下午了。

VI. *Three American Vices*

TO the Chinese, therefore, with the fine philosophy that"Nothing matters to a man who says nothing matters,"Americans offer a strange contrast. Is life really worth all the bother, to the extent of making our soul a slave to the body? The high spirituality of the philosophy of loafing forbids it. The most characteristic advertisement I ever saw was one by an engineering firm with the big words: "Nearly Right Is Not Enough."The desire for one hundred percent efficiency seems almost obscene. The trouble with Americans is that when a thing is nearly right, they want to make it still better, while for a Chinese, nearly right is good enough.

The three great American vices seem to be efficiency, punctuality and the desire for achievement and success. They are the things that make the Americans so unhappy and so nervous. They steal from them their inalienable right of loafing and cheat them of many a good, idle and beautiful afternoon. One must start out with a belief that there are no catastrophes in this world, and that besides the noble art of getting things done, there is a nobler art of leaving things undone. On the whole, if one answers letters promptly, the result is about as good or as bad as if he had never answered them at all. After all, nothing happens, and while one may have missed a few good appointments, one may have also avoided a few unpleasant ones. Most of the letters are not worth answering, if you keep them in your drawer for three months; reading them three months afterwards, one might realize how utterly futile and what a waste of time it would have been to answer them all. Writing letters really can become a vice. It turns our writers into fine promotion salesmen and our college professors into good efficient business executives. In this sense, I can understand Thoreau's contempt for the American who always goes to the post office.

Our quarrel is not that efficiency gets things done and very well done, too. I always rely on American water-taps, rather than on those made in China, because American water-taps do not leak. That is a consolation. Against the old contention, however, that we must all be useful, be efficient, become officials and have power, the old reply is that there are always enough fools left in the world who are willing to be useful, be busy and enjoy power, and so somehow the business of life can and will be carried on. The only point is who are the wise, the loafers or the hustlers? Our quarrel with efficiency is not that it gets things done, but that it is a thief of time when it leaves us no leisure to enjoy ourselves and that it frays our nerves in trying to get things done perfectly. An American editor worries his hair gray to see that no typographical mistakes appear on the pages of his magazine. The Chinese editor is wiser than that. He wants to leave his readers the supreme satisfaction of discovering a few typographical mistakes for themselves. More than that, a Chinese magazine can begin printing serial fiction and forget about it halfway. In America it might bring the roof down on the editors but in China it doesn't matter, simply because it doesn't matter. American engineers in building bridges calculate so finely and exactly as to make the two ends come together within one-tenth of an inch. But when two Chinese begin to dig a tunnel from both sides of a mountain, both come out on the other side. The Chinese's firm conviction is that it doesn't matter so long as a tunnel is dug through, and if we have two instead of one, why, we have a double track to boot. Provided you are not in a hurry, two tunnels are as good as one, dug somehow, finished somehow and if the train can get through somehow. And the Chinese are extremely punctual, provided you give them plenty of time to do a thing. They always finish a thing on schedule, provided the schedule is long enough.

The tempo of modern industrial life forbids this kind of glorious and magnificent idling. But worse than that, it imposes upon us a different conception of time as measured by the clock, and eventually turns the human being into a clock himself. This sort of thing is bound to come to China, as is evident, for instance in a factory of twenty thousand workers. The luxurious prospect of twenty thousand workers coming in at their own sweet pleasure at all hours is, of course, somewhat terrifying. Nevertheless, this is what makes life so hard and hectic. A man who has to be punctually at a certain place at five o'clock has the whole afternoon from one to five ruined for him already. Every American adult is arranging his time on the pattern of the schoolboy—three o'clock for this, five o'clock for that, six-thirty for change of dress; six-fifty for entering the taxi and seven o'clock for emerging into a hotel room. It just makes life not worth living.

And Americans have now come to such a sad state that they are booked up not only for the following day, or the following week, but even for the following month. An appointment three weeks ahead of time is a thing unknown in China. And when a Chinese receives an invitation card, happily he never has to say whether he is going to be present or not. He can put down on the invitation list"coming"if he accepts, or"thanks"if he declines, but in the majority of cases the invited party merely writes the word"know,"which is a statement of fact that he knows of the invitation and not a statement of intention. An American or a European leaving Shanghai can tell me that he is going to attend a committee meeting in Paris on April 19,1938, at three o'clock and that he will be arriving in Vienna on May 21st by the seven o'clock train. If an afternoon is to be condemned and executed, must we announce its execution so early? Cannot a fellow travel and be lord of himself, arriving when he likes and taking departure when he likes? But above all, the American's inability to loaf comes directly from his desire for doing things and in his placing action above being. We should demand that there be character in our lives as we demand there be character in all great art worthy of the name. Unfortunately, character is not a thing which can be manufactured overnight. Like the quality of mellowness in wine, it is acquired by standing still and by the passage of time. The desire of American old men and women for action, trying in this way to gain their self-respect and the respect of the younger generation, is what makes them look so ridiculous to an Oriental. Too much action in an old man is like a broadcast of jazz music from a megaphone on top of an old cathedral. Is it not sufficient that the old people are something? Is it necessary that they must be forever doing something? The loss of the capacity for loafing is bad enough in men of middle age, but the same loss in old age is a crime committed against human nature.

Character is always associated with something old and takes time to grow, like the beautiful facial lines of a man in middle age, lines that are the steady imprint of the man's evolving character. It is somewhat difficult to see character in a type of life where every man is throwing away his last year's car and trading it in for the new model. As are the things we make, so are we ourselves. In 1937 every man and woman look 1937, and in 1938 every man and woman will look 1938. We love old cathedrals, old furniture, old silver, old dictionaries and old prints, but we have entirely forgotten about the beauty of old men. I think an appreciation of that kind of beauty is essential to our life, for beauty, it seems to me, is what is old and mellow and well-smoked.

Sometimes a prophetic vision comes to me, a beautiful vision of a millennium when Manhattan will go slow, and when the American"go-getter"will become an Oriental loafer. American gentlemen will float in skirts and slippers and amble on the sidewalks of Broadway with their hands in their pockets, if not with both hands stuck in their sleeves in the Chinese fashion. Policemen will exchange a word of greeting with the slow-devil at the crossings, and the drivers themselves will stop and accost each other and inquire after their grandmothers' health in the midst of traffic. Some one will be brushing his teeth outside his shopfront, talking the while placidly with his neighbors, and once in a while, an absent-minded scholar will sail by with a limp volume rolled up and tucked away in his sleeve. Lunch counters will be abolished, and people will be lolling and lounging in soft, low armchairs in an Automat, while others will have learned the art of killing a whole afternoon in some cafe. A glass of orange juice will last half an hour, and people will learn to sip wine by slow mouthfuls, punctuated by delightful, chatty remarks, instead of swallowing it at a gulp. Registration in a hospital will be abolished, "emergency wards" will be unknown, and patients will exchange their philosophy with their doctors. Fire engines will proceed at a snail's pace, their staff stopping on the way to gaze at and dispute over the number of passing wild geese in the sky. It is too bad that there is no hope of this kind of a millennium on Manhattan ever being realized. There might be so many more perfect idle afternoons.

/ Chapter Eight   **The Enjoyment of the Home** /

第八章

家庭之乐

一. 趋近生物观念

依我看来，不论哪一种文明，它的最后测验即是它能产生何种形式的夫妻父母。除了这个严峻而又简单的问题之外，文明的他种成就，如：艺术、哲学、文学和实际生存，都退到无关重要的地位。

对于中国费尽心力以东西文明做比较的人们，我每每用这句话给他们当做一服清凉剂，并且极有效验，这是使我很得意的。研究西方生活和学术的学生，不论远渡重洋或在本国做研究，他们对于西方的灿烂成就，从医学、地质学、天文学，到摩天的大厦、优美的汽车公路和天然色彩的照相机，自然觉得目迷五色，极可惊异。他们必转着热烈羡慕或自惭不如他人的念头，也许两样念头都有，于是一种反抗自卑的意念油然而生，能使他不知不觉地努力替东方文明辩护，甚至斥摩天的大厦和优美的汽车公路为无用之物。——不过我还没有听见过斥照相机为废物的话——这种状态是很可怜的，使他失去了合理地和旁观地衡量东西两方优劣的资格。他在这种被自惭不如别人的思想所烦扰和炫惑的时候，实应给他一个定心丸，使他的心平静下来。

我所建议的这种测验，能扫除文明和文化中的一切不必要的事物，而有使人类归于平等，将一切人类都置在一个简单又明白的方程式之下的奇效。于是文明的其他一切成就都可被认为是促进产生优良夫妻父母的方法。人类之中，百分之九十有夫妻关系，百分之百是人子，而婚姻和家庭确是人类生活中最亲密的部分，所以能产生优良夫妻父母的文明，实造成一种较快乐的人类生活，因此是一种较高级的文明。这是很明显的。和我们同居的男子或女人的素质，较之他们完成的工作重要得多。所以凡是女子，也应对可以给她一个较好的丈夫之文明表示感激。这种事物都是相对成就的，因此理想的夫妻父母无时无地不有之。欲有优良夫妻父母的最好方法或者是优生学，可以使我们节省许多教导他们的辛劳。反之，凡是轻视家庭或揿之于低下地位的文明，往往产生较为低劣的子女。

我承认我渐渐趋于生物主义。但我本属于生物，世上男女也都属于生物，不管我们是否愿意，终免不了是个生物，所以趋于生物主义那句话，其实也是多说的。我们因生物性而快乐，因生物性而发怒，因生物性而有志愿，因生物性而信神或爱好和平，虽然我们自己或许还没有觉得是如此。我们既是生物，自不能逃避出生、吃母奶、婚嫁和生育等事。每个男人都是妇人所生，每个男人（除了少数之外）都须和一个妇人共过一生去做小孩的父亲。每个女人也都是妇人所生，每个妇人（除了少数之外）也都须和一个男人共过一生，生育小孩。中间也有几个不愿意做父母，这等于花木之不肯生子以传它们的种；但是没有一个人能不要父母而生，也正如花木之不能不要种子而生。因此我们就得到生命中最紧要的相互关系，就是男人、女人、小孩三者之间的相互关系那桩事实，而生命哲学除非是讨论这个必须的相互关系，即不能称为适当的哲学，或不成其为哲学。

但单是男女之间的关系还嫌不够。这关系必须生出婴孩，否则便不能称为完备。所以无论哪一代的文明，绝无理由剥夺男人女人有婴孩的权利。我知道目前曾发生一个真正难题，有许多男女不肯结婚，另有许多人虽结婚，但因这样或那样理由不肯有婴孩。据我的意见，不论他们所持的是何种理由，凡是男女不遗留子女而离开这世界，实在是犯了一件对于自身的大罪。如若他们的不生育是因为身体关系，那么他们的身体已是退化或有差错。如若是因为婚姻的程度过高，那么这过高的婚姻程度就有不合理的地方。如若是因为一种谬误的个人主义哲学，那么个人主义哲学必是错的。最后，如若是因为整个的社会组织，那么这整个社会组织是不对的。待到二十一世纪，我们对于生物科学已有较高的认识，能更了解我们之为生物时，男女们大概就会见到这个真理。我深信二十世纪将为生物学世纪，正如十九世纪之为自然科学世纪。等到人们更能了解自己，而觉悟到对于造化所赋予的天性即使争斗也是徒然时，他们就会更加重视这类简单智慧。从瑞士心理学家荣格劝告有钱的病人回到乡间去饲养鸡、鸭、小孩和栽种萝卜那件事，我们看到这种生物学的和医学的智慧已有发展的征兆。这类有钱的女性病人，她所犯的弊病就在未能顺着生物性发挥本能，或是她们的发挥程度过于低下。

自有历史以来，人们从来没有学着去和女人共同生活。最奇怪的事是，尽管如此，从来没有一个人能完全脱离女人而生活。一个人如觉悟他绝不能没有母亲而生到这世界上来时，便不会轻蔑女人。从出生到死亡，他的四围没有一天没有女人，如母亲、妻子、女儿等。即使他不娶亲，也免不了和诗人华兹华斯一般依赖他妹妹过日子，或和诗人斯本塞一般依赖他的管家婆。没有一种哲学能拯救他的灵魂，如同他不能和母亲姐妹们建立相当的关系，如若他甚至不能和"管家婆"建立适当关系，那么他简直不能算人。

凡是未能和女人达到适当关系，而又走着道德歧路的人，如王尔德之类，实在有些可怜。他们一面喊着男人万难和女人共同生活，但一方面又说男人不能无女人而生活。这样看来，由一个印度故事的著者，直到二十世纪的王尔德，中间虽已经过四千余年，但是人类的智慧好似没有分寸进步。因为那印度著者正抱着和王尔德同样的心理。据这本印度故事所载，上帝创造女人时，系采取花的美丽，鸟的歌音，虹霓的彩色，风的柔态，水的笑容，羊的温柔，狐的狡猾，云的难于捉摸和雨的变幻无常，将它们交织成一个女人，而拿她送给男人做妻子。印度亚当很快乐，他俩便在这美丽的世界中自在游行。几天之后，印度亚当跑去向上帝说："请你将这女人带走，我实在不能和她过下去了。"上帝答应他的请求，将夏娃带了回去。于是亚当觉得很寂寞，依旧不能快乐。几天之后，他又到上帝那里说："你所创造的这个夏娃，仍请你收了回去，我发誓不能和她过下去。"上帝于无限智慧之中仍然顺从了他。等到亚当第四次走来说没有了那个女伴不能生活时，上帝虽允了他的请求，但要他答应以后绝不改变心肠，不论甘苦，以后都和她永远过下去，尽他俩的智力在这个世上共度生活。我以为这幅景象，直到如今并没有什么根本改变。

I. On Getting Biological

IT has seemed to me that the final test of any civilization is, what type of husbands and wives and fathers and mothers does it turn out? Besides the austere simplicity of such a question, every other achievement of civilization–art, philosophy, literature and material living–pales into insignificance.

This is a dose of soothing medicine that I have always given to my countrymen engaged in the head-racking task of comparing Chinese and Western civilizations, and it has become a trick with me, for the medicine always works. It is natural that the Chinese student of Western life and learning, whether in China or studying abroad, is dazzled by the brilliant achievements of the West, from medicine, geology, astronomy to tall skyscrapers, beautiful motor highways and natural-color cameras. He is either enthusiastic about such achievements, or ashamed of China for not having made such achievements, or both. An inferiority complex sets in, and in the next moment you may find him the most arrogant, chauvinistic defender of the Oriental civilization, without knowing what he is talking about. Probably as a gesture, he will condemn the tall skyscrapers and the beautiful motor highways, although I haven't yet found one that condemns a good camera. His plight is somewhat pathetic, for that disqualifies him for judging the East and the West sanely and dispassionately. Perplexed and dazzled and harrassed by such thoughts of inferiority, he has great need of what the Chinese call a medicine for"calming the heart"to allay his fever.

The suggestion of such a test as I propose has the strange effect of leveling all mankind by brushing aside all the non-essentials of civilization and culture and   bringing all under a simple and clear equation. All the other achievements of civilization are then seen as merely means toward the end of turning out better husbands and wives and fathers and mothers. Insofar as ninety per cent of mankind are husbands or wives and one hundred per cent have parents, and insofar as marriage and the home constitute the most intimate side of a man's life, it is clear that that civilization which produces better wives and husbands and fathers and mothers makes for a happier human life, and is therefore a higher type of civilization. The quality of men and women we live with is much more important than the work they achieve, and every girl ought to be grateful for any civilization that can present her with a better husband. Such things are relative, and ideal husbands and wives and fathers and mothers are to be found in every age and country. Probably the best way to get good husbands and wives is by eugenics, which saves us a great deal of trouble in educating wives and husbands. On the other hand, a civilization which ignores the home or relegates it to a minor position is apt to turn out poorer products.

I realize that I am getting biological. I am biological, and so is every man and woman. There is no use saying,"Let's get biological,"because we are so whether we like it or not. Every man is happy biologically, angry biologically, or ambitious biologically, or religious or peace-loving biologically, although he may not be aware of it. As biological beings, there is no getting around the fact that we are all born as babies, suck at mothers' breasts and marry and give birth to other babies. Every man is born of a woman, and almost every man lives with a woman through life and is the father of boys and girls, and every woman is also born of a woman, and almost every woman lives with a man for life and gives birth to other children. Some have refused to become parents, like trees and flowers that refuse to produce seeds to perpetuate their own species, but no man can refuse to have parents, as no tree can refuse to grow from a seed. So then we come to the basic fact that the most primary relationship in life is the relationship between man and woman and the child, and no philosophy of life can be called adequate or even called philosophy at all unless it deals with this essential relationship.

But the mere relationship between man and woman is not sufficient; the relationship must result in babies, or it is incomplete. No civilization has any excuse for depriving a man or woman of his or her right to have babies. I understand that this is a very real problem at present, that there are many men and women today who don't get married, and many others who, after getting married, refuse to have babies for one reason or another. My point of view is, whatever the reason may be, the fact of a man or woman leaving this world without children is the greatest crime he or she can commit against himself or herself. If sterility is due to the body, then the body is degenerate and wrong; if it is due to the high cost of living, then the high cost of living is wrong; if it is due to a too high standard of marriage, then the too high standard of marriage is wrong; if it is due to a false philosophy of individualism, then the philosophy of individualism is wrong; and if it is due to the entire fabric of social system, then the entire fabric of social system is wrong. Perhaps men and women of the twenty-first century will come to see this truth when we have made better progress in the science of biology and there is a better understanding of ourselves as biological beings. I am quite convinced that the twentieth century will be the century of biology, as the nineteenth century was the century of comparative natural science. When man comes to understand himself better and realizes the futility of warring against his own instincts, with which nature has endowed him, man will appreciate more such simple wisdom. We see already signs of this growing biological and medical wisdom, when we hear the Swiss psychologist Jung advise his rich women patients to go back to the country and raise chickens, children and carrots. The trouble with rich women patients is that they are not functioning biologically, or their biological functioning is disgracefully low-grade.

Man has not learned to live with woman, since history began. The strange thing is that no man has lived without a woman, in spite of that fact. No man can speak disparagingly of woman if he realizes that no one has come into this world without a mother. From birth to death, he is surrounded by women, as mother, wife and daughters, and even if he does not marry, he has still to depend on his sister, like William Wordsworth, or depend on his housekeeper, like Herbert Spencer. No fine philosophy is going to save his soul if he cannot establish a proper relationship with his mother or his sister, and if he cannot establish a proper relationship even with his housekeeper, may God have pity on him!

There is a certain pathos in a man who has not arrived at a proper relationship with woman and who has led a warped moral life, like Oscar Wilde, who still exclaims, "Man cannot live with a woman, nor can he live without her!" So that it seems human wisdom has not progressed an inch farther between the writer of a Hindu tale and Oscar Wilde at the beginning of the twentieth century, for that writer of the Hindu tale of the Creation expressed essentially the same thought four thousand years ago. According to this story of the Creation, in creating woman, God took of the beauty of the flowers, the song of the birds, the colors of the rainbow, the kiss of the breeze, the laughter of the waves, the gentleness of the lamb, the cunning of the fox, the waywardness of the clouds and the fickleness of the shower, and wove them into a female being and presented her to man as his wife. And the Hindu Adam was happy and he and his wife roved about on the beautiful earth. After a few days, Adam came to God and said, "Take this woman away from me, for I cannot live with her." And God listened to his request and took Eve away. Adam then became lonely and was still unhappy, and after a few days he came to God again and said, "Give me back my woman, for I cannot live without her." Again God listened to his request and returned him Eve. After a few days again, Adam came to God and asked, "Please take back this Eve that Thou has created, for I swear I cannot live with her." In His infinite wisdom God again consented. When finally Adam came a fourth time and complained that he could not live without his female companion, God made him promise that he was not going to change his mind again and that he was going to throw in his lot with her, for better and for worse, and live together on this earth as best they knew how. I do not think the picture has essentially changed much, even today.

二. 独身主义——文明的畸形产物

采取这种简单而自然的生物性观点，包含两种冲突：第一，个人主义和家庭的冲突；第二，富有智力的无生殖哲学和天性的较有热情的哲学的冲突。因为个人主义和崇拜智力往往能蒙蔽一个人，使他看不见家庭生活之美丽。两者比较起来，尤以后者为更可恶。一个相信个人主义者向着它的合理后果而进行，尚不失为一个具有理解力的生物。但专一相信冷静头脑，而毫不知有热情心肠者，简直是个呆子。因为家庭的集体性，就其为一个社会单位而论，尚有可以替代的物事，但是配偶天性和父母天性之失灭，是无从弥补的。

在这起点，我们不能不假定人类不能单独无伴地生活于这世界而得到快乐，必须和近旁的一群人做伴，而成一个范围较大的我。这个我的范围并不限于本身身体轮廓之内，而实在是向外伸展到他的心灵和社会活动所以能达到之处。不论在哪一个时代或国度里，不论在什么政体下，一个人所真正爱好的生活绝不和当时的国家或时代同其广泛，而必仅限于他所熟识的人和所做的活动那个较小的范围之内。此即所谓较大的我。他生活、活动于这个社会单位之内，而为其中的一个生物。这种单位可以是一个教区，一个学堂，一个监狱，一家商店，一个秘密社会或一个慈善机关。这类单位有时可以替代家庭，甚至完全取而代之。宗教本身或一场广大的政治运动有时也可以占尽一个人的心力，使他抛弃一切。但在这许多团体中，仍只有家庭是自然的、具有生物性的实在的、可以使人们满意而有意义的生存单位，因为人们在出生时即已置身于家庭之中，并且将终身如此，所以家庭于他是自然的。因为嫡血的关系，导致人们对于较大的我实在是一件看得见的实物的观念，所以家庭是生物性实在的。一个人如不能在这个自然的团体生活中获得成功，则他在以外的团体生活中，大概也难于期望有所成就。孔子说："弟子入则孝，出则悌，谨而信，泛爱众，而亲仁。行有余力，则以学文。"离了这个人所视为重要的团体生活之外，人们必须有一个相当的异性分子和谐地辅佐他，方能使他有完美的表现，完美的尽职，将他的个性发展到最高的程度。

女人具有比男人更深的生物性感觉，所以很明了这一点的中国女孩儿都潜意识地羡慕红裙花轿；西方的女孩儿也同样地羡慕结婚网纱和结婚钟。大自然所赋予女人的母性根深蒂固，所以不易于被人造的文明毁灭。我毫不怀疑地相信大自然创造女人，尤其期望她做母亲，而不仅做一个配偶，因此所赋予的心灵和道德本质，都是诱导之于母亲任务之类，而于母性中获得其真正解释和和谐，如现实主义、判断力，遇事不厌求详，怜爱弱小，乐于施助，较强烈的兽性爱憎，较厉害的个人偏见和感情用事和对事物的一般的个人眼光。所以哲学如果离弃了大自然的本意，不计及这个母性（即女人整个生存之具有支配力的特点和中心解释）而要想使女人快乐，实已走入歧途。因此，在未受教育和受过合理教育的女人中，这个母性是从不强自压制的，它萌芽于儿童时期，渐渐强盛而达到充足于成熟时期。但在男人中，这个父性在三十五岁之前大概都隐而不显，或至少须等到子女已经五岁方能感觉。我想二十五岁的少年大概不会想到将要做父亲的事。这时他只知道爱上一个女子，无意之间生下一个孩子就丢开了，等他的妻子去一心照顾。总要到三十岁之后，才能一旦觉得自己已有了一个可以携带到公共场所炫耀于人前的孩子而感觉到他的父性。二十余岁的少年对于孩子的观念大多视为有些可笑，但除了觉得有些可笑之外，便不再加以思索。至于一个有了孩子或将有孩子的女人，这就成为她一生中最严重的一桩事情，甚至变更她的整个生命，变换她的性情和嗜好。女人一到怀孕将产，便似进了另一个世界，从此她能认清自己生命的使命和到世上生存的目的，而毫无疑惑。她知道有人需要她，所以即发挥她的效能。我曾看见过最娇生惯养的中国富家女郎，于她的小孩病中变为异常伟大，目不交睫地整个月服侍下去。在大自然的配合中，无须如此的父性，所以并不给他，因为男人也好似雄鸭雄鹅一般，除了种子之外，对于子嗣的其他事情均毫不关心。所以一个女人如若生命的中心主动力得不到表现和发挥的机会，即在心理上受到最大的痛苦。美国容忍那么许多很可爱的女人无辜地失去嫁人的机会，因此，如有人向我称赞美国的文明对于女子是怎样仁慈，我简直不相信。

我相信美国婚姻所以调整失当，大多是由于这类女人的母性和男人的父性参差过甚所致。美国人的所谓情感不成熟性，除了这个生物性事实外，没有其他的解释。男人因在青年时代过惯了过于放浪的生活，这种社会制度使他们缺乏负责思想的天然节制，然而女人因了较深大的母性，仍是具有的。大自然如若未曾赋予女人以充分的镇静性去应付将做母亲的心理预备，事情就将不可收拾。所以大自然就如此做去。贫穷人家的子弟由于艰苦的环境，已将负责思想深印于脑筋之中，只剩下那些生活放浪的富家儿郎，在崇拜青春和纵容青年的国度里，于理想的情形中发展成为情感上和社会上的"低能儿"。

说来说去，我们所关切的实在只是如何去度一个快乐生活的问题。一个人除非在外部生活的浅薄成熟之余，能触动内心个性的发条，使它得到合于常规的发挥之外，别无求得生活快乐之道。独身主义在个人事业的形式上成为理想目标时，不但带着个人主义的色彩，并也带着愚拙的智力主义色彩。因了后者的理由，这种独身主义应由我们唾弃。我常疑心立誓不娶不嫁的男女，由于已经变成无用的智力主义者，不肯更变心肠，他们都已被外部的成就所蒙蔽，误信他们以属于人类而言，能从家庭之替代物中得到快乐，或从能使他们满意的智力艺术或职业兴趣中得到快乐。

我认为他们是错误的。这种个人主义的现象：不婚嫁，无子息，拟从事业和个人成熟之中寻求充足满意生活的替代物和阻止虐待牲畜，在我看来，都是很愚笨可笑的。在心理方面而言，这颇仿佛几个老处女因在马戏团里边看见老虎背上有几条鞭痕，引起疑心，而拟控诉马戏团老板虐待老虎一般。这种抗议用非其地，是母性的畸形发挥。试想真正的老虎会在乎打几鞭子吗？这种老处女是在盲目地摸索一个生命中的位置，而又自以为是地想旁人承认她们为合理。

政治文学和艺术的成熟所给予成功者的报酬，不过是些空心的智力上的喜悦，但眼看自己的儿女长成人，其愉快出于衷心而何等实在。著作家和艺术家，有几个能在老年时对于自己的作品感觉满意？其中大多数无非视之为消遣中的偶然产物，或借以维持生活的工作而已。据说赫伯特·斯宾塞（Herbert Spencer，英国著名的"社会达尔文主义之父"）在临终的前几天，将他所著的《综合哲学》（The Synthelic Philosophy）十八巨册放在膝上，当他觉到其沉重时，颇有这分量如若换上一个孙儿岂不更好的感触。聪明的伊里亚岂不是愿意将他所著的论文去兑换一个梦想中的儿女吗？人造粮、人造乳油、人造棉花，已够讨人嫌，如再加上人造儿童，岂不更可悲吗？约翰·洛克菲勒（John D. Rockfeller）的慈善施舍遍及全世界，受益的人不计其数，他对这些自然感到一种道德上的审美的满意，但同时我深信如此的满意是异常淡薄和脆弱的，在高尔夫球棍一挥之际即能完全忘却，而真实的不会遗忘的满意，仍在于他的儿子。

从另一方面看来，世人的快乐大部分在于能否寻到一种值得用毕生心力去做的工作，即心爱的工作。现在男女人所做的职业，我很疑心有百分之九十属于非其所好。我们常听人夸说："我很爱我的工作。"但这句话是否言出于衷颇是一个问题。我们从没有听人说："我爱我的家。"因为这是当然的，是不言而喻的。普通职业人士每天走进他们的办公室时，基本上是抱着和中国妇人之对生育儿女一般"人人如此我又何能例外"的心理。个个都在说我爱我的工作，这句话如出之于电梯司机、电话接线女服务员或牙科医生之口，显然是一句谎言；如系出于编辑人、地产经理人或证券经纪人，则尤其属于违心之论。我以为除了到北极去的探险家和试验室中的科学家专心致力于发明之外，人们对于他的工作充其量也不能超过颇感兴趣、颇合性情，而总够不上"爱"字。对工作的爱，万不能比拟母亲对儿女的爱。许多人常因对自己的职业发生厌恶而屡次改业，但从没有一个母亲会对养育教导儿女这桩毕生工作发生疑义。成功的政治家会放弃他的政治成就，编辑人会舍弃他所出版的杂志成就，飞行家会放弃飞行成就，拳击家会放弃比拳场成就，优伶会放弃舞台，但从没有听见过一个母亲不论是成功或失败会放弃她的母职。她觉得自己是必不可少的人物，在生命中已有了一个地位，坚信她的职务没有一个人能够代替，这信念比希特勒自信只有他能够拯救德国还要深切。除了这桩自知已在生命中得到切实地位的满意以外，还有什么东西能给人以更大的快乐？我敢说执业的人当中，也许有百分之五居然能得到合于自己性情的工作，而能够爱好，但百分之百的父母都觉得抚育自己的儿女乃是生活目标中最深入最切身的部分。因此可知一个女人自然在做母亲的天职中比做一个建筑师更易于获得真正的快乐，而大自然也绝不使她失望。所以婚姻岂不是最宜于女人的职业吗？

我知道女人未始没有这种感觉，不过没有形之于口罢了。现在经我一说破，而又明说家庭的重担终究必须由女人去肩挑，女人听了或许要受到一些惊惧，但这个确是我的本意和题旨。我们等着看罢，究竟哪个待女人好些？因为我们所关切者不过在于女人的快乐，不是社会成就上的快乐，而是自我的快乐。即从相宜或合格的观点上讲起来，我以为没有几个银行经理能像女人般适宜于其职务。我们常听见说才力不及的主任，才力不及的营业经理，才力不及的银行家，才力不及的总理，但难得听见才力不及的母亲，所以女人天然合宜于母职的这桩事，她们自己都知道，并也愿意去做。现在的美国女大学生似乎已有离开谬误理想的趋向，已渐渐知道用合理的眼光去视察生命，而公然宣布她们愿意嫁人了。我理想中的女人爱好化妆品也爱好数学，是个柔媚的女性，而不单是个女性。让她们去调脂弄粉，如尚有余力，则让她们如孔子所说一般去从事数学。

我在上边所说的这些是以普通男女的一般理想而论。女人中也和男人一般有才能杰出者，世上的真正进步即依赖他们的创造才能而成功。我希望一般女人能把婚姻当做理想职业去生育小孩，或者洗一些碗盏。我也希望一般男人放弃了艺术而单去做赚钱养家的工作：如理发、擦鞋、捉贼、补锅、饭店侍者等。生育小孩教养他们，当心他们，使他们成长为有用聪明的国民，这个职务势需有人去担任，而男人显然不但不能生育，即使叫他去抱小孩或替小孩洗澡也手足无措，所以我只好期望女人去担任了。这两类工作：教养小孩和理发擦鞋或替人开门，哪类高尚一些？我也不能确定。但我总以为丈夫既在那里替别人开门，则他的妻子又何必憎嫌洗碗盏，从前立柜台的都是男人，现在被大批女人跑进来占据了这位置，而将男人挤了出去做替人开门的工作。如女人认为这是较为高尚的工作，我们很欢迎她们来做，但须知在生活寻求的方式上，工作无所谓高低之别。一旦女人在衣帽室里收付衣帽，未必高尚于替丈夫补袜子，其间的分别，不过在于补袜子者和这袜子所有人有荣辱与共的关系，而衣帽管理人是没有的。我们自然希望这袜子所有人是一个良好丈夫而值得他妻子的服侍，但人们也不应该悲观到凡是丈夫的袜子都是不值一补的地步，男人也未必都是不肖的。最重要的一点就是：教养儿童之神圣工作的家庭生活于女人太为低贱这个普通假说，实不能称为合理的社会态度。只有在女人、家庭和母职未能受着相当尊重的文化中，才有这种观念。

II. Celibacy a Freak of Civilization

THE taking of such a simple and natural biological viewpoint implies two conflicts, first, the conflict between individualism and the family, and second, a deeper conflict between the sterile philosophy of the intellect and the warmer philosophy of the instinct. For individualism and worship of the intellect are likely to blind a man to the beauties of home life, and of the two, I think the first is not so wicked as the second. A man believing in individualism and carrying it to its logical consequences can still be a very intelligent being, but a man believing in the cold head as against the warm heart is a fool. For the collectivism of the family as a social unit, there can be substitutes, but for the loss of the mating and paternal-maternal instincts, there can be none.

We have to start with the assumption that man cannot live alone in this world and be happy, but must associate himself with a group around him and greater than himself. Man's self is not limited by his bodily proportions, for there is a greater self which extends as far as his mental and social activities go. In whatever age and country and under whatever form of government, the real life that means anything to a man is never co-extensive with his country or his age, but consists in that smaller circle of his acquaintances and activities which we call the"greater self."In this social unit he lives and moves and has his being. Such a social unit may be a parish, or a school, or a prison, or a business firm, or a secret society or a philanthropic organization. These may take the place of the home as a social unit, and sometimes entirely displace it. Religion itself or sometimes a big political movement may consume a man's whole being. But of all such groups, the home remains the only natural and biologically real, satisfying and meaningful unit of our existence. It is natural because every man finds himself already in a home when he is born, and also because it remains with one for life; and it is biologically real because the blood relationship lends the notion of such a greater self a visible reality. One who does not make a success of this natural group life cannot be expected to make a success of life in other groups. Confucius says, "The young should learn to be filial in the home and respectful in society; they should be conscientious and honest, and love all people and associate with the kindly gentlemen. If after acting on these precepts, they still have energy left, let them read books." Apart from the importance of this group life, man expresses and fulfils himself fully and reaches the highest development of his personality only in the harmonious complementing of a suitable member of the other sex.

Woman, who has a deeper biological sense than man, knows this. Subconsciously all Chinese girls dream of the red wedding petticoat and the wedding sedan, and all Western girls dream of the wedding veil and wedding bells. Nature has endowed women with too powerful a maternal instinct for it to be easily put out of the way by an artificial civilization. I have no doubt that nature conceives of woman chiefly as a mother, even more than as a mate, and has endowed her with mental and moral characteristics which are conducive to her role as mother, and which find their true explanation and unity in the maternal instinct—realism, judgment, patience with details, love of the small and helpless, desire to take care of somebody, strong animal love and hatred, great personal and emotional bias and a generally personal outlook on things. Philosophy, therefore, has gone far astray when it departs from nature's own conception and tries to make women happy without taking into account this maternal instinct which is the dominant trait and central explanation of her entire being. Thus with all uneducated and sanely educated women, the maternal instinct is never suppressed, comes to light in childhood and grows stronger and stronger through adolescence to maturity, while with man, the paternal instinct seldom becomes conscious until after thirty-five, or in any case until he has a son or daughter five years old. I do not think that a man of twenty-five ever thinks about becoming a father. He merely falls in love with a girl and accidentally produces a baby and forgets all about it, while his wife's thoughts are occupied with nothing else, until one day in his thirties he suddenly becomes aware that he has a son or daughter whom he can take to the market and parade before his friends, and only then does he begin to feel paternal. Few men of twenty or twenty-five are not amused at the idea of their becoming a father, and beyond amusement, there is little thought spent on it, whereas having a baby, even anticipating one, is probably the most serious thing that ever comes to a woman's life and changes her entire being to the point of affecting a transformation of her character and habits. The world becomes a different world for her when a woman becomes an expectant mother. Thenceforth she has no doubt whatever in her mind as to her mission in life or the purpose of her existence. She is wanted. She is needed. And she functions. I have seen the most pampered and petted only daughter of a rich Chinese family growing to heroic stature and losing sleep for months when her child was ill. In nature's scheme, no such paternal instinct is necessary and none is provided for, for man, like the drake or the gander, has little concern over his offspring otherwise than contributing his part. Women, therefore, suffer most psychologically when this central motive power of their being is not expressed and does not function. No one need tell me how kind American civilization is to women, when it permits so many nice women to go unmarried through no fault of their own.

I have no doubt that the maladjustment in American marriages is very largely due to this discrepancy between the maternal instinct of women and the paternal instinct of men. The so-called"emotional immaturity"of American young men can find no other explanation than in this biological fact; the men being brought up under a social system of over-pampering of youth do not possess the natural check of responsible thinking which the girls have through their greater maternal instinct. It would be ruinous if nature did not provide women with sufficient soberness when they are physiologically ready to become mothers, so nature just does it. Sons of poor families have responsible thinking drilled into their system by harder circumstances, thus leaving only the pampered sons of rich families, in a nation which worships and pampers youth, in an ideal condition for developing into emotional and social incompetents.

After all, we are concerned only with the question:"How to live a happy life?"and no one's life can be happy unless beyond the superficial attainments of the external life, the deeper springs of his or her character are touched and find a normal outlet. Celibacy as an ideal in the form of"personal career"carries with it not only an individualistic, but also a foolishly intellectualistic taint, and is for the latter reason to be condemned. I always suspect the confirmed bachelor or unmarried woman who remains so by choice of being an ineffectual intellectualist, too much engrossed with his or her own external achievements, believing that he or she can, as a human being, find happiness in an effective substitute for the home life, or find an intellectual, artistic or professional interest which is deeply satisfying.

I deny this. This spectacle of individualism, unmarried and childless, trying to find a substitute for a full and satisfying life in"careers"and personal achievements and preventing cruelty to animals has struck me always as somewhat foolish and comical. It is psychologically symptomatic in the case of old maiden ladies trying to sue a circus manager for cruelty to tigers because their suspicion has been aroused by whipmarks on the animals' backs. Their protests seem to come out of a misplaced maternal instinct, applied to a wrong species, as if true tigers ever thought anything of a little whipping. These women are vaguely groping for a place in life and trying very hard to make it sound convincing to themselves and to others.

The rewards of political, literary and artistic achievement produce in their authors only a pale, intellectual chuckle, while the rewards of seeing one's own children grow up big and strong are wordless and immensely real. How many authors and artists are satisfied with their accomplishments in their old age, and how many regard them as more than mere products of their pastime, justifiable chiefly as means of earning a living? It is said that a few days before his death, Herbert Spencer had the eighteen volumes of The Synthetic Philosophy piled on his lap and, as he felt their cold weight, wondered if he would not have done better could he have a grandchild in their stead. Would not wise Elia have exchanged the whole lot of his essays for one of his "dream children"? It is bad enough to have ersatz-sugar, ersatz-butter and ersatz-cotton, but it must be deplorable to have ersatz-children! I do not question that there was a moral and aesthetic satisfaction to John D. Rockefeller in the feeling that he had contributed so much to human happiness over such a wide area. At the same time, I do not doubt that such a moral or aesthetic satisfaction was extremely thin and pale, easily upset by a stupid stroke on the golf course, and that his real, lasting satisfaction was John D., Jr.

To look at it from another aspect, happiness is largely a matter of finding one's life work, the work that one loves. I question whether ninety per cent of the men and women occupied in a profession have found the work that they really love. The much vaunted statement, "I love my work," I suspect, must always be taken with a grain of salt. One never says, "I love my home," because it is taken for granted. The average business man goes to his office in very much the same mood as Chinese women produce babies: everybody is doing it, and what else can I do? "I love my work," so says everybody. Such a statement is a lie in the case of elevator men and telephone girls and dentists, and a gross exaggeration in the case of editors, real estate agents and stock brokers. Except for the Arctic explorer or the scientist in his laboratory, engaged in the work of discovery, I think, to like one's work, finding it congenial, is the best we can hope for. But even allowing for the figure of speech, there is no comparison between one's love of his work and the mother's love of her children. Many men have doubts about their true vocation, and shift from one to another, but there is never a doubt in a mother's mind concerning her life work, which is the taking care and guiding of the little ones. Successful politicians have thrown up politics, successful editors have thrown up magazine work, successful aviators have given up flying, successful boxers have given up the ring, and successful actors and actresses have given up the stage, but imagine mothers, successful or unsuccessful, giving up motherhood! It is unheard of. The mother has a feeling that she is wanted; she has found a place in life, and has the deep conviction that no one in the world can take her place, a conviction more profound than Hitler's that he must save Germany. And what can give man or woman a greater, deeper happiness than the satisfaction of knowing he or she has a definite place in life? Is it not common sense to say that whereas less than five per cent are lucky enough to find and be engaged in the work they love, a hundred per cent of parents find the work of looking after their children the most deep and engrossing of their life motives? Is it not true, therefore, that the chance of finding real happiness is surer and greater for a woman if she is engaged as a mother rather than as an architect, since nature never fails? Is it not true that marriage is the best profession for women?

My feminist readers must have sensed this all along, and bit by bit have begun trembling with rage as I grew more and more enthusiastic about the home, knowing that the cross of the home eventually must be borne by women. Such is exactly my intention and my thesis. It remains to be seen who is kinder to women, for it is with women's happiness alone that we are concerned, happiness not in terms of social achievements, but in terms of the depths of personal being. Even from the point of view of fitness or competency, I have no doubt that there are fewer bank presidents really fit for their jobs than women fit for mothers. We have incompetent department chiefs, incompetent business managers, incompetent bankers, and incompetent presidents, but we rarely have incompetent mothers. So then women are fit for motherhood, they want it and they know it. I understand that there has been a swing in the right direction away from the feminist ideal among the American college girls of today, that the majority of them are able to look at life sanely enough to say openly that they want to get married. The ideal woman for me is one who loves her cosmetics along with her mathematics, and who is more feminine than feminist. Let them have their cosmetics, and if they still have energy left, as Confucius would say, let them play with mathematics also.

It is to be understood that we are talking of the average ideal of the average man and woman. There are distinguished and talented women as there are distinguished and talented men, whose creative ability accounts for the world's real progress. If I ask the average woman to regard marriage as the ideal profession and to bear babies and perhaps also wash dishes, I also ask the average man to forget the arts and just earn the family bread, by cutting hair or shining shoes or catching thieves or tinkering pots or waiting at tables. Since some one has to bear the babies and take care of them and see them safely through measles and raise them to be good and wise citizens, and since men are entirely ineffectual in bearing babies and frightfully awkward in holding and bathing them, naturally I look to the women to do the job. I'm not so sure which is the nobler work—comparing the averages—raising babies or cutting people's hair or shining people's shoes or opening doors at department stores. I don't see why women have to complain about washing dishes if their husbands have to open doors for strangers at department stores. Men used to stand behind counters, and now girls have rushed in to take their place behind counters while the men open the doors, and they are welcome to it, if they think it is a nobler kind of work. Considered as a means of living, no work is noble and no work is ignoble. And I am not so sure that checking men's hats is necessarily more romantic than mending husbands' socks. The difference between the hat-check girl and the sock darner at home is that the sock darner has got a man over whose destiny it is her privilege to preside, while the hat-check girl hasn't. It is to be hoped, of course, that the wearer of the socks is worth the woman's labor, but it would be also unwarranted pessimism to lay down as a general rule that his socks are not worth her mending. Men are not all quite as worthless as that. The important point is that the general assumption that home life, with its important and sacred task of raising and influencing the young of the race, is too low for women can hardly be called a sane social attitude, and it is possible only in a culture where woman and the home and motherhood are not sufficiently respected.

三．性的吸引

　　在女权和增进女子社会权利的门面背后，即便美国也未能把应有的地位给予女人。我希望我这印象是不对的，我也希望在女权增进的当中，男人的中世纪骑士精神没有减退。因为这两件事并非一定联系而行，骑士精神或真心敬重女人精神，并非即是听任女人自由花钱，听任女人自由行动，听任女人担任执行公务，听任女人投票选举之谓。在我这种生于老旧世界，抱有老旧观念的人看来，世上的事情都有切己和不切己之别。美国女人在不切己的事情上，其地位已远胜于旧世界的姐妹，但在真正切己的事情上，依旧相同，没有寸进。骑士精神在美国也未见得较明显于欧洲。美国女人能真正发挥权力之处依旧是在她传统的区域，即家庭中，她在这区域中，以服务安琪儿的精神做她的主人。我曾见过这样的安琪儿，但只有在家庭的神圣私生活中才能见到，只在一个女人飘然往来于厨房客室之间，做专心爱好家庭的主妇中才能见到。她在这里放射出一种灿烂的光明，如移置到办公室去时，便是不合适而不可思议的了。

　　这只不过因为女人穿了飘逸的长裙比穿着短小的公事服更为动人，更为娇美。这或许不过是我的想象，问题的要点只在：女人居家正如鱼之居水，而一穿上公事服之后，男人就会拿她们当做相类的同事，即可以任意批评。但一等到她们在办公时间之外穿上了飘飘长裙，男人即自然放弃和她们竞争的心思，而只会坐着向她们张口呆看了。一进公事范围，女人更守纪律，对于日常例行公事也更优胜，但一出公事房，例如在喜事茶会等处，同事彼此遇到时，女人即较为活泼，较会发挥本能。她们会随口劝告男同事甚至她们的经理，快去理发，或告诉他们到哪里去买最好的医治脱发药水。在公事房中女人说话时极低声下气，但在公事房之外，即以权威者的态度发言。

从一个男人的观点坦白说起来——其实也无须讳言——我以为公共场所中有了女人，实在使生活上例如办公室中和街上增添不少动人的点缀和自然的礼貌。这于男人很有益处，可以使办公室中的语声变为柔和，色彩较为鲜艳，办公桌上也可以较为整齐。我又以为造化所赋予的性吸引力，或性吸引力的欲望，其程度至今没有变更。不过美国的男人享受较大的艳福，就因为在性诱引上，美国女人较别国人如中国女人更致力于获得男人的欢心。我的结论是：西方的人们过于着重性，而过于轻视女人。

美国女人所费于整理头发的时间长久，和旧时中国女人差不多。她们对于修饰较为公开、恒久和不择地而行，对于节食、运动、按摩和寻求保持美态的方法较为热烈。早晨在床上做两腿起落的运动以收束腰身，较为严切奉行。在一个中国女子已经抛弃脂粉的年龄，在美国依然会抹粉扑脂，色染头发。她们于生发水和香水上花费较多的钱，所以美国的化妆品如白天雪花膏、夜用雪花膏、无色雪花膏、搽粉之前所用的冷膏、面霜、防止毛孔涨大的冷霜、柠檬冷霜、避日炙冷油、去皱纹冷油、鱼油，以及各种稀奇古怪的香油，多至不可胜计，销路多而广。也许美国女人有较多的时间和钱可以耗费。她们或者穿衣服以讨男人的欢喜，脱衣服以讨自己

的欢喜，或两者都有之，都说不定。中国女人没有这样厉害，也许是中国
女人没有得到这许多种化妆品的机会的缘故。但对于女人想要吸引男人的
欲望，我实在不敢说各种族之间有什么不同。以前的中国女人也因极力想
讨男人的欢喜，将她的双足缠成三寸金莲，不过现在已经解放，而改为穿
高跟皮鞋了。我虽然不是先知，但我敢预言，在不久的将来，中国女人也
必于每天早晨在床上做几分钟腰腿运动，以讨她的丈夫或自己的欢喜。然
而很显明的事实在这里：现在的美国女人是更致力于身体上的性吸引力，
并专在性吸引力方面讲究她们的衣饰，以讨男人的欢喜。其结果是：在公
园里或街上所见的一般女人，大多具有较为苗条的身材，都穿着较为讲究
的衣裳。这确须感谢女人所费于保持其曲线的努力，因而使男人获得很大
的愉快，但据我的想象，她们在神经上必因此受很重大的影响。我所说的
性吸引力，均系对母性吸引力，或对女人的一般吸引力而言。我疑心现代
文明的这一种状态，已把它的特质深印于现代恋爱和婚姻上了。

　　艺术使现代男人有了性的意识。对于这点，我毫无疑义。先前是艺
术，后来变为商业性的利用，将女人的全体直到最后一条曲线和最后一根
染色的脚趾完全开拓起来。我从未见过女人的肢体经过这样的商业性开
拓，并且很奇怪何以美国女人竟肯这样驯服地听任去将她们的肢体开拓到
这个地步，在东方人的心目中，这种对女人身体的商业性开拓和尊重女人
的观念绝不能并立。艺术家称之为美，看戏的观众称之为艺术，只有戏馆
老板和经理直称之为性吸引力，而男人们也就因此得到很好的消遣。在
这男人所创造，男人所统治的社会中，女人可以剥光了去供商业性的开
拓，而男人除了歌舞团团员之外，很少裸身露体的时候，这实是一种奇特
的现象。在戏台上我们能看到女人几乎不穿着什么，但男人依然都是衣冠
整齐。如若这个世界是女人统治的，那么我们当然就要看见裸身露体的男

人，而女人都穿上长裙了。艺术家对于男女人的身体构造同样研究，但似乎终没有法子可以将男人的身体美化为商业性的用场。戏院以裸体为号召，但大多是剥去了女人的衣服以吸引男人，而绝不剥去男人的衣服以吸引女人。就是在较为高尚的表演中，虽然是艺术和道德并重，也总是让女人艺术化，男人道德化，而从不坚持女人道德化，男人艺术化（在歌舞杂耍表演中，男演员大都偏重于滑稽，即跳舞的时候也是如此，而观众尚认为是艺术的）。商业广告都抓住这一点，千方百计地利用。所以现在的人们如要知道什么是"艺术"，只须买一本杂志，将里边的广告看一遍，便能一目了然。其结果使女人的脑中深印下极其深刻的"女人必须艺术化"的印象，甚至下意识地默认了这个原则，自愿地忍饥节食、运动、按摩，严格遵守一切纪律，以求对美的世界有所贡献。心地不很明白的人看了，几乎要认做女人除了利用性吸引力外，竟没有其他抓住男人的方法了。

我认为这种过于注重性吸引力，造成一种对女人整个天性的不成熟和不充足的观念，影响恋爱和婚姻的性质，也引起了对这两件事的谬误或不充足的见解。因此，女人被认为只是一个可能的伴侣，而不是家庭的主持人物。女人是妻，也是母。但因现在如此注意于性，以致伴侣的意想取代了为母意想的地位。不过我仍坚持女人只在为母时才能达到她的最崇高身份。如若一个女人竟拒绝为母，她便立刻丧失大部分的尊严和庄重，而有成为玩具的危险。我以为一个女人，不论在法律上的身份如何，只要有了子女，便可视之为妻；而如若没有子女，即使是妻，也只能视做姘妇。子女使姘妇抬高身份，而无子女使妻降级。现代女人有许多不愿生育，理由也很明显，无非恐惧怀孕将妨碍她的苗条身段罢了。

好色的天性对于增进人生的生趣有相当助力，但行之过分不利于女人。保持性吸引力的意念当然是一种加在女人神经上的压力，而男人没有。这也是太不公平，因为人们如过于重视美丽及青春，中年女人便陷于对灰白的头发和不可挽回的光阴做必不能胜的奋斗。中国某诗人早已提醒我们说，青春之泉是无稽之谈，无人能系住光阴不让它前进。因此中年女人保持做性吸引力的努力，即等于和年龄赛跑，太不合理。只有幽默能补救这个处境。如若明知和老年白发无从争斗，何不就认白发为美丽？朱杜（明朝女道士朱桂英）的诗：

> 白发新添数百茎，
> 几番拔尽白还生；
> 不如不拔由他白，
> 哪得功夫与白争。

这整个情形既不自然且太不平，于为母者和年龄较大的女人不平允。重量拳击家将锦标保持了数年之后，势必不得不将宝座让给较为年轻的后进；跑马场里得锦标的马，过了数年，自不得不将第一位置让给年龄较小的马；同样，年龄较大的女人，绝不能战胜年轻的少女。同类相争，其实何必。中年女人想在性吸引力上和少女争高下是愚拙危险而又无望的举动。女人所应重视者尚有比性更进一层的东西。谈情说爱的动作，大部分当然以身体的吸引力为根据，较为成熟的男女自应视之为过去的事情，而不必再斤斤计较了。

我们知道人类是动物中最喜欢表示爱情的动物，但除了表示爱情的天性外，还同样具有一种有力的父母天性，结果产生了人类家庭生活。多数动物大概都和人类一般的具有爱情和父母天性，但人类家庭生活的起点好像是发源于长臂猿。不过在一种过分矫饰的文化中，人类受了艺术上不断

的性的刺激如电影和戏院之类，他们的家庭天性便有屈服于表示爱情天性
的危险。在如此的文化中，家庭理想的必需性常易于被人们忘却，尤其又
有一种个人主义观念的潮流，所以在这种社会中，我们对婚姻即有了一种
新奇的意见，视为不过是许多次接吻，而于结婚的钟声中结束；对女人也
有了一种新奇的观念，视为不过是男人的伴侣，而不是为母者，因此理想
的女人须具有完美相称的身体和动人的体态。但在我看来，一个女人最美
丽的时候是在她立在摇篮前面；最恳切最庄严的时候是在她怀抱婴儿或扶
着四五岁小孩行走；最快乐的时候则如我所见的一幅西洋画像一般，是在
拥抱一个婴儿睡在枕上。或许我对母道有一些迷信，但我们中国人有一些
心理上的迷信是并无妨碍的。我以为我对女人的见解并非由于迷信母道，
实是由中国式家庭理想之影响。

III. On Sex Appeal

BEHIND the facade of woman's rights and increased social privileges for women, I always think woman is not given her due even in modern America. Let us hope that my impression is incorrect, and that with the increase of woman' s rights, chivalry has not decreased. For the two things do not necessarily go together, chivalry or true respect for women and allowing women to spend money, to go where they please, to hold executive jobs and to vote. It has seemed to me (a citizen of the Old World with the Old World outlook) that there are things which matter and things which don't, and that American women are far ahead of their Old World sisters in all things that don't matter, and remain very much in the same situation in all things that do. Anyway, there is no clear index of a greater chivalry in America than in Europe. What real authority the American woman does exercise is still from her traditional old throne–the hearth–over which she presides as the happy ministering angel. I have seen such angels, but only in the sanctity of a private home, where a woman glides along in the kitchen or in the parlor, true mistress of a home consecrated to family love. Somehow she suffuses a radiance which would be unthinkable or out of place in an office.

Is it merely because woman is more charming and more graceful in a chiffon dress than in a business jacket, or is it merely my imagination? The gist of the matter seems to lie in the fact that women at home are like fish in water. Clothe women in business jackets and men will regard them as co-workers with the right to criticize, but let them float about in georgette or chiffon one out of the seven office hours in the day and men will give up any idea of competing with them, and will merely sit back and wonder and gasp. Submitted to business routine, women are disciplined quite easily and make better routine workers than men, but the moment the office atmosphere is changed, as when as a business staff meet at a wedding tea, and you will find that women immediately come into their own by advising their men colleagues or their boss to get a haircut, or where to get the best lotion for curing dandruff. In the office, women talk with civility; outside the office, they talk with authority.

Frankly speaking from a man' s point of view—there is no use in pretending to speak otherwise—I think that the appearance of women in public has added greatly to the charm and amenities of life, life in the office and in the street, for the benefit of men; that voices in the offices are softer, colors gayer and the desks neater. I think also that not a whit of the sexual attraction or desire for sexual attraction provided by nature has changed, but that in America, men are having a grander time because American women are trying harder to please the men than, for instance, Chinese women, so far as attention to sex appeal is concerned. And my conclusion is that in the West, people think too much of sex and too little of women.

Western women spend almost as much time fixing their hair as Chinese women used to do; attend to their make-up more openly, constantly and ubiquitously; diet, exercise, massage and read advertisements for keeping the figure more assiduously; kick their legs up and down in bed to reduce their waistline more religiously; lift their faces and dye their hair, at an age at which no Chinese women ever think of doing such a thing. They are spending more money, not less, on lotions and perfumes, and there is a bigger business in beauty aids and day creams, night creams, vanishing creams, foundation creams, face creams, hand creams, pore creams, lemon creams, suntan oils, wrinkle oils, turtle oils, and every conceivable variety of perfumed oil. Perhaps it is simply because American women have more time and more money to spend. Perhaps they dress to please men and undress to please themselves, or the other way round, or both. Perhaps the reason is merely that Chinese women have fewer available modern beauty aids, for I hesitate very much to draw a distinction between races when it comes to woman's desire to attract men. Chinese women were trying hard enough to please men by binding their own feet half a century ago, and now they have gayly capered their way from their"bow shoes"into high heels. I am not usually a prophet, but I can say with prophetic conviction that in the immediate future, Chinese women will be having their morning ten minutes of kicking their legs up and down in order to please their husbands or themselves. Yet the obvious fact is there: American women at present seem to be trying harder to please the men by spending more thought on their bodily sex appeal and dress with better understanding of sex appeal. The net result is that women as a whole, as seen in the parks and in the streets, have better figures and are better dressed, thanks to the continuous tremendous daily efforts of women to keep their figure—to the great delight of men. But I imagine how it must wear on their nerves. And when I speak of sex appeal, I mean it in contrast to motherhood appeal, or woman's appeal as a whole. I suspect this phase of modern civilization has stamped its character on modern love and marriage.

Art has made the modern man sex conscious. I have no doubt about it. First art and then commercial exploitation of the woman's body, down to its last curve and muscular undulation and the last painted toe-nail. I have never seen every part of a woman's body so completely exploited commercially, and find it hard to understand how American women have submitted so sweetly to this exploitation of their bodies. To an Oriental, it is hard to square this commercial exploitation of the female body with respect for women. Artists call it beauty, theater-goers call it art, only producers and managers honestly call it sex appeal, and men generally have a good time. It is typical of a man-made and man-ruled society that women are stripped for commercial exploitation and men almost never, outside a few acrobats. On the stage, one sees women nearly undressed, while the men still keep their morning coats and black ties; in a woman-ruled world, one would certainly see the men half undressed while the women kept their skirts. Artists study male and female anatomy equally, but somehow find it difficult to turn their study of the male body beautiful to commercial account. The theater strips to tease, but generally strips the women to tease the men, and does not strip the men to tease the women. Even in the higher-class shows, where they try to be both artistic and moral, people allow the women to be artistic and the men to be moral, but never insist on the women being moral and the men artistic. (All men actors in vaudeville shows merely try to be funny, even in dancing, which is supposed to be"artistic.") The commercial advertisements pick up the theme and play it up in endless variations, so that today all a man needs to do when he wants to be"artistic"is to take a copy of a magazine and run through the advertising section. The result is, the women themselves are so impressed with the duty of being artistic that they unconsciously accept the doctrine and starve themselves or submit to massage and rigorous discipline, in order to contribute toward a more beautiful world. The less clear-minded are almost led to think that their only way of getting a man and holding him is by sex appeal.

I consider that this over-emphasis on sex appeal involves an adolescent and inadequate view of the entire nature of woman, with certain consequences upon the character of love and marriage, whose conception becomes also false or inadequate. Woman is thus more thought of as a mating possibility than as a presiding spirit over the hearth. Woman is wife and mother both, but with the emphasis on sex as such, the notion of a mate displaces the notion of the mother, and I insist that woman reaches her noblest status only as mother, and that a wife who by choice refuses to become a mother immediately loses a great part of her dignity and seriousness and stands in danger of becoming a plaything. To me, any wife without children is a mistress, and any mistress with children is a wife, no matter what their legal standing is. The children ennoble and sanctify the mistress, and the absence of children degrades the wife. It is a truism that many modern women refuse to have babies because pregnancy would spoil their figures.

The amorous instinct has its proper contribution to make to the enrichment of life, yet it can be overdone to the detriment of woman herself. The strain of keeping up sex appeal necessarily falls upon the nerves of women and not of men. It is also unfair, for by placing a premium upon beauty and youth, middle-aged women are confronted with the hopeless task of fighting their gray hair and time's course. A Chinese poet has already warned us that the fountain of youth is a hoax, that no man can yet"tie a string to the sun"and hold back its course. Middle-aged woman's effort to keep up sex appeal thus becomes an arduous race with the years, which is quite senseless. Only humor can save the situation. If there is no use carrying on a hopeless fight against old age and white hair, why then not call the white hair beautiful? So sings Chu Tu:

> I've gained white hairs, some hundreds, on my head.
> As often as they're plucked, still more grow in their stead.
> Why not stop plucking, then, and let the white alone?
> Who has the time to fight against the silvery thread?

The whole thing is unnatural and unfair. It is unfair to the mother and older women, because as surely as a heavyweight champion must hand over his title in a few years to a younger challenger and an old champion horse must yield in a few years to a younger horse, so must the old women fight a losing battle against the younger women, and after all they are all fighting against their own sex. It is foolish, dangerous, and hopeless for middle-aged women to meet younger women on the issue of sex appeal. It is also foolish because there is more to a woman than her sex, and while wooing and courtship are necessarily largely based on physical attraction, maturer men and women should have outgrown it.

Man, we know, is the most amorous animal in the zoological kingdom. Besides this amorous instinct, however, there is an equally strong parental instinct, resulting in the human family life. The amorous and paternal instincts we share in common with most of the animals, but the beginnings of a human family life seem to be found among the gibbons. There is a danger, however, of the amorous instinct subjugating the family instinct in an oversophisticated culture surrounding man with constant sexual stimuli in art, the movies and the theatre. In such a culture, the necessity of the family ideal can be easily forgotten, especially when in addition there is a current of individualistic ideas. In such a society, therefore, we get a strange view of marriage, as consisting of eternal kissing, generally ending with the wedding bells, and a strange view of woman, chiefly as man's mate and not as mother. The ideal woman, then, becomes a young woman with perfect physical proportions and physical charm, whereas for me, woman is never more beautiful than when she is standing over a cradle, never more serious and dignified than when she is holding a baby in her breast and leading a child of four or five years by the hand, and never more happy than, as I have seen in a Western painting, when she is lying in bed against a pillow and playing with a baby at her breast. Perhaps I have got a motherhood complex, but that is all right because psychological complexes never do a Chinese any harm. Any suggestion of an Oedipus complex or father-and-daughter complex, or of a son-and-mother complex, in a Chinese always seems to me ridiculous and unconvincing. I suggest that my view of woman is not due to a motherhood complex, but is due to the influence of the Chinese family ideal.

四. 中国式家庭理想

我颇以为《创世记》中"创造"一节应该从头写过。中国小说《红楼梦》里边的才子贾宝玉是一个极富感情的柔性男人，最喜和女人为伴，万分崇拜他许多姐妹的美色，而常常自恨是个男人。他曾说，女人是水做的，而男人是泥做的。其理由是：女人都伶俐聪明，娇媚可爱，而男人都愚蠢粗鲁，面目可憎。如若《创世记》的作者换了贾宝玉，心地和他一样明白，则《创世记》必不是这样写法。上帝抓了一把泥土，捏成一个人形，从鼻孔吹一口气进去，亚当就此造成，但是亚当渐渐燥裂，泥土松碎，一片片掉落下来，所以上帝又取了一些水和将进去，使泥土凝结。这种掺入亚当生命的水，就是夏娃，亚当的生命非有这水不能完成。我以为婚姻的特别意义至少如此。女人是水，男人是泥土，水掺入泥土而使之成形，泥土盛了这水而有形质，水即于流动生活中而有了具体。

元朝名画家赵孟頫的妻子管夫人也是一位著名画家，早已引用过这个泥土和水的譬喻。当夫妻俩都在中年的时候，孟对她的爱情似乎减退，想纳一个妾，管夫人即作了下面这一首小令，使她的丈夫看了非常感动，便打消了纳妾的念头。

你侬我侬，

忒煞情多，

情多处，

热如火。

把一块泥，

捻一个你，

塑一个我。

将咱两个，

一齐打破，

用水调和，

再捻一个你，

再塑一个我。

我泥中有你，

你泥中有我；

与你生同一个衾，

死同一个椁。

中国的社会和生活都是组织于家庭制度基础上的，人所共知，这个制度决定并润色整个中国式生活的模型。但这个家庭生活的理想是从何而来的呢？这个问题从来没有人提出过。因为中国人都视为理所当然，而外国自觉不够资格去问这句话。把家庭制度作为一切社会和政治生活的基础，大家都知道孔夫子曾给予一个哲学的根基。他非常注重夫妻关系，认为是一切人类关系的根基，也注重孝顺父母，每年祭扫祖墓、崇拜祖先，设立祖先祠。

中国的祖先崇拜，曾被某些作家称为一种宗教，我相信在某种程度上是很对的。至于它的非宗教方面就在于它很少超自然的成分，不涉及神怪，所以崇拜祖先不妨和信仰基督仙佛或回教神道同时并行。崇拜祖先所用的礼仪造成一种宗教的形式，非常自然而且合理，因为凡属信念是不能没有表现方式的。照这种情形而论，我以为对着一块长约十五英寸的长方木牌表示尊意，其尊敬程度和英国把英王肖像印在邮票之上并没有什么高下之分。第一，中国人对于这种祖先之灵并不十分视同神道，而不过当他如在世的老长辈一般侍奉，并不向他祈求福佑，也不求他治病，并不像普通的崇拜者和被崇拜之间必有一种施必望报的情形；第二，这种崇拜仪式不过是对已死的祖先表示敬意的典礼，不过借这一天使全家团聚一次，并纪念祖先对这家庭所贻的世泽。这种仪式充其量如同替长辈做一次小规模的生日庆典，和平常替父母做寿，和美国的母亲节并没有什么分别。

基督教士不许中国信徒参加崇拜祖先的仪节，其唯一反对理由，是因为祭祖时大家都须跪地磕头，认为这违反"十诫"中的第一条。这是基督教士太缺乏谅解的表征之一。中国人的膝盖不若西方人的膝盖那样宝贵，中国人都向他们的皇帝、官府磕头，新年都向在世的父母磕头，视为常事，所以中国人的膝弯较为柔曲，而跪在神主牌之前磕几个头，也不会使他即记得变为一个崇信异端的人。城市村镇中的中国信徒却因此和一般的社团生活相隔绝，不能去参加大众节日的欢聚，也不便捐助这些节日的戏份儿，所以中国的基督信徒几乎不与本族的人相往来。

这种对于一己的家庭的虔敬和神秘性义务的感觉，有时确也能变成一种很深的宗教态度，例如十七世纪的儒者颜元在老年的时候，独自出外周历天下，找寻他的哥哥。因为自己没有儿子，所以希望寻到他的哥哥和一个侄子，以便传宗接代。他是河北人，笃信儒宗，专事力行。他的哥哥失踪已经多年，他忽然厌弃教读生活，如奉神召一般，决计出外寻兄。他连哥哥的影踪都不知道，盲目找寻，这是何等艰难的事情。况且这个时期正值明朝覆亡、全国混乱的时候，遍地伏葬，旅行极为危险，但他不顾一切，冒险前行，所到之处都贴下招纸，悬赏找寻。他走了一千余里的路程，经过中国北部数省，直到数年之后，他走某处时，被他的侄子看见了他手中所拿伞柄上刻着的姓名，知道是他的叔父，方将他引导到自己家中。那时他的哥哥已死，但他的目的仍算达到，因为有一个侄子可继香火了。

孔子极为推崇孝道，其理由何在？没有人能够知道。据吴经熊博士在某篇论文所说，是因为孔子乃是一个没有父亲的人，所以他的心理作用无非也和名歌《家，甜蜜的家》（Home, Sweet Home，约翰·霍华德·佩恩所写）的作者其实从来没有享过家庭幸福完全一样。如若孔子幼时父亲尚在，则他的父职概念不至于会这样深刻远到。再如若他已成年，而他的父亲尚在世，则结果恐怕更坏于此，如此他即有机会可以看到父亲的弱点，而会觉得力行纯孝未必是件容易做到的事情了。总之，他出世的时节，父亲已经故世，并且不知道葬在哪里。他是一个私生子，他的母亲从来没有告诉过他父亲是谁。他的母亲死后，他就将母亲的遗体葬在"五父之衢"，这当中或许含一些故意亦未可知。后来居然有一个年老妇人将他父亲的葬处告诉了他，他方将母亲的灵柩迁去合葬。

这一个巧妙的假说有怎样的价值，我们不必苛求，但中国文学中对于家庭理想的必须，确实举出不少的理由。它是以一个人还不是一个单位，而只是家庭单位中的一分子为出发点。由"生活潮流"假说（这是我所题的名称）所具的生活观念为之支持，而由认力行天性为道德和政治的最后目标的哲理证之为正当。

　　家庭制度的理想和个人主义的理想显然不能并立。一个人终究不能完全独自过一生，照这样的个人观念，太缺乏实在性。一个人如若不认他是一个人子，一个兄弟，一个父亲或一个朋友，则我们当他是个什么？如此的个人将成为形而上的抽象物。中国人的心理都偏于生物思想，所以他们对于人类自然先想到他在生物上的关系。因此家庭即自然成为人生中的生物性单位，婚姻也成为一件家庭事件，而不单是个人事件。

　　和这种西方的个人主义和国家主义对照的就是家庭理想。在这种理想中，人并不是个人，而被认是家庭的一分子，是家庭生活巨流中的一个必需分子，这就是我所谓"生活潮流"假说的意义。人类生活就整个而论，可以为多种不同的种族生活潮流。但人们所能直接触到和看到的只有家庭中的生活潮流。东西两方都有家庭如大树这个譬喻，每个人的生命不过是这大树上的一枝，借着树干而生存，尽他协助树干滋长下去的本分。所以就我们所见，人类生活显然是一种生长或连续作用。在这当中，每个人都在家庭历史上有一番作为，尽他对于整个家庭的义务，不过成绩有优劣，有些替家庭争到光荣，有些使家庭蒙受恶名。

　　家庭意识和家庭荣誉的感觉，或许就是中国人人生中唯一的团体精神或团体意识。家庭中每个分子因须振其家声，必须好好地做人，而不得贻羞于家族，他应该像一个球员一般将球推向前去。"败家子"不但是个人之耻，也是全家之羞，正如一个球员失足而被对方抢去球一般。凡去考试而金榜题名的好像是一个获得胜利的球员，光荣不但属于他个人，也属于他的一家。考中状元或一个三甲进士的人光被全族，使全族的人甚至连亲戚和同乡都得到精神上的兴奋和实质的权益。即使在一二百年之后，乡人尚会夸说某某年本乡怎样出过状元。从前人中了状元或进士，全家全乡都庆祝，荣归挂匾，大家何等欢欣兴奋，觉得荣耀非凡，人人有份，和这个相较起来，现代学校毕业接受文凭时是何等冷静缺乏意趣啊！

　　在这一幅家庭生活的景象内，其变化和色彩有很大的伸缩余地。人们须经过童年、成人和老年这几个时期。先由别人养育他，再由他去养育别人，最后于老年时重复由别人侍奉他。起先他尊奉别人，受别人的指挥，等到成人以后，他便渐渐地受人尊奉，指挥别人。更重要的，女人置身于家庭之中，使这幅景象增加不少色彩。女人在连续不断的家庭生活中不单是个装饰品或玩具，也不单是一个妻，实是这株家庭大树一个关系生存和必需的分子。因为使连续成为可能者即是女人，而家庭中各个支派的盛衰也是以所娶来媳妇的体质心性为依归。一个聪明的家长于选择媳妇时，必注意她的出身是否清白，正如园丁对于接果树的枝必须加以选择一般。很有些人认为一个人的生活，尤其是家庭生活，或苦或乐，都以他所娶的妻为定，而未来家庭的整个性质也由此而决定。子孙体格的强弱，性情的优劣，完全以媳妇的体格性情为依归，因此产生一种根据于遗传性的无形的、界限不分明的优生学，注重于门第的高低，实则无非是家长对于未来媳妇的体格姿色和教育的一种取舍标准。普通的标准大都着重于她的家教。依传统的说法，最好的媳妇应出于勤俭书礼之家。有时候家长发现所娶的媳妇不贤惠时，便要咒骂亲家的家教不良。所以为父母者又多出了一种教养女儿将来成为好媳妇而不致贻羞家声的责任，例如不会烹饪，不会做年糕之类，都须视为没有家教。

　　依照这种家族制度里边的"生命潮流"假说，永生几乎是可以看到可以触到的东西了。一个祖父看见孙儿背了书包上学校，便觉得他好似已在这个小孩之中重新生活。他用手去抚摸这个小孩时，即感到这就是自己的血肉。他的生命不过是家庭大树上的一枝，或永远向前流去的潮流中一部分，所以他虽死也是快乐的。因此中国的家长所关切的事情就是亲见男婚女嫁，视为比将来自己所葬的坟墓或所睡的棺材更重要。他必须亲眼看见媳妇或女婿是怎样的人，方始放心，如若都是很好并且满意的，他便可以含笑而逝，一无遗恨了。

这种生命概念的结果使一个人对任何物事都有一种伸长的见解，而不再认生命为始于个人，止于个人。球队当中虽有一两个守卫中途退出，但他们的位置即刻有人填补，球赛依旧可以继续下去，成败也因而变换了性质。中国的生活理想是：做人须无愧于祖宗，并且无愧于自己的好儿子。中国官员在辞职时每每引用下面这两句老话：

有子万事足，无官一身轻。

一个人最不幸的遭遇或许就是儿子不肖，不能维持家声、保持家产。一个富翁看见他的儿子好赌，就觉得半生辛劳所积聚的家产不能保持。如果儿子失败，这失败便是绝对的。在另一方面，一个眼光远大的寡妇，如有一个五岁的好孩子，她就能含辛茹苦、历尽艰难去教养他。中国历史中这种守节抚孤的女人很多很多，期望经历多年苦况之后儿子成人、飞黄腾达。蒋介石就是这类事件的一个最近榜样。他幼时和寡母常受邻人的欺侮，但寡母因有这个儿子，终愿奋志抚养。寡居的母亲，由于富有实际见解，常常教养出才德俱优的大人物，这桩情形使我觉得单以抚育儿童而言，父亲竟是不必要的。寡妇的笑声最响，因为她总是末了一个发出笑声的人。

因此，生活在这种家庭方式之中是令人满意的，因为人生的生物性各方面都已顾到。此即孔子所关切的事情。依孔子的见解，政治的最终理想无疑是生物性的。他说"老者安之，少者怀之""内无怨女，外无旷夫"。这话不单是对于枝节问题的一种表白，而实是政治的最后目标，所以尤堪注意。此即人性学者所谓的达情哲理。孔子意欲使一切人类天性都得到满足，以为必须如此方能使人在满意的生活中得到道德的和平，而只有道德的和平方是真正的和平。这是一种政治理想，其目的在于使政治成为不必要，因为这种和平是发于人类本心、极为稳固的和平。

IV. The Chinese Family Ideal

I rather think that the Genesis story of the Creation needs to be rewritten all over again. In the Chinese novel Red Chamber Dream, the boy hero, a sentimental mollycoddle very fond of female company and admiring his beautiful female cousins intensely and all but sorry for himself for being a boy, says that,"Woman is made of water and man is made of clay,"the reason being that he thinks his female cousins are sweet and pure and clever, while he himself and his boy companions are ugly and muddle-headed and bad-tempered. If the writer of the Genesis story had been a Paoyü and knew what he was talking about, he would have written a different story. God took a handful of mud, molded it into human shape and breathed into its nostrils a breath, and there was Adam. But Adam began to crack and fall to pieces, and so He took some water, and with the water He molded the clay, and this water which entered into Adam's being was called Eve, and only in having Eve in his being was Adam's life complete. At least that seems to me to be the symbolic significance of marriage. Woman is water and man is clay, and water permeates and molds the clay, and the clay holds the water and gives its substance, in which water moves and lives and has its full being.

The analogy of clay and water in human marriage was long ago expressed by Madame Kuan, wife of the great Yüan painter Chao Mengfu and herself a painter and teacher at the Imperial Court. When in their middle age Chao's ardor was cooling, or anyway when he was thinking of taking a mistress, Madame Kuan wrote the following poem, which touched his heart and changed his mind:

> *'Twixt you and me*
> *There's too much emotion.*
> *That's the reason why*
> *There's such a commotion!*
> *Take a lump of clay,*
> *Wet it, pat it,*
> *And make an image of me,*
> *And an image of you.*
> *Then smash them, crash them,*
> *And add a little water.*
> *Break them and re-make them*
> *Into an image of you,   .*
> *And an image of me.*
> *Then in my clay, there's a little of you.*
> *And in your clay, there's a little of me.*
> *And nothing ever shall us sever;*
> *Living, we'll sleep in the same quilt,*
> *And dead, well be buried together.*

It is a well-known fact that Chinese society and Chinese life are organized on the basis of the family system. This system determines and colors the entire Chinese pattern of life. Whence came this family ideal of life? It is a question that has seldom been asked, for the Chinese seem to take it for granted, while foreign students do not feel competent to enter into the question. Confucius is reputed to have provided the philosophical foundation for the family system as the basis of all social and political life, with its tremendous emphasis on the husband-wife relationship as the foundation of all human relationships, on filial piety toward parents, annual visits to ancestral graves, ancestor worship, and the institution of the ancestral hall.

Chinese ancestor worship has already been called a religion by certain writers, and I believe this to a very great extent is correct. Its non-religious aspect is the exclusion or the much less significant place of the supernatural element. The supernatural is left almost untouched, and ancestor worship can go side by side with belief in a Christian, a Buddhist, or a Mohammedan god. The rituals of ancestor worship provide a form of religion and are both natural and justifiable because all beliefs must have an outward symbol and form. As it is, I do not think the respect paid to square wooden tablets about fifteen inches long, inscribed with the names of ancestors, is more religious or less so than the use of the picture of the King on a British postage stamp. In the first place, these ancestral spirits are conceived less as gods than as human beings, continuing to be served as they were in their old age by their descendants. There is no prayer for gifts and no prayer for cure of sickness, and none of the usual bargaining between the worshiper and the worshiped. In the second place, this ceremony of worship is no more than an occasion for pious remembrance of one's departed ancestors, on a day consecrated to family reunion and reflections of gratitude on what the ancestor has done for the family. At best, it is only a poor substitute for celebrating the ancestor's birthday when he was alive, but in spirit it differs in no way from the celebration of a parent's birthday, or of Mother's Day in America.

The only objection which led Christian missionaries to forbid Chinese converts to participate in the ceremonies and communal feasting and merrymaking of ancestor worship is that the worshipers are required to kneel down before the ancestral tablets, thus infringing upon the first of the

Ten Commandments. This is about the most flagrant instance of lack of understanding on the part of the Christian missionaries. Chinese knees are not quite as precious as Western knees, for we kneel down before emperors, magistrates and before our own parents on New Year's Day, when they are living. Consequently Chinese knees are naturally more flexible, and one doesn't become a heathen more or less by kneeling before an inscribed wooden tablet, resembling a calendar block. On the other hand, Chinese Christians in the villages and towns are forced to cut themselves off from the general community life by being forbidden to participate in the general feasting and merrymaking, or even to contribute money toward the theatrical performances usual on such an occasion. The Chinese Christians, therefore, practically excommunicate themselves from their own clan.

There is hardly a question that, in many cases, this feeling of piety and of mystic obligation toward one's own family actually amounted to a deeply religious attitude. We have, for instance, the case of Yen Yüan, one of the greatest Confucianist leaders in the seventeenth century, who in his old age started out on a pathetic journey of search for his brother, in the hope that his brother might be found to have a son, since he himself had none. This follower of Confucianism, who believed in conduct more than in knowledge, was living in Szechuen. His brother had been missing for years. Tired of teaching the doctrines of Confucius, one day he felt what among missionaries would be regarded as a"divine call" to search for this lost brother. The situation was practically hopeless. He had no idea where his brother might be, or even if he were living. Travel was a highly perilous undertaking in those days, and the country was in disorder because of the collapse of the Ming regime. Still this old man set out on this truly religious journey, with no better means of locating his brother than pasting placards on city gates and inns wherever he went. Thus he traveled from Western China to the northeastern provinces, covering over a thousand miles, and only after years of desperate search, was he brought to the home of his brother through the latter's son recognizing his name on an umbrella left standing against a wall while he was in a public privy. His brother was then dead, but he achieved his goal, which was to find a male descendant for his ancestor's family.

Why Confucius laid such emphasis on filial piety nobody knows, but it has been suggested by Dr. John C. H. Wu in an illuminating essay that the reason was that Confucius was born without a father. The psychological reason is therefore similar to that of the writer of Home, Sweet Home, who never knew a home in his entire life. Had Confucius' father been living when he was a child, the idea of fatherhood could not have been invested with such romantic glamour, and if his father had been living after he grew up, the result might have been still more disastrous. He would have been able to see his father's foibles, and he might have found the precept of absolute piety somewhat difficult to live up to. Anyway his father was dead when he was born, and not only that, but Confucius did not even know where his father's grave was. He had been born out of wedlock, and his mother refused to tell him who his father was. When his mother died, he buried her (cynically, I suppose) at the"Road of the Five Fathers,"and only after he had found out the location of his father's grave from an old woman, did he provide for the burial of his parents together at another place.

We have to let this ingenious theory stand for what it is worth. But for the necessity of the family ideal, there is no lack of reasons in Chinese literature. It starts out with a view of man not as an individual, but as a member of a family unit, is backed by the view of life which I may call the"stream-of-life" theory, and justified by a philosophy which regards the fulfillment of man's natural instincts as the ultimate goal of morals and politics.

The ideal of the family system is necessarily dead set against the ideal of personal individualism. No man, after all, lives as an individual completely alone, and the idea of such an individual has no reality to it. If we think of an individual and regard him as neither a son, nor a brother, nor a father, nor a friend, then what is he? Such an individual becomes a metaphysical abstraction. And being biologically minded as the Chinese are, they naturally think of a man's biological relationships first. The family then becomes the natural biological unit of our existence, and marriage itself becomes a family affair, and not an individual affair.

In place of this individualism and nationalism of the West, there is then the family ideal in which man is not regarded as an individual but as a member of a family and an essential part of the great stream of family life. That is what I mean by the "stream-of-life" theory. Human life as a whole may be regarded as consisting of different racial streams of life, but it is the stream of life in the family that a man feels and sees directly. In accordance with both a Chinese and Western analogy, we speak of the "family tree," and every man's life is but a section or a branch of that tree, growing upon the trunk and contributing by its very existence to its further growth and continuation. Human life, therefore, is inevitably seen as a growth or a continuance, in which every man plays a part or a chapter in the family history, with its obligations toward the family as a whole, bringing upon itself and upon the family life shame or glory.

This sense of family consciousness and family honor is probably the only form of team spirit or group consciousness in Chinese life. In order to play this game of life as well as, or better than, another team, every member of the family must be careful not to spoil the game, or to let his team down by making a false move. He should, if possible, try to bring the ball further down the field. A derelict son is a shame to himself and to his family in exactly the same sense as a quarterback who makes a fumble and loses the ball. And he who comes out on top in the civil examinations is like a player who makes a touchdown. The glory is his own and at the same time that of his family. The benefits of one's becoming a chuangyüan ("No. 1" in the imperial examinations), or even a third-class chinshih, are both sentimentally and materially shared by members of his immediate family, his relatives, his clan, and even his town. For a hundred or two hundred years afterwards, the townspeople will still boast that they produced a chuangyüan in such and such a reign. In comparison with the family and town rejoicing when a man got a chuangyüan or chinshih and came home to place a golden-painted tablet of honor high upon his ancestral hall, with his mother probably shedding tears and the entire clan feeling themselves honored by the great occasion, the getting of a college diploma today is a pretty dull and tame affair.

In this picture of the family life, there is room for the greatest variety and color. Man himself passes through the stages of childhood, youth, maturity and old age: first being taken care of by others, then taking care of others, and in old age again being taken care of by others; first obeying and respecting others, and later being obeyed and respected in turn in proportion as he grows older. Above all, color is lent to this picture by the presence of women. Into this picture of the continuous family life comes woman, not as a decoration or a plaything, nor even essentially as a wife, but as a vital and essential part of the family tree—the very thing which makes continuity possible. For the strength of any particular branch of a family depends so much upon the woman married into the home and the blood she contributes to the family heritage. A wise patriarch is pretty careful to select women of sound heritage, as a gardener is careful to select the proper strain for grafting a branch. It is pretty well suspected that a man's life, particularly his home life, is made or unmade by the wife he marries, and the entire character of the future family is determined by her. The health of one's grandchildren and the type of family breeding that they are going to receive (upon which great emphasis is laid) depend entirely upon the breeding of the daughter-in-law herself. Thus there is a kind of amorphous and ill-defined eugenic system, based on belief in heredity and often placing great emphasis on menti (literally"door and home"or lineage or family standing), but in any case based on standards of desirability in the health, beauty and breeding of the bride as seen by the eyes of the parents or grandparents of the family. In general, the emphasis is upon family breeding (in the same sense that a Westerner would choose a girl from a"good home"), representing the fine old traditions of thrift, hard work, good manners and civility. And when sometimes a parent discovers to his sorrow that his son has married a worthless daughter-in-law with no manners, he always secretly curses the other family for not training their daughter better. Hence upon the mother and father devolves the duty of training their daughters so that they shall not be ashamed of them when they marry into another household—as, for instance, when they do not know how to cook or how to make a good New Year pudding.

According to the stream-of-life theory as seen in the family system, immortality is almost visible and touchable. Every grandfather seeing his grandchild going to school with a satchel feels that truly he is living over again in the life of the child, and when he touches the child's hand or pinches his cheeks, he knows it is flesh of his own flesh and blood of his own blood. His own life is nothing but a section of the family tree, or of the great family stream of life flowing on forever, and therefore he is happy to die. That is why a Chinese parent's greatest concern is to see that his sons and daughters are properly married before he dies, for that is an even more important concern than the site of his own grave or the selection of a good coffin. For he cannot know what kind of life his children are going to have until he sees with his own eyes what type of girls and men his sons and daughters marry, and if the daughters-in-law and sons-in-law look pretty satisfactory, he is quite willing to"close his eyes without regret"on his deathbed.

The net result of such a conception of life is that one gets a lengthened outlook on everything, for life is no more regarded as beginning and ending with that of the individual. The game is continued by the team after the center or the quarterback is put out of action. Success and failure begin to take on a different complexion. The Chinese ideal of life is to live so as not to be a shame to one' s ancestors and to have sons of whom one need not be ashamed. A Chinese official when resigning office often quotes the line:

Having sons, I am content with life;
Without office, my body is light.

The worst thing that can happen to a man, probably, is to have unworthy sons who cannot"maintain the family glory"or even the family fortune. The millionaire father of a gambling son sees his fortune dispersed already, the fortune that he has taken a life time to build up. When the son fails, the failure is absolute. On the other hand, a farsighted widow is able to endure years of misery and ignominy and even persecution, if she has a good boy of five. Chinese history and literature are full of such widows who endured all kinds of hardships and persecutions, but who lived for the day when their sons should do well and prosper, and perhaps even become prominent citizens... The success of widows in giving their children a perfect education of character and morals, through woman's generally more realistic sense, has often led me to think that fathers are totally unnecessary, so far as the upbringing of children is concerned. The widow always laughs the loudest because she laughs last.

Such an arrangement of life in the family, then, is satisfying because a man's life in all its biological aspects is well taken care of. That, after all, was Confucius' chief concern. The final ideal of government, as Confucius conceived it, was curiously biological:"The old shall be made to live in peace and security, the young shall learn to love and be loyal, that inside the chamber there may be no unmarried maids, and outside the chamber there may be no unmarried males."This is all the more remarkable because it is not merely a statement of a side issue, but of the final goal of government. This is the humanist philosophy known as tach'ing, or"fulfillment of instincts."Confucius wanted to be pretty sure that all our human instincts are satisfied, because only thus can we have moral peace through a satisfying life, and because only moral peace is truly peace. It is a kind of political ideal which aims at making politics unnecessary, because it will be a peace that is stable and based upon the human heart.

五. 乐享余年

　　据我的见解，中国的家族制度大概是一种对老者和幼者的个别准备的布置。因为童、幼、老三个时期须占到人生岁月之半，所以幼者和老者都应当使他们过满意的生活。其中幼者虽因不知人事而比较不会自己当心自己，但对于物质的享用需要不如老者那么深切。小孩对于物质供给的缺乏往往不太有感觉，所以贫苦人家的孩子常和富家的孩子一样快乐。他因没有鞋穿而赤脚，但在他未始不是一种舒适，而在老者，赤足便觉得十分难受了。这是因为幼童都较为充满生气。他有时虽也知道忧虑，但一会儿便会忘却。他不像老者那般，并没有钱财观念，有时也会收藏几张香烟里边的赠品券，但他的目的不是积财，而是想去调换一支气枪。老年人便与此不同，而去收藏自由公债了。这两种收藏举动在意趣上是不能比较的。其理由是：因为幼童不像成人那样受过生活的压迫，个人习嗜尚没有形成。他喝咖啡并不一定非某种牌子不可，无非是有什么吃什么；他并没有什么种族偏见，思想和概念都尚没有固定的轨道。所以老者比幼童更需要他人的帮助这事好似很奇怪，但其原因是老者的恐惧心较为明显，欲望较为无限制而已。

中国人在上古时代已有优视老年人的意识，这种意识我以为可以比拟西方的骑士精神和优视女人习惯。其实这种举动也可以称为武士精神。孟子所说"颁白者不负载于道路矣"，即表示一种优良政治的最后目标。孟子又列述世上四种最困苦的人为鳏、寡、孤、独。他说，第一第二两种应由一种政治经济的安排方法使他们男婚女嫁，各将其偶。他对于孤儿的处置没有提起，但当时已有养老院，而育婴堂也是各时代都有的。不过人人知道养老院和育婴堂终不足以替代家庭一般的感觉，只有家庭能给老年和幼童以一种相当满意的供给。小孩子自有父母爱护他们，毋庸细说。不过晚辈对于长辈的孝养，正如中国的俗谚"水往低处流"一般，不像长辈爱小辈那么自然，必须由文化去培植出来。一个自然人必会爱他的子女，但只有受过文化洗礼的人才会孝养父母、敬爱老年。这个教训到现在已成为大众所公认的原理，并且据有些学者说来，能得孝养父母的机会已成一种权利，而为人所渴望的了。父母病的时候未能亲侍汤药，死的时候未能送终，已被中国人视为终身莫大的遗憾；官员到了五六十岁尚不能迎养父母，于官署中晨昏定省，已被认为犯了一种道德上的罪名，而本人对于亲友和同僚也必定要时常设法解释不能迎养的理由。从前有一个人，因回到家里时父母死了，即不胜悲憾，说了下面这两句话：

树欲静而风不息，子欲养而亲不在。

我们应该可以假定如果人们能过一种诗意的生活，他就会拿晚年当做他一生中最快乐的时代，非但不再畏惧老年，反将希望这个时期早些来临，当它是一生中一个最美好最快乐的时期，而时常来先预备去享受它。我将东西两方的生活拿来做比较的时候，觉得两者之间虽有许多不同的地方，但绝对不相同的实在只有对于老年的态度这一点。这态度在东西两方绝对不同，而且区别分明，毫无折中调和的余地。两方对于性，对于女人，对于工作娱乐和成就，在态度上虽是不同，但都不过是相对的。例如：中国的夫妻之间的关系和西方的夫妻关系，根本上没有什么很大分别。父母和子女之间的关系也是如此。此外如对于个人的自由、民主制度、人民和统治者之间的关系等观念，实在也并没有什么极大的不同之处。但对于老年一事的态度，两方的态度竟绝对不同，所持的见地竟绝对相反。这一点在向人询问年龄和说出自己的年龄时，就可以极明白地看出来。中国习惯在拜访生人时，问过尊姓大名之后，接下来必问他贵庚。如对方很谦虚地回说只有二十三或二十八岁，问者必以"前程远大后福无量"一语去安慰他。但那人如回说已经三十五或三十八岁，则问者便会表现尊敬的态度，而赞他好福气。总之所回报的年龄越高，所受到的尊敬越深。如答话的已经五六十岁，则问者必低声下气地以晚辈自居，表示极端的尊敬。所以凡是年老的人，可能的话，都应该到中国去居住。因为在那里哪怕是白发龙钟的乞丐，讨起饭来也比别人容易些。中年的人常希望快些过他的五十岁生日。得意的商人和官员常大做四十岁生日，但是五十岁生日，即所谓年已半百，更为人所重视。以后每隔十年必做一次寿，六十岁生日比五十岁更快活，七十岁生日比六十岁更快活，如能做八十岁生日，更将被人视为得天独厚。颔下留起长须来，是祖父一辈人的特权。没有到这资格的人，如还没有孙子或年龄未过五十者，如若留须，常会被人背后讥笑。因此，年轻的人也都喜学做老成持重，抱着和老年人相同的见解。刚从中学毕业的少年书生，已在那里写"青年应知"和"青年应读"等类的文章，并以为父母者的态度而讨论青年的堕落问题了。

我们如了解中国人之如何珍视老年，便能明了为什么中国人都喜欢倚老卖老，自认为老。第一，照中国的礼貌，只有长者有发言的权利，年轻的人只许静听，所以中国有"少年用耳不用口"那句老话。凡有年龄较高的人在座时，年轻的人只许洗耳恭听。世人大都欢喜发言而受人听，因此，在中国必须到相当的年龄才有发言权利，使人期望早些达到老年，以便无论到什么地方都可以多说几句话。这种生活程序之中，人人须循序而进，每个人都有同等达到老年的机会，而没有一个人能躐等超前。当一个父亲教训他的儿子时，如若祖母走来插口，那做父亲的便须停口、谨敬恭听。这时他当然很羡慕那祖母的地位。年老的人能说："我所走过的桥比你所走过的街还要多几条。"因此，以经验而言，年轻的人在长者面前没有发言的权利，只能洗耳恭听，这是很公允的。

我虽然已很熟悉西方的生活，并很明白西方人对于老年的态度，但有时所听见的话仍使我非常诧异，很出我的意料。这种使我奇异的态度，常有所遇。我曾听见过一位年老的妇人说，她已有几个孙儿女，其中以长孙儿使她受到的感触最大，她的意思是长孙儿已如此长大，将反映她自己的年龄之高。我很明白美国人最恨别人说他已老，但意料不到他们的畏惧心竟会到这个地步。五十岁以下的人大都希望旁人视他为依然年富力强，这很在意中，但是一个头发已经花白的老妇人，在旁人提到她的年龄时尚要顾左右而言他，实在使我觉得出乎意料。当我在让一位老者先走进电梯或公共汽车时，心中自不免有认为他已老的意思，但我总不敢形之于口。有一天遇到这样一件事时，我无意间说了出来，不料那位很尊严的老者坐下去时，竟会向坐在他下手的太太用着讥笑的口气说我："这年轻的人，竟以为他比我年纪轻得多啊！"

这种情形太缺乏意识，使我不解其所以然。我很谅解年轻和中年未嫁的女人因为保爱其青春，所以不愿意将年纪告诉旁人。中国女郎达到二十二岁而尚未出嫁或定亲时，也常常感到一些恐惧。岁月很忍心地按部就班地消逝，一刻也不肯停留，女人常常怕被岁月遗弃，如在公园晚间园门关时不及出去而被关闭在里边一般。因此常常有人说，女人一生中最长的一年是二十九岁，直可以延长到三四年之久而依然是二十九岁。但除了这种情形以外，隐瞒年龄便属毫无意思。在旁人的眼光中，人非老何以能够聪明？年轻的人对于生命婚姻和真有价值的事物能知道些什么？我很谅解。因为西方生活的整个模型都过于重视青春，所以不论男女都不敢将自己的年龄告诉他人。一个年纪四十五岁的女书记，其实很富于精力，办事效能很高，但是她将年龄一旦说破，便将为了不可解的理由被人认为毫无用处，无怪她为了要保全饭碗起见，而不能不隐瞒年龄。这种生活的模型和对于青春的过于重视，都太缺乏意识，照我看来，竟毫无意义。这种情形显然是职业生活所造成，因为我深信在敬老上，家庭胜于办公室。除非美国人民渐渐觉得憎嫌工作效能和成就，上述的情形竟是无可避免的。我颇以为等到美国的为父者能视家庭而不是办公室为他生活中的理想处所，能公然如中国父亲一般泰然自若地告诉旁人他已有一个好儿子可以继续他的事业，并且觉得受其奉养很可夸耀时，他便会期望这种快乐时期的来临，在尚未到五十岁的时候，即要屈指计算，好像等得不耐烦了。

美国身体康健的老年人常对人说他尚年轻，而旁人也说他年轻，但实在的意义是说他康健，这真是一种语言上的不幸。老年健壮是人生的莫大幸运，但改称之为健壮年轻便将减削意义，使原来很完美的东西变为不完美了。实在说起来，这世界上再没有比一个健壮而智慧的老者更美丽的，有着红润的面颊，雪白的头发，以通晓世故的态度，用和蔼的口气，谈着做人的道理，中国人很明白这一点，所以画起老翁来总是红面白须，视之为人世终极快乐的象征。中国人所画的寿星，美国人大概也看见过的，他那高高的额角，红红的面孔，雪白的长须，笑容可掬的样子，是何等生动。他手抚长须，悠然自得，何等庄严，令人起敬。因为从没有人对他的智慧发生疑问，所以他极端自信。因为他见惯了人世的忧苦，所以极仁慈。对于富有生气的老者，我们每每说他们是老当益壮，像利奥德·乔治（David Lloyd Geroge）这样的人，我们每每称他为"老姜"，意即姜桂之性，越老越辣。

我在美国几乎连白须老者的影子也看不到，他们好似结了伴躲避我。我在美国已那么久了，只有一次在新泽西州看见过一个略具白须老者样子的人。这或者是保安剃刀的成绩，其可惜和愚笨正如中国北方的农民将各处山上的树木一起砍伐净尽，弄得美丽的青山都变成秃顶光皮不相上下。美国尚有一处宝藏需待他们去发现，这就是美丽和智慧的宝藏，美国人民发现时方能觉得这宝藏是何等赏心悦目。飘飘长髯的山姆大叔已不复可见，因为他已用保安剃刀将长须剃去，变成一个双颧高耸，双颊凹瘪，戴着一副牛角框眼镜，透出炯炯目光的滑稽样子了。这一变立刻使他失去了旧日的庄严伟大，那是何等可惜！我对最高法院问题所取的态度（这问题其实和我并不相干），完全系以爱好查尔斯·埃文思·休斯（Charles Evans Hughes，曾任美国首席大法官）的面貌而决定的。他简直已是美洲硕果仅存的伟大老人，试问此外还有别个吗？为了优待起见，自应让他退休，但如果说他已衰老不堪任事，在我看来竟是绝大的侮辱。他的面貌是雕刻家所认为最合理想的。

美国的老人依旧要如年轻人一般的忙劳，显然是个人主义推行得太过分所致。他以自立为荣，而以依赖晚辈为耻。美国宪法曾替人民规定下许多应享的权利，但不应遗漏了老年人应由其子女赡养这一条。因为这也是由服役而产生的一种权利和义务。为父母者在子女幼小时何等辛劳，子女小有病痛必整日整夜地服侍，换下来的尿布每天必须洗涤，须费二十余年的工夫方能完成教养，使他们可以出去应世做事。他们即费了这大的辛苦，到了老年时，应该由他们的子女赡养并受人尊敬，尚有拒绝不给予他们的道理吗？在普通方式的家庭生活中，凡是人都先受父母的教养，后来则接下去教养自己的子女，最后则受子女的赡养，程序极为自然，其间没有个人自傲的余地。中国人因为他们对生活的概念是完全以家庭中互助为基础，所以并没有个人独立的意识，因此，到了老年受子女们的赡养时，也不觉得有什么可耻的地方，反而将因有子女赡养他们而自己觉得欣幸。中国人的生存目的也仅此而已。

西方人则不然，他宁可住在底层有餐厅的旅馆中，出于"大公无私"的愿望，不愿为子女所累，不愿去干涉他们的家庭生活。他其实有干涉的权利，这种干涉即使将使子女们不愉快，但确属十分自然。因为一切生活，尤其是家常生活，本是一种节制课程。试想人在幼时，岂不都受父母的干涉吗？操行主义者以为子女须离开父母，在这种思想中，我们看到不干涉的逻辑。父母曾为我们费过一番极大的辛劳，如若我们在他们老而无能时尚不能容忍他们，则我们在家庭中尚能容忍什么人？一个人无论如何须学习自制，否则连婚姻也失去效力。试想骨肉的情爱奉侍，岂是旅馆仆役所能代替的吗？

中国人对于年老父母的躬亲奉侍概念，系完全根据于"有恩必报"的理由。一个人从朋友方面所受到的恩惠都可以用数字计算，但父母的养育之恩绝不是数字所能记录。中国的教孝论文中，一再提起洗尿布，这件事使轮到自己做父母时觉得有意义。所以为了报答起见，父母年老时，为子女者岂不应好好地侍奉，视其所好，每天以精美的膳食供养吗？为子女者尽孝道不是一件很容易的事情，不单是像医院看护服侍一个陌生病人一般，但求尽职就能算数的。以下是屠羲时所著《童子礼》中的一节。这篇文字从前小学生都当做教科书读，中间详述子女应该怎样对父母尽其孝道：

> 夏月侍父母，常须挥扇于其侧，以清炎暑及驱逐蚊蝇。冬月，则审察衣被之厚薄，炉火之多寡，时为增益；并候视窗户罅隙，使不为风寒所侵，务期父母安乐方已。
>
> 十岁以上，侵晨先父母起，梳洗毕，诣父母榻前，问夜来安否？如父母已起，则就房先作揖，后致问，问毕，乃一揖退。昏时，候父母将寝，则拂席整衾以待，已寝，则下帐闭户而后息。

因此在中国，哪个不期望做老人、做父母或祖父母？但其中实有一种佳趣，因此，中国内地的老年人都还牢守这个思想。最重要的一点是：凡人不能不老，如果他足够长寿，他当然渴望这样。愚拙的个人主义似乎假定个人可以在抽象的境地中生存，可以实际独立。如若舍弃这个思想，便会承认我们必须如此计划我们的生活方式，以使人生的极乐时期出现在老年之时，而并不在知识未充分的青年时期。因为我们如若取持和此相反的态度，则我们将于不知不觉之间和光阴做必不能获胜的竞赛，对于未来永远怀着一种恐惧，深怕它的莅临。一个人绝不能不老，凡自己以为不老的人，都是在那里自欺欺人。人类不能和大自然相对抗，何不安于由此而老呢？生命的交响曲，其终点处应是伟大的和平晴朗，物质舒适和精神上的满足，而不是破锣破鼓的刺耳响声。

V. On Growing Old Gracefully

THE Chinese family system, as I conceive it, is largely an arrangement of particular provision for the young and the old, for since childhood and youth and old age occupy half of our life, it is important that the young and the old live a satisfactory life. It is true that the young are more helpless and can take less care of themselves, but on the other hand, they can get along better without material comforts than the old people. A child is often scarcely aware of material hardships, with the result that a poor child is often as happy as, if not happier than, a rich child. He may go barefooted, but that is a comfort, rather than a hardship to him, whereas going barefooted is often an intolerable hardship for old people.   This comes from the child's greater vitality, the bounce of youth. He may have his temporary sorrows, but how easily he forgets them. He has no idea of money and no millionaire complex, as the old man has. At the worst, he collects only cigar coupons for buying a pop-gun, whereas the dowager collects Liberty Bonds. Between the fun of these two kinds of collection there is no comparison. The reason is the child is not yet intimidated by life as all grown-ups are. His personal habits are as yet unformed and he is not a slave to a particular brand of coffee, and he takes whatever comes along. He has very little racial prejudice and absolutely no religious prejudice. .His thoughts and ideas have not fallen into certain ruts. Therefore, strange as it may seem, old people are even more dependent than the young because their fears are more definite and their desires are more delimited.

Something of this tenderness toward old age existed already in the primeval consciousness of the Chinese people, a feeling that I can compare only to the Western chivalry and feeling of tenderness toward women. If the early Chinese people had any chivalry, it was manifested not toward women and children, but toward the old people. That feeling of chivalry found clear expression in Mencius in some such saying as,"The people with grey hair should not be seen carrying burdens on the street,"which was expressed as the final goal of a good government. Mencius also described the four classes of the world's most helpless people as:"the widows, widowers, orphans and old people without children."Of these four classes the first two were to be taken care of by a political economy which should be so arranged that there would be no unmarried men and women. What was to be done about the orphans Mencius did not say, so far as we know, although orphanages have always existed throughout the ages, as well as pensions for old people. Every one realizes, however, that orphanages and old age pensions are poor substitutes for the home. The feeling is that the home alone can provide anything resembling a satisfactory arrangement for the old and the young. But for the young, it is to be taken for granted that not much need be said, since there is natural paternal affection. "Water flows downwards and not upwards,"the Chinese always say, and therefore the affection for parents and grandparents is something that stands more in need of being taught by culture. A natural man loves his children, but a cultured man loves his parents. In the end, the teaching of love and respect for old people became a generally accepted principle, and if we are to believe some of the writers, the desire to have the privilege of serving their parents in their old age actually became a consuming passion. The greatest regret a Chinese gentleman could have was the eternally lost opportunity of serving his old parents with medicine and soup on their deathbed, or not to be present when they died. For a high official in his fifties or sixties not to be able to invite his parents to come from their native village and stay with his family at the capital, "seeing them to bed every night and greeting them every morning," was to commit a moral sin of which he should be ashamed and for which he had constantly to offer excuses and explanations to his friends and colleagues. This regret was expressed in two lines by a man who returned too late to his home, when his parents had already died:

> *The tree desires repose, but the wind will not stop;*
> *The son desires to serve, but his parents are already gone.*

It is to be assumed that if man were to live this life like a poem, he would be able to look upon the sunset of his life as his happiest period, and instead of trying to postpone the much feared old age, be able actually to look forward to it, and gradually build up to it as the best and happiest period of his existence. In my efforts to compare and contrast Eastern and Western life, I have found no differences that are absolute except in this matter of the attitude towards age, which is sharp and clear-cut and permits no intermediate positions. The differences in our attitude towards sex, toward women, and toward work, play and achievement are all relative. The relationship between husband and wife in China is not essentially different from that in the West, nor even the relationship between parent and child. Not even the ideas of individual liberty and democracy and the relationship between the people and their ruler are, after all, so very different . But in the matter of our attitude toward age, the difference is absolute, and the East and the West take exactly opposite points of view. This is clearest in the matter of asking about a person's age or telling one's own. In China, the first question a person asks the other on an official call, after asking about his name and surname is, "What is your glorious age?" If the person replies apologetically that he is twenty-three or twenty-eight, the other party generally comforts him by saying that he has still a glorious future, and that one day he may become old. But if the person replies that he is thirty-five or thirty-eight, the other party immediately exclaims with deep respect, "Good luck!"; enthusiasm grows in proportion as the gentleman is able to report a higher and higher age, and if the person is anywhere over fifty, the inquirer immediately drops his voice in humility and respect. That is why all old people, if they can, should go and live in China, where even a beggar with a white beard is treated with extra kindness. People in middle age actually look forward to the time when they can celebrate their fifty-first birthday, and in the case of successful merchants or officials, they would celebrate even their forty-first birthday with great pomp and glory. But the fifty-first birthday, or the half-century mark, is an occasion of rejoicing for people of all classes. The sixty-first is a happier and grander occasion than the fifty-first and the seventy-first is still happier and grander, while a man able to celebrate his eighty-first birthday is actually looked upon as one specially favored by heaven. The wearing of a beard becomes the special prerogative of those who have become grandparents, and a man doing so without the necessary qualifications, either of being a grandfather or being on the other side of fifty, stands in danger of being sneered at behind his back. The result is that young men try to pass themselves off as older than they are by imitating the pose and dignity and point of view of the old people, and I have known young Chinese writers graduated from the middle schools, anywhere between twenty-one and twenty-five, writing articles in the magazines to advise what "the young men ought and ought not to read," and discussing the pitfalls of youth with a fatherly condescension.

This desire to grow old and in any case to appear old is understandable when one understands the premium generally placed upon old age in China. In the first place, it is a privilege of the old people to talk, while the young must listen and hold their tongue. "A young man is supposed to have ears and no mouth," as a Chinese saying goes. Men of twenty are supposed to listen when people of thirty are talking, and these in turn are supposed to listen when men of forty are talking. As the desire to talk and to be listened to is almost universal, it is evident that the further along one gets in years, the better chance he has to talk and to be listened to when he goes about in society. It is a game of life in which no one is favored, for everyone has a chance of becoming old in his time. Thus a father lecturing his son is obliged to stop suddenly and change his demeanor the moment the grandmother opens her mouth. Of course he wishes to be in the grandmother's place. And it is quite fair, for what right have the young to open their mouth when the old men can say, "I have crossed more bridges than you have crossed streets!"What right have the young got to talk ?

In spite of my acquaintance with Western life and the Western attitude toward age, I am still continually shocked by certain expressions for which I am totally unprepared. Fresh illustrations of this attitude come up on every side. I have heard an old lady remarking that she has had several grandchildren, but, "It was the first one that hurt."With the full knowledge that American people hate to be thought of as old, one still doesn't quite expect to have it put that way. I have made allowance for people in middle age this side of fifty, who, I can understand, wish to leave the impression that they are still active and vigorous, but I am not quite prepared to meet an old lady with gray hair facetiously switching the topic of conversation to the weather, when the conversation without any fault of mine naturally drifted toward her age. One continually forgets it when allowing an old man to enter an elevator or a car first; the habitual expression"after age"comes up to my lips, then I restrain myself and am at a loss for what to say in its place. One day, being forgetful, I blurted out the usual phrase in deference to an extremely dignified and charming old man, and the old man seated in the car turned to his wife and remarked jokingly to her, "This young man has the cheek to think that he is younger than myself! "

The whole thing is as senseless as can be. I just don't see the point. I can understand young and middle-aged unmarried women refusing to tell their age, because there the premium upon youth is entirely natural. Chinese girls, too, get a little scared when they reach twenty-two and are unmarried or not engaged. The years are slipping by mercilessly. There is a feeling of fear of being left out, what the Germans call a Torschlusspanik, the fear of being left in the park when the gates close at night. Hence it has been said that the longest year of a woman' s life is when she is twenty-nine; she remains twenty-nine for three or four or five years. But apart from this, the fear of letting people know one's age is nonsensical. How can one be thought wise unless one is thought to be old? And what do the young really know about life, about marriage and about the true values? Again I can understand that the whole pattern of Western life places a premium on youth and therefore makes men and women shrink from telling people their age. A perfectly efficient and vigorous woman secretary of forty-five is, by a curious twist of reasoning, immediately thought of as worthless when her age becomes known. What wonder that she wants to hide her age in order to keep her job? But then the pattern of life itself and this premium placed upon youth are nonsensical. There is absolutely no meaning to it, so far as I can see. This sort of thing is undoubtedly brought about by business life, for I have no doubt there must be more respect for old age in the home than in the office. I see no way out of it until the American people begin somewhat to despise work and efficiency and achievement. I suspect when an American father looks upon the home and not the office as his ideal place in life, and can openly tell people, as Chinese parents do, with absolute equanimity that now he has a good son taking his place and is honored to be fed by him, he will be anxiously looking forward to that happy time, and will count the years impatiently before he reaches fifty.

It seems a linguistic misfortune that hale and hearty old men in America tell people that they are"young,"or are told that they are"young"when really what is meant is that they are healthy. To enjoy health in old age, or to be"old and healthy,"is the greatest of human luck, but to call it"healthy and young"is but to detract from that glamour and impute imperfection to what is really perfect. After all, there is nothing more beautiful in this world than a healthy wise old man, with"ruddy cheeks and white hair,"talking in a soothing voice about life as one who knows it. The Chinese realize this, and have always pictured an old man with"ruddy cheeks and white hair' as the symbol of ultimate earthly happiness. Many Americans must have seen Chinese pictures of the God of Longevity, with his high forehead, his ruddy face, his white beard–and how he smiles! The picture is so vivid. He runs his fingers through the thin flowing beard coming down to the breast and gently strokes it in peace and contentment, dignified because he is surrounded with respect, self-assured because no one ever questions his wisdom, and kind because he has seen so much of human sorrow. To persons of great vitality, we also pay the compliment of saying that"the older they grow, the more vigorous they are,"and a person like David Lloyd George would be referred to as"Old Ginger,"because he gains in pungency with age.

On the whole, I find grand old men with white beards missing in the American picture. I know that they exist, but they are perhaps in a conspiracy to hide themselves from me. Only once, in New Jersey, did I meet an old man with anything like a respectable beard. Perhaps it is the safety razor that has done it, a process as deplorable and ignorant and stupid as the deforestation of the Chinese hills by ignorant farmers, who have deprived North China of its beautiful forests and left the hills as bald and ugly as the American old men's chins. There is yet a mine to be discovered in America, a mine of beauty and wisdom that is pleasing to the eye and thrilling to the soul, when the American has opened his eyes to it and starts a general program of reclamation and reforestation. Gone are the grand old men of America! Gone is Uncle Sam with his goatee, for he has taken a safety razor and shaved it off, to make himself look like a frivolous young fool with his chin sticking out instead of being drawn in gracefully, and a hard glint shining behind horn-rimmed spectacles. What a poor substitute that is for the grand old figure! My attitude on the Supreme Court question (although it is none of my business) is purely determined by my love for the face of Charles Evans Hughes. Is he the only grand old man left in America, or are there more of them? He should retire, of course, for that is only being kind to him, but any accusation of senility seems to me an intolerable insult. He has a face that we would call"a sculptor's dream."

I have no doubt that the fact that the old men of America still insist on being so busy and active can be directly traced to individualism carried to a foolish extent. It is their pride and their love of independence and their shame of being dependent upon their children. But among the many human rights the American people have provided for in their Constitution, they have strangely forgotten about the right to be fed by their children, for it is a right and an obligation growing out of service. How can any one deny that parents who have toiled for their children in their youth, have lost many a good night's sleep when they were ill, have washed their diapers long before they could talk and have spent about a quarter of a century bringing them up and fitting them for life, have the right to be fed by them and loved and respected when they are old? Can one not forget the individual and his pride of self in a general scheme of home life in which men are justly taken care of by their parents and, having in turn taken care of their children, are also justly taken care of by the latter? The Chinese have not got the sense of individual independence because the whole conception of life is based upon mutual help within the home; hence there is no shame attached to the circumstance of one's being served by his children in the sunset of one's life. Rather it is considered good luck to have children who can take care of one. One lives for nothing else in China.

In the West, the old people efface themselves and prefer to live alone in some hotel with a restaurant on the ground floor, out of consideration for their children and an entirely unselfish desire not to interfere in their home life. But the old people have the right to interfere, and if interference is unpleasant, it is nevertheless natural, for all life, particularly the domestic life, is a lesson in restraint. Parents interfere with their children anyway when they are young, and the logic of non-interference is already seen in the results of the Behaviorists, who think that all children should be taken away from their parents. If one cannot tolerate one's own parents when they are old and comparatively helpless, parents who have done so much for us, whom else can one tolerate in the home? One has to learn self-restraint anyway, or even marriage will go on the rocks. And how can the personal service and devotion and adoration of loving children ever be replaced by the best hotel waiters?

The Chinese idea supporting this personal service to old parents is expressly defended on the sole ground of gratitude. The debts to one's friends may be numbered, but the debts to one's parents are beyond number. Again and again, Chinese essays on filial piety mention the fact of washing diapers, which takes on significance when one becomes a parent himself. In return, therefore, is it not right that in their old age, the parents should be served with the best food and have their favorite dishes placed before them? The duties of a son serving his parents are pretty hard, but it is sacrilege to make a comparison between nursing one's own parents and nursing a stranger in a hospital. For instance, the following are some of the duties of the junior at home, as prescribed by T'u Hsishih and incorporated in a book of moral instruction very popular as a text in the old schools:

In the summer months, one should, while attending to his parents, stand by their side and fan them, to drive away the heat and the flies and mosquitoes. In winter, he should see that the bed quilts are warm enough and the stove fire is hot enough, and see that it is just right by attending to it constantly. He should also see if there are holes or crevices in the doors and windows, that there may be no draft, to the end that his parents are comfortable and happy.

A child above ten should get up before his parents in the morning, and after the toilet go to their bed and ask if they have had a good night. If his parents have already gotten up, he should first curtsy to them before inquiring after their health, and should retire with another curtsy after the question. Before going to bed at night, he should prepare the bed, when the parents are going to sleep, and stand by until he sees that they have fallen off to sleep and then pull down the bed curtain and retire himself.

Who, therefore, wouldn't want to be an old man or an old father or grandfather in China?

This sort of thing is being very much laughed at... but there is a charm to it which makes any old gentleman inland cling to it... The important point is that every man grows old in time, if he lives long enough, as he certainly desires to. If one forgets this foolish individualism which seems to assume that an individual can exist in the abstract and be literally independent, one must admit that we must so plan our pattern of life that the golden period lies ahead in old age and not behind us in youth and innocence. For if we take the reverse attitude, we are committed without our knowing to a race with the merciless course of time, forever afraid of what lies ahead of us—a race, it is hardly necessary to point out, which is quite hopeless and in which we are eventually all defeated. No one can really stop growing old; he can only cheat himself by not admitting that he is growing old. And since there is no use fighting against nature, one might just as well grow old gracefully. The symphony of life should end with a grand finale of peace and serenity and material comfort and spiritual contentment, and not with the crash of a broken drum or cracked cymbals.

全两册
中英双语

The Importance
of Living

生活的
艺术

林语堂 著

越裔 译

下

湖南文艺出版社
HUNAN LITERATURE AND ART PUBLISHING HOUSE

博集天卷
CS·BOOKY

THE IMPORTANCE OF LIVING

By Lin Yutang

This edition arranged with Curtis Brown Group Ltd.

through Andrew Nurnberg Associates International Limited

著作权合同登记号：图字18-2016-127

图书在版编目（CIP）数据

生活的艺术 = The Importance of Living：全两册：汉英对照/林语堂著；越裔译. — 长沙：湖南文艺出版社，2017.6（2021.8重印）

ISBN 978-7-5404-8081-3

Ⅰ. ①生… Ⅱ. ①林… ②越… Ⅲ. ①人生哲学—汉、英 Ⅳ. ①B821

中国版本图书馆CIP数据核字（2017）第095669号

上架建议：名家经典·文化

SHENGHUO DE YISHU: QUAN LIANG CE: HAN-YING DUIZHAO

生活的艺术：全两册：汉英对照

作　　者：林语堂
译　　者：越　裔
出 版 人：曾赛丰
责任编辑：薛　健　刘诗哲
监　　制：蔡明菲　邢越超
特约策划：王　维
特约编辑：蔡文婷
版权支持：辛　艳　张雪珂
营销支持：文刀刀　周　茜
封面设计：棱角视觉
版式设计：李　洁
出　　版：湖南文艺出版社
　　　　　（长沙市雨花区东二环一段508号　邮编：410014）
网　　址：www.hnwy.net
印　　刷：三河市兴博印务有限公司
经　　销：新华书店
开　　本：880mm×1270mm　1/32
字　　数：688千字
印　　张：28
版　　次：2017年6月第1版
印　　次：2021年8月第3次印刷
书　　号：ISBN 978-7-5404-8081-3
定　　价：88.00元（全两册）

若有质量问题，请致电质量监督电话：010-59096394
团购电话：010-59320018

/ Chapter Nine **The Enjoyment of Living** / ────

第九章

生活的享受

一. 安卧眠床

　　我好像终将成为一个走方式的哲学家，但这也是无法的事，一般的哲学好似都属于一种将简单的事情弄成令人难懂的学科，但我的心目中能想象到一种相反的哲学，即是将烦难的事情化成简单的科学。一般的哲学中虽用物质主义、人性主义、超凡主义、多元主义、什么主义的冗长字眼，但我终以为这类学说未必能比我的哲学更精深。人类的生活终不过包括吃饭、睡觉、朋友间的离合、接风、饯行、哭笑、每隔两星期左右理一次发、植树、浇花、仁望邻人从他的屋顶掉下来等类的平凡事情。大学哲学家用深奥的字句来描写这类简单的生活状态，无非是一种遮掩概念的极端缺乏和模糊的技巧而已。所以，哲学实已渐渐趋近于使人类对于自己的事情更加不懂。哲学目前的成就仅是：越加解说，越加使人模糊。

　　安睡眠床艺术的重要性，能感觉的人至今甚少，这是很令人惊异的。我的意见以为，世上所有的重要发明，不论科学的或哲学的，其中十有九桩都是科学家或哲学家在清晨二点到五点之间，蜷卧于床上时忽然得到的。

　　有些人在白天睡觉，有些人则在晚间睡觉。这里的所谓“睡觉”也做说谎解（按：英文中的 Lying 一字做安睡解，也做说谎解）。我觉得凡是同意我深信安睡眠床是人生最大乐事之一者都是诚实人，而不信者都是说谎人。他们简直是在白天说谎。提倡道德者，幼儿园教师，读《伊索寓言》者，即属于这一类的人。至于和我一般肯坦白承认安卧眠床艺术理应有意识地培植者，则尽是诚实的人，都是宁可阅读不含道德教训的故事如《爱丽丝梦游仙境》（Alice in Wonderland）之类者。

安睡卧床，对身体和心灵，究竟有什么意义呢？在身体上，这是和外界隔绝而独隐。人在这个时候，是将其身体置放于最宜于休息、和平，以及沉思的姿势。安睡易有一种适宜和舒服的感觉。生活大艺术家孔子从来是"寝不尸"，即不要像僵尸一般地挺睡，而必须蜷腿侧卧。我也觉得蜷腿睡在床上，是人生最大乐事之一。两臂的安置也极关重要，须十分适宜，方能达到身体上的极度愉快和心灵上的极度活泼。我深信最适宜的姿势不是平卧床上，而是睡在斜度约在三十度的软木枕头上，两臂或一臂搁在头的后面。用这种姿势，不论哪一个诗人都能写出不朽的佳作，不论哪一个哲学家都能改革人类思想，不论哪一个科学家都有划时代的新发明。

寂静和沉思的价值，能感觉到的人很少，这是令人惊奇的。安卧眠床艺术，其意义不单是令人在整天的劳苦工作之后，在和人相见谈话、无意义地说了许多废话之后，在哥哥姐姐遇事必要矫正以便保护你升到天堂、致使你的神经极受刺激之后，得到身体上的休息。还有更进一层的意义。这艺术如果加以相当培植，可以成为一种心灵上的大扫除。有许多生意人，办公桌上安着三架电话机，片刻不停地一天忙到晚，还自己觉得非此不可，引以为慰，但他实在不知道倘若在半夜后或清晨间安睡在床上做一小时的沉思，反而可以赚进加倍的钱。一个人即使睡到八点钟方起身，那又有什么关系？他如在洗脸刷牙之前，先在床上悠闲地吸几支香烟，将这一天所要做的事情计划一下，而不要匆忙地起身，则对他的益处将不仅以倍数计算。这时候他穿着宽大的睡衣，以最舒服的姿势睡在床上，没有紧狭的内衣，牵扳的背带，窒息的硬领，也没有很重的皮鞋束缚他，使他那白天势必失去自由的足趾也得到了解放舒适，这时，他的生意头脑方能真正运用。因为一个人的头脑，只在他的足趾自由时，方是真正自由的；只在头脑自由时，他方有真正思想的可能。在如此舒适的境地中，他能思量昨天的成就和失败，并将当日的事情分其轻重，决定进行。一个商人不妨先预备好一切，到十点钟时再走进办公室去。这较胜于在九点钟，或甚至在八点三刻时，即像奴隶监工一般地赶到办公室，就像中国人的所谓"无事忙"。

对于思想家、发明家、"概念家"（有想法的人），在床上一小时的安睡，其所助犹不止于此。一个著作家在这种姿势中，能比整天坐在写字台前得到更多的论文或小说资料。因为这时节，他完全不受电话、来客和日常小节的烦扰。他好似从一片玻璃或一挂珠帘中看到人间的生活，而现实世界的周围好似悬着一圈云彩，使它增添了一种神奇的美丽。这时他所看见的，不是生硬的生活，而已变为一幅比生活更真实的画像，如倪云林或米芾的名画一般。

睡在床上，所以有益于人的，理由大概如下：一个人睡在床上时，他的筋肉静息，血液的流行较为平顺有节，呼吸较为调匀，视觉听觉和脉系神经也几乎完全静息，造成一种身体上的静态，所以能使心思集中，不论于概念或感觉都更为纯粹。就是在感觉方面，例如嗅觉和听觉，也是在这个时候最为锐敏。所以好的音乐须卧而听之。李笠翁于他所著的"杨柳"篇中说：人们须在清晨未起身时，卧听鸟的叫声。我们在清晨苏醒后，睡在床上听百鸟的鸣声，这其实是何等美丽的境界啊！百鸟的鸣声就是在大多城市中也可以听到，不过我敢说，能够感觉到的人很少罢了。以下所述，即某天清晨我在上海寓所听到而记下来的：

这天，我在一宵好梦之后，于五点钟时醒来，即听到一阵极为悦耳的声音。最初所听见的是高低不一的厂家笛声。稍停是一阵远远的马蹄"嘚嘚"声，大概是几个骑马的印度巡捕在愚园路上经过。在寂静中，我所享受的美的愉快更胜于勃拉姆斯的交响曲。又过一阵，即来了一阵细碎的鸟鸣声。可惜我对鸟类没有什么研究，所以不知道叫的是什么鸟，但我的享受相同。

同时，自然还有别的声音。有几个外国青年，大概是在外面狂欢了一宵，这时回家敲后门。一个清道夫在打扫隔壁的弄堂，扫帚的"唰唰"声清晰可闻。忽然之间，大概是一只野鸭在天空一声长唳，悠扬不绝。六点二十五分左右，我听见沪杭甬火车隆隆之声自远而来，到极司非而路车站停止。隔壁房中有一两个小孩的啼叫声。此后各处渐有人声，一刻增多一刻，因而知道各处已在那里渐渐上市了。我自己的屋中，仆人也一一起身，即听见开窗和铁钩插上去的声音，轻轻的咳嗽声，轻轻的足声，杯盘碗盏声，忽然又有一个小孩呼妈妈声。

这些就是那天早晨我在上海寓所听到的音乐会之《协奏曲》。

那年的春天，我所最爱听的就是鸸鹋的鸣声。它们在互相叫唤之中共有四个音阶（即 do. mi：re-：- ti）。其中 re 的延长约三拍，在第三拍的中间突然中断，接上一个低的 ti 音。这种鸣声，我在南方的山中时常听到。最有趣的是，每天清晨一只雄鸟必先在我寓所附近的树上叫起来，随后雌鸟即在离开约百码以外之处以鸣声相答。它们的鸣声的快慢有时也有参差，似乎是因心境的变迁，有时则末一短音不叫出来。那地方各种鸟鸣声种类不一，但鹧鸪的鸣声最动人。各种鸟鸣声悦耳异常，除以音乐比拟之外，实在不能用字句形容。据我所能辨别者，其中有百灵鸟、喜鹊、啄木鸟和鸽子。每天早晨，老鸦的叫声最迟，理由大概是如李笠翁所说，别种鸟类多畏惧人类的猎枪和儿童的石子，所以它们必在清早人类尚未起身之前出来奏它们的音乐，以免被人类打断，而老鸦并不畏惧，所以它们起身较迟。

I. On Lying in Bed

IT seems I am destined to become a market philosopher, but it can't be helped. Philosophy generally seems to be the science of making simple things difficult to understand, but I can conceive of a philosophy which is the science of making difficult things simple. In spite of names like "mate rialism," "humanism," "transcendentalism," "pluralism," and all the other longwinded "isms," I contend that these systems are no deeper than my own philosophy. Life after all is made up of eating and sleeping, of meeting and saying good-by to friends, of reunions and farewell parties, of tears and laughter, of having a haircut once in two weeks, of watering a potted flower and watching one's neighbor fall off his roof, and the dressing up of our notions concerning these simple phenomena of life in a kind of academic jargon is nothing but a trick to conceal either an extreme paucity or an extreme vagueness of ideas on the part of the university professors. Philosophy therefore has become a science by means of which we begin more and more to understand less and less about ourselves. What the philosophers have succeeded in is this: the more they talk about it, the more confused we become.

It is amazing how few people are conscious of the importance of the art of lying in bed, although actually in my opinion nine-tenths of the world's most important discoveries, both scientific and philosophical, are come upon when the scientist or philosopher is curled up in bed at two or five o'clock in the morning.

Some people lie in the daytime and others lie at night. Now by"lying"I mean at the same time physical and moral lying, for the two happen to coincide. I find that those people who agree with me in believing in lying in bed as one of the greatest pleasures of life are the honest men, while those who do not believe in lying in bed are liars and actually lie a lot in the daytime, morally and physically. Those who lie in the daytime are the moral uplifters, kindergarten teachers and readers of Aesop's Fables, while those who frankly admit with me that a man ought to consciously cultivate the art of lying in bed are the honest men who prefer to read stories without a moral like Alice in Wonderland.

Now what is the significance of lying in bed, physically and spiritually? Physically, it means a retreat to oneself, shut up from the outside world, when one assumes the physical posture most conducive to rest and peace and contemplation. There is a certain proper and luxurious way of lying in bed. Confucius, that great artist of life,"never lay straight" in bed"like a corpse,"but always curled up on one side. I believe one of the greatest pleasures of life is to curl up one's legs in bed. The posture of the arms is also very important, in order to reach the greatest degree of aesthetic pleasure and mental power. I believe the best posture is not lying flat on the bed, but being upholstered with big soft pillows at an angle of thirty degrees with either one arm or both arms placed behind the back of one's head. In this posture any poet can write immortal poetry, any philosopher can revolutionize human thought, and any scientist can make epoch-making discoveries.

It is amazing how few people are aware of the value of solitude and contemplation. The art of lying in bed means more than physical rest for you, after you have gone through a strenuous day, and complete relaxation, after all the people you have met and interviewed, all the friends who have tried to crack silly jokes, and all your brothers and sisters who have tried to rectify your behavior and sponsor you into heaven have thoroughly got on your nerves. It is all that, I admit. But it is something more. If properly cultivated, it should mean a mental house-cleaning. Actually, many business men who pride themselves on rushing about in the morning and afternoon and keeping three desk telephones busy all the time on their desk, never realize that they could make twice the amount of money, if they would give themselves one hour's solitude awake in bed, at one o'clock in the morning or even at seven. What does it matter even if one stays in bed at eight o'clock? A thousand times better that he should provide himself with a good tin of cigarettes on his bedside table and take plenty of time to get up from bed and solve all his problems of the day before he brushes his teeth. There, comfortably stretched or curled up in his pyjamas, free from the irksome woolen underwear or the irritating belt or suspenders and suffocating collars and heavy leather boots, when his toes are emancipated and have recovered the freedom which they inevitably lose in the daytime, the real business head can think, for only when one's toes are free is his head free, and only when one's head is free is real thinking possible. Thus in that comfortable position, he can ponder over his achievements and mistakes of yesterday and single out the important from the trivial in the day's program ahead of him. Better that he arrived at ten o'clock in his office master of himself, than that he should come punctually at nine or even a quarter before to watch over his subordinates like a slave driver and then"hustle about nothing,"as the Chinese say.

But for the thinker, the inventor and the man of ideas, lying quietly for an hour in bed accomplishes even more. A writer could get more ideas for his articles or his novel in this posture than he could by sitting doggedly before his desk morning and afternoon. For there, free from telephone calls and well-meaning visitors and the common trivialities of everyday life, he sees life through a glass or a beaded screen, as it were, and a halo of poetic fancy is cast around the world of realities and informs it with a magic beauty. There he sees life not in its rawness, but suddenly transformed into a picture more real than life itself, like the great paintings of Ni Yünlin or Mi Fu.

Now what actually happens in bed is this. When one is in bed the muscles are at rest, the circulation becomes smoother and more regular, respiration becomes steadier, and all the optical, auditory and vaso-motor nerves are more or less completly at rest, bringing about a more or less complete physical quietude, and therefore making mental concentration; whether on ideas or on sensations, more absolute. Even in respect to sensations, those of smell or hearing for example, our senses are the keenest in that moment. All good music should be listened to in the lying condition. Li Liweng said in his essay on "Willows" that one should learn to listen to the birds at dawn when lying in bed. What a world of beauty is waiting for us, if we learn to wake up at dawn and listen to the heavenly concert of the birds! Actually there is a profusion of bird music in most towns, although I am sure many residents are not aware of it. For instance, this was what I recorded of the sounds I heard in Shanghai one morning:

This morning I woke up at five after a very sound sleep and listened to a most gorgeous feast of sounds. What woke me up were the factory whistles of a great variety of pitch and force. After a while, I heard a distant clatter of horse's hoofs; it must have been cavalry passing down Yuyuen Road; and in that quiet dawn it gave me more aesthetic delight than a Brahms symphony. Then came a few early chirps from some kind of birds. I am sorry I am not proficient in birdlore, but I enjoyed them all the same.

There were other sounds of course—some foreigner's "boy," presumably after a night
of dissipation outside, appeared at about half past five and began to knock at someone's
back door. A scavenger was then heard sweeping a neighboring alleyway with the swish-
swash of his bamboo broom. All of a sudden, a wild duck, I suppose, would sail by in
the sky, leaving echoes of his kung-tung in the air. At twenty-five past six, I heard the
distant rumble of the engine of the Shanghai-Hangchow train arriving at the Jessfield
Station. There were one or two sounds coming from the children in their sleep in the
next room. Life then began to stir and a distant hum of human activities in the near
and distant neighborhood gradually increased in volume and intensity. Downstairs in
the house itself, the servants had got up, too. Windows were being opened. A hook was
being placed in position. A slight cough. A soft tread of footsteps. A clanging of cups and
saucers. And suddenly the baby cried, "Mamma! "

This was the natural concert I heard that morning in Shanghai.

Throughout the whole spring that year my greatest delight was to
listen to a kind of bird probably called a quail or partridge in English.
Its lovecall consists of four notes (do. mi: re—:—. ti,) the re lasting
two or three beats and ending in the middle of a beat, followed by
an abrupt, staccato ti in the lower octave. It is the song I used to hear
in the mountains in the south. The most beautiful part of it was that
a male bird would start the call on top of a tree about twenty yards
from my place early at dawn, and a female bird would counterpoint
it at a distance of about a hundred yards. Then once in a while there
would be a slight variation, a quickening as it were of tempo and of
the bird's heart, and the last staccato note would be left out. This bird-
song stands out preeminently among those of others, of which there is
a great profusion. I am at a loss to describe these songs except by resort
to musical notation, but I know they include the songs of orioles and
magpies and woodpeckers, and the cooing of pigeons. The sparrows seem
to wake up later, and the reason, I suppose, must be as our great epicure-

poet Li Liweng gave it. The other birds have to sing early because they are continually afraid of men's guns and children's stones during the day. These birds, therefore, can sing at ease only before this insufferable human species wake up from their sleep. As soon as men wake up, the birds can never finish their song at ease. But the sparrows can, because they are not afraid, and therefore they can sleep longer.

二. 坐在椅中

我向来以喜欢躺在椅中出名，所以我要写一段坐椅法的哲学。朋友之中，喜欢躺在椅中者不止我一个，但不知如何单是我出名，至少在中国文艺界中是如此。我认为在这现代世界中，我并不是唯一的好躺椅中者，而人之说我者，也有些言过其实。这件事的经过是如此，某年我刊发了一本《论语》(Analects Fortnightly) 杂志。其中，我颇力辩所谓吸烟之害并无其事。杂志当中虽没有刊载卷烟广告，但文字中很多称赞尼古丁的美德的话，因此传了开去，说我一天到晚不做事情，只是躺在椅中吸雪茄。我虽屡次否认，并极力声明我实在是中国最勤于做事者之一，但传说之词依然风行一时，甚至成为我是被人憎恶的有闲知识阶级之一的证据。两年后，又因我刊行了一种注重通俗文章的杂志，于是更落实我是一个懒鬼。当时我因看不惯流行文章的体裁过于迟钝、不忠实和虚诞，认为还是旧式私塾命十二三岁的孩子做"救国"和"恒心之德"等类题目的文章的遗毒，故而以为必须提倡一种坦白通俗的文体，方能解放中国文章，使之脱去陈腐的桎梏。但我于不经意之间，将通俗文体写成潇洒文体……因而我被认为是中国懒惰成家中最懒惰者之一，"在这国难时期中，更为杀无赦"。

我承认时常躺在朋友家客室里的椅中，但别人何尝不如此？如若沙发椅不是为了躺躺而设，则何必有沙发椅？如若二十世纪的男女都必须正襟危坐，则现代的客室中何必摆着那种沙发椅？而我们极应该坐在挺硬的红木椅子上，身量较矮的妇女，须两脚悬空地挂着。

其实躺在椅中这件事也有一种哲学。古人和今人的坐法之不同，其起因即在于对恭敬的注重与否。古人的坐，以态度恭敬为主，今人则以舒服为主。两者之间有一种哲理上的冲突。因为依照古人的见解（五十年前尚是如此），舒服即罪恶，耽于舒服即趋于失敬。这一点奥尔德斯·赫胥黎在他讨论"舒适"（Comfort）那篇文章里已讲得很明白。赫氏所说西方的封建社会阻止了躺椅的产生，直到近时这句话和中国的情形正完全相同。今日凡自认是在朋友之列的，坐在他的房中时，尽可把两足高高地搁在他书桌上而不必有所顾忌，这是熟不拘礼，并不是失敬。不过这种行为，如在老辈面前，当然是要被斥为不当的。

道德和建筑与室内陈设之间，有一种我们寻常所意料不到的密切关系。赫胥黎指出西方女人因为怕看见自己的肉体，不常洗澡，因而使现代式白瓷澡盆的发明迟延了数百年之久。当我们认识儒教的公私行为都以恭敬为主时，我们就能了然旧式的中国木器为什么制成那种样子。我们在红木椅子上，只有挺起背脊笔直地坐着，就因为这是社会所公认的唯一合适的坐法。中国皇帝的宝座，坐时并不舒服，如叫我去坐，就是五分钟也不愿意。英王的宝座也是如此。克利奥帕特拉出外时，总是斜躺在睡椅上，令人抬着行走。她敢如此，就因为她没有受过孔子的教训。这种样子如被孔子看见，当然也要像对付原壤一般"以杖叩其胫"了。在儒家社会中，不论男女都应该恭身正容，至少在正式场合中应该如此。在这种时节，如有人将腿脚略为跷起，便立刻会被人视作村野失礼。事实上，最恭敬的姿势如在谒见长官时，坐的时节应斜欠着身子，将臀部搁在椅子的边沿上，才算知礼。儒家古训和中国建筑之间也有密切的关系，但这里姑且不论。

我们应该感谢十八世纪末叶和十九世纪初叶的浪漫派运动，它打破了古礼的传统思想，方使"舒服"这件事不再被人认为罪恶。另一方面，除了浪漫运动之外，又因对人类心理有了进一步的认识，于是对人生也产生了一种较为真切的态度。这种态度的改变，使人们不再视戏剧为淫猥，不再视莎士比亚作"野蛮人"（barbarian，伏尔泰称他为"喝醉的野蛮人"）外，也使女人的浴衣、清洁的澡盆、舒服的躺椅和睡椅得以出现，并使生活和文章有了一种较为真切、较为亲热的体裁。在这种意义上，我喜欢躺在椅中的习惯和我拟想将一种亲热自由潇洒的文体导入中国杂志界中的企图之间，确有一种联系存在着。

如若我们承认舒服并不是一种罪恶，则我们也须承认我们在朋友家的客室中以越舒服的姿势坐在躺椅上，越是在对这个朋友表示最大的恭敬。简括地说，客人能自己找寻舒服，实是在招待上协助主人，使他减少烦虑。试看多少做主人者每每为能否使客人舒服自在而担忧啊！所以我坐时，每每将一只脚高搁在茶桌或就近的家具上面，以协助做主人者，因而使其余的客人也可以趁此机会抛弃他们假装出来的尊严态度。

关于坐卧器具的舒服比较，我已发明了一个公式。这公式可以用简单字句表达如下：椅子越低，坐时越加舒服。有许多人坐在朋友家的某种椅中觉得异常舒适，即因为这个理由。当我尚未发明这个公式前，我以为室内装潢家对于一张椅子如何可以使坐者得到最高度的舒适，其高度阔度和斜度之间必有一种数学的公式。但自从我的公式发明之后，我即知道这事其实比较简单。如将中国红木椅子脚锯去数寸，坐时即立刻较为舒服，如再锯去一些，必更为舒服。这种情形的合理结论当然是，最舒服的姿势就是平躺在床上。这岂不简单？

从这个基本原则，我们即能演绎出一个附则，即我们倘因坐在一把太高又不便将脚锯去的椅子上而觉得不大舒服时，只须在椅子的前面找一个搁脚的地方，以减少我们的腰部平线和着脚处的距离，也即等于减低椅子的高度。我所最常利用的一个极普通方法就是，将写字台的屉斗拉一只出来搁脚。但这条附则应该怎样聪明地实施，须视各人的常识了。别人说我一天之中倒有十六小时醒着的时间是躺在椅中的，为了化解这个误会，我当说明我也能在写字台或打字机前很耐心地坐上三小时。我所要使人明白的是：松弛我们的肌肉，不一定是一件罪恶，但我并没有说我们可以一天到晚松弛我们的肌肉，或如此办法是最合卫生的姿势。我的原意并不如此，人类的生活终须由工作和游憩循环为用，即紧张和松弛相替为用。男人的脑力和工作能力也如女人的身体一般，每月有一种循环式变迁。威廉·詹姆斯说，脚踏车的链子如若绷得太紧，即有碍于转动的顺利，人类的心力也正相同。无论什么事情终是个习惯问题，人体内有一种调节的无穷能力。日本人惯于盘腿坐在地上，我颇疑心如叫他们改坐椅上，他们即易于犯腿抽筋的毛病。我们只有借着将工作时间中完全挺直的姿势和工作完毕后躺在睡椅中的舒服姿势循环变换，方能成就生活的最高智慧。

至于妇女，坐着的时候如若眼前没有搁脚的地方，则可把两腿蜷搁在睡椅上，你们应知道这是一个最惹人爱的姿势。

II. On Sitting in Chairs

I want to write about the philosophy of sitting in chairs because I have a reputation for lolling. Now there are many lollers among my friends and acquaintances, but somehow I have acquired a special reputation for lolling, at least in the Chinese literary world. I contend that I am not the only loller in this modern world and that my reputation has been greatly exaggerated. It happened like this. I started a magazine called the Analects Fortnightly, in which I consistently tried to disprove the myth of the harmfulness of smoking. In spite of the fact that we did not have cigarette advertisements in our magazine, I wrote and published essay after essay praising the virtues of Lady Nicotine. Somehow, therefore, a legend developed that I was a man doing nothing the whole day but lolling idly on a sofa smoking a cigar, and in spite of my disclaimers and my protests that I am one of the hardest working men in China, the legend had got about and was constantly used as an evidence of my belonging to the hateful leisure-class intelligentsia. Two years after, the situation was aggravated by the fact that I started another magazine devoted to the familiar essay. Bored stiff by the dilatory, hypocritical and pompous style of Chinese editorials, which is the result of the method of teaching school composition a generation ago, making young boys of twelve or thirteen write essays on"The Salvation of the Country"and"The Virtue of

Persistence,"I saw that the introduction of a more familiar style of writing was the means of emancipating Chinese prose from the straitjacket of Confucian platitudes. It happened, however, that I translated the phrase"familiar style"by a Chinese phrase meaning"leisurely style."... And now I have the indisputable reputation of being the most leisurely of all leisurely writers in China and therefore the most unforgivable,"while we are living in this period of national humiliation."

I admit that I do loll about in my friends' drawing-rooms, but so do the others. What are armchairs for anyway, except for people to loll in? If gentlemen and ladies of the twentieth century were supposed to sit upright all the time with absolute dignity, there should be no armchairs at all in the modern drawing-room, but we should all be sitting on stiff redwood furniture, with most ladies' feet dangling about a foot from the ground.

In other words, there is a philosophy about lolling in chairs. The mention of the word"dignify"explains exactly the origin of the difference in the styles of sitting between the ancient and the modern people. The ancient people sat in order to look dignified, while modern people sit in order to be comfortable. There is a philosophic conflict between the two, for, according to ancient notions only half a century ago, comfort was a sin, and to be comfortable was to be disrespectful. Aldous Huxley has made this sufficiently plain in his essay on"Comfort."The feudalistic society which made the rise of the armchair impossible until modern days, as described by Huxley, was exactly the same as that which existed in China up to a generation ago. Today any man who calls himself another's friend must not be afraid to put his legs on top of a desk in his friend's room, and we take that as a sign of familiarity instead of disrespect, although to put one's legs on top of a desk in the presence of a member of the older generation would be a different matter.

There is a closer relationship between morals and architecture and interior decoration than we suspect. Huxley has pointed out that Western ladies did not take frequent baths because they were afraid to see their own naked bodies, and this moral concept delayed the rise of the modern white-enameled bathtub for centuries. One can understand why in the design of old Chinese furniture there was so little consideration for human comfort only when we realize the Confucian atmosphere in which people moved about. Chinese redwood furniture was designed for people to sit upright in, because that was the only posture approved by society. Even Chinese emperors had to sit on a throne on which I would not think of remaining for more than five minutes, and for that matter the English kings were just as badly off. Cleopatra went about inclining on a couch carried by servants, because apparently she had never heard of Confucius. If Confucius should have seen her doing that, he would certainly have"struck her shin with a stick,"as he did to one of his old disciples, Yüan Jang, when the latter was found sitting in an incorrect posture. In the Confucian society in which we lived, gentlemen and ladies had to hold themselves perfectly erect, at least on formal occasions, and any sign of putting one's leg up would be at once construed as a sign of vulgarity and lack of culture. In fact, to show extra respect, as when seeing official superiors, one had to sit gingerly on the edge of a chair at an oblique angle, which was a sign of respect and of the height of culture. There is also a close connection between the Confucian tradition and the discomforts of Chinese architecture, but we will not go into that now.

Thanks to the romantic movement in the later eighteenth and earlier nineteenth centuries, this tradition of classical decorum has broken down, and to be comfortable is no longer a sin. On the other hand, a more truthful attitude towards life has taken its place, due as much to the romantic movement as to a better understanding of human psychology. The same change of attitude which has ceased to regard theatrical amusements as immoral and Shakespeare as a"barbarian,"has also made possible the evolution of ladies' bathing costumes, clean bathtubs and comfortable armchairs and divans, and of a more truthful and at the same time intimate style of living and writing. In this sense, there is a true connection between my habit of lolling on a sofa and my attempt to introduce a more intimate and free and easy writing into modern Chinese journalism.

If we admit that comfort is not a sin, then we must also admit that the more comfortably a man arranges himself in an armchair in a friend's drawing-room, the greater respect he is showing to his host. After all, to make oneself at home and look restful is only to help one's host or hostess succeed in the difficult art of hospitality. How many hostesses have feared and trembled for an evening party in which the guests are not willing to loosen up and just be themselves. I have always helped my hosts and hostesses by putting a leg up on top of a tea table or whatever happened to be the nearest object, and in that way forced everybody else to throw away the cloak of false dignity.

Now I have discovered a formula regarding the comparative comfort of furniture. This formula may be stated in very simple terms: the lower a chair is, the more comfortable it becomes. Many people have sat down on a certain chair in a friend's home and wondered why it is so cozy. Before the discovery of this formula, I used to think that students of interior decoration probably had a mathematical formula for the proportion between height and width and angle of inclination of chairs which conduced to the maximum comfort of sitters. Since the discovery of this formula, I have found that it is simpler than that. Take any Chinese redwood furniture and saw off its legs a few inches, and it immediately becomes more comfortable; and if you saw off another few inches, then it becomes still more comfortable. The logical conclusion of this is, of course, that one is most comfortable when one is lying perfectly flat on a bed. The matter is as simple as that.

From this fundamental principle, we may derive the corollary that when we find ourselves sitting in a chair that is too high and can't saw its legs off, all we have to do is seek some object in the foreground on which we can rest our legs and therefore theoretically decrease the difference between the levels of our hips and our feet. One of the commonest devices that I use is to pull out a drawer in my desk and put my feet on it. But the intelligent application of this corollary I can leave to everybody's common sense. To correct any misunderstanding that 1 am lolling all the lime for sixteen waking hours of the day, I must hasten to explain that I am capable of sitting doggedly at a desk or in front of a typewriter for three hours. Just because I wish to make it clear that relaxation of our muscles is not necessarily a crime, I do not mean that we should keep our muscles relaxed all the time, or that this is the most hygienic posture to be assumed all the time. That is far from my intention. After all human life goes in cycles of work and play, of tension and relaxation. The male brain energy and capacity for work goes in monthly cycles just like women's bodies. William James said that when the chains of a bicycle are kept too tight, they are not conducive to the easiest running, and so with the human mind. Everything, after all, is a matter of habit. There is an infinite capacity in the human body for adjustments. Japanese who have the habit of sitting with crossed legs on the floor are liable to get cramps, I suspect, when they are made to sit on chairs. Only by alternating between the absolutely erect working posture of office hours and the posture of stretching ourselves on a sofa after a hard day's work can we achieve that highest wisdom of living.

A word to the ladies: when there is nothing in the immediate foreground on which you can rest your feet, you can always curl up your legs on a sofa. You never look more charming than when you are in that attitude.

三. 谈话

　　"与君一夕话，胜读十年书。"这是一位中国学者和他一个朋友谈了一次话以后的一句赞语，这句话中含有不少真理。"一夕话"现在已成为一种口头语，以代表和朋友所做的一次愉快谈话，不论是已过的或期望的。中国有两三种著作，其书名即《一夕谈》（A Night's Talk）或《山中一夕谈》（A Night's Tlak in the Mountains），书的性质和英文的"周末文集"（Weekend omnibus）相似。这种和朋友的一夕快谈，是人生难得遇到的。因为正如李笠翁所说，凡是真正的智者都拙于言谈，而善谈者又罕是智者。所以在高山的寺院中忽然发现了一位深解人生的高士，同时是善谈者，则其愉快自不亚于一位天文家发现一颗新行星，或一位生物学家发现一种新植物了。

　　现在有许多人都以为围炉聚谈或坐桶聚谈的谈话艺术，已因今日商业生活的动率而丧失掉。我以为动率对于这事确也有些关系。不过谈话艺术的毁灭，实开端于家庭改为没有火炉的公寓，而由汽车的影响完成这桩毁灭工作。这动率是完全不合的，因为谈天这桩事只在一群富有闲适精神的人当中，写意、心平气和、幽默自然的时候方能办得到。因为"说话"和

"谈话"之间显然有分别，这两个中国名词已表示得很明白。在谈话的时候所说的话，天南地北，较为琐碎，态度较为闲适，而没有办公事时那种煞有介事。商务信和文友之间通信也有着相类的区别。我们可以和任何人都说话或谈公事，但不是和任何人可以做一夕之谈的。所以我们如若得到一个能真正谈天的朋友，则其愉快实不下于读一本名著，更不用说亲耳听见他的语音，亲眼看见他的动作的乐趣。这种快乐的谈天，我们有时得之于老友的重逢或回溯当年的谈话中，有时则在夜晚间火车的吸烟室中，或旅行时的旅舍中。所谈的话，狐鬼、神怪、独裁、卖国，谈言微中，料及未来，也是常事。这种谈天，过后可以长在心头，一世不忘。

当然夜间是最宜于谈天的时候，因为白天的谈天总好似缺乏夜间那种魔力。至于谈天的地点我以为毫无关系。在十八世纪式的"沙龙"（即客厅）中，可以谈关于文学或哲学的闲天，但在农家木桶的旁边也未尝不可以谈。或在风雨之夕的航船中，对河船上的灯光微映水波，而卧听船夫闲谈当地的一个女子怎样被选去做皇后娘娘的故事。这类谈天之所以悦人者，实在于所得的乐趣因地点时间和谈者而各不相同。我们所以能牢记不忘，有时因为谈天的时候是正在桂子飘香、秋月悬空的佳景下；有时因为是正在风雨之夕，一炉柴火之前；有时因为是正坐在一个高亭之中，远眺河中船只往来，而有一只船忽因潮流过激而侧翻的时候；或是在清晨坐在车站候车室中的时候。这种眼前即景常和所谈的天联系一起，因而使我们永不能忘。如若在室内，谈者或是两三人或是六七人，老陈微醉，老秦有些伤风鼻塞，都可以使这夕的谈天增添趣味。人生是限制于月不常圆，花不常好，良朋不能常聚之中的，所以我们做这类简单的乐趣，我想不至于为造物主所忌吧？

依常例而言，好的谈天等于一篇好的通俗文章。两者之间的体裁和资料都相仿。如狐狸精、苍蝇、英国人的古怪脾气、东西方文化的异点、塞纳河旁的书摊、成衣铺中色迷迷的学徒、各国元首政治家和军人的逸事、储藏佛手的方法等，都是极好极相宜的谈天资料。它之所以类似文章，即在体裁的通俗。所谈的题目尽可以严肃重大，如本国情形的惨苦混乱，或疯狂的政治概念潮流之下文明的没落，剥夺人民的自由、人类的尊严，甚至剥夺人类快乐的终点，或关涉真理和公平的大问题等，均无不可。不过意见的发表总是出之以一种偶然的、闲适的和亲切的态度。因为在文明的当中，不论我们对强夺我们的自由怎样恼恨，我们至多只许用我们的舌头和笔尖，以轻描淡写的字句来表示感想。至于充分发挥真情感的激烈言论，只可以在少数几个知己朋友之间，私下发泄一下子。所以要做一次真正的谈天，其必要条件是一间关上门的屋子，几个知己的朋友，旁边没有我们所不愿意看见的人，那时，方能悠闲地发表我们的意见。

这种真正的谈天有异于政治上的交换意见，正如一篇优美通俗的文章之有异于政治家的宣言。这类政治家的宣言虽表现着较为高尚的情感，例如：对于民主制度的意见、服务的愿望、穷人的福利问题、精忠报国、崇高的理想主义、酷爱和平、保证维持国交、绝不贪图权位金钱或名誉等动人听闻的说话，但其中终免不了带着些令人远而避之的气息，正如我们畏避一个打扮过分、胭脂水粉搽得太浓的妖娆女人一般。反之，我们在听到一次真正有趣的谈天，或读到一篇优美的通俗文章时，便如面对着一个在河边洗衣的乡村少女，穿着极淡雅的布衣服，头发或者有一绺拖在前面，身上的纽扣或许有一粒未曾扣上，其天真烂漫的姿态自然令人见而生爱，这也就是西方女人特意穿着便服所想要模仿的动人姿态。凡是有趣的谈天和优美的文章，都必然具有这种天然的动人之处。

所以谈天的适当方式应是亲密的，毫无顾忌的。在座的人谈到出神时，都已忘却身处何地，也不再想到身上穿的是什么衣服，谈言吐语，一举一动都是任性为主。而所谈的，也是忽而东忽而西，想着便谈，并无一定的题目。我们只有在知己朋友相遇、肯互相倾吐肺腑时，方能真正地谈天，而谈时各人也是任性坐卧，毫无拘束，一个将两脚高高地搁在桌上，一个坐在窗槛上，一个坐在地板上，将睡椅上的垫子搬下来当褥子用。因为必须在手足都安放在极舒服的位置，全部身体感受舒适时，我们的心地方能安闲舒适，此即前人所谓：

眼前一笑皆知己，
举座全无碍目人。

这些都是真正谈天的必要条件。谈时不择题目，想到便谈，天南地北，越去越远，既无秩序，也无定法，随意所之，所以谈到兴尽之时，也就欢然而散。

这就是谈天和空闲的联系关系，也就是谈天和散文之勃兴的联系关系。因为我相信一个国家的真正优美散文必须在谈天一道已经发展成为一项艺术的地步方能产生。这个情形在中国和希腊文的发展中最为显明。我以为孔子之后的数百年中，思想的活动产生所谓"九家"的学说，其起源即因当时有一群学者，平生唯以说话为事，所以即发展成一种文化的背景。这种发展，除此之外，实在说不出其他的理由。当时列国有五位豪富的公子，都以慷慨好客著名一时。每人的家中都聚着食客数千人，例如：齐国的孟尝君，家中养着珠履之客三千人。其人数既如此众多，则当时你谈我说，议论纷纷的情形，也就可想而知。这类人的说话，在传于后世的《列子》《淮南子》《战国策》和《吕氏春秋》诸书中，可以得其大概。《吕氏春秋》据说实是吕不韦的门下所著，而不过用他的名义（这和英国十六世纪与十七世纪时代的作家著了书用赞助人之名义发表的情形相似）。这部书中已经发展了一种善处人生的概念，大意是不善处人生，不如不生活。此外还有一群长于说辞的纵横家，列国君王常利用他们到邻国去下说辞，或去挽回一次危局，或去劝说退兵解围，或去说合联盟，而他们也大都能成功而返。这群纵横家或学者都是长于口才，善于譬喻，他们的言论很多记载于《战国策》中。从这种自由而智巧的言论中产生了几位大哲学家，以"为我主义"著名的杨朱，以"现实主义"著名的韩非子（他和马基维里Machiaevelli相似但较为温和）和以敏捷辩论著名的大外交家晏子。这些都可以证实我的假说。

公元前三世纪末叶，楚国李园把他的才貌双全的妹妹献给楚相春申君。这桩事就是当时社会生活很文明的一个榜样。后来春申君又将这女子献与楚王，以致楚国渐渐衰弱，为秦始皇所灭。

昔者楚考烈王相春申君吏李园。园女弟女环谓园曰："我闻王老无嗣，可见我与春申君，我欲假于春申君，我得见春申君，径得见王矣！"园曰："春申君贵人也。千里之佐，吾何托敢言？"女环曰："即不见我，汝求谒于春申君才人，告远道客，请归待之。彼必问汝，汝家何运道客者。因对曰：'园有女弟，鲁相闻之，使使者来求之园。'才人使告园者。彼必有问：'女弟何能？'对曰：'能鼓琴，读书通一经。'故彼必见我。"

园曰："诺。"明日辞春申君："才人有远道客，请归待之。"春申君果问："汝家等远道客？"对曰："园有女弟，鲁相闻之，使使求之。"春申君曰："何能？"对曰："能鼓琴，读书通一经。"春申君曰："可得见乎？明日使待于离亭。"园曰："诺。"既归告女环曰："吾辞于春申君，许我明日夕待于离亭。"女环曰："园宜先供待之。"

　　春申君到，驰人呼女环，女环至，大纵酒。女环鼓琴，曲未终，春申君大悦，留宿……

　　这就是当时受过教育的女子和闲适的文士的社会背景，因而使中国的散文也有了第一次重要的发展。当时有善说辞、通文才、娴于音乐的女子，使男女共处的社会中有着社交的、美术的和文学的动机交织，点缀着社会的性质和气象当然是贵族化的，因为，相国是常人很难于见到的贵官，但他在知道一个女子娴于音乐擅长文才时，便也渴于一见了。这就是古代中国文人和哲学家所度的闲适生活，而当时的一切著作，也不过是彼此谈话的产物而已。

　　只有在有闲的社会中，谈话艺术方能产生，这是很显明的；只有从谈话艺术中，优美通俗的文章方能产生，这是同样显明的。一般说起来，谈话艺术和优美通俗文章的艺术在人类文明进步史中产生的时间比较迟。因为人类的心灵必须先经过一种敏锐微妙技巧的发展，方能达此地步。而要发展这些，则又非生活有闲不可。文化本身的进步，实是有赖于空闲的合理利用，而谈话不过是其中的方式之一罢了。一天忙到晚的生意人，吃了晚饭就睡觉，鼾声如牛者，是绝不能有所助于文化的。

　　一个人的空闲，有时是环境所迫，而不是自我的，许多文学杰作都是在环境所迫的空闲中完成。因此我们如遇到一个极有希望的文学天才，看见他虚糜时间于社交或写作流行的政治论文时，对待他的最好方法是将他关进监狱去。因为我们须记得《周易》，一部讨论人生变迁的哲学巨著，即是周文王被囚在羑里时写成。而中国的历史杰作《史记》，也是司马迁被囚在狱中写成的。古代许多著名的作家大都因为宦途不达，屈在下僚，

或是伤心国是，转变生活而产生了他们的文学或艺术杰作。元朝何以产生
这许多名画家和词曲家？清初何以能产生名画家石涛和八大山人？即由于
这个理由。激于耻为夷狄之民的爱国思想，使他们致一生心力于艺术和学
问。石涛实是中国最伟大的画家之一，但因清朝皇帝对于这班心不臣服的
艺术家有意埋没，所以他名不甚著，西方人知道的很少。此外还有很多科
举考试名落孙山的人也发愤而致力于创作，例如：施耐庵之著《水浒传》
和蒲留仙之著《聊斋》。

《水浒传》的序文中（金圣叹本），有一段形容朋友谈天之乐的绝妙
文字：

> 吾友毕来，当得十有六人，然而毕来之日为少；非甚风雨而尽不来之
> 日亦少。大率日以六七人来为常矣。吾友来，亦不便饮酒，欲饮则饮，欲
> 止则止，各随其心，不以酒为乐，以谈为乐也。吾友不谈及朝廷，非但安分，
> 亦以路遥传闻为多，传闻之言无实，无实即唐丧唾津矣。亦不及人过失者，
> 天下之人本无过失，不应吾诋诬之也。所发之言，不求惊人，人亦不惊。
> 未尝不欲人解，而人卒亦不能解者，事在性情之际，世人多忙，未曾常闻也。

《水浒传》即在如此环境和情感中产生的，而所以能产生，即因他懂
得享受空闲。

希腊的散文早年也是在同样的空闲社会背景中产生的。希腊思想的清
明，散文体裁的简洁，显系空闲谈艺术造成。柏拉图以《会话》为其书
名，即能证明此点。在《宴饮》一篇中，我们看见一群希腊文士斜躺在地
上，在美酒鲜果和美少年的氛围中欢笑谈天。因为这种人已养成了谈天的
艺术，所以他们的思想才能如此清朗，文体如此简洁，与现代作家的夸大

迂腐恰成一种对比。这方面希腊人显然已学会了用轻描淡写的态度去应付哲学问题。希腊哲学家动人的闲谈气象，好谈天的欲望，对聆听有趣味的谈天的重视和对谈天的适当环境的选择，都在《斐德若》一篇序文中描写得很分明。

柏拉图在他的《理想国》一篇中，并不像现代作家用"人类文明从它发展的各个连续梯阶观察起来，乃是一种从多种生殖变化为纯一生殖之动力的运动"。或诸如此类令人费解的话头开场，而只说："昨天我和亚里士多德的儿子格劳可到比雷埃夫斯去拜女神，同时想去看看他们将怎样庆祝这个节日，因为这尚是第一次举行。"早年中国哲学家的气象，即思想最活泼最有力时代的气象也可以从希腊人的画像中看得到，在这种画像中几个希腊人偶然齐集在一起，如《宴饮》一篇中所描写的，讨论一个伟大的悲剧作家是否同时必是一个伟大的喜剧作家。集会的气氛中，交织着严肃轻快和恶意的敏捷应对。旁人嘲弄苏格拉底的酒量，但他仍是旁若无人地坐在那里，欲饮即斟酒而饮，欲止即止。他口若悬河地谈了一整夜，直谈到除了阿里斯托芬和阿迦松（Agathon）之外，其余听者都已沉沉睡去。后来连那二人也倦极睡去，只剩下他自己一人，方起身离开筵席，走到教授室去洗了一个澡，即又精神焕发了。希腊的哲学即是在这种善意的谈论之气氛中所产生的。

毫无疑义的，我们在高尚地谈天时，须有几个女子夹杂在座中，以使这谈天具有必不可少的轻俏性。谈天如缺乏轻俏性和愉快性，即变为沉闷乏味，而哲学本身也就变为缺乏理智，和人生相隔离了。不论在哪一个国家，不论在哪一个时代，凡是注意了解生活艺术的文化者，都一致发展欢迎女子加入为点缀的习尚。雅典在伯里克利时代即是如此，十八世纪法国沙龙时代也是如此。中国男女之间虽禁止交际，但是历代文士都渴欲女子

加入他们的谈天，在晋宋明三朝之中，当清谈艺术最为流行的时候，都有
许多才女如谢道蕴、朝云、柳如是等掺杂中间。因为，中国人虽对于自己
的老婆力主贤德，回避男子，但自己免不了极想和有才的女子为友。因此
中国的文学史中，几乎随时能发现才女名妓的踪迹。男子谈天之时，渴望
女子加入以调剂精神，乃是一种普遍的愿望。我曾碰到过几位德国女子，
她们能从下午五点钟谈到晚间十一点钟。我曾碰到过几位英美女子，她们
熟习经济学使我不胜惊异，因为这种学问是我所不敢研究而自认无望的。
无论如何，即使一时没有能和我对于卡尔·马克思和恩格斯学说做辩论的
女子，如若座中杂坐几位善于听人谈论、心地玲珑的女子，实可以使在座
者精神格外兴奋。我觉得座中面对玲珑的女子，实胜于和一个满脸笨相的
人谈天。

III. On Conversation

"TALKING with you for one night is better than studying books for ten years,"–this was the comment of an old Chinese scholar after he had had a conversation with another friend. There is much truth in that statement, and today the phrase"a night's talk"has become a current expression for a happy conversation with a friend at night, either past or anticipated. There are two or three books which resemble an English"weekend omnibus,"bearing the title A Night's Talk, or A Night's Talk in the Mountains. Such a supreme pleasure as a perfect conversation with a friend at night is necessarily rare, for as Li Liweng has pointed out, those who are wise seldom know how to talk, and those who talk are seldom wise. The discovery of a man up in a mountain temple, who really understands life and at the same time understands the art of conversation, must therefore be one of the keenest pleasures, like the discovery of a new planet by an astronomer or of a new variety of plant by a botanist.

People today are complaining that the art of conversation around a fireplace or on cracker barrels is becoming lost, owing to the tempo of business life today. I am quite sure that this tempo has something to do with it, but believe also that the distortion of the home into an apartment without a log fire began the destruction of the art of conversation, and the influence of the motor car completed it. The tempo is entirely wrong, for conversation exists only in a society of men imbued in the spirit of leisure, with its ease, its humor, and its appreciation of light nuances. For there is an evident distinction between mere talking and conversation as such. This distinction is made in the Chinese language between shuohua (speaking) and t'anhua (conversation), which implies the discourse is more chatty and leisurely and the topics of conversation are more trivial and less business-like. A similar difference may be noted between business correspondence and the letters of literary friends. We can speak or discuss business with almost any person, but there are very few people with whom we can truly hold a night's conversation. Hence, when we do find a true conversationalist, the pleasure is equal to, if not above, that of reading a delightful author, with the additional pleasure of hearing his voice and seeing his gestures. Sometimes we find it at the happy reunion of old friends, or among acquaintances indulging in reminiscences, sometimes in the smoking room of a night train, and sometimes at an inn on a distant journey. There will be chats about ghosts and fox-spirits, mixed with amusing tales or impassioned comments on dictators and traitors, and sometimes before we know it, light is shown by a wise observer and conversationalist on things taking place in a certain country which are a premonition of the impending collapse or change in a regime. Such conversations remain among the memories that we cherish for life.

Of course, night is the best time for conversation, because there is a certain lack of glamour in conversations during the daytime. The place of conversation seems to me entirely unimportant. One can enjoy a good conversation on literature and philosophy in an eighteenth-century salon, or sitting on barrels at a plantation of an afternoon. Or it may be on a windy or rainy night when we are traveling in a river boat and the lantern lights from boats anchoring on the opposite bank of the river cast their reflections into the water and we hear the boatmen tell us stories about the girlhood of the Queen. In fact, the charm of conversation lies in the fact that the circumstances in which it takes place, the place, the time and the persons engaged in it, vary from occasion to occasion. Sometimes we remember it in connection with a breezy moonlight night, when the cassia flowers are in bloom, and sometimes we associate it in memory with a dark and stormy night when a log fire is glowing on the hearth, and sometimes we remember we were sitting on top of a pavilion watching boats coming down the river, and perhaps a boat was overturned by the swift current, or again, we were sitting in the waiting room of a railroad station in the small hours of the morning. These pictures associate themselves indelibly with our memory of those particular conversations. There were perhaps two or three persons in the room, or perhaps five or six; maybe old Chen was slightly drunk that night, or old Chin had a cold in the nose and spoke with a slight twang, adding to the particular flavor of that evening. Such is human life that "the moon cannot always be round, the flowers cannot always look so fine and good friends cannot always meet together,"and I do not think the gods will be jealous of us when we engage in such simple pastimes.

As a rule, a good conversation is always like a good familiar essay. Both its style and its contents are similar to that of the essay. Such topics as fox-spirits, flies, the strange ways of Englishmen, the difference between Oriental and Occidental culture, the bookstalls along the Seine, a nymphomaniac apprentice in a tailor shop, anecdotes of our rulers, statesmen and generals, the method of preserving"Buddha's Fingers"(a variety of citron)—these are all good and legitimate topics of conversation. The point it has most in common with the essay is its leisurely style. However weighty and important the topic may be, involving reflections on the sad change or state of chaos of one's own country, or the sinking of civilization itself under a current of mad political ideas, depriving man of liberty, human dignity and even the goal of human happiness, or even involving moving questions of truth and justice, still such ideas are expressed in a casual, leisurely and intimate manner. For in civilization, however a man chafes and is angry at the robbers of our liberty, we are allowed only to express our sentiments by a light smile around our lips or at the tip of our pen. Our really impassioned tirades, in which we give full reins to our sentiments, may be heard only by a few of our most intimate friends. Hence the requisite condition of a true conversation is that we are able to air our views at leisure in the intimacy of a room with a few good friends and with no people around whom we hate to look at.

It is easy to see this contrast between the true genre of conversation and other kinds of polite exchange of opinion by referring to the similar contrast between a good familiar essay and the statements of politicians. Although there are a good deal more noble sentiments expressed in politicians' statements, sentiments of democracy, desire for service, interest in the welfare of the poor, devotion to the country, lofty idealism, love of peace and assurances of unfailing international friendship, and absolutely no suggestions of greed for power or money or fame, yet there is a smell about it which puts one off at a distance, like an over-dressed and over-painted lady. On the other hand, when we hear a true conversation or read a good familiar essay, we feel that we have seen a plainly dressed country maiden washing clothes by the riverbank, with perhaps her hair a little disheveled and one button loose, but withal charming and intimate and likable. That is the familiar charm and studied negligence aimed at in a Western woman's negligée. Some of this familiar charm of intimacy must be a part of all good conversations and all good essays.

The proper style of conversation is, therefore, a style of intimacy and nonchalance, in which the parties engaged have lost their self-consciousness and are entirely oblivious of how they dress, how they speak, how they sneeze, and where they put their hands, and are equally indifferent as to which way the conversation is drifting. We can engage in a true conversation only when we meet our intimate friends and are prepared to unburden our hearts to each other. One of them has put his feet on a neighboring table, another is sitting on a window sill, and still another is sitting on the floor, upholstered

by a cushion which he has snatched from the sofa, thus leaving one-third of
the sofa seat uncovered. For it is only when your hands and feet are relaxed
and the position of your body is at ease that your heart can be at ease also. It
is then that:

Before my face are friends who know my heart,
And at my side are none who hurt my eyes.

This is the absolutely necessary condition of all conversation worthy
of the name of an art. And since we do not care what we are talking about,
the conversation will drift further and further, without order and without
method, and the company break up, happy of heart.

Such is the connection between leisure and conversation and the
connection between conversation and the rise of prose that I believe the
truly cultured prose of a nation was born at a time when conversation had
already developed as a fine art. This we see most clearly in the development
of Chinese and Greek prose. I cannot otherwise imagine an explanation for
the vitality of Chinese thought in the centuries following Confucius, giving
birth to the so-called" Nine Schools of Thought,"than the development of
a cultured background, in which there was a class of scholars whose business
was only to talk. For confirmation of my theory, we find there were five great
rich noblemen, noted for their generosity, chivalry and fondness for guests.
All of them had thousands of scholarly guests at their homes, as for instance
Mengch'ang of Ch'i Kingdom who was reputed to have three thousand
scholars, or"guests"wearing"pearled shoes,"being"fed"at his home. One can
imagine the conversational hubbub that was going on in those houses. The
content of the conversation of scholars of those days is today reflected in the
books of Liehtse, Huainantse, Chankuots'eh and Lülan. It is noteworthy that
with respect to the last one, which was a book admittedly written by Lü's
guests but published in his name (in a sense similar to the" patrons"of English

sixteenth-century and seventeenth-century authors), there was already developed the idea of the art of living well, in the formula that it would be better to live well or not at all. There was besides a class of brilliant sophists or professional talkers, who were engaged by the different warring states and sent out as diplomats to avert a crisis or persuade a hostile army to retreat from the walls of a besieged city, or to bring about an alliance, as so many did. Such professional sophists were always distinguished by their wit, their clever parables and their general persuasive power. The conversations or clever arguments of such sophists are preserved for us in the book Chankuots'eh. From such an atmosphere of free and playful discussion arose some of the greatest names in philosophy, Yang Chu, noted for his cynicism, Hanfeitse, noted for his realism (similar to Machiaevelli's, but more tempered), and the great diplomat Yentse, noted for his wit.

An example of the cultured social life existing in the third century B. C., toward the end of this period, may be seen in the record of how a certain scholar by the name of Li Yüan succeeded in presenting his accomplished sister to the court of a rich patron in the Kingdom of Ch'u. The patron in turn secured the favor of the King for this girl, which was eventually responsible for the destruction of the Kingdom of Ch'u by the conquering army of the First Emperor of Ch'in, who united the Chinese Empire.

Formerly there was Li Yüan, serving as a subordinate of Prince Ch'unshen, the Prime Minister of the King of Ch'u. Li had a sister by the name of Nühuan who spoke to him one day, "I hear that the King is without an heir. If you will present me to the Prime Minister, through him I will be able to see the King." "But the Prime Minister is a high official," replied her brother. "How dare I mention it to him?" "You just go and see him," said his sister, "and then tell him that you have to come home because a noble guest has arrived. He will then ask you who is the noble guest and you can reply that you have a sister, that the Prime Minister of the Kingdom of Lu has heard of her reputation and has sent a delegate to ask for her from you, and that a messenger from home has just brought the news. He will then ask, what can your sister do? And you will reply that I can play on the ch'in, can read and write, and have mastered one of the classics. He is certain to send for me that way.

Li then promised to do as she said, and the next morning, after seeing the Prime Minister, he said, "A messenger from home told me that there is a guest from a distant country, and I must return to receive him." The Prime Minister Ch'unshen then actually asked him, "Who is this noble guest from a distant country?" And Li replied, "I have a sister, and the Prime Minister of the Kingdom of Lu has heard of her reputation and has sent a delegate to ask for her." "May I see her?" asked the Prime Minister. "Ask her to come and meet me at the Li Pavilion." "Yes sir," replied Li, and he returned and told his sister that the Prime Minister expected to see her the next evening at the Li Pavilion. "You must go there yourself first in order to be there when I arrive," said the girl.

The Prime Minister then arrived at the time and asked to see Nühuan. She was presented and they drank a great deal. Nühuan played on the ch'in, and before her song was finished, the Prime Minister was greatly pleased and asked her to stay there for the night...

This then, was the social background of cultivated ladies and leisurely scholars, which produced for us the first important development of prose in China. There were ladies who could talk and read and write and play on a musical instrument, making for that peculiarly light mixture of social, artistic and literary motives that was always found in a society where men and women mixed together. It was undoubtedly aristocratic in character and atmosphere, for the Prime Minister Ch'unshen was difficult to see, but when he heard of a lady who could play on a musical instrument and had mastered one of the classics, he insisted on seeing her. This was then the life of leisure which the early Chinese sophists and philosophers lived. The books of these early Chinese philosophers were nothing but the results of leisurely conversation among these scholars.

It is clear that only in a society with leisure can the art of conversation be produced, and it is equally clear also that only when there is an art of conversation can there be good familiar essays. In general, both the art of conversation and the art of writing good prose came comparatively late in the history of human civilization, because the human mind had to develop a certain subtlety and lightness of touch, and this was possible only in a life of leisure. (...) The enjoyment of leisure cannot be a sin, but on the other hand the progress of culture itself depends on an intelligent use of leisure, of which conversation is only one form. Business men who are busy the whole day and immediately go to bed after supper, snoring like cows, are not likely to contribute anything to culture.

Sometimes this"leisure"is enforced upon one and does not come of one's own seeking. All the same, many good works of literature have been produced in an atmosphere of enforced leisure. When we see a literary genius with great promise, dispersing his energy in futile social parties or writing essays on current politics, the best and kindest thing we can do to him is to shut him up in jail. For we must remember that it was in prison that the King Wen wrote his Book of Changes, a classic of philosophy on the changes of human life, and Ssema Ch'ien wrote his masterpiece, Shihchi (conventionally spelled Shiki), the best history ever written in the Chinese language. Sometimes the authors were defeated in their ambitions for a political career, or the political situation was too discouraging, and great works of literature or of art were produced. That is the reason why we had such great Yüan painters and Yüan dramatists during the Mongol regime and such great painters as Shih T'ao and Pata Shanjen... Shih T'ao is undoubtedly one of the very greatest painters China has ever produced, and the fact that he is not generally known in the West is due to an accident and to the fact that the Manchu emperors were not willing to give credit to these artists not in sympathy with their rule. Other great writers who had failed in the imperial examination, began to sublimate their energy and turn to creation, as in the case of Shih

Naian who gave us All Men Are Brothers and P'u Liuhsien who gave us Strange Stories From A Chinese Studio.

We have in the preface to All Men Are Brothers, attributed to Shih, one of the most delightful descriptions of the pleasure of conversation among friends:

When all my friends come together to my house, there are sixteen persons in all, but it is seldom that they all come. But except for rainy or stormy days, it is also seldom that none of them comes. Most of the days, we have six or seven persons in the house, and when they come, they do not immediately begin to think; they would take a sip when they feel like it and stop when they feel like it, for they regard the pleasure as consisting in the conversation, and not in the wine. We do not talk about court politics, not only because it lies outside our proper occupation, but also because at such a distance most of the news is based upon hearsay; hearsay news is mere rumour, and to discuss rumours would be a waste of our saliva. We also do not talk about people's faults, for people have no faults, and we should not malign them. We do not say things to shock people and no one is shocked; on the other hand, we do wish people to understand what we say, but people still don't understand what we say. For such things as we talk about lie in the depths of the human heart, and the people of the world are too busy to hear them.

It was in this kind of style and with this kind of sentiment that Shih's great work was produced, and this was possible because they enjoyed leisure.

The rise of Greek prose took place clearly in the same kind of a leisurely social background. The lucidity of Greek thought and clearness of the Greek prose style clearly owe their existence to the art of leisurely conversation, as is so clearly revealed in the title of Plato's Dialogues. In the"Banquet"we see a group of Greek scholars inclining on the ground and chatting merrily along in an atmosphere of wine and fruit and beautiful boys. It was because these people had cultivated the art of talking that their thought was so lucid and their style so clear, providing a refreshing contrast to the pomposity and pedantry of modern academic writers. These Greeks evidently had learned to handle the topic of philosophy lightly. The charming conversational atmosphere of the Greek philosophers, their desire for talking, the value they placed upon hearing a good talk and the choice of surroundings for conversations were beautifully described in the introduction to "Phaedrus." This gives us an insight into the rise of Greek prose.

Plato's "Republic" itself does not begin, as some of the modern writers would have it, with some such sentence as, "Human civilization, as seen through its successive stages of development, is a dynamic movement from heterogeneity to homogeneity," or some other equally incomprehensible rot. It begins rather with the genial sentence: "I went down yesterday to the Piraeus, with Glauco, the son of Aristo, to pay my devotion to the goddess; and desirous, at the same time, to observe in what manner they would celebrate the festival, as they were now to do it for the first time." The same atmosphere that we find among the early Chinese philosophers when thinking was most active and virile, we find in the picture of Greek men, gathered to discuss the topic whether a great writer of tragedies should or should not be also a great writer of comedies, as described in "The Banquet." There was an atmosphere of mixed seriousness and gaiety and friendly repartee. People were making fun of Socrates' drinking capacity, but there he sat, drinking or stopping as he liked, pouring a cup for himself when he felt like it, without bothering about others. And thus he talked the whole night out until everybody in the company fell asleep except Aristophanes and Agathon. When he had thus talked everybody to sleep and was thus the only one awake, he left the banquet and went to Lyceum to have a morning bath, and passed the day as fresh as ever. It was in this atmosphere of friendly discourse that Greek philosophy was born.

There is no question but we need the presence of women in a cultured conversation, to give it the necessary frivolity which is the soul of conversation. Without frivolity and gaiety, conversation soon becomes heavy and philosophy itself becomes foolish and a stranger to life. It has been found in all countries and in all ages that, whenever there was a culture interested in the understanding of the art of living, there always developed a fashion of welcoming women in society. This was the case of Athens in the time of Pericles, and it was so in the eighteenth-century French salons. Even in China, where mixed company was tabooed, Chinese men scholars still demanded the presence of women who could join in their conversation. In the three dynasties, Chin, Sung and Ming, when the art of conversation was cultivated and became a fashion, there always appeared accomplished ladies, like Hsieh Taoyün, Ch'oayün, Liu Jushih and others. For although Chinese men demanded that their wives be virtuous and abstain from seeing men, they did not on that account cease to desire the company of talented women themselves. Chinese literary history after all was very much mixed up with the lives of professional courtesans. The demand for a touch of feminine charm in a company during conversation is a universal demand. I have met German ladies who can talk from five o'clock in the afternoon till eleven at night, and I have come across English and American ladies who frighten me by their familiarity with economics, a subject that I despair of ever having the courage to study. But it seems to me, even if there are no ladies around who can debate with me on Karl Marx and Engels, conversation is always pleasantly stimulated when there are a few ladies who know how to listen and look sweetly pensive. I always find it more delightful than talking to stupid-looking men.

四．茶和交友

我以为从人类文化和快乐的观点论起来，人类历史中的杰出新发明，其能直接有力地有助于我们享受空闲、友谊、社交和谈天者，莫过于吸烟、饮酒、饮茶的发明。这三件事有几样共同的特质：第一，它们有助于我们的社交；第二，这几件东西不至于一吃就饱，可以在吃饭的中间随时吸饮；第三，都是可以借嗅觉去享受的东西。它们对于文化的影响极大，所以餐车之外另有吸烟车，饭店之外另有酒店和茶馆，而至少在中国和英国，饮茶已经成为社交上一种不可少的规矩。

烟酒茶的适当享受，只能在空闲、友谊和乐于招待之中发展出来。因为只有富于交友心、择友极慎、天然喜爱闲适生活的人士，方有圆满享受烟酒茶的机会，如将乐于招待心除去，这三种东西便毫无意义。享受这三件东西，也如享受雪月花草一般，须有适当的同伴。中国的生活艺术家最注意此点，例如：看花须和某种人为伴，赏景须有某种女子为伴，听雨最好在夏日山中寺院内躺在竹榻上。总括起来，赏玩一样东西时，最紧要的是心境。我们对每一种事物，各有一种不同的心境，不适当的同伴，常会败坏心境。所以生活艺术家的出发点就是，如更想要享受人生，第一个必要条件即是和性情相投的人交朋友，须尽力维持这友谊，如妻子要维持其丈夫的爱情一般，或如一个下棋名手宁愿跑一千里的长途去会见一个同志一般。

所以气氛是重要的东西。我们必须先对文士的书室的布置和它的一般的环境有了相当的认识，方能了解他怎样享受生活。第一，他们必须有共同享受这种生活的朋友，不同的享受须有不同的朋友。和一个勤学而含愁思的朋友共去骑马，即非其类，正如和一个不懂音乐的人去欣赏一次音乐表演一般。因此，某中国作家明陈继儒《小窗幽记》曾说过：

赏花须结豪友，观妓须结淡友，登山须结逸友，泛舟须结旷友，对月须结冷友，待雪须结艳友，捉酒须结韵友。

他对各种享受选定了不同的适当游伴之后，还须去找寻适当的环境。所住的房屋，布置不必一定讲究，地点也不限于风景幽美的乡间，不必一定需一片稻田方足供他的散步，也不必一定有曲折的小溪以供他在溪边的树下小憩。他所需的房屋极其简单，只需："有屋数间，有田数亩，用盆为池，以瓮为牖，墙高于肩，室大于斗，布被暖余，藜藿饱后，气吐胸中，充塞宇宙……凡静室，须前栽碧梧，后种翠竹。前檐放步，北用暗窗，春冬闭之，以避风雨，夏秋可开，以通凉爽。然碧梧之趣，春冬落叶，以舒负暄融和之乐，夏秋交荫，以蔽炎烁蒸烈之气。"一个人可以"筑室数楹，编槿为篱，结茅为亭。以三亩荫竹树栽花果，二亩种蔬菜。四壁清旷，空诸所有。蓄山童灌园薙草，置二三胡床着亭下。挟书剑，伴孤寂，携琴弈，以迟良友"。

到处充满着亲热的空气：

　　吾斋之中，不尚虚礼。凡入此斋，均为知己。随分款留，忘形笑语。不言是非，不侈荣利。闲谈古今，静玩山水。清茶好酒，以适幽趣。臭味之交，如斯而已。

　　在这种同类相引的气氛中，我们方能满足色香声的享受，吸烟饮酒也在这个时候最为相宜。我们的全身便于这时变成一种盛受器械，能充分享受大自然和文化所供给我们的色声香味。我们好像已变为一把优美的小提琴，正待由一位大音乐家来拉奏名曲了。于是我们"月夜焚香，古桐三弄，便觉万虑都忘，妄想尽绝。试看香是何味，烟是何色，穿窗之白是何影，指下之余是何音，恬然乐之而悠然忘之者，是何趣，不可思量处，是何境？"

　　一个人在这种神清气爽、心气平静、知己满前的境地中，方真能领略到茶的滋味。因为茶须静品，而酒须热闹。茶之为物，性能引导我们进入一个默想人生的世界。饮茶之时有儿童在旁哭闹，或粗蠢妇人在旁大声说话，或自命通人者在旁高谈国是，即十分败兴，也正如在雨天或阴天去采茶一般糟糕。因为必须天气清明的清早，当山上的空气极为清新，露水的芬芳尚留于叶上时，所采的茶叶方称上品。照中国人说起来，露水实在具有芬芳和神秘的功用，和茶的优劣很有关系。照道家的返自然和宇宙之能生存全恃阴阳二气交融的说法，露水实在是天地在夜间和融后的精英。至今尚有人相信露水为清鲜神秘的琼浆，多饮即能致人兽长生。德昆西（De Quincey，英国著名散文家和批评家）说的话很对，他说："茶永远是聪慧的人们的饮料。"但中国人更进一步，视它为风雅隐士的珍品。

　　因此，茶是凡间纯洁的象征，在采制烹煮的手续中，都须十分清洁。采摘烘焙，烹煮取饮之时，手上或杯壶中略有油腻不洁，便会使它丧失美味。所以也只有在眼前和心中毫无富丽繁华的景象和念头时，方能真正地享受它。和妓女作乐时，当然用酒而不用茶。但一个妓女如有了品茶的资格，便可以跻于诗人文士所欢迎的妙人儿之列了。苏东坡曾以美女喻茶，但后来，另一个持论家，《煮泉小品》的作者田艺蘅即补充说，如果定要以茶去拟女人，则唯有麻姑仙子可比拟。至于"必若桃脸柳腰，宜驱屏之销金幔中，无俗我泉石"。又说："啜茶忘喧，谓非膏粱纨绮可语。"

据张源《茶录》所说："其旨归于色香味，其道归于精燥洁。"如果要体味这些质素，静默是一个必要的条件；也只有"以一个冷静的头脑去看忙乱的世界"的人，才能够体味出这些质素。自从宋代以来，一般喝茶的鉴赏家认为一杯淡茶才是最好的东西，当一个人专心思想的时候，或是在邻居嘈杂、仆人争吵的时候，或是由面貌丑陋的女仆侍候的时候，自会很容易地忽略了淡茶的美妙气味。同时，喝茶的友伴不可多，因为"饮茶以客少为贵。客众则喧，喧则雅趣乏矣。独啜曰幽；二客曰胜；三四曰趣；五六曰泛；七八曰施"。《茶疏》的作者许次纾说："若巨器屡巡，满中泻饮，待停少温，或求浓苦，何异农匠作劳，但需涓滴；何论品赏？何知风味乎？"

因为这个理由，因为要顾到烹时的合度和洁净，有茶癖的中国文士都主张烹茶须自己动手。如嫌不便，可用两个小童为助。烹茶须用小炉，烹煮的地点须远离厨房，而近饮处。茶童须受过训练，当主人的面前烹煮。一切手续都须十分洁净，茶杯须每晨洗涤，但不可用布揩擦。童儿的两手须常洗，指甲中的污腻须剔干净。"三人以上，止爇一炉，如五六人，便当两鼎，炉用一童，汤方调适，若令兼作，恐有参差。"真正鉴赏家常以亲自烹茶为一种殊乐。中国的烹茶饮茶方法不像日本那样过分严肃和讲规则，而仍属一种富有乐趣而又高尚重要的事情。实在说起来，烹茶之乐和

饮茶之乐各居其半，正如吃西瓜子，用牙齿咬瓜子壳之乐和吃瓜子肉之乐
实各居其半。

　　茶炉大都置在窗前，用硬炭生火。主人很郑重地扇着炉火，注视着水
壶中的热气。他用一个茶盘，很整齐地装着一个泥茶壶和四个比咖啡杯小
一些的茶杯，再将贮茶叶的锡罐安放在茶盘的旁边，随口和来客谈着天，
但并不忘了手中所应做的事。他时时顾看炉火，等到水壶中渐发沸声后，
就立在炉前不再离开，更加用力地扇火，还不时要揭开壶盖望。那时壶底
已有小泡，名为"鱼眼"或"蟹沫"，这就是"初滚"。他重新盖上壶盖，
再扇上几扇，壶中的沸声渐大，水面也渐起泡，这名为"二滚"。这时已
有热气从壶口喷出来，主人也就格外注意。到将届"三滚"、壶水已经沸
透之时，他就提起水壶，将小泥壶里外一浇，赶紧将茶叶加入泥壶，泡出
茶来。这种茶如福建人所饮的"铁观音"，大都泡得很浓。小泥壶中只可
容水四小杯，茶叶占去其三分之一的容隙。因为茶叶加得很多，所以一泡
之后即可倒出来喝了。这一道茶已将壶水用尽，于是再灌入凉水，放到炉
上去煮，以供第二泡之用。严格说起来，茶在第二泡时为最妙。第一泡譬
如一个十二三岁的幼女，第二泡为年龄恰当的十六女郎，而第三泡已是少
妇了。照理论上说起来，鉴赏家认第三泡的茶为不可复饮，但实际上，享
受这个"少妇"的人仍很多。

　　以上所说是我本乡一种泡茶方法的实际素描，这个艺术是中国北方人
所不晓的。在中国的一般人家中，所用的茶壶都较大。至于一杯茶，最好
的颜色是清中带微黄，而不是英国茶那样的深红色。

　　我们所描写的当然是指鉴赏家的饮茶，而不是店铺中的以茶奉客。这
种雅举并非普通人所能办到，也并非人来人往、论碗解渴的地方所能办
到。《茶疏》的作者许次纾说得好："宾朋杂沓，止堪交钟觥筹；乍会泛交，

仅须常品酬酢。惟素心同调，彼此畅适，清言雄辩，脱略形骸，始可呼童篝火，吸水点汤，量客多少，为役之烦简。"而《茶解》作者所说的就是此种情景："山堂夜坐，汲泉煮茗。至水火相战，如听松涛。倾泻入杯，云光滟潋。此时幽趣，故难与俗人言矣。"

凡真正爱茶者，单是摇摩茶具，已经自有其乐趣。蔡襄年老时已不能饮茶，但他每天必烹茶以自娱，即其一例。又有一个文士周文甫，每天自早至晚，必在规定的时刻自烹自饮六次，他极宝爱他的茶壶，死时甚至以壶为殉。

因此，茶的享受技术包括下列各节：第一，茶味娇嫩，茶易败坏，所以整治时，须十分清洁，须远离酒类香类一切有强味的事物和身带这类气息的人；第二，茶叶须贮藏于冷燥之处，在潮湿的季节中，备用的茶叶须贮于小锡罐中，其余则另贮大罐，封固藏好，不取用时不可开启，如若发霉，则须在文火上微烘，一面用扇子轻轻挥扇，以免茶叶变黄或变色；第三，烹茶的艺术一半在于择水，山泉为上，河水次之，井水更次，水槽之水如来自堤堰，因为本属山泉，所以很可用得；第四，客不可多，且须文雅之人，方能鉴赏杯壶之美；第五，茶的正色是清中带微黄，浓的红茶即不能不另加牛奶、柠檬、薄荷或他物以调和其苦味；第六，好茶必有回味，大概在饮茶半分钟后，当其化学成分和津液发生作用时，即能觉出；第七，茶须现泡现饮，泡在壶中稍稍过候即会失味；第八，泡茶必须用刚沸之水；第九，一切可以混杂真味的香料，须一概摒除，至多只可略加些桂皮或花，以合有些爱好者的口味而已；第十，茶味最上者，应如婴孩身上一般的"奶花香"。

据《茶疏》之说，最宜于饮茶的时候和环境是这样：

饮　时：

心手闲适　披咏疲倦　意绪棼乱　听歌闻曲　歌罢曲终　杜门避事
鼓琴看画　夜深共语　明窗净几　洞房阿阁　宾主款狎　佳客小姬　访友
初归　风日晴和　轻阴微雨　小桥画舫　茂林修竹　课花责鸟　荷亭避暑
小院焚香　酒阑人散　儿辈斋馆　清幽寺观　名泉怪石

宜　辍：

作字　观剧　发书柬　大雨雪　长筵大席　阅卷帙

人事忙迫　及与上宜饮时相反事

不宜用：

恶水　敝器　铜匙　铜铫　木桶　柴薪　麸炭　粗童　恶婢　不洁巾
帨　各色果实香药

不宜近：

阴屋　厨房　市喧　小儿啼　野性人　童奴相哄　酷热斋舍

IV. On Tea and Friendship

I do not think that, considered from the point of view of human culture and happiness, there have been more significant inventions in the history of mankind, more vitally important and more directly contributing to our enjoyment of leisure, friendship, sociability and conversation, than the inventions of smoking, drinking and tea. All three have several characteristics in common: first of all, that they contribute toward our sociability; secondly, that they do not fill our stomach as food does, and therefore can be enjoyed between meals; and thirdly, that they are all to be enjoyed through the nostrils by acting on our sense of smell. So great are their influences upon culture that we have smoking cars besides dining cars, and we have wine restaurants or taverns and tea houses. In China and England at least, drinking tea has become a social institution.

The proper enjoyment of tobacco, drink and tea can only be developed in an atmosphere of leisure, friendship and sociability. For it is only with men gifted with the sense of comradeship, extremely select in the matter of forming friends and endowed with a natural love of the leisurely life, that the full enjoyment of tobacco and drink and tea becomes possible. Take away the element of sociability, and these things have no meaning. The enjoyment of these things, like the enjoyment of the moon, the snow and the flowers, must take place in proper company, for this I regard as the thing that the Chinese artists of life most frequently insist upon: that certain kinds of flowers must be enjoyed with certain types of persons, certain kinds of scenery must be associated with certain kinds of ladies, that the sound of raindrops must be enjoyed, if it is to be enjoyed fully, when lying on a bamboo bed in a temple deep in the mountains on a summer day; that, in short, the mood is the thing, that there is a proper mood for everything, and that wrong company may spoil the mood entirely. Hence the beginning of any artist of life is that he or anyone who wishes to learn to enjoy life must, as the absolutely necessary condition, find friends of the same type of temperament, and take as much trouble to gain and keep their friendship as wives take to keep their husbands, or as a good chess player takes a journey of a thousand miles to meet a fellow chess player.

The atmosphere, therefore, is the thing. One must begin with the proper conception of the scholar's studio and the general environment in which life is going to be enjoyed. First of all, there are the friends with whom we are going to share this enjoyment. Different types of friends must be selected for different types of enjoyment. It would be as great a mistake to go horseback riding with a studious and pensive friend, as it would be to go to a concert with a person who doesn't understand music. Hence as a Chinese writer expresses it:

For enjoying flowers, one must secure big-hearted friends. For going to sing-song houses to have a look at sing-song girls, one must secure temperate friends. For going up a high mountain, one must secure romantic friends. For boating, one must secure friends with an expansive nature. For facing the moon, one must secure friends with a cool philosophy. For anticipating snow, one must secure beautiful friends. For a wine party, one must secure friends with flavor and charm.

Having selected and formed friends for the proper enjoyment of different occasions, one then looks for the proper surroundings. It is not so important that one's house be richly decorated as that it should be situated in beautiful country, with the possibility of walking about on the rice fields, or lying down under shady trees on a river bank. The requirements for the house itself are simple enough. One can "have a house with several rooms, grain fields of several mow, a pool made from a basin and windows made from broken jars, with the walls coming up to the shoulders and a room the size of a rice bushel, and in the leisure time after enjoying the warmth of cotton beddings and a meal of vegetable soup, one can become so great that his spirit expands and fills the entire universe. For such a quiet studio, one should have wut'ung trees in front and some green bamboos behind. On the south of the house, the eaves will stretch boldly forward, while on the north side, there will be small windows, which can be closed in spring and winter to shelter one from rain and wind, and opened in summer and autumn for ventilation. The beauty of the wut'ung tree is that all its leaves fall off in spring and winter, thus admitting us to the full enjoyment of the sun's warmth, while in summer and autumn its shade protects us from the scorching heat." Or as another writer expressed it, one should "build a house of several beams, grow a hedge of chin trees and cover a pavilion with a hay-thatch. Three mow of land will be devoted to planting bamboos and flowers and fruit trees, while two mow will be devoted to planting vegetables. The four walls of a room are bare and the room is empty, with the exception of two or three rough beds placed in the pavilion. A peasant boy will be kept to water the vegetables and clear the weeds. So then one may arm one's self with books and a sword against solitude, and provide a ch'in (a stringed instrument) and chess to anticipate the coming of good friends. "

An atmosphere of familiarity will then invest the place.

"In my studio, all formalities will be abolished, and only the most intimate friends will be admitted. They will be treated with rich or poor fare such as I eat, and we will

chat and laugh and forget our own existence. We will not discuss the right and wrong
of other people and will be totally indifferent to worldly glory and wealth. In our leisure
we will discuss the ancients and the moderns, and in our quiet, we will play with the
mountains and rivers. Then we will have thin, clear tea and good wine to fit into the
atmosphere of delightful seclusion. That is my conception of the pleasure of friendship. "

In such a congenial atmosphere, we are then ready to gratify our senses,
the senses of color and smell and sound. It is then that one should smoke and
one should drink. We then transform our bodies into a sensory apparatus
for perceiving the wonderful symphony of colors and sounds and smells
and tastes provided by Nature and by culture. We feel like good violins
about to be played on by master violinists. And thus "we burn incense on a
moonlight night and play three stanzas of music from an ancient instrument,
and immediately the myriad worries of our breast are banished and all our
foolish ambitions or desires are forgotten. We will then inquire, what is the
fragance of this incense, what is the color of the smoke, what is that shadow
that comes through the white papered windows, what is this sound that arises
from below my fingertips, what is this enjoyment which makes us so quietly
happy and so forgetful of everything else, and what is the condition of the
infinite universe?"

Thus chastened in spirit, quiet in mind and surrounded by proper
company, one is fit to enjoy tea. For tea is invented for quiet company as wine
is invented for a noisy party. There is something in the nature of tea that leads
us into a world of quiet contemplation of life. It would be as disastrous to
drink tea with babies crying around, or with loud-voiced women or politics-
talking men, as to pick tea on a rainy or a cloudy day. Picked at early dawn on
a clear day, when the morning air on mountain top was clear and thin, and
the fragrance of dews was still upon the leaves, tea is still associated with the
fragrance and refinement of the magic dew in its enjoyment. With the Taoist
insistence upon return to nature, and with its conception that the universe

is kept alive by the interplay of the male and female forces, the dew actually stands for the "juice of heaven and earth" when the two principles are united at night, and the idea is current that the dew is a magic food, fine and clear and ethereal, and any man or beast who drinks enough of it stands a good chance of being immortal. De Quincey says quite correctly that tea "will always be the favorite beverage of the intellectual," but the Chinese seem to go further and associate it with the high-minded recluse.

Tea is then symbolic of earthly purity, requiring the most fastidious cleanliness in its preparation, from picking, frying and preserving to its final infusion and drinking, easily upset or spoiled by the slightest contamination of oily hands or oily cups. Consequently, its enjoyment is appropriate in an atmosphere where all ostentation or suggestion of luxury is banished from one's eyes and one's thoughts. After all, one enjoys sing-song girls with wine and not with tea, and when sing-song girls are fit to drink tea with, they are already in the class that Chinese poets and scholars favor. Su Tungp'o once compared tea to a sweet maiden, but a later critic, T'ien Yiheng, author of Chuch'üan Hsiaop'in (Essay On Boiling Spring Water) immediately qualified it by adding that tea could be compared, if it must be compared to women at all, only to the Fairy Maku, and that, "as for beauties with peach-colored faces and willow waists, they should be shut up in curtained beds, and not be allowed to contaminate the rocks and springs." For the same author says, "One drinks tea to forget the world's noise; it is not for those who eat rich food and dress in silk pyjamas."

It must be remembered that, according to Ch'alu, "the essence of the enjoyment of tea lies in appreciation of its color, fragrance and flavor, and the principles of preparation are refinement, dryness and cleanliness." An element of quiet is therefore necessary for the appreciation of these qualities, an appreciation that comes from a man who can "look at a hot world with a cool head." Since the Sung Dynasty, connoisseurs have generally regarded a cup of pale tea as the best, and the delicate flavor of pale tea can easily pass unperceived by one occupied with busy thoughts, or when the neighborhood is noisy, or servants are quarreling, or when served by ugly maids. The company, too, must be small. For, "it is important in drinking tea that the guests be few. Many guests would make it noisy, and noisiness takes away from its cultured charm. To drink alone is called secluded; to drink between two is called comfortable; to drink with three or four is called charming, to drink with five or six is called common; and to drink with seven or eight is called [contemptuously] philanthropic." As the author of Ch'asu said, "to pour tea around again and again from a big pot, and drink it up at a gulp, or to warm it up again after a while, or to ask for extremely strong taste would be like farmers or artisans who drink tea to fill their belly after hard work; it would then be impossible to speak of the distinction and appreciation of flavors. "

For this reason, and out of consideration for the utmost rightness and cleanliness in preparation, Chinese writers on tea have always insisted on personal attention in boiling tea, or since that is necessarily inconvenient, that two boy servants be specially trained to do the job. Tea is usually boiled on a separate small stove in the room or directly outside, away from the kitchen. The servant boys must be trained to make tea in the presence of their master and to observe a routine of cleanliness, washing the cups every morning (never wiping them with a towel), washing their hands often and keeping their fingernails clean. "When there are three guests, one stove will be enough, but when there are five or six persons, two separate stoves and kettles will be required, one boy attending to each stove, for if one is required to attend to both, there may be delays or mix-ups." True connoisseurs, however, regard the personal preparation of tea as a special pleasure. Without developing into a rigid system as in Japan, the preparation and drinking of tea is always a performance of loving pleasure, importance and distinction. In fact, the preparation is half the fun of the drinking, as cracking melon-seeds between one's teeth is half the pleasure of eating them.

Usually a stove is set before a window, with good hard charcoal burning. A certain sense of importance invests the host, who fans the stove and watches the vapor coming out from the kettle. Methodically he arranges a small pot and four tiny cups, usually smaller than small coffee cups, in a tray. He sees that they are in order, moves the pewter-foil pot of tea leaves near the tray and keeps it in readiness, chatting along with his guests, but not so much that he forgets his duties. He turns round to look at the stove, and from the time the kettle begins to sing, he never leaves it, but continues to fan the fire harder than before. Perhaps he stops to take the lid off and look at the tiny bubbles, technically called"fish eyes"or"crab froth,"appearing on the bottom of the kettle, and puts the lid on again. This is the"first boil." He listens carefully as the gentle singing increases in volume to that of a"gurgle,"with small bubbles coming up the sides of the kettle, technically called the"second boil."It is then that he watches most carefully the vapor emitted from the kettle spout, and just shortly before the"third boil"is reached, when the water is brought up to a full boil,"like billowing waves,"he takes the kettle from the fire and scalds the pot inside and out with the boiling water, immediately adds the proper quantity of leaves and makes the infusion. Tea of this kind, like the famous"Iron Goddess of Mercy,"drunk in Fukien, is made very thick. The small pot is barely enough to hold four demi-tasses and is filled one-third with leaves. As the quantity of leaves is large, the tea is immediately poured into the cups and immediately drunk. This finishes the pot, and the kettle, filled with fresh water, is put on the fire again, getting ready for the second pot. Strictly speaking, the second pot is regarded as the best; the first pot being compared to a girl of thirteen, the second compared to a girl of sweet sixteen, and the third regarded as a woman. Theoretically, the third infusion from the same leaves is disallowed by connoisseurs, but actually one does try to live on with the"woman. "

The above is a strict description of preparing a special kind of tea as I have seen it in my native province, an art generally unknown in North

China. In China generally, tea pots used are much larger, and the ideal color of tea is a clear, pale, golden yellow, never dark red like English tea.

Of course, we are speaking of tea as drunk by connoisseurs and not as generally served among shopkeepers. No such nicety can be expected of general mankind or when tea is consumed by the gallon by all comers. That is why the author of Ch'asu, Hsü Ts'eshu, says,"When there is a big party, with visitors coming and coming, one can only exchange with them cups of wine, and among strangers who have just met or among common friends, one should serve only tea of the ordinary quality. Only when our intimate friends of the same temperament have arrived, and we are all happy, all brilliant in conversation and all able to lay aside the formalities, then may we ask the boy servant to build a fire and draw water, and decide the number of stoves and cups to be used in accordance with the company present."It is of this state of things that the author of Ch'achieh says,"We are sitting at night in a mountain lodge, and are boiling tea with water from a mountain spring. When the fire attacks the water, we begin to hear a sound similar to the singing of the wind among pine trees. We pour the tea into a cup, and the gentle glow of its light plays around the place. The pleasure of such a moment cannot be shared with vulgar people. "

In a true tea lover, the pleasure of handling all the paraphernalia is such that it is enjoyed for its own sake, as in the case of Ts'ai Hsiang, who in his old age was not able to drink, but kept on enjoying the preparation of tea as a daily habit. There was also another scholar, by the name of Chou Wenfu, who prepared and drank tea six times daily at definite hours from dawn to evening, and who loved his pot so much that he had it buried with him when he died.

The art and technique of tea enjoyment, then, consists of the following: first, tea, being most susceptible to contamination of flavors, must be handled throughout with the utmost cleanliness and kept apart from wine, incense, and other smelly substances and people handling such substances. Second, it must be kept in a cool, dry place, and during moist seasons, a reasonable quantity for use must be kept in special small pots, best made of pewterfoil, while the reserve in the big pots is not opened except when necessary, and if a collection gets moldy, it should be submitted to a gentle roasting over a slow fire, uncovered and constantly fanned, so as to prevent the leaves from turning yellow or becoming discolored. Third, half of the art of making tea lies in getting good water with a keen edge; mountain spring water comes first, river water second, and well water third; water from the tap, if coming from dams, being essentially mountain water and satisfactory. Fourth, for the appreciation of rare cups, one must have quiet friends and not too many of them at one time. Fifth, the proper color of tea in general is a pale golden yellow, and all dark red tea must be taken with milk or lemon or peppermint, or anything to cover up its awful sharp taste. Sixth, the best tea has a "return flavor" (hueiwei), which is felt about half a minute after drinking and after its chemical elements have had time to act on the salivary glands. Seven, tea must be freshly made and drunk immediately, and if good tea is expected, it should not be allowed to stand in the pot for too long, when the infusion has gone too far. Eight, it must be made with water just brought up to a boil. Nine, all adulterants are taboo, although individual differences may be allowed for people who prefer a slight mixture of some foreign flavor (e.g., jasmine, or cassia). Eleven, the flavor expected of the best tea is the delicate flavor of "baby's flesh. "

In accordance with the Chinese practice of prescribing the proper moment and surrounding for enjoying a thing, Ch 'asu, an excellent treatise on tea, reads thus:

Proper moments for drinking tea :
When one's heart and hands are idle.
Tired after reading poetry.
When one's thoughts are disturbed.
Listening to songs and ditties.

When a song is completed.

Shut up at one's home on a holiday.

Playing the ch'in and looking over paintings.

Engaged in conversation deep at night.

Before a bright window and a clean desk.

With charming friends and slender concubines.

Returning from a visit with friends.

When the day is clear and the breeze is mild.

On a day of light showers.

In a painted boat near a small wooden bridge.

In a forest with tall bamboos.

In a pavilion overlooking lotus flowers on a summer day.

Having lighted incense in a small studio.

After a feast is over and the guests are gone.

When children are at school.

In a quiet, secluded temple.

Near famous springs and quaint rocks.

Moments when one should stop drinking tea :

At work. Watching a play.

Opening letters.

During big rain and snow.

At a long wine feast with a big party.

Going through documents.

On busy days.

Generally conditions contrary to those enumerated in the above section.

Things to be avoided :

Bad water.

Bad utensils.

Brass spoons.

Brass kettles.

Wooden pails (for water).
Wood for fuel (on account of smoke).
Soft charcoal.
Coarse servant.
Bad-tempered maid.
Unclean towels.
All varieties of incense and medicine.
Things and places to be kept away from:
Damp rooms.
Kitchens.
Noisy streets.
Crying infants.
Hotheaded persons.
Quarreling servants.
Hot rooms.

五．淡巴菰和香

　　现在的世人，分为吸烟者和不吸烟者两类。吸烟者确然使不吸烟者有些讨厌，但这种讨厌不过是属于物质性质，而不吸烟者之讨厌于吸烟者是精神上的。不吸烟者之中，当然也有对吸烟者采取不干涉态度的人，为妻者之中，当然也有容许其丈夫在床上吸烟的，这种夫妻，显然是在婚姻上获有圆满结果的佳偶。但颇也有人以为不吸烟者在道德上较为高尚，以为他们具有一种可以傲人的美德，而不知他们即已因此丧失了人类的最大乐趣之一。我很愿意承认吸烟是道德上的一个弱点，但在另一方面，一个没有道德弱点的人，也不是可以全然信任的，他惯于持严肃的态度，从不做错误的事情，他的习惯一般是有规则的，举动较为近于机械性，智能时常控制其心情。我很欢喜富于情理的人，同样憎嫌专讲理智的人。因为这个理由，我踏进人家的屋子而找不到烟灰缸时，心中便会惊慌，觉得不自在。这种屋子中，往往过于清洁有秩序，椅垫从不随意乱摆，主人也必是极严肃毫无情感的人。这将使我也不能不正襟危坐，力持礼貌，因而失去了一切舒适。

　　这种毫无错误，正直而无感情，毫无诗意的人们，从不会领略吸烟在道德上和精神上的裨益。但是我们这批吸烟者，每每被人从道德而不是艺术方面加以攻击。所以，第一步我也须从道德方面加以辩护，而以为吸烟者的道德在大体上高于不吸者。口含烟斗者是最合我意的人，这种人都较为和蔼，较为恳切，较为坦白，又大都善于谈天。我总觉得我和这般人必能彼此结交相亲。我对萨克雷所说的话极表同情。他说，烟斗从哲学家的口中引出智慧，也封闭愚拙者的口，使之缄默；它能产生一种沉思的、富有意思的、仁慈的和无虚饰的谈天风格。

　　吸烟者的手指当然较为污秽，但只要他心有热情，这又何妨？无论如何，沉思的、富有意思的、仁慈的和无虚饰的谈天风格终究是罕遇之物，所以，需付一笔巨大的代价去享受它，也是值得的。最重要的一点是：口含烟斗的人都是快乐的，而快乐终是一切道德效能中之最大者。梅金（W. Maggin）说："吸雪茄的人，从没有自杀者。"更确凿有据的事情是，吸管烟的人从不会同自己的太太吵嘴，其理由很显明，因为口含烟斗的人，绝不能同时高声叫骂。我从来没有见过如此的人。当一个人吸着管烟时，语音当然很低，一个吸烟的丈夫发怒时，他的办法就是立刻点一支卷烟或一斗管烟吸起来，显出一些抑郁的神气，但这种神情不久即能消灭，因为他的怒气已有了发泄之处。即使他有意想把怒容维持下去，以表示他发怒的正当，或表示他受了侮辱，但事实上绝不能持续。因为烟斗中的烟味是如此和润悦性，以致他所贮着的怒气早已在无意间，跟着一口一口喷出来的烟消逝了。所以聪明的妻子，当她看见丈夫快要发怒时，就应该赶紧拿烟斗塞在丈夫的口中，说："得了，不必再提。"这个方法万试万灵。为妻者或许不能平抑丈夫的发怒，但烟斗是从不失败的。

　　从一个吸烟者的短期戒烟中所经验的忽忽若有所失的感觉，最足以显出吸烟的艺术和实际的价值。每个吸烟者一生之中，免不了在欠少思量的时候忽有想和"尼古丁女士"脱离关系的尝试。但经过一番和缥缈的良心责备争斗之后，他必又重新恢复他的理智。我有一次，也很欠思量地戒烟三个星期，但后来终究为良心所驱使而重新登上正当的途径。从此我就立誓不再起叛逆之心，立誓在她的神座前做一个终身的敬信崇拜者，直到我年老无能，或许落入一个属于节制会的太太手中而失去了自主的权力时。因为到了这种老年无能时期，一个人对于自己的一切行动当然无须再负责任了。但只要我的自主力和道德观念一日存在，则我必一日不做背叛的尝试。这个有功效的新发明所供给精神上的动力和道德上的安宁观念是怎样伟大，我们如若拒绝它，岂不是不可赦的不道德行为吗？因为按照英国大生物化学家霍尔丹（Haldane）的说法，吸烟是人类历史中四大发明之一，曾于人类文化上留下一种很深的生物性影响。

在我这次做懦夫的三个星期中，我竟会故意拒绝一件我明知具有巨大提升灵魂力量的东西。其经过实在极为可耻的。现在我已恢复了理智。在清明中回想这件事时，我正不解当时这种道德的不负责任行为何以会维持这般的久。我在这痛苦的三个星期中，内心日夜交战着，如要将这段经过描写出来，恐怕用三千句荷马（Homer）体的诗，或一百五十页小字的散文尚且写不尽哩。当时我的动机其实很可笑。我不解对宇宙中的人类而言，为什么不能吸烟？对这句问话，我现在实在找不出答语。我猜想当一个人只为了求一些克服抵抗力的乐趣，借此以消磨他的道德动力的暂时剩余，因而想做一种违反本性的举动时，这种不合情理的意旨或许就会在胸中产生。除了这个理由之外，我实在想不出我为什么会突然很愚蠢地决意戒烟。换句话说，当时我实在和许多人们耽于瑞典式体操一样——为体操而体操，所费的力对于社会一无用处。我当时的举动，其实不过是如此一种道德上的枉费力量罢了。

在最初的三天中，我当然觉得很无聊不自在。食道的上部尤其难受。为了消除这种不自在，我特地吃些重味的薄荷橡皮糖、福建茶和柠檬糖，居然在第三天即消灭了这种不快的感觉。但这不过是属于身体方面的，所以克服极其容易，而且照我事后想起来，实是这次争斗中最卑鄙的部分。倘若有人以为这已经包括这种卑鄙战争的全局，则他简直是在那里胡说八道。他们忘却了吸烟是一种精神上的行为，凡是对于吸烟的精神上的意义毫无了解之人，可不必来妄论这件事情。三天之后，我已踏进第二个梯阶，真正的精神上的交战也开始发生。我顿时觉得眼前金星乱碰。由这次的经验，我即发现世上实有两种吸烟人，而其中一种实在不能算为真正吸烟者，在这种人之中并没有这第二个梯阶。我因此方恍然知道为什么有许多人能毫不费力地戒除烟癖。他们之能摒除烟习如丢弃一支用旧的牙刷一般容易，即表明他们其实尚没有学会吸烟。有许多人还称赞他们的意志力坚强，但其实他们并不是真正的吸烟者，也从没有学会吸烟。对于这一种人，吸烟不过是一种身体上的行为，如每天早晨的洗脸刷牙一般——只是一种身体的兽性的习惯，而并不具有灵魂上获得满足的质素。我很疑惑这种迁就事实的人们是否能有一天调和他们的灵魂，而达到大诗人雪莱或肖邦（Chopin）所描写的境地，这种人于戒烟时并不感觉有什么不自在，或许觉得和自己那不进烟酒的太太共读《伊索寓言》是更为快乐一些的。

但在我们这种真正吸烟者，另外有一个烟酒不入的太太或爱读《伊索寓言》的丈夫所不能梦想其万一的问题。在我们，不久就显然知道这个举动不但是委屈自己，而且实在是毫无意义。见识和理智不久便会反抗而诘问："一个人为了那一种社会的、政治的、道德的、生理的或经济的理由，而须有意识地用自己的意志力去阻抑自己去企求那种完备的精神安乐，那种深切富有幻想的认识和充分反响的创造力的境地？"——这种境地是圆满享受和友人围炉聚谈，或阅读一本古书时心中发生真正热情，或动笔著作时文思佳句有节奏地泉涌出来所必需的境地。在这种时节，一个人天然觉得伸手去拿一支烟是道德上最正当的举动，倘若去拿一块橡皮糖塞在口中以为替代便是一种罪恶。此处我当略举一两个我所经验的实例。

我的朋友某君从北平来探望我。我们阔别已经三十年。当同在北平时，我们时常促膝而坐，抽烟谈天，消磨晚间的时光，所谈者大多是政治、哲学和现代艺术等题目。此次久别重逢，自然有不少甜蜜的回忆，于是我们又随便谈天，谈谈以前在北平时所知道的许多教授、诗人和奇人。每谈到有趣味的话时，我心里屡次想到伸手去拿卷烟，但刚站了起来，便又强自抑制地缩回坐下，我的朋友则边吸边谈，十分恬然自得。我就告诉他，我已戒烟了，为了自尊起见，实在不愿当着他的面前破戒。我嘴里虽如此说，但心底里实在觉得很不自在，使我在知己相对应该两情融洽、心意交流时，很不应该地装出冷淡富于理智的样子。所以这次谈天，大部分皆是我的朋友在说话，而我好似只有半个人在场。后来我的朋友告辞去了，我好似做了一次凶残的争斗，虽借着意志力获得了胜利，但自己深知实在不快乐。数日之后，这朋友在旅途中写了一封信给我说，我已不是从前那富于热情、狂放不羁的人，并说，或许因上海的环境不良，似致如此。那天晚上，我没有抽烟的过失，直到现在，我尚不能宽恕自己。

……

自此之后，我的良心渐渐啃蚀我的灵魂。因为我曾自问，没有想象的思想将成为什么东西？想象这东西哪里能够附在不吸烟者已经修剪的灰色翅膀上飞行？因此，某天下午，我即去探望一位女友，我已预备在这天回头。当时室中只有我们主客两人，显然可以促膝而谈。女主人手中正拿一支已燃着的烟卷，另一只手则拿着一个卷烟罐，斜着身躯，以极娇媚的态度向着我。我知道时机到了，所以我就缓缓地伸手向罐内取了一支，明白这一个举动已使我从道德堕落的妄举中脱身出来。

我回家之后，立刻叫小童去买一听"绞盘"牌卷烟。我的写字台右边有一条焦痕，那是因为我习惯将香烟头放在这个老地方而留下的痕迹。据我的计算，这焦痕大概需七八年的功夫方能烧穿这二寸厚的台面。因为我这次戒烟的间断，这焦痕竟许久没有加添深度，这使我看了很负疚。现在好了，我照旧很快乐地把烟头放在原处，而烧炙台面的工作也能照常进行了。

中国文学中，提到淡巴菰的好处者很少，不像称赞酒类那么随处可见。因为吸烟的习惯是直到十六世纪方始由葡萄牙水手传到中国的。我曾查遍这个时代以后的中国文学著作，但可称为有价值的赞美言词实在稀若麟毛。称赞淡巴菰的抒情诗显然须如牛津大学般的文人方能著得出来。但中国人对于嗅觉也极灵敏，他们能领略茶酒食物之味即是一个证据。所以他们在淡巴菰未曾传入中国之前，另已发展了一种焚香的艺术。中国文学中提到这件事时，都视之为类于茶酒雅物。远在中国治权伸张到印度支那的汉朝时代，由南方所进贡的香料即已为宫中和贵人的家中所焚用。讨论生活起居的书籍，其中必有一部分讲香料种类、质地和焚法。屠隆所著的《考盘余事》一书中，有一段焚香之趣的描写：

香之为用，其利最薄。物外高隐，坐语道德，焚之可以清心悦神。四更残月，兴味萧骚，焚之可以畅怀舒啸。晴窗塌帖，挥尘闲吟，温灯夜读，焚以远辟睡魔。谓古伴月可也。红袖在侧，秘语谈私，执手拥炉，焚以薰心热意。谓古助情可也。坐雨闭窗，午睡初足，就案学书，啜茗味淡，一炉初热，香蔼馥馥撩人。更宜醉筵醒客，皓月清宵，冰弦曳指，长啸空楼，苍山极目，未残炉热，香雾隐隐绕帘。又可祛邪辟秽，随其所适，无施不可。品其最优者，伽南止矣。第购之甚艰，非山家所能卒办。其次莫若沉香。沉有三等，上者气太厚，而反嫌于辣；下者质太枯，而又涉于烟；唯中者约六七分一两，最滋润而幽甜，可称妙品。煮茗之余，即乘茶炉火便，取入香鼎，徐而蒸之。当斯会心景界，俨居太清宫与上真游，不复知有人世矣。噫，快哉近世焚香者，不博真味，徒事好名，兼以诸香合成斗奇争巧，不知沉香出于天然，其幽雅冲澹，自有一种不可形容之妙。

冒辟疆在所著的《影梅庵忆语》中，描写他和爱姬董小宛的闺房之乐，屡次提到焚香之趣。中间有一节说：

姬每与余静坐香阁，细品茗香。宫香诸品淫，沉水香俗。俗人以沉香著火上，烟朴油腻，顷刻而灭。无论香之性情未出，即著怀袖皆带焦腥。沉香坚致而纹横者，谓之"横隔沉"，即回种沉香内革沉横纹者是也，其香特妙。又有沉水结而未成，如小笠大菌，名"蓬莱香"。余多蓄之，每慢火隔纱，使不见烟，则阁中皆如风过伽楠（佛教寺院的通称），露沃蔷薇，热磨琥珀，酒倾犀礜之味，久蒸枲枕间，和以肌香，甜艳非常，梦魂俱适。

V. On Smoke and Incense

THE world today is divided into smokers and non-smokers. It is true that the smokers cause some nuisance to the non-smokers, but this nuisance is physical, while the nuisance that the nonsmokers cause the smokers is spiritual. There are, of course, a lot of non-smokers who don't try to interfere with the smokers, and wives can be trained even to tolerate their husbands' smoking in bed. That is the surest sign of a happy and successful marriage. It is sometimes assumed, however, that the non-smokers are morally superior, and that they have something to be proud of, not realizing that they have missed one of the greatest pleasures of mankind. I am willing to allow that smoking is a moral weakness, but on the other hand, we must beware of the man without weaknesses. He is not to be trusted. He is apt to be always sober and he cannot make a single mistake. His habits are likely to be regular, his existence more mechanical and his head always maintains its supremacy over his heart. Much as I like reasonable persons, I hate completely rational beings. For that reason, I am always scared and ill at ease when I enter a house in which there are no ash trays. The room is apt to be too clean and orderly, the cushions are apt to be in their right places, and the people are apt to be correct and unemotional. And immediately I am put on my best behavior, which means the same thing as the most uncomfortable behavior.

Now the moral and spiritual benefits of smoking have never been appreciated by these correct and righteous and unemotional and unpoetic souls. But since we smokers are usually attacked from the moral, and not the artistic side, I must begin by defending the smoker's morality, which is on the whole higher than that of the non-smokers. The man with a pipe in his mouth is the man after my heart. He is more genial, more sociable, has more intimate indiscretions to reveal, and sometimes he is quite brilliant in conversation, and in any case, I have a feeling that he likes me as much as I like him. I agree entirely with Thackeray, who wrote:"The pipe draws wisdom from the lips of the philosopher, and shuts up the mouths of the foolish; it generates a style of conversation contemplative, thoughtful, benevolent, and unaffected."

A smoker may have dirtier finger-nails, but that is no matter when his heart is warm, and in any case a style of conversation contemplative, thoughtful, benevolent, and unaffected is such a rare thing that one is willing to pay a high price to enjoy it. And most important of all, a man with a pipe in his mouth is always happy, and after all, happiness is the greatest of all moral virtues. W. Maggin says that"no cigar smoker ever committed suicide,"and it is still truer that no pipe smoker ever quarrels with his wife. The reason is perfectly plain: one cannot hold a pipe between one's teeth and at the same time shout at the top of one's voice. No one has ever been seen doing that. For one naturally talks in a low voice when smoking a pipe. What happens when a husband who is a smoker gets angry, is that he immediately lights a cigarette, or a pipe, and looks glum. But that will not be for long. For his emotion has already found an outlet, and although he may want to continue to look angry in order to justify his indignation or sense of being insulted, still he cannot keep it up, for the gentle fumes of the pipe are altogether too agreeable and soothing, and as he puffs the smoke out, he also seems to let out, breath by breath, his stored-up anger. That is why when a wise wife sees her husband about to fly into a fit of rage, she should gently stick a pipe in his mouth and say, "There! forget about it!" This formula always works. A wife may fail, but a pipe never.

The artistic and literary value of smoking can best be appreciated only when we imagine what a smoker misses when he stops smoking for a short period. Every smoker has, in some foolish moment, attempted to abjure his allegiance to Lady Nicotine, and then after some wrestling with his imaginary conscience, come back to his senses again. I was foolish enough once to stop smoking for three weeks, but at the end of that period, my conscience irresistibly urged me onto the right road again. I swore I would never relapse, but would keep on being a devotee and a worshiper at her shrine until my second childhood, when I might conceivably fall prey to some Temperance Society wives. When that unhappy old age arrives, one is of course not responsible for one's actions. But so long as I have a modicum of will-power and moral sense left, I shall not attempt it again. As if I had not seen the folly of it all—the utter immorality of trying to deny oneself the spiritual force and sense of moral well-being provided by this useful invention. For according to Haldane, the great English bio-chemist, smoking counts as one of the four human inventions in the history of mankind that have left a deep biologic influence on human culture.

The story of those three weeks, when I played the coward to my own better self and willfully denied myself something that I knew to be of great soul-uplifting force, was indeed a disgraceful one. Now that I can look back upon it in a matter-of-fact and rational way, I can hardly understand at all how that fit of moral irresponsibility lasted so long. If I were to detail my spiritual Odyssey by day and by night during those three weeks in the Joycean manner, I am sure it would fill three thousand good Homeric lines in verse, or a hundred and fifty closely printed pages in prose. Of course, the object, to begin with, was ridiculous. Why, in the name of the human race and the universe, should one not smoke? I cannot answer now. But such unreasonable moods do come to a man sometimes, when one, I suppose, wishes to do

something against the grain just for the pleasure of overcoming resistance
and in this way use up his momentary excess of moral energy. Beyond this,
I cannot account for my sudden and unholy resolution to cut out smoking.
In other words, I was giving myself a moral test much in the same manner as
people indulge in Swedish gymnastics—movement for the sake of movement,
without actually accomplishing any work useful to society. It was apparently
this kind of moral luxury that I was giving myself, and that was all.

Of course, in the first three days, I felt a queer sinking sensation
somewhere along the alimentary canal, especially in the upper part of it.
To relieve that queer sensation, I took Double-Mint chewing gum, good
Fukien tea, and Montesserat Lime-Fruit Pastilles. I conquered and killed
that sensation in exactly three days. This was the physical, and therefore the
easiest and, to my mind, the most contemptible part of the battle. People who
think that herein lies the whole of the unholy struggle against smoking have
no idea of what they are talking about. They forget that smoking is a spiritual
act, and those who have no idea of the spiritual significance of smoking ought
never to meddle with the affair. After three days, I encountered the second
stage, when the real spiritual battle began. Scales fell from my eyes and I saw
that there were two races of smokers, one of which never deserved the name
at all. For these people, the second stage never existed. I began to understand
why we hear of the"easy conversions"of many smokers who seem to have
given up their smoking without any struggle at all. The fact that they could
stop such a habit as easily as they could throw away an old toothbrush shows
that they have never really learned to smoke at all. People credit them with
a"strong will-power,"whereas the fact is these people are never true smokers
and have never been so in their lives. For them, smoking is a physical act, like
the washing of their faces and brushing of their teeth every morning—a mere
physical, animal habit without any soul-satisfying qualities. I doubt whether
this race of matter-of-fact people would ever be capable of tuning up their
souls in ecstatic response to Shelley's"Skylark"or Chopin's"Nocturne."These

people miss nothing by giving up their smoke. They are probably happier reading Aesop's Fables with their Temperance wives.

For us true smokers, however, a problem existed, of which neither the Temperance wives nor their Aesop-reading husbands have even an inkling. For us, the injustice to oneself and the senselessness of it all soon became apparent. Good sense and reason soon began to revolt against it and ask: for what reason, social, political, moral, physiological or financial, should one consciously use one's will-power to prevent oneself from attaining that complete spiritual well-being, that condition of keen, imaginative perception, and full, vibrant creative energy—a condition necessary to our perfect enjoyment of a friend's conversation by the fireside, or to the creating of real warmth in the reading of an ancient book, or to that bringing forth of a perfect cadence of words and thought from the mind that we know as authorship? At such moments, one instinctively feels that reaching out for a cigarette is the only morally right thing to do, and that sticking a piece of chewing gum in the mouth instead would be criminally wicked. Of such moments, I can tell only a few here.

My friend B—had arrived from Peiping and called on me. We had not seen each other for three years. At Peiping, then called Peking, we used to chat and smoke the evenings out, discussing politics and philosophy and modern art. And now he had come, and we were engaged in the fascinating task of rambling reminiscences. We discussed the whole bunch of professors, poets and cranks we used to know in Peiping. At every pointed remark, I was mentally reaching out for a cigar, but instead inhibited myself and only rose up and sat down again. My friend, on the other hand, was rattling along amidst his cigar fumes in perfect contentment. I had told him that I had given up smoking, and I had enough self-respect not to break down right in his presence. But down in my heart, I knew I was not at my best, and was only unjustly making myself look coldly rational, when I wished to partake of the full communion of the two souls with a complete surrender of the emotions. The conversation went on, somewhat onesidedly, with half of myself there, and then my friend left. I had stuck it all out somewhat grimly. By that fiction of "willpower," I had "won," but I knew only that I was unhappy. A few days later, my friend wrote me on his voyage that he had found me not the old, vibrant, ecstatic self, and suggested that perhaps living in Shanghai had something to do with it. To this day, I have not forgiven myself for failing to smoke that night.

......

After this, my conscience began to gnaw upon my soul. For, I said to myself, what was thought without imagination, and how could imagination soar on the clipped wings of a drab, non-smoking soul? Then one afternoon, I visited a lady. I was mentally prepared for the re-conversion. Nobody else was in the room, and we were apparently going to have a real tête-a-tête. The young lady was smoking with one arm resting on her crossed knee, slightly inclined forward, and looking wistful and in her best style. I felt the moment had arrived. She offered the tin, and I took one firmly and slowly from it, knowing that by that act I had recovered from my temporary fit of moral degradation.

I came back, and at once sent my boy to buy a tin of Capstan Minum. On the right side of my desk, there was a regular mark, burnt in by my habitually placing burning cigarette ends there. I had calculated that it would take somewhere between seven and eight years to burn through the two-inch desk top, and had regretted to observe that, after my last disgraceful resolution, it was going to remain at about half a centimeter. It was with great delight, therefore, that I had the pleasure of placing my burning cigarette on that old mark again, where it is happily at work now, trying to resume its long journey ahead.

In contrast with wine, there is comparatively little praise of tobacco in Chinese literature, because smoking as a habit was introduced by Portuguese sailors as late as the sixteenth century. I have ransacked the entire Chinese literature after that period, but have found only a few scattered insignificant lines, quite unworthy of the fragrant weed. An ode in praise of tobacco evidently has to come from some undergraduate of Oxford. The Chinese people, however, always had a very high sense of smell, as is evident in their appreciation of tea and wine and food. In the absence of tobacco, they had developed the art of burning incense, which in Chinese literature was always classified in the same category and mentioned in the same breath, with tea and wine. From the earliest time, as far back as the Han Dynasty when the Chinese Empire extended its rule to Indo-China, incense brought as tribute from the South began to be used at court and in rich men's homes. In books on the art of living, sections have always been devoted to a discussion of the varieties and quality and preparation of incense. In the chapter on incense in the book K'aop'an Yüshih, written by T'u Lung, we have the following description of the enjoyment of incense:

The benefits of the use of incense are manifold. High-minded recluse scholars, engaged in their discussion of truth and religion, feel that it clears their mind and pleases their spirit when they burn a stick of incense. At the fourth watch of the night, when the solitary moon is hanging in the sky, and one feels cool and detached toward life, it emancipates his heart and enables him to whistle leisurely. When one is examining old rubbings of calligraphy before a bright window, or leisurely singing some poetry with a fly-whip in his hand, or when one is reading at night in the lamp light, it helps to drive away the Demon of Sleepiness. You may therefore call it "the ancient companion of the moon." When a lady in red pyjamas is standing by your side, and you are holding her hand around the incense burner and whispering secrets to each other, it warms your heart and intensifies your love. You may therefore call it "the ancient stimulant of passion." Or when one has waked up from his afternoon nap and is sitting before a closed window on a rainy day and practising calligraphy and tasting the mild flavor of tea, the burner is just getting warm and its subtle fragrance floats about and encircles our bodies. Even better still is it when one wakes up from a drinking party and a full moon is shining upon the clear night, and he moves his fingers across the strings or makes a whistle in an empty tower, with the green hills in the distance in full sight, and the half-visible smoke from the remaining embers floats about the door screen. It is also useful for warding off evil smells and the malicious atmosphere of a swamp, useful anywhere and everywhere one goes. The best in quality is chianan, but this is difficult to obtain, not accessible to a man living in the mountains. The next best is aloeswood or eaglewood, which is of three grades. The highest grade has too strong a smell, tending to be sharp and pungent; the lowest grade is too dry and also too full of smoke; the middle grade, costing about six or seven cents an ounce, is most soothing and fragrant and can be regarded as exquisite. After one has boiled a pot of tea, he can make use of the burning charcoals and put them in the incense container and let the fire heat it up slowly. In such a satisfying moment, one feels like being transported to the heavenly abode in the company of the immortals, entirely oblivious of human existence. Ah, indeed great is the pleasure! People nowadays lack the appreciation of true fragrance and go in for strange and exotic names, trying to outdo one another by having a mixture of different kinds, not realizing that the fragrance of aloeswood is entirely natural, and that the best of its kind has an indescribable subtlety and mildness.

Mao Pichiang in his Reminiscences of My Concubine, describing the art of life of this rich poet and his accomplished and understanding mistress, gives various descriptions of their enjoyment of incense, of which the following is one:

My concubine often sat quietly with me in her fragrant bedchamber to sample or judge famous incense. The so-called "palace incense" is seductive in quality, while the popular way of preparing aloeswood is vulgar. The ordinary people often set aloeswood right on the fire and its fragrant fume is soon put out by the burning resin. Thus not only has its fragrance failed to be brought out, but it also leaves a smoky, choking odor behind around one's body. The hard quality with horizontal grains called hengkoch'en, has a superb fragrance, being one of the four kinds of aloeswood, but distinguished by having horizontal fibers. There is another variety of this wood, known as p'englaihsiang, which is the size of a mushroom and conically shaped, being not yet fully grown. We kept all these varieties, and she burned them on top of fine sand over a slow fire so that no actual smoke was visible. Its subtle perfume permeated the chamber like the smell of chianan wood wafted by a breeze, or like that of dew-bedecked roses, or of a piece of amber rubbed hot by friction, or of fragrant liquor being poured into a horn cup. When bedding is perfumed by this method, its fragrance blends with that of the woman's flesh, sweet and intoxicating even in one's dreams.

六．酒令

我生平不善饮酒，所以实在不配谈酒。我的酒量不过绍兴三杯，有时只喝了一杯啤酒便会觉得头脑昏昏然。这显是限于天赋，无从勉强。所以善于饮茶吸烟者，未必同时善于饮酒。我有几个朋友酒量极好，但一吸雪茄不到半支，便会头晕。我则除去睡眠时间之外，几乎没有一个小时不吸烟，而从不觉得有什么不舒服，但酒不能多饮。李笠翁曾很坚决地记录他的意见说：善饮茶者必不好酒，掉过来也是如此。李笠翁是一个茶鉴赏家，但承认并不善饮酒。所以我最乐于在我所合意的中国著作家中，搜寻口说好饮酒而实在不善于酒的人。从他们的著作中，找寻这类自承的事实颇费一些时间，但终被我找到好几个，如：李笠翁、袁子才、王渔洋和袁中郎。他们都爱酒，但实不善饮。

我虽然没有饮酒资格，但不能就将这个题目置而不论，因为这样东西，比之别物更有所助于文学，也如吸烟在早已知道吸烟之术的地方一般，能有助于人类的创作力，得到极持久的效果。饮酒之乐，尤其是中国文学中常提到的所谓"小饮"之乐，起初我总视为神秘，不能了解，直到一位美丽的上海女士在她半醉之时，以灿花妙舌畅论酒的美德后，方感到所描写的乐境必是真实不虚。"一个人在半醉时，说话含糊，喋喋不休，这是至乐至适之时。"她说。在这时节，一种扬扬得意的感觉，一种排除一切障碍力量的自信心，一种加强的锐感和一种好像介于现实和幻想之间的创作思想力，好似都已被提升到比平时更高的行列。这时使人具着一种创作所必需的自信和解放动力。在下文论及艺术时，我们便能了解，这种自信的感觉和脱离规矩及技巧羁绊的感觉是怎样息息相关。

有人说，现代欧洲独裁者如此危害人情，即因他们都是不饮酒的人。这个想法很聪明。我在阅读过去数年的流行文字中，觉得一九三七年六月份《哈珀杂志》(Harper) 所载查尔斯·弗格森 (Charles W. Ferguson) 所著《独裁者不饮酒》(Dictators Don't Drink) 那篇文字最为切当诙谐，富有见识，其思想很可取，而且文章流利。我很想完全引用，但因不便，故只得略为引证几句。弗格森思想的起点是："斯大林、希特勒、墨索里尼都是严肃有节的模范。这些用现代方式行使暴虐行为的人，这些人民的新式统治者，都是希望出人头地的有志青年所足以奉为圭臬的典型。他们之中，不论哪一个都是良好的女婿和丈夫。他们足以代表福音传教士所认为模范道德的理想人物。……希特勒不食肉，不饮酒，不吸烟。他在这种闷人的美德之外，再加上更进一步、更可著称的克欲德行。墨索里尼在饮食方面较像一匹马。但他用了坚强不屈的勇气摒绝醇酒，不过偶尔喝一杯淡酒——只要不足以妨碍他征服一个民族的国家大计就是了。斯大林很俭朴地住在一所三间房的公寓屋子中，衣着朴素，食品粗糙，对于白兰地，只如鉴赏家般沾唇尝尝而已。"但是这种事实使我们从中能看出些什么呢？"这些事实是否指出人类现在是处于一小群本性整饬的，过分自谓正直的，很倔强地自认为德行完备的人们的掌握中，以致变为十分危险。因此，如能劝诱他来做一次哄然热闹的畅饮，世界的大部分便会立刻改观而有所进步。……有瑕疵的人绝不会成为一个危险的独裁者，他的无上尊严念头必会立刻破碎。他必以为已在他的子民之前铸了大错，因而受了挫辱。他将降为民众当中一个——最低微的当中一个——这种经验可以调和他那种难堪的自大心。"这位作家以为尚能预订一次国际"鸡尾酒"（Cocktail）会，专请这班特别领袖来畅饮一回，以平静他们的意气，则第二天早晨，"他们决然已经不是今日的超人、世上的特种人物，将一变而为寻常人物，能如最低微的人们一般感觉痛苦，具有如常人一般而不是半神道一般的处事心胸了。"

我所以反对独裁者，就因为他们不近人情。因为不近人情者总是不好的。不近人情的宗教不能算是宗教；不近人情的政治是愚笨的政治；不近人情的艺术是恶劣的艺术；而不近人情的生活也就是畜类式生活。这种是否近人情的试验，是普遍可以适用于各界人类和各种系统的思想。人类所能期望的最高理想，不应是一具德行的陈列箱，而应是只去做一个和蔼可亲、近情理的人。

中国人能以饮茶之术教西方人，而西方人能以饮酒之术教中国人。当一个中国人踏进一家美国酒店，看见贴有五光十色标签的酒瓶时，必觉得眼花缭乱。因为他在本国所看见的无非是绍兴酒而已。除了绍兴酒之外，虽尚有六七种酒，如药酒和麦米所酿的高粱酒等，但总不过这几种。中国人尚没有发展以不同的酒类配供不同的菜肴的技巧。但绍兴酒非常普遍，各处都有。绍兴本乡甚至在一个女孩儿出嫁时，嫁妆之中至少有一坛二十年陈的美酒。"花雕"之名即由此而得，因为这种坛子的外面都是画着花的。

中国人极讲究饮酒的时机和环境，这一点即弥补了酒类缺少花色的缺点。饮酒应有饮酒时的心胸，所以有人分别酒茶之不同说："茶如隐逸，酒如豪士。酒以结友，茶当静品。"作家列举饮酒时应具的心胸和最适当的地点说："法饮宜舒，放饮宜雅，病饮宜小，愁饮宜醉，春饮宜庭，夏饮宜郊，秋饮宜舟，冬饮宜室，夜饮宜月。"

又说："凡醉，各有所宜。醉花宜昼，袭其光也；醉雪宜夜，清其思也；醉得意宜唱，宜其和也；醉将离宜击钵，壮其神也；醉文人宜谨节奏章程，畏其侮也；醉俊人宜益觥盂，加旗帜，助其烈也；醉楼宜暑，资其清也；醉水宜秋，泛其爽也。此皆审其宜，考其景；反此，则失饮之人矣。"（出自《小窗幽记》）

在我的心目中，中国人对于酒的态度和酒席上的行为，一部分是难于了解应该斥责的，而一部分是可加赞美的。应该斥责的部分就是强行劝酒以取乐。这类事我在西方的社交中似乎没有看见过。在席的人，凡是稍能饮酒者，必以酒量自豪，而总以为别人不如他自己，于是即有强行劝酒，希望灌醉别人的举动。但劝酒时，总是出之以欢乐友谊的精神，其结果即引起许多大笑声和哄闹声，也使这次欢会增出不少的兴趣。宴席到了这种时候，情形极为有趣。客人好似都已忘形，有的高声唤添酒，有的走来走去和别人调换位，所有的人到了这时都已沉浸于狂欢之中，甚至也无所谓主客之别。这种宴席到了后来，必以豁拳行令斗酒为归宿。各人都必用尽心机以能胜对方为荣，还须时时防对方的取巧作弊。其中的快乐，大约即在这种竞争精神当中。

中国的食酒方式当中，可以赞美的部分就在声音的喧哗。在一家中国菜馆中吃饭，有时使人觉得好像是置身于一次足球比赛中。这些具有美妙韵节如同足球比赛时助威呐喊一般的嘈杂声音，究竟是因何而发的呢？其答语就是豁拳。豁拳的方法是，两人同时伸出几个手指，一面即各由口中高声喊猜两方手指加起来的总数，猜着者为胜。所喊的一二三四等数字，都有极雅致的代表名词：如"七巧""八马"或"八仙过海"之类。豁拳伸指时，双方必须在快慢上和谐合拍，因此嘴里的喊声也随之而生出高低快慢、抑扬顿挫的韵调，如音乐中的节拍一般。还有些人在上下句喊声的中间插入一种如音乐过门一般的句子。所以这种喊声可以连续有节拍地接下去，直到两人之中有一个胜了，由输者喝完事先所约定的杯酒时，方暂时停顿一下子。这种豁拳并不只是盲目胡猜，须极注意对方伸数的习惯，而立刻加以极敏捷的推测。其兴趣完全看豁拳者是否高兴，豁时音调是否迅速合拍。

　　我们到此，方能算是对中国的酒宴有了真正的认识。因为下述的酒席情形使我们明了何以中国的宴集为时如此之久？菜肴为什么如此之多？上菜为什么如此之慢？一个人坐到酒席上去，并不是专为了吃菜饮酒，也作乐，须一面做富有意趣的游戏如讲故事、说笑话、猜谜、行令等。这种筵席其实好似一种口令游戏的集会，每隔五六分钟上一道菜，以便客人放松脑筋，进一些酒菜。这办法有两种功效。第一，这种用嘴叫喊的游戏，无疑可以使喝下去的酒易于从身体内发泄出来；第二，这种席面每每延长一小时之久，其时吃下去的东西一部分已经消化，所以会越吃越饿。默不作声，实在是吃东西时一种恶习，这是不道德的，因为它是不合卫生的。有些在中国的西方人，如若依旧疑惑中国人是一种略带拉丁色彩的快乐民族，仍认中国人是静默沉着、缺乏情感的人类，只须去看一看中国人请客吃饭时的情形，便会知道自己的认识错误。因为中国人只有在这个时候，方露出他的天生性格和完备的道德。中国人如若不在饮食之时找些乐趣，尚有什么时候可以找寻乐趣呢？

　　中国人的文虎很著名，不过各种酒令知者尚少。他们以酒为罚，从中发明了不少劝客饮酒的游戏。大多数中国小说都忠实地记录每次酒席上所供的菜肴，也描写各种联句和诗酒令，每每占去书中许多篇幅。中国女子所欢迎的小说《镜花缘》中，曾描写许多种通文的女子间所行的酒令（内中包括声韵学的酒令），好似这就是故事的主题。

最简单的酒令是射覆，其方法是选取两个字，截头弃尾，然后将剩余的部分联成一字，请对方去猜截去的部分。例如 Humdrum 和 Drumstick 两字。第一字的尾部和第二字的头部都是 Drum，现把这 Drum 字节截去，而将其余部分联起来，成为 Hum-stick 字，请对方去猜这截去的字节。又如 Acorn 和 Cornstarch 两字，将 Corn 截去，联成 A-starck，请对方去猜这截去的字节。照正式的猜法，猜者不许直接说出所猜的字节，而应另外加上一个头尾，成为两个不同的字，然后说出来。例如射第二个覆时，射者应在 Corn 这字节的前后都加上另外一个字节如 Popcorn 和 Corner，而举出 Pop-er 这样的答案，行此令者固然一听即能了然其是否射着，但边坐的人仍茫然不解。有时答案虽不和出令者心中的字眼相符，但如其较为切贴，则出令者也须认为射着。行此令时，两方可以同时出令请对方去射。这种射有时极简单，有时极深奥。如 A-ound 所截去的字节之为 Prou。至于如 Cam-ephant，则一望而知截去的字节是 el。如学者们即不妨用极深奥的字眼，如历史的名词，或取自莎士比亚的剧本，或巴尔扎克的小说中的人名。

以文字为游戏的酒令，种类多至不能胜计。最流行的一种即联句，由第一个人吟一句诗，即令第二个人联上一句。这种联句极为有趣，联下去时，后来的诗句竟会离题万里，不知所云。联句大都以人物或景色为题，各人挨着次序联一句或两句。要点是前后的诗句必须押韵，如若座中都是熟读"四书五经"的人，令官往往有用"女儿羞""女儿乐""女儿悲"等为题，而请众人集句联吟。唐诗和曲牌名是酒令中常用的材料，有时并也限用切于曲牌名的药名和花名。为了使英美人易于了解起见，这里当举几个英文名字为例，如切于妇女用品的花名有 Queen-Anne's lace, fox-glove 等。这类文字上的假借比拟是否可能，须视这种文字中所用以形容花树药草的字眼是否也通用于人类美丽的形容而定，例如英国人的姓可假借以隐射歌曲之题名（如 Rockefeller 可指 Sit down, You're rocking the boat 或 Whitehead 可指 Silver threads among the gold）。其比拟是否切贴，视人之才智而异。至于这种游戏的乐趣，则是在于其自然和想象丰富中，而且不必一定需饱学之士才会行的。英文名字中如 Lugwell，Sitwell 和 Frankfurlir 等，极易于用以隐射滑稽的意义（如 Frankfurlir 一字我以为可以隐射 Not cold, not pig），学校中的学生可以利用教师的姓名为资料，而行出各种极有趣的酒令。

比较复杂的酒令，行时须用令筹。中国小说《兰花梦》中曾记载着下述这个酒令：其令筹分为三组，以六种人在六种地点做六种事，错综配合为游戏。

　　六种人是：纨绔子　老僧　佳人　屠夫　妓女　叫花

　　六种地点是：官道　方丈　闺阁　大街　红楼　坟茔

　　六种所做的事是：骑马　念经　刺绣　打架　调情　睡觉

每人随意从这三组中各抽一筹，而将人地事配合起来，往往成为极滑稽可笑的事情。例如：老僧在闺阁中调情，妓女在坟茔中念经，叫花在红楼中睡觉，屠夫在官道上刺绣，佳人在方丈中打架，等，都可以拿来当报纸的绝妙标题。待各人将筹抽定，即以所配合的人事为题，令各人说一句五言诗，一个曲牌名，再加一句诗经以咏之，总以意思贴切者为上。

所以一次宴集，时间延长到两小时以上，很不足为奇。宴集的目的，不是专在吃喝，而是在欢笑作乐。因此在席者以半醉为最上，其情趣正如陶渊明这弹无弦的琴，因为好饮之人所重者不过是情趣而已。因此，一个人虽不善饮，也可享酒之趣。"世有目不识丁之人而知诗趣者，世有不能背诵经文之人而知宗教之趣者，世有滴酒不饮之人而识酒趣者，世有不识石之人而知画趣者。"像这些，都是诗人、圣贤、饮者和画家的知己。

VI. On Drink and Wine Games

I am no drinker and am therefore totally unqualified to talk of wines and liquors. My capacity is three cups of shaohsing rice wine, and I am even capable of getting tipsy on a mere glass of beer. This is evidently a matter of natural gift, and the gifts of drinking tea and wine and smoking do not seem to go together. I have found among my friends great drinkers who get sick before they go through half a cigar, while I smoke every waking hour of the day without any appreciable effect, but am not very good with liquors. Anyway, Li Liweng has put down on record his sworn opinion that great drinkers of tea are not fond of wine, and vice versa. Li himself was a great tea connoisseur, but confessed that he had no pretentions to being a drinker of wine at all. It is therefore my special delight and comfort to discover so many distinguished Chinese authors that I like who had really but a small capacity for wine, and who said so. It has taken me some time to collect these confessions from their letters or other writings. Li was one, Yüan Tsets'ai, Wang Yüyang, and Yüan Chunglang were others. All of them, however, were people who had"the sentiment for wine"without having an actual capacity for it.

In spite of my disqualification, I still cannot ignore this topic, because more than anything else, it has made an important contribution to literature, and in the same measure as smoking, wherever the custom of smoking was known, it has greatly helped man's creative power, with considerable lasting results. The pleasure of wine drinking, especially in what the Chinese call"a little drink,"so constantly met with in Chinese literature, had always seemed a mystery to me, until a beautiful Shanghai lady, when half-drunk herself dilated upon its virtues with such convincing power that I finally thought the condition described must be real."One just babbles along and babbles along in the state of halfdrunkenness, which is the best and happiest state,"she said. There seems to be a sense of elation, of confidence in one's power to overcome all obstacles and a heightened sensibility, and man's power of creative thinking, which seems to lie in the borderland of fact and fancy, is brought up to a higher pitch than at normal times. There seems to be a force of self-confidence and emancipation, which is so necessary at the creative moment. The importance of this sense of confidence and emancipation from mere rules and technique will be made quite clear when we come to the section on art.

There is a wise thought in the suggestion that the modern dictators of Europe are so dangerous to humanity because they don't drink. In my reading of current literature of the past year, I have come across no better and wiser and wittier writing than an article by Charles W. Ferguson on"Dictators Don't Drink"in Harper's for June, 1937. The thought is worth pursuing, and the writing is so good throughout that I feel tempted to quote it in full but have to refrain from doing so... The question is what do these facts signify for us?"Do they indicate that we are today in the grip of a coterie of men essentially smug, disastrously self-righteous, grimly aware of their tremendous rectitude, and hence so dangerous that the world at large would be better off if it could entice them on a roaring drunk?"..."No man could be a dangerous dictator with a hang-over. His sense of God-almighty-ness would be wrecked. He would feel himself to have been gross and humiliated in the presence of his subjects. He would have become one of the masses—one of the lowest of them—and the experience would have done something to his insufferable conceit."The writer thinks that should there be an international cocktail party, attended only by these chosen leaders, in which"the main object would merely be to fry the dignitaries as smoothly and as quickly as possible,"the next morning,"Far from being the irreproachable supermen of today, the world's best would have become ordinary fellows, afflicted like their meanest followers, and perhaps in a frame of mind to grapple with matters as men and not as demigods. "

The reason I don't like dictators is that they are inhuman, and anything which is inhuman is bad. An inhuman religion is no religion, inhuman politics is foolish politics, inhuman art is just bad art, and the inhuman way of life is the beast's way of life. This test of humanness is universal and can be applied to all walks of life and all systems of thought. The greatest ideal that man can aspire to is not to be a show-case of virtue, but just to be a genial, likable and reasonable human being.

While the Chinese can teach the Westerners about tea, Westerners can teach the Chinese about wine. A Chinese is easily dazzled by the variety of bottles and labels when he enters an American wine-shop for wherever he goes, in his own country, he sees shaohsing, again skaohsing, and nothing but shaohsing. There are six or seven other varieties, and there are distilled liquors from millet, the kaoliang, besides the class of medicinal wines, but the list is soon exhausted. The Chinese have not developed the nicety of serving different drinks with different courses of food. On the other hand, the popularity of shaohsing is such at the place giving its name to this wine, that there as soon as a girl is born, her parents make a jar of wine, so that by the time she marries, she is sure to have at least a jar of wine about twenty years old as part of her trousseau. Hence the name huatiao, the proper name for this wine, which means"florally decorated, "from the jar decoration.

This lack in the variety of wine they make up for by greater insistence on the proper moment and surrounding for drinking. The feeling for wine is essentially correct. The contrast between wine and tea is expressed in the form that"tea resembles the recluse, and wine resembles the cavalier; wine is for good comradeship, and tea is for the man of quiet virtue."Specifying the proper moods and places for drink, a Chinese writer says,"Formal drinking should be slow and leisurely, unrestrained drinking should be elegant and romantic; a sick person should drink a small quantity, and a sad person should drink to get drunk. Drinking in the spring should take place in a courtyard, in summer in the outskirts of a city, in autumn on a boat and in winter in the house, and at night it should be enjoyed in the presence of the moon. "

Another writer says,"There is a proper time and place for getting drunk. One should get drunk before flowers in the daytime, in order to assimilate their light and color; and one should get drunk in snow in the night-time, in order to clear his thoughts. A man getting drunk when happy at success should sing, in order to harmonize his spirit; and a man getting drunk at a farewell party should strike a musical tone, in order to strengthen his spirit. A drunk scholar should be careful in his conduct, in order to avoid humiliations; and a drunk military man should order gallons and put up more flags, in order to increase his military splendor. Drinking in a tower should take place in summer, in order to profit from the cool atmosphere; and drinking on the water should take place in autumn, in order to increase the sense of elated freedom. These are proper ways of drinking in respect of mood and scenery, and to violate these rules is to miss the pleasure of drinking."

The Chinese attitude toward wine and behavior during a wine feast is partly incomprehensible or reprehensible to me, and partly commendable. The reprehensible part is the custom of getting pleasure out of forcing a man to drink beyond his capacity. I am not aware that such a practice exists or is common in Western society. It is usual among drinkers to place a mystic value upon the mere quantity of drinking, whether by oneself, or by those in the company. No doubt there is a certain hilarity connected with it and such urging is done in a playful or friendly spirit, resulting generally in a lot of noise and hubbub and confusion, which adds to the fun of the occasion. It is beautiful to look at when the company reach a state when they all forget themselves and the guests shout for more wine or leave or exchange their seats, and nobody remembers who is the host and who are the guests. It usually degenerates into a drinking match, played with great pride and subtlety and finesse and always with the desire to see the other fellow under the table. One has to be on the lookout for foul play, and guard against the other party's underhanded tactics. Probably the fun lies there, in the spirit of contest.

The commendable side of Chinese drinking lies in the noise. Eating at a Chinese restaurant sometimes makes one imagine one is attending a football match. How is the volume of noise produced and whence come those noises with beautiful rhythm resembling cheers and yells at a football match? The answer lies in the custom of"guessing fingers,"in which each party puts up a number of fingers simultaneously with his opponent and shouts the number of the total sum of fingers that he guesses will be put out by both parties. The numbers,"one, two, three, four,"and so on, are given in poetic, polysyllabic phrases like"seven stars"(Ch'ich'iao, the constellation"Dipper"), or"eight horses"or"eight immortals crossing the sea."The necessity for perfectly timed and simultaneous action in putting out the fingers forces the phrases into definite musical beats or bars, into which the varying syllables have to be compressed, and these are accompanied during the interval by a set introductory phrase occupying another musical bar, and the song is carried on without interruption rhythmically until one party makes a correct conjecture and the other party has to drink a full cup, large or small, or two or three, as previously agreed. Guessing at the total is not mere blind conjecture, but is based on observation of the opponent's habit of sequence or alternation of numbers and demands some quick thinking. The fun and swing of the game depends entirely upon the speed and uninterrupted rhythm of the players.

We have come to the real point about the conception of a wine party, for this alone gives a satisfactory explanation of the length of a Chinese feast, its number of courses and its method of service. One does not sit down at a feast to eat, but to have a good time, provided by telling of stories and jokes and all kinds of literary puzzles and poetic games between the serving of different dishes. The party looks more like a time for oral games, punctuated every five or seven or ten minutes with the appearance of a dish on the table and a bite or two by the company. This produces two effects: first, the vociferousness of oral games undoubtedly helps to let the spirituous liquors evaporate from the system, and secondly, by the time one comes to the end of a feast lasting over an hour, some of the food is already digested, so that the more one eats, the more hungry he becomes. Silence after all is a vice during eating; it is immoral because it is unhygienic. Any foreigner in China who has lasting doubts about the Chinese being a gay and happy people with a touch of Latin gaiety, who still clings to the preconceived notion that the Chinese people are silent, sedate and unemotional, should watch them while eating, for then the Chinaman is in his natural element and his moral perfections are complete. If the Chinaman does not have a good time when he is eating, when does he have a good time?

Famous as the Chinese are for their puzzles, their wine games are less well-known. With wine as forfeit, a great variety of games have been invented as excuses for drinking. All Chinese novels dutifully record the names of dishes served at a dinner, and equally dutifully describe the contests of poetry which have no difficulty in filling an entire chapter. The feminist novel Chinghuayüan describes so many games among the literary girls (including games in phonetics), as to seem to make these the main theme of the story.

The simplest game is shehfu, in which a syllable forming the beginning of one word and the end of another is concealed by joining the other syllables into a word and the player has to guess at the missing syllable. Thus"drum" being the syllable common to"humdrum"and"drum stick,"the puzzle is given in the combination "hum-stick"and the other party is to supply the missing syllable. Or given"a-starch,"the other person is to find the missing middle syllable"corn"in"acorn-corns tarch." Properly played the person who has guessed at the middle syllable is not to declare it, but to form a counter-puzzle with the syllable"corn"and simply answer"pop-er"("pop- corn-corner"), or" pop-muffin,"from which the original maker of the puzzle is able to tell whether he has got the correct answer, while it remains a mystery to the rest of those present. Sometimes an answer not originally intended, but even better than the one the maker has in mind, has to be accepted. Both parties can set syllable-puzzles for each other to solve at the same time. Some puzzles are simple and some are carefully concealed, as"a-ounce"for the missing syllable"pron,"while"cam-ephant"can be easily detected to contain"el"in"camel-el ephant."Rare and difficult words might be used, and in the practice of scholars, rare historical names might be used, taxing one's scholarship: e.g., names from one of Shakespeare's plays or Balzac's novels.

Variations of literary games are infinite. One popular among scholars is for each person in turn to say a doggerel line of seven words for the other person to follow up with another rhymed line, the poem as a whole degenerating into pure nonsense at the end. Lines usually begin with some comment on some object or person in view, or the scenery. Every person is to say two lines, the first one completing a couplet begun by the preceding person, and the second leading off a new couplet for the successor to finish. The first line sets the rhyme, and the third, fifth, seventh (and so on) lines must keep to it. In the milieu of scholars, by whom every name and sentence from the Four Books or the Book of Poetry has been memorized by heart, demands may be made by the toastmaster for apt quotations illustrating a topic, (e.g.,"Girl shy","Girl happy","Girl cries"). Names of popular ditties, and lines from T'ang poems are often included. Or the party may be required to give names of medicines or flowers that answer to the description in a given title of a popular tune, or, to make the matter appear simpler in English, to give names of medicines or flowers that refer to an article pertaining to women; e. g., Queen Anne's lace, fox-glove, etc. Possibility of such combinations depends on the beauty of names given to flowers, medicines, trees, etc. in a language. English family names might, for instance, be given to call up names of popular songs, (as an instance,"Rockefeller"may suggest" Sit Down, You're Rocking the Boat,"and"Whitehead"may suggest"Silver Threads Among the Gold"). The aptness of such juxtapositions depends on one's ingenuity, and the fun of such games lies in spontaneity and fanciful, but not necessarily learned, associations. Names like"Tugwell, ""Sitwell" and"Frankfurter" can easily be made to serve any humorous purpose (for the last I suggest"Non-cold Not-Pig"). College students can have a good time making wine games out of their professors' names.

More elaborate games require specially designed chips. In the novel An Orchid's Dream, one finds, for instance, a description of the following game. Three sets of chips (which may be made of paper) contain the following combination of six persons doing six things in six places:

Dandy	*goes horse-riding*	*thoroughfare*
Abbot	*says prayers*	*abbot's room*
Lady	*embroiders*	*Lady's chamber*
Butcher	*fights*	*streets*
Courtesan	*flirts*	*red-light district*
Beggar	*sleeps*	*cemetery*

Chips drawn from the three sets by a person may form the weirdest combinations: thus,"Abbot flirts in a lady's room,""Courtesan says prayers in a cemetery,""Beggar sleeps in the red-light district,""Butcher embroiders in a thoroughfare,""Lady fights in an Abbot's room,"etc., all of which would make good newspaper headlines. With some such situation as the main theme, each person is to give a five-word line from a poem, followed by the name of a song, and concluded by a line from the Book of Poetry, the whole to make an apt description of the situation.

It is no wonder therefore that a wine feast easily lasts two hours. The object of a dinner is not to eat and drink, but to join in merrymaking and to make a lot of noise. For that reason, he who drinks half drinks best. Like the poet T'ao Yüanming playing upon a stringless instrument, for the drinker the sentiment is the thing. And one may enjoy the sentiment for wine without the capacity to drink it. "There are people who cannot read a single word, but have the sentiment for poetry; people who cannot repeat a single prayer but have the sentiment for religion; people who cannot touch a drop but have the sentiment for wine; and people who do not understand a thing about rocks, but have the sentiment for painting." It is such people who are fit company for poets, saints, drinkers and painters.

七．食品和药物

　　我们如把对于食品的观点范围放大些，则食品之为物，应该包括一切可以滋养我们身体的物品，正如对于房屋的观点放大起来，即应包括一切关于居住的事物。因为我们都属于动物类，所以不能不吃食以维持生命。我们的生命并不在上帝的掌握中，而在厨子的掌握中。因此中国绅士都优待他们的厨子，因为厨子实在掌着予夺他们生活享受之大权。中国之为父母者——我猜想西方人也是如此——大都善视其儿女的奶妈，因为他们知道儿女的健康，完全依赖奶妈的性情、快乐和起居。为了同样的理由，我们自然也应善待职司喂养我们的厨子，如若我们也和留意儿女们一般留意我们的身体健康。如若一个人能在清晨未起身时，很清醒地屈指算一算一生之中究竟有几件东西使他得到真正的享受，则一定以食品为第一。所以倘要试验一个人是否聪明，只要去看他家中的食品是否精美，便能知道了。

　　现代城市生活之节奏是如此紧张，致使我们一天更比一天无暇去顾到烹调和滋养方面的事情。一个同时是著名记者的主妇，绝不能埋怨将罐头汤和罐头豆供给她的丈夫。不过一个人如若只为了工作而进食，而不是为了须进食而工作，实在可说是不合情理的生活。我们须对己身施行仁慈慷慨，方会对别人施行仁慈和慷慨。一个女人即使极致力于市政事业，极致力于改进一般的社会情形，但自己只能在一副两眼煤气灶上煮饭烧菜，每顿只有十分钟的吃饭时间，这于她又有什么益处？她如遇到孔子，定被休回娘家，一如孔子因太太失于烹调，而将她休掉一般。

孔子之妻究竟是被休，还是因受不了丈夫的种种苛求而自己逃回娘家，其中的事实不很明了。在孔子，"食不厌精，脍不厌细"，"不得其酱不食，割不正不食；色恶不食，臭恶不食"。我敢断定孔太太对于这些要求纵是能忍受，但是有一天她买不到新鲜的食物，不得已命她的儿子鲤到店铺里去买些酒和熟食以供餐，孔子即说："沽酒市脯不食。"到这时，她除了整一整行李，弃家逃走之外，还有什么办法？这个孔子之妻的心理设想，是我所创造出来的。但孔子对于这位可怜的太太所立下的许多严厉规条，确是明明白白地列在《论语》中，有籍可稽。

中国人对于食物，向来抱一种较为广泛的见解。所以对于食品和药物并不加以很明显区别。凡有益于身体者都是药物，也都是食物。现代科学直到上一世纪，方始知道食事在医疗上的重要。现时的医院中都已聘有经验丰富的食事专家是一件可喜的事情。但如若各医院的当局肯更进一步，将这班食事专家送到中国去受一下训练，或许就会减少玻璃瓶的使用。古代医学作家孙思邈（六世纪）说："谓其医者先晓病源，知其所犯，先以食疗，不瘥，然后用药。"元代太医院某大夫忽思慧，于一三三〇年著了一本中国的第一部完整的饮食卫生和食疗专书《饮膳正要》，认食物为基本的养生法。他在序文中说：

　　善摄生者，薄滋味，省思虑，节嗜欲，戒喜怒，惜元气，简言语，轻得失，破忧沮，除妄想，远好恶，收视听，勤内顾。不劳神，不劳形，神形既安，病患何由而致也？故善养性者，先饥而食，食勿令饱，先渴而饮，饮勿令过。食欲数而少，不欲顿而多。盖饱中饥，饥中饱，饱则伤肺，饥则伤气，若食饱，不得便卧，即生百病。

所以这本烹调书，也和其他的中国烹调书一般，实等于一本药方书。

你如向上海河南路走一遭，去那里看看卖中国药物的铺子，竟难于断言这种铺子里边究竟是药物多于食物，还是食物多于药物。你在那里可以看见桂皮和火腿，虎筋和海狗肾及海参，鹿茸和麻菇及蜜枣并排地阵列在一处，这都是有益于身体的，都是富于滋养的。此外如虎骨木瓜酒，显然也难于区别其究竟是食物还是药物。中国补药不像西药般用次磷酸盐三毫升，砒二厘所合成，是一件可喜的事情。生地炖童鸡即是一碗绝妙的补药。这完全是由于中国药物使用法，西药大都以丸或片为式，而中国药物大都为汤式。而且中国药的配制方法和寻常的汤相同，是用许多味不同的药物合煮而成。中国的汤药，其中药物往往多至七八十种，都是君臣相齐，以滋补和加强身体的整体为主，而不专在于治疗某一部分的病患。因为中国的医学在基本上和最新的西方医学见解相合，认为当一个人患肝病时，并不单是肝部而实是全体都有病患。总而言之，药之为物，其效用不过在于以增强生机力为原则，使其对于人身非常复杂的器官、液汁和内分泌系统自然发生作用，而让身体增加抵抗疾病的力量，自己去治疗其患处。

中国医生对于病人并不给予阿司匹林片，而是给他喝一大碗药茶以取汗，所以将来的病人，或许不必再吃金鸡纳片（奎宁），而只需喝一碗加些金鸡纳皮的冬菇甲鱼汤。现代医院的食事部分势须加以扩充，到了将

来，医院本身大概将变成一个类似疗养院的大菜馆。最后，我们必将达到
认为健康和疾病二者有交互作用的地步。到那时，人类即会因预防疾病而
进食，而不再是为医治疾病而吃药了。这一点目下尚未为西方人所充分注
意，因为西方人尚只知有病时去找医生，而不知道在未病时即去找医生。
待达到这个程度时，滋补药物和治病药物之间的区别将废除。

　　所以，我们对于中国人的药食不分应该庆贺，这个观念使他们的药物
减少药性，而使食物增加其可食性。饕餮之神在人类刚有历史时代即已出
现似乎有一种象征的意义，我们现在发现这神道，远在古代即已是铸像家
和雕刻家所爱塑造的目标。我们身体中都有饕餮的精神，这使我们的药方
书类似烹饪书，使我们的烹饪书类似药方书，并使植物学和动物学不可能
发展为自然科学的一支。因为中国的科学家看见一条蛇、一只猢狲、一
条鳄鱼或一个驼峰时，他始终只是想去尝尝它们的滋味。真正的科学好奇
心，在中国不过是一种烹饪艺术的好奇心而已。

　　因为在野蛮部落中药物和法术往往混为一谈，因为道家专心于养生之
道和寻求长生的方法，因此我们的食物无形中受着他们的支配，在上文已
提及的那部元朝太医院大夫所著的食谱中有许多章即专讲如何长生，如何
免病的。道家最尊信大自然，所以偏向于蔬类的花果和食物，把含露的鲜
莲视为高人的无上食品。这里边便有诗意和出世思想的交织。据他们的意
见，如若可能，单吸所含的露更好。这类食物包括松子、葛粉、藕粉之
类，都是道家所认为足以助人长生的仙品，因为它们都是性能清心醒脾的
东西，一个人在吃莲子时，心中不可怀有俗念如女色等事。似药物而常为
人所吃食，以为足以助人长生的食品有：天门冬、生地、枸杞子、白术、
黄精，尤其是人参和黄芪等物为贵品。

　　中国的药方书可供西方科学研究以广大的研讨领域。西方医学直到上

一世纪方始发现动物肝之为食物，具有补血的功用，但在中国极早就拿这种东西为老年人的补食。我颇疑心当一个西方屠夫宰一口猪时，大概将腰子肚子大小肠（肠中显然满含着胃汁）猪血骨髓和脑子一并弃去，而不知这些实是含有最丰富的滋养料部分。现在已渐渐有人发现骨是人体内血中的红血球制造处，这不免使我可惜为什么羊骨、猪骨、牛骨都被随手丢弃，而不拿来熬一碗美味的汤，岂不是虚耗有价值的食物吗？

西方的食物中，有几种是我所爱吃的。第一，我当提到蜜露瓜，因为以"蜜露"为名是很近于中国式的。在古代如能有人拿一串葡萄送给一个道家，则他大概以为已经得到了可致长生不老的仙药。因为道家所欲求的，都是奇花异果的特别滋味。以番茄汁为食品，应为二十世纪西方大发明之一。因为中国也像一世纪前的西方人，尚认番茄是不适于食用的东西。其次是芹菜生吃法，好似中国人爱吃爽脆物品如笋之类。芦笋在未青的时候很好吃，但中国人尚不知道。最后，我当承认我极爱吃英国式红烧牛肉和其他红烧物。不论哪一种食品，只要在新鲜之时，在它的本处烹煮出来，总是好吃的。美国家庭所供的美国式菜肴很合我的胃口，但是在纽约的大旅馆中，我从来没有尝到过美味佳肴。也不能全怪旅馆或菜馆，即在中国，除非预定或特别烹煮，也是难于得到美味的。

另一方面，欧美的烹调法实有极显著的缺点。他们于饼类点心和糖果上一日进步千里，但在菜肴上仍是过于单调，不知变化。一个人只要在旅馆公寓或轮船上连吃三个星期的饭食，吃来吃去无非是皇帝鸡、牛排、羊排、菲力（里脊）这几样菜，便会胃口完全倒尽。西方的烹调对于烧煮蔬菜类更为幼稚：第一，所用的蔬菜类太少；第二，只知放在水中白煮；第三，总是煮得过了度，以致颜色暗淡、烂糟糟的。菠菜从不好好地烹煮，以致儿童见了就讨厌。因为他们总是把菠菜煮成烂糟糟的，而不知用油盐在极热的锅中煎炒，在未烂之前起锅是最可口的吃法。莴苣用同样的烧法也极可口，在烧这类蔬菜时，第一应注意的是煎炒不可过久。鸡肝已被西方人认为美味，铁排羊腰也认做佳肴，但仍有不少食物未经他们试吃。这就是西方菜肴缺乏花色的原因。炸鸡肫和鸡肝用椒盐蘸吃，乃是中国人常吃的菜。烧鲤鱼头连着面颊和颔下的脂肉是佳肴之一。猪肚是我爱吃的，牛肚有一部分也很好吃。如以肚子下面，或将肚子加在别种汤中一滚即起锅离火，其爽脆不下于生芹菜。蜗牛（单用其嘴部厚的部分）是法国人很爱吃的美味，中国亦然，在滋味及耐嚼上和鲍鱼及扇贝颇为相似。

西方的汤类，花色稀少有两个原因：第一，不懂拿荤素之品混合在一起烹煮，其实只需五六种作料如虾米、冬菇、笋、冬瓜、猪肉等间花配合，便能煮出数十甚至百种的好汤来。冬瓜汤是西方菜肴中所没有的。其实，这种瓜如用各种方法煮起来，再加入一些虾米屑，乃是夏天里一样最可口的菜。第二，汤的种类缺少是由于不知尽量利用海产。扇贝在西方只知炸了吃，而不知干的扇贝实是做汤的最佳作料。鲍鱼也是如此。西菜中虽有蛤蜊浓汤这个名目，但我从来没有吃着其中的蛤蜊味道。又如虽有甲鱼汤，但汤中从来看不到甲鱼肉。真正的甲鱼汤应该煮得极浓，乃中国广东菜中的美味，有时则加入鸡鸭掌，在一起同煮。浙江绍兴人有一样佳肴名为"醉大转弯"（the big corners），其中的材料即鸡腿鸡翅，因为这两件东西都是肉中夹着筋和皮，所以十分耐嚼好吃。我所认为最美味可口的汤即蛤蜊鲫鱼汤。凡是用介蛤之类所做的汤，其要点是不可过于油腻。

下面我将从李笠翁所著的《闲情偶寄》中引用一段论蟹的文字，以为中国人对于食物见解的例证：

予于饮食之美，无一物不能言之，且无一物不穷其想象，竭其幽渺而言之。独于蟹螯一物，心能嗜之，口能甘之，无论终身一日，皆不能忘之。至其可嗜可甘与不可忘之故，则绝口不能形容之。此一事一物也者，在我则为饮食中之痴情，在彼此为天地间之怪物矣。

予嗜此一生。每岁于蟹之未出时，即储钱以待。因家人笑予以蟹为命，即自呼其钱为"买命钱"。自初出之日始，至告竣之日止，未尝虚负一夕，缺陷一时。同人知予癖蟹，招者饷者，皆于此日，予因呼九月十月为"蟹秋"。……向有一婢，勤于事蟹，即易其名为"蟹奴"，今亡之矣。

蟹乎！蟹乎！汝与吾之一生，殆相终始者乎？

李笠翁对于蟹如此称美，其理由即蟹完全具有食物必备的三种美：色、香、味。李氏的见解也就是现代大多数中国人的见解，不过中国人所称美的蟹，只限于淡水中所产之一种。

在我个人，食物哲学大概可以归纳为三事，即新鲜，可口，火候适宜。高手厨师如若没有新鲜的作料，也做不出好菜。他们大概都能使你知道烹调的良否一半在于办作料。十七世纪的大诗人和享乐家袁子才在著作中述及他的厨师：他是一个极高尚自重的人，如若作料不是新鲜，即使强迫他，也不肯动手烹煮的。这厨师的脾气很坏，但他因为主人知味，所以依旧能久于其职。四川现在有一位年纪很大的高手厨师，要请他来做一次菜很费事，须提前一星期预约，以便他有充分买办作料的时间，须完全听他自择菜肴，而不许点菜。

普通人都知道凡是新鲜食品都是好吃的。这种知识使力不足以雇高手厨师的人，也有着享用美味的机会。在享受的供给上，依赖大自然实胜于依赖文化。为了这个理由，凡家里有菜园或居住乡间的人，虽然没有高手厨师，也自必能够享受种种美食。为了同样的理由，食物必须在其产地吃过之后，方能评断其美恶。但对一个不懂买办新鲜食品的主妇，或单是吃冷藏食物即觉得满意的人，对他讲何以享受美味实是徒然的。

食物的口味在酥嫩爽脆上，完全是火候关系。中国的菜馆因为有特备的炉子，所以能做出普通家庭中所不能烹煮的菜肴。至于滋味上，食物可以分为两类：第一，是专以本味见长的食物，这类菜肴中，除了盐或酱油之外，不可加入别的作料；第二，是必须配以别样作料方有滋味的食物。例如：鳜鱼和鲥鱼都宜清炖，方显其本味；较肥的鱼如鲱鱼，则加酸菜豆腐烹煮更为好吃。美国的豆粟羹是各味调和的一个好例子。世间有许多食品好像都是为调味而出，必须和别种食品合烧，方显其至美之味。笋烧猪肉是一种极可口的配合，肉借笋之鲜，笋则以肉而肥。火腿似乎最宜于甜吃。我住在上海时的厨子有一样拿手好菜，即用火腿和蜜枣为酿的番薯饼。木耳、鸭蛋汤和南乳烧纽约龙虾都属佳肴。专为调味而设的食品甚多。如：麻菇、笋、榨菜等都属于此类。

此外则有一种中国所视为珍品而本身没有味道的食物，这类食物都须借别样作料的调和配合，方成好菜。中国最贵重的食品，本身都具有三种特质，即无色、无臭和无味，如鱼翅、燕窝、银耳。这三种食品都是含胶质的东西，都是无色、无臭、无味，其所以成佳肴，全在用好汤去配合。

VII. On Food and Medicine

A broader view of food should regard it essentially as including all things that go to nourish us, just as a broader view of house should include everything pertaining to living conditions. As we are all animals, it is but common sense to say that we are what we eat. Our lives are not in the lap of the gods, but in the lap of our cooks. Hence every Chinese gentleman tries to befriend his cook, because so much of the enjoyment of life lies within his power to give or to take away as he sees fit. Chinese, and I suppose Western, parents always try to befriend the wet-nurse and treat her royally, because they realize that the health of their baby depends on the temper and happiness and general living conditions of the wet-nurse. Pari passu, we should give our cooks who feed us the same royal treatment, if we care as much for our own health as we care for that of our babies. If a man will be sensible and one fine morning, when he is lying in bed, count at the tips of his fingers how many things in this life truly give him enjoyment, invariably he will find food is the first one. Therefore it is the invariable test of a wise man whether he has good food at home or not.

The tempo of modern city life is such that we are giving less and less time and thought to the matter of cooking and feeding. A housewife who is at the same time a brilliant journalist can hardly be blamed for serving her husband with canned soup and beans. Nevertheless, it is a pretty crazy life when one eats in order to work and does not work in order to eat. We need a certain kindness and generosity to ourselves before we learn kindness and generosity to others. What good does it do a woman to do some muckraking for the city and improve general social conditions, if she herself has to cook on a two-burner range and allow ten minutes for eating her meal? Confucius undoubtedly would divorce her, as he divorced his wife for failure in good cooking.

The story is not exactly clear as to whether Confucius divorced her or she just had to run away in order to flee from the demands of this fastidious artist of life. For him "rice could never be white enough and mince meat could never be chopped fine enough." He refused to eat "when meat was not served with its proper sauce," "when it was not cut square, ""when its color was not right" and "when its flavor was not right." I am quite sure that even then his wife could have stood it, but when one day, unable to find fresh food, she sent her son Li to buy wine and cold meat from some delicatessen and be through with it, and he announced that he "would not drink wine that was not homemade, nor taste meat that was bought from the shops," what else could she do except pack up and run away? This insight into the psychology of Confucius' wife is mine, but the severe conditions that he imposed upon his poor wife stands there today in the Confucian classics.

Taking then the broader view of food as nourishment, the Chinese do not draw any distinction between food and medicine. What is good for the body is medicine and at the same time food. Modern science has only in the last century come to realize the importance of diet in curing diseases, and happily today all modern hospitals are well equipped with trained dietitians. If modern doctors would carry it a step further, and send these dietitians to be trained in China, they might have less use for their glass bottles. An early medical writer, Sun Ssemiao (sixth century, A. D.), says: "A true doctor first finds out the cause of the disease, and having found that out, he tries to cure it first by food. When food fails, then he prescribes medicine."Thus we find the earliest existing Chinese book on food, written by an imperial physician at the Mongol Court in 1330, regards food essentially as a matter of regimen for health, and makes the introductory remarks:

He who would take good care of his health should be sparing in his tastes, banish his worries, temper his desires, restrain his emotions, take good care of his vital force, spare his words, regard lightly success and failure, ignore sorrows and difficulties, drive away foolish ambitions, avoid great likes and dislikes, calm his vision and his hearing, and be faithful in his internal regimen. How can one have sickness if he does not tire his spirits and worry his soul? Therefore he who would nourish his nature should eat only when he is hungry and not fill himself with food, and he should drink only when he is thirsty and not fill himself with too much drink. He should eat little and between long intervals, and not too much and too constantly. He should aim at being a little hungry when well-filled and being a little well-filled when hungry. Being well-filled hurts the lungs and being hungry hurts the flow of vital energy.

This cook book, like all Chinese cook books, therefore reads like a pharmacopoeia.

Walking down Honan Road in Shanghai and passing through the shops selling Chinese medicine, one might find it hard to decide whether they sell more medicine than food or more food than medicine. For there one finds cinnamon bark standing side by side with ham, tiger's tendons and beaver kidneys along with sea slugs, and horns of young deer along with mushrooms and Peiping dates. All of them are good for the body and all of them nourish us. The distinction between food and medicine is positively impossible in the case of a bottle of"tiger-tendon and quince wine."Happily a Chinese tonic does not consist of three grams of hypophosphate and .02 grains of arsenic. It consists of a bowl of black-skinned chicken soup, cooked with rehmannia lutea. This is due entirely to the practice of Chinese medicine, for while

Western medicines are taken in pills or tablets, Chinese medicines are served as stews and literally called"soups."And Chinese medicine is conceived and prepared in the same manner as ordinary soup, with proper regard for mixing of the flavors and ingredients. There are anywhere from seven or eight to twenty ingredients in a Chinese stew, so designed as to nourish and strengthen the body as a whole, and not to attack the disease solely. For Chinese medicine essentially agrees with the most up-to-date Western medicine in thinking that, when a man's liver is sick, it is not his liver alone but the entire body that is sick. After all, all that medicine can do comes down to the essential principle of strengthening our vital energy, through acting on this most highly complicated system of organs and fluids and hormones called the human body, and letting the body cure itself.

Instead of giving patients aspirin tablets, Chinese doctors would therefore ask them to take large bowls of medicinal tea to produce perspiration. And instead of taking quinine tablets, the patients of the future world might conceivably be required to drink a bowl of rich turtle soup with mushrooms, cooked with pieces of cinchona bark. The dietetics department of a modern hospital will have to be enlarged, and the future hospital itself will very nearly resemble a sanatorium-restaurant. Eventually we have to come to a conception of health and disease by which the two merge into each other, when men eat in order to prevent disease instead of taking medicine in order to cure it. This point is not stressed enough in the West, for the Westerners go to see a doctor only when they are sick, and do not see him when they are well. Before that time comes, the distinction between medicine which nourishes the body and medicine which cures disease will have to be abolished.

We have, therefore, to congratulate the Chinese people on their happy confusion of medicine and food. This makes their medicine less of a medicine, but makes their food more of a food. There seems to be a symbolic significance in the fact that the God of Gluttony appeared even in our semi-historical period, the God T'aot'ieh being found today as a favorite motif among our earliest bronze and stone sculptures. The spirit of T'aot'ieh is in us. It makes our pharmacopoeias resemble our cook books and our cook books resemble a pharmacopoeia, and it makes the rise of botany and zoology as branches of the natural science impossible, for the Chinese scientists are thinking all the time of how a snake, a monkey, or a crocodile's flesh or a camel's hump would taste. True scientific curiosity in China is a gastronomic curiosity.

With the confusion of medicine and magic, found in all savage tribes, and with the Chinese Taoists making the"nourishment of life"and the search after immortality or long life their central object, we find that food and medicine often lie in their hands. In the Imperial Cook Book of the Mongol Dynasty referred to above, Yinshan Chengyao, there are chapters devoted to preserving long life and warding off disease. With the Taoist passionate devotion to Nature, the tendency is always to emphasize fruits and food of a vegetarian nature. There is a sort of combination of poetry and Taoist detachment from life, regarding the eating of fresh lotus seeds with their delicate flavor born of the dew as the height of a scholar's refined pleasure. He would drink the dew itself, if he could. In this class belong the seeds of pine trees, arrowroot and Chinaroot which are all regarded as tending toward long life, because they clarify one's heart and purify one's soul. One is not supposed to have mortal desires, like the desire for women, when eating a lotus seed. More like medicine and constantly taken as part of one's food and highly valued for prolonging life are asparagus lucidus, rehmannia lutea, lycium chinense, atratylis ovata, polygonatum giganteum, and particularly ginseng and astragalus hoantely.

The Chinese pharmacopoeia offers an immense field waiting for Western scientific research. Western medicine has only within the past decade discovered the blood-building value of liver, while Chinese have all along regarded it as an important tonic for old people. I have a suspicion that when a Western butcher kills a pig, he throws away all the parts that have the greatest nourishing food value—kidneys, stomachs, intestines (which must be full of gastric juice), blood, bone marrow and brains. It is beginning to be discovered that the bone is the place where the red corpuscles of one's blood are manufactured, and I cannot help thinking the throwing away of mutton bones and pig's bones and cow's bones without stewing them into a fine soup is a terrific waste of food value.

There are many Western foods that I like, and first of all I must mention the honeydew melon, because its suggestion of dew is so Chinese. Also, if an ancient Chinese Taoist were given a grapefruit, he might imagine that he had discovered the elixir of immortality, for it was the exotic flavor of strange and unknown fruits that the Taoists were looking for. Tomato juice must be ranked as one of the greatest Western discoveries in the twentieth century, for the Chinese, like the Westerners of a century ago, used to consider tomatoes not fit for eating. Next comes the eating of raw celery, which comes nearest to the Chinese idea of eating food for texture, as with bamboo shoots. Asparagus is fine, when it is not green, but it is not unknown in China. Finally I must confess to a great liking for English roast beef, and for all roasts. Every food is good when cooked and tasted in its own country and in its proper season. I have always liked American food when it is served in American homes, but have never yet tasted food that impressed me as good in the best hotels of New York. The fault is not due to hotels or restaurants, for even in Chinese restaurants, it is impossible to get good food unless with long notice and unless it is prepared with individual care.

On the other hand, there are glaring deficiencies in American and European cuisine. Far ahead in bakery and the making of sweets and desserts, Western cuisine strikes one as being pretty dull and insipid and extremely limited in variety. After eating in any hotel or boarding house or steamship for three weeks, and after one has had chicken à la king, prime ribs of beef and lamb chops and filet for the thirteenth time, the food begins to pall on one's palate. The most undeveloped branch of Western cooking is that of preparing vegetables. In the first place, vegetables are extremely limited in variety; in the second place, they are merely boiled in water; and in the third place, they are always over-cooked until they lose their color and look mushy. Spinach, the calamity of all children, is never cooked properly; it is cooked until it becomes mushy, whereas if it is fried on a very hot pan with oil and salt and taken away before it loses its crispness, it is one of the most palatable of foods. Lettuce prepared in the same manner is also delightful, the only consideration being not to allow it to stand in the pan for too long. Chicken liver is considered a delicacy in the West, and even grilled lamb kidneys, but there is a large number of foods of the same class which have not even been experimented upon. This explains the lack of variety in Western food. Fried chicken gizzard, along with fried chicken liver, dipped in salt, is among the commonest dishes in China. Carp's head, with its delicate flesh around the cheeks and jowl, is served as a special dish of great delicacy. Pig's tripe is my favorite food, and for that matter, certain parts of the ox's tripe. It makes a delicious soup with noodles, or it may be thrown into boiling soup over an extremely hot fire and immediately taken out, so that it has a crispness almost like that of raw celery. Large snails (meaning only the thick covering at their mouth) are a delicacy much sought after in France, and they are also a delicacy in China. In taste and texture and resistance to the teeth, they are practically the same as abalones and scallops.

The lack of variety of soups is due to two causes. First, the lack of experiment on mixtures of vegetables with meat. By combinations and permutations, five or six ingredients, like dried shrimp, mushroom, bamboo-shoot, melon, pork, etc. , can give a hundred varieties of different soups. Melon soup is unknown in the West, and yet, made with different varieties and prepared with a dash of dried shrimp, it is one of the most delicious dishes in summer. Secondly, the lack of variety in soup is due to failure to make full use of sea food. Scallops are always fried in the West, but dried scallops are one of the most important elements for making good soup, and so is abalone. As for clam chowder, I never smell the clam in it, and one, of course, never sees turtle flesh in turtle soup. A real turtle soup, cooked until it is sticky on the lips, is one of the favorite Cantonese dishes, sometimes being prepared with the webbed feet of ducks or geese. The Shaohsing people of Chekiang have a favorite dish called"the big corners,"consisting of the wings and legs of chicken, because there is a happy combination of skin and tendon and meat in chicken wings and feet. The best soup I have tasted, however, is the soup of bastard carp and small softshell clams combined. The test of soup generally made from shell food is that it should not be oily.

As an instance of Chinese feeling about food, I may quote here from Li Liweng's essay on"Crabs"in the section on food in his Art of Living.

There is nothing in food and drink whose flavor I cannot describe with the utmost understanding and imagination. But as for crabs, my heart likes them, my mouth relishes them, and I can never forget them for a year and a day, but find it impossible to describe in words why I like them, relish them, and can never forget them. Ah, this thing has indeed become for me a weakness in food, and is in itself a strange phenomenon of the universe. All my days I have been extremely fond of it. Every year before the crab season comes, I set aside some money for the purpose and because my family say that"crab is my life"I call this money"my life ransom."From the day it appears on the market to the end of the season, I have never missed it for a night. My friends who know this weakness of mine always invite me to dinner at this season, and I therefore call October and November"crab autumn. "... I used to have a maid quite devoted to attending to the care and preparation of crabs and I called her"my crab maid."Now she is gone! O crab! my life shall begin and end with thee!

The reason that Li finally gave for his appreciation of crab was that it was perfect in the three requisites of food—color, fragrance and flavor. Li's feeling about crabs is quite generally shared by Chinese of all classes today, the kind eaten being from freshwater lakes.

For me, the philosophy of food seems to boil down to three things: freshness, flavor and texture. The best cook in the world cannot make a savory dish unless he has fresh things to cook with, and any good cook can tell you that half the art of cooking lies in buying. Yüan Tsets'ai, the great epicure and poet of the seventeenth century, wrote beautifully about his cook as a man carrying himself with great dignity who absolutely refused to cook a dish ordered unless the thing was in its best season. The cook had a bad temper, but confessed that he continued to serve the poet because the latter understood flavor. Today there is a cook over sixty years old in Szechuen who must be courteously invited to prepare a dinner for some special occasion, and who must be given a week's notice to collect and buy things and must be left entirely free to be the sole lord and judge of the menu to be served.

For the common people who cannot afford expensive cooks, there is comfort in the knowledge that anything tastes good in its season, and that it is always better to depend upon nature than upon culture to furnish us with the greatest epicurean delights. For this reason, people who have their own garden or who live in the country may be quite sure that they have the best food, although they may not have the best cook. For the same reason, food ought to be tasted in its place of origin, before any judgment can be pronounced upon it. But for a wife who does not know how to buy fresh food or a man who is willing to put up with cold storage foods, any discussion of epicurean values is futile.

The texture of food, as regards tenderness, elasticity, crispness and softness, is largely a matter of timing and adjusting the heat of the fire. Chinese restaurants can produce dishes not possible in the home because they are equipped with a fine oven. As for flavor, there are clearly two classes of food, those that are best served in their own juice, without adulteration except salt or soya-bean sauce, and those that taste best when they are combined with the flavor of another food. Thus, in the case of fish, fresh mandarin fish or trout should be prepared in its natural juice to get its full flavor, while more fatty fish like the shad tastes best with Chinese pickled beans. The American succotash is an example of the perfect combination of tastes. There are certain flavors in nature which seem to be made for each other and reach their highest degree of delectability only in combination with each other. Bamboo shoot and pork seem to make a perfect pair, each borrowing its fragrance from the other and lending it in return its own. Ham somehow combines well with the sweet flavor, and one of the proudest dishes of my cook in Shanghai is ham with rich Peking golden dates, steamed together in a casserole. So does black tree fungus combine perfectly with duck's egg in soup, and New York lobster combine with Chinese pickled bean-curd sauce (nanju). In fact, there is a large class of eatables whose chief function seems to be to lend their flavor to others—mushroom, bamboo shoots, Szechuen tsats'ai, etc. And there is a large class of food, most valued by the Chinese, which have no flavor of their own, and depend entirely on borrowing from others.

The three necessary characteristics of the most expensive Chinese delicacies are that they must be colorless, odorless, and flavorless. These articles are shark-fin, bird's nest and the"silver fungus."All of them are gelatinous in quality and all have no color, taste or smell. The reason why they taste so wonderful is because they are always prepared in the most expensive soup possible.

八．几件奇特的西俗

东方文明和西方文明的一个重大不同就是，在行相见礼时，西方人以手互握，而东方人握自己的手（即拱手）。我以为一切可笑的西俗当中以握手为最。我虽然是一个极前进的人，也能领略西方的艺术文学、美国丝袜、巴黎香水，甚至英国战舰的好处，但终不能懂极前进的欧洲人何以竟会听任这个握手的野蛮习俗存留到今日。我相信欧洲中必有人私下很反对这个习俗，如有许多人反对同样可笑的戴帽和戴硬领习俗一样，但他们并无成就，因为旁人总认他们为小题大做，对这种小节不值得费心思。我是一个极注意小节的人，因为我是中国人，所以比欧洲人更憎恶这个西方习俗。我和人相见或辞别时，宁可照中国古礼对人拱手作揖。

我们当然都知道这个习俗也和另一个脱帽习俗一般，还是欧洲野蛮时代的产物。这类习俗都是起于武士道盛行的中古时代，那时的绿林豪客、英雄武士遇到非敌人时都须除去面具或头盔以示友态或善意。现在的人已不戴什么面具或头盔，若还用这个姿势，岂不可笑？但野蛮时代所遗留的习俗，每每为人所不肯委弃，例如决斗之风至今犹存。

　　我出于卫生的和许多别的理由，反对这个习俗。握手是人类彼此之间
的一个接触方式，握时的姿势和表情各自不同，各类不一。别出心裁的美
国大学生写毕业论文时，大可以"各种握手式的时间和动作之研究"为
题，以严肃的体裁讨论其握时的轻重，时间的久暂，是否带着幽默性，对
方有无感应等，进而研究不同性别者握手时的种种变态，身体长度的关
系。因为高矮之不同，所以握时的姿势亦就各自不同。因职业和阶级不
同，皮肤的颜色也如何不同等。此处并可附上几幅图像和表格。这篇论文
如若做得充分深奥冗长，则我敢保他博士头衔唾手可得。

　　现在可以谈谈卫生上的反对理由。居住上海的西方人说我们的铜元是
微生物的寻常集合所，所以碰都不敢碰，但是在街上随便和张三李四握手
时，并不觉得什么。这实属不合逻辑。因为你怎能知道这张三李四的手没
有摸过你所畏如蛇蝎的铜元呢？更坏的是，有时你或许遇到一个咳嗽时常
用手帕掩着口部以示卫生，但露出已患肺病气色的人竟也伸出手来和你相
握。在这一点上，中国的习俗实较为科学化，因为中国人不过是握了自己
的手拱拱而已。我不知道中国这个习俗从何而起，但从医学卫生的观点说
起来，我们不能否认它的长处。

　　此外对于握手还有感受上和心理上的反对理由。当你将一只手伸出去
时，就等于听人宰割。对方可以由着自己的意握得或轻或重，或久或暂。
手是人体上感觉最敏锐的器官之一，极易感觉压力，辨其轻重。例如：第
一，你所遇到的或许是青年会式 Y.M.C.A（基督教青年会）握手。对方一
手拍你的肩膀，另一手则握着你的手，重重地抖动一下，直抖得你浑身骨
头都几乎脱节。如若这青年会书记也是一位棒球名手（往往如此），竟
可以使被握者啼笑皆非。这种款式的握手，再加上他的坦直、好自我主张
态度，简直是等于向你说："听着，你现在已在我的掌握中，你必须买一

张下次开会时的入场券，或答应买一份舍伍德·埃迪（Sherwood Eddy，青年会作者）的小册带回去，我方能放你。"遇到这种情形时，我无非是赶紧掏出皮夹来。

我如挨次列述下去，可以举出许多种轻重不同的握手，从那种漠然无诚意因而毫无意思的握手，到那种伸伸缩缩、微微颤抖、表示畏惧的握手。最后还有那种态度高雅的社会交际花，和人握手时，不过微微伸出指尖，好似不过让你看看她那染色指甲的握手。所以从这种人身的接触，很可以看得出双方关系的深浅亲疏。有几位小说家以为从握手的款式，可以看出其人的性格，如：独断的、退缩的、不诚实的、懦弱的和令人畏惧的重手，都是能一见而辨的。但我极愿省去这种遇到人时即须分析其个性的麻烦，或从他用力的轻重当中，去揣度对我的感情增减的麻烦。

脱帽的习俗更为没有意思，这里面包括许多种极不通情理的礼节。例如女人在礼拜堂，或下午室内茶会时都须戴帽子。这个礼拜堂内须戴帽子的习俗是否和第一世纪小亚细亚的习俗有关系？我不知道，但我颇疑心它其实是起源于盲从圣保罗在礼拜堂中男子不应戴帽而女子须遮没其头部的教训。倘若是如此，则这个习俗简直是基于西方人所弃绝已久的男女不平等的亚洲哲学，这岂不是可笑的矛盾吗？电梯中有女人时，男子须脱去帽子，也是一件可笑的事情，简直没有理由可言。这一点，电梯不过是走廊的延长部分，男子既然无须在走廊中脱帽，则何以须在电梯中脱帽？凡是偶然戴着帽子在走廊中行走的人，如若仔细想一想，当即能知道这极没有意义。第二点，电梯和别种行具如汽车之类实在没有什么合于逻辑的分别，一个有良心感觉的人既无须在和女人同坐一辆汽车之中时脱去帽子，又何须禁止他于同样情形时在电梯中戴帽子呢？

　　总而言之，我们的世界尚是一个缺乏理性的世界，没有一处地方不看到人类的愚钝，从现代国际关系的愚钝直到现代教育制度的愚钝。人类的聪明虽足以发明无线电，但不足以制止战争，将来也是如此。所以我对于许多小节的愚钝，宁可听其自然，而不过旁观暗笑罢了。

VIII. Some Curious Western Customs

ONE great difference between Oriental and Occidental civilizations is that the Westerners shake each other's hands, while we shake our own. Of all the ridiculous Western customs, I think that of shaking hands is one of the worst. I may be very progressive and able to appreciate Western art, literature, American silk stockings, Parisian perfumes and even British battleships, but I cannot see how the progressive Europeans could allow this barbarous custom of shaking hands to persist to the present day. I know there are private groups of individuals in the West who protest against this custom, as there are people who protest against the equally ridiculous custom of wearing hats or collars. But these people don't seem to be making any headway, being apparently taken for men who make mountains of molehills and waste their energy on trivialities. I am one of these men who are always interested in trivialities. As a Chinese, I am bound to feel more strongly against this Western custom than the Europeans, and prefer always to shake my own hands when meeting or parting from people, according to the time-honored etiquette of the Celestial Empire.

Of course, everyone knows this custom is the survival of the barbaric days of Europe, like the other custom of taking off one's hat. These customs originated with the medieval robber barons and chevaliers, who had to lift their visors or take off their steel gauntlets to show that they were friendly or peacefully disposed toward the other fellow. Of course it is ridiculous in modern days to repeat the same gestures when we are no longer wearing helmets or gauntlets, but survivals of barbaric customs will always persist, as witness, for instance, the persistence of duels down to the present day.

I object to this custom for hygienic and many other reasons. Shaking hands is a form of human contact subject to the finest variations and distinctions. An original American graduate student could very well write a doctorate dissertation on a "Time-and-motion Study of the Varieties of Hand-Shaking," reviewing it, in the approved fashion, as regards pressure, duration of time, humidity, emotional response, and so forth, and further studying it under all its possible variations as regards sex, the height of the person concerned (giving us undoubtedly many "types of marginal differences"), the condition of the skin as affected by professional work and social classes, etc. With a few charts and tables of percentages, I am sure a candidate would have no difficulty in getting a Ph.D., provided he made the whole thing sufficiently abstruse and tiresome.

Now consider the hygienic objections. The foreigners in Shanghai, who describe our copper coins as regular reservoirs of bacteria and will not touch them, apparently think nothing of shaking hands with any Tom, Dick or Harry in the street. This is really highly illogical, for how are you to know that Tom, Dick or Harry has not touched those coppers which you shun like poison? What is worse is, sometimes you may see a consumptive-looking man who hygienically covers his mouth with his hands while coughing and in the next moment stretches his hand to give you a friendly shake. In this respect, our celestial customs are really more scientific, for in China, each of us shakes his own hand. I don't know what was the origin of this Chinese custom, but its advantage from a medical or hygienic point of view cannot be denied.

Then there are aesthetic and romantic objections to handshaking.

When you put out your hand, you are at the mercy of the other person, who is at liberty to shake it as hard as he likes and hold it as long as he likes. As the hand is one of the finest and most responsive organs in our body, every possible variety of pressure is possible. First you may have the Y. M. C. A. type of handshaking; the man pats you on the shoulder with one hand and gives you a violent shake with the other until all your joints are ready to burst within you. In the case of a Y. M. C. A. secretary who is at the same time a baseball player with a powerful grip, and the two often go together, his victim often does not know whether to scream or to laugh. Coupled with his straightforward self-assertive manner, this type of handshaking practically seems to say,"Look here, you are now in my power. You must buy a ticket for the next meeting or promise to take back with you a pamphlet by Sherwood Eddy before I'll let your hand go."Under such circumstances I am always very prompt with my pocketbook.

Coming down the scale, we find different varieties of pressure, from the indifferent handshake which has utterly lost all meaning, to that kind of furtive, tremulous, retiring handshake which indicates that the owner is afraid of you, and finally to the elegant society lady who condescends to offer you the very tip of her fingers in a manner that almost suggests that you look at her red-painted fingernails. All kinds of human relationships, therefore, are reflected in this form of physical contact between two persons. Some novelists profess that you can tell a man's character from his type of handshake, distinguishing between the assertive, the retiring, the dishonest and the weak and clammy hands which instinctively repel one. I wish to be spared the trouble of analyzing a person's moral character every time I have to meet him, or reading from the degree of his pressure the increase or decrease of his affection towards me.

More senseless still is the custom of taking off one's hat. Here we find all kinds of nonsensical rules of etiquette. Thus a lady should keep her hat on during church service or during afternoon tea indoors. Whether this custom of wearing hats in church has anything to do with the customs of Asia Minor in the first century A. D. or not, I do not profess to know, but I suspect it comes from a senseless following of St. Paul's injunction that women should have their heads covered in church while men should not, being based thus on an Asiatic philosophy of sexual inequality which the Westerners have so long repudiated. For the men, there is that ridiculous custom of taking off one's hat in an elevator when there are ladies in it. There can be absolutely no defense for this meaningless custom. In the first place, the elevator is but a continuation of the corridor, and if men are not required to take off their hats in a corridor, why should they be made to do so in a lift? Any one would see the utter senselessness of it all, if he happens to pass from one floor to another in the same building with a hat on. In the second place, the elevator cannot by any logical analysis be distinguished from other types of conveyance, the motor car, for instance. If a man can, with a free conscience, keep his hat on while driving in a motor car in the company of ladies, why should he be forbidden from doing the same in a lift?

All in all, this is a very crazy world of ours. But I am not surprised. After all, we see human stupidity around us everywhere, from the stupidity of modern international relations to that of the modern educational system. Mankind may be intelligent enough to invent the radio and wireless telephones, but mankind is simply not intelligent enough to stop wars, nor will ever be. So I am willing to let stupidity in the more trivial things go by, and content merely to be amused.

九．西装的不合人性

虽然西装已经风行于土耳其、埃及、印度、日本和中国，虽然西装已经成为全世界外交界的普遍服装，但我仍依恋着中国衣服。常有许多好友问我为什么不改穿西装？他们问到这句话，能算是我的知己吗？这等于问我为什么用两足直立。凑巧这两件事正是有相互关系的。下文可以说明我所穿的是世上最合人性的衣服，更何必举出什么理由来呢？凡喜欢在家中穿着土著式长袍，或穿着浴衣拖鞋在外面走来走去的人，何需举出不裹扎于令人窒息的硬领、马甲、腰带、臂箍、吊袜带中的理由。西装的尊严，其基础也未必稳固于大战舰和柴油引擎的尊严，并不能在审美的、道德的、卫生的或经济的立场上给予辩护。它所占的高位，完全不过是出于政治的理由。

我所取的态度是矫情的吗？或是我中国哲学已有进步的象征吗？我以为都不然。我取这个态度，富于思想的同辈中国人都和我同感。中国的绅士都穿中国衣服。此外如名成利就的中国高士、思想家、银行家，有许多从来没有穿过西装，有许多则于政治、金融或社会上获得成就立刻改穿中装。他们会立刻回头，因为他们已经知道自己的地位稳固无虑，无须再穿上一身西装，以掩饰他们的浅薄英文智识或他们的低微本能。上海的绑匪绝不会去绑一个穿西装的人，因为他们明知这种人是不值一绑的。你可知道中国现在穿西装者是怎样一些人吗？大学生，赚百元一月薪俸的小职员，到处去钻头觅缝的政治家，党部青年，暴发户，愚人，智力薄弱的人。最后，当然还有那亨利·溥仪，俗极无比地题上一个外国名字，穿上一身西装，还要加上一副黑眼镜，单是这身装束，已足使他丧失一切复登大雅之堂的机会。即使日本天皇拿出全部兵力来帮助他，也不会中用。因为或许可以用种种谎话去欺骗中国人，但绝无法使他们相信一个穿西装戴墨镜的家伙是他们的皇帝。溥仪一日穿着西装，一日用"亨利"为名，则一日不能安坐皇位，而只合优游于利物浦的船坞中罢了。

中装和西装在哲学上不同之点就是，后者意在显出人体的线形，而前者意在遮隐之。但人体在基本上极像猢狲的身体，所以普通应该是越少显露越好。试想甘地只围着一条腰裙时是个什么样子？西装之为物，只有不识美丑者方会说它好。其实呢，"完美的体形世上很少"这句话也是迂腐之谈。你只要到纽约游戏场去一趟，便能看到人的体形是如何美丽。但美点的显露，并不是穿了西装使人一望而知其腰围是三十二英寸或在三十八英寸的说法。一个人何必一定要被人一望而知他的腰围是在三十二英寸呢？如若是一个颇为肥胖的人，何必一定被人知道他腰围的大小，而不能单单自己明白呢？

因此，我也相信年在二十到四十之间、身材苗条的女人和身体线形没有被现代不文明生活所毁损的儿童，确是穿西装较为好看。但是叫所有男女不分美丑，都把身体线形显露于别人的眼前，又是另一句话了。女人穿了西式晚礼服的优雅好看，实不是东方的成衣匠所能梦想到的。但一个四十多岁的肥胖妇人，穿了露出背脊的礼服出现于戏院中，其刺目也是西方所特有的景象。对于这样的妇人，中国衣服实较为优容，也和死亡一般使大小美丑一律归于平等。

所以，中国衣服是更为平等的。以上都是关于审美方面的讨论，以下可以谈谈卫生和常识方面的理由。凡是头脑清楚的人，大概都不会矫说硬领——红衣主教黎塞留和瓦尔特·罗利爵士（Sir Walter Raleigh，近代英国批评家、作家）时代的遗物——是一种助于健康的东西。即在西方，也有许多富于思想的人屡次表示反对。西方女人的衣服已在这一点上得到了许多以前所不许享受的舒适。但是男人的颈项，依旧被所有受过教育的人当做丑恶猥亵、不可见人的部分，认为须遮隐起来，正和腰围大小应尽量显露成一个反比例。这件可恶的服饰，使人在夏天不能适当地透气，在冬天不能适当地御寒，并一年到头使人不能做适当的思想。

从硬领以下，竟是一大篇连续不断同样厉害的加害人类常情的记录。能发明霓虹灯和狄赛尔柴油引擎的聪明西方人，何以竟会缺乏常识到这个地步，至于桎梏人的全身，而仅仅留出一个头部的自由，实令人不解。种种服饰的不近人情无庸一一细说——例如紧绷在身上的内衣裤，妨碍了身体的透气自由，马甲使人连背脊都弯转不来，背带或腰带使人在饥饱不同的时候没有宽紧的分别。其中最不合情理的是马甲，凡略略研究过人体线形者都知道人的胸背两部，除了在身体笔挺的时候之外，是绝不能同时平直的。凡穿过硬胸衬衫的人都从经验知道当身体向前屈时，衬衫必定拱起来。但马甲是假定人的胸背随时都是平直而裁制的，因此令人须将身体时时笔直地挺着。实际上绝没有人能始终维持这个姿势，于是马甲的边沿尖角都因起了绉痕，时时触刺人身。但若是一个肥胖的人，这马甲简直是画了一个突出的弧形线，尽头之处触出空中，再由腰带和裤子接续着向下面渐渐弯去。人类诸发明中，还有比这个更离奇的事物吗？无怪现在已有人在那里发起一种裸体运动，以反抗这束缚人体的离奇东西了。

如若人是四足动物，则腰带还有情理可说，因为还可以如马的肚带一般宽紧随心。但人类已经改为直立的地位，这腰带依旧是假定人是四足动物而制的，正如腹膜肌肉一般根据四足地位完全将它的重量系于背脊骨上。这种不合理的生理配置，使孕妇易于流产——在兽类并无此弊——而男人的腰带也因易于向下脱落，不能不束得极紧，结果是妨碍脏腑的自由活动。

我深信西方人对于身外之物有了更大的进步后，必有一天会对本身所用的物件更费一些心思，而于衣服这件事上变为较近情理。西方男人为了对这件事不肯费力革新，已受了长久的重罪，但西方女人早已于衣服一道上，达到简单和近于情理的地步了。我深信在远期中——近期中尚办不到———男人必会以直立姿势为根据，而终究发展出一种合于情理的衣服，如女人所已经达到的一般。一切累赘的衣带必会被废弃，而男人的衣服必将改为很自然合适地悬挂于两肩的制法；衣服的肩部必不再塞上许多棉花垫高起来，必将改为比现式更舒适像便袍一般的样子。据我的眼光，那时男女服色的区别，必只在男人穿裤，而女人穿裙。至于上半身，则将以同样舒适自在为基本原则了。男人的颈项将和女人一般自由，马甲将被废弃。男人的外褂也将如时下女人的外褂一般的穿法，也将像现在的女人一般以不穿外褂的时候为多。

这个改革当然包括更改衬衫的制法，衬衫将不再是单为穿在里面之用。它的颜色必改为较深的，而可以穿在外面。所用的材料将改为从最薄的绸到最厚的呢绒，以合时令，并改良式样，以求更为美观。外褂可穿可不穿，将以天气的寒暖而不以虚文为标准。因为这一种服色将成为不论到哪里都可以穿着的衣服。为了解除令人难受的腰带和背带，衬衫和裤子将连在一起，穿时只须像现在的女衣一般从头上套进去。腰部的宽紧可以看情形做得大些或小些，以适应身体的肥瘦。

　　就是现在式样的男服，也未尝不可以保持原式略加改良，即能将腰带或背带废除。它的整个原则是全部重量须悬挂到两肩上去，并均匀分配开来，而不应该借着约束之力紧系于肚皮上面。男人的腰部束缚须加解放，人们须在能领略这个原则时，男人的衣服方有渐渐改变为宽松的长袍的可能。我们现在倘以废除马甲为改革的第一步，只须将衬衫如儿童衣服一般用纽扣扣在裤子上。等到衬衫渐渐变成外衣时，我们即可以改用较好的材料做衬衫，采取和裤子同料同色或相配的颜色。又如我们倘不便把马甲马上废除，则可把马甲和裤子连在一起，以保持其形式。马甲的背部应该改为两条阔带子。此外，马甲即使不和裤子连在一起，腰带或背带也有弃置的方法。我们只须在马甲的反面前后钉上六条带子，前四后二，扣在裤子上，就可以把裤子系住了。因为扣带是在马甲的里面，而马甲是遮在裤子的上面，所以形式上将和现在的束腰带或用背带并无两样。改革一旦实行，人们觉得现在的衣服式样实在不合情理时，他们便会逐步改进，废除马甲，将上下衣裤做成仿佛现在的工人裤子，但较为好看的式样。

　　在适合时令的调节上，中国衣服也是显而易见的最近情理。穿西装不论寒暑表低到零摄氏度以下，或高到一百摄氏度以上，总是限于一身内衣裤，一件衬衫，一件外衣，连或不连马甲，但中国衣服可以加减随心。据一个故事说：有一个中国妇人，看见她儿子打一个喷嚏，即替他加上一件袍子；打两个喷嚏，再加一件；打第三个喷嚏，再加一件。这是西方做母亲的人所办不到的。她到儿子打第三个喷嚏时，恐怕就要手足无措，而只有去请教医生之一法了。我不能不信中国民族所以能够不被肺痨和肺炎所灭尽，全靠那一件棉袍的力量。

IX. The Inhumanity of Western Dress

IN spite of the popularity of Western dress among the modern Turks, Egyptians, Hindus, Japanese and Chinese, and in spite of its universality as the official diplomatic costume in the entire world, I still cling to the old Chinese dress. Many of my best friends have asked me why I wear Chinese instead of foreign dress. And those people call themselves my friends! They might just as well ask me why I stand on two legs. The two happen to be related, as I shall try to show. Why must I give a reason for wearing the only"human"dress in the world? Need anyone who in his native garb practically goes about the house and outside in his pyjamas and slippers give reasons why he does not like to be encased in a system of suffocating collars, vests, belts, braces and garters? The prestige of the foreign dress rests on no more secure basis than the fact that it is associated with superior gunboats and Diesel engines. It cannot be defended on esthetic, moral, hygienic or economic grounds. Its superiority is simply and purely political.

Is my attitude merely a pose, or symptomatic of my progress in knowledge of Chinese philosophy? I hardly think so. In taking this attitude, I am supported by all the thinking persons of my generation in China. The Chinese dress is worn by all Chinese gentlemen. Furthermore, all the scholars, thinkers, bankers and people who made good in China either have never worn foreign dress, or have swiftly come back to their native dress the moment they have"arrived"politically, financially or socially. They have swiftly come back because they are sure of themselves and no longer feel the need for a coat of foreign appearance to hide their bad English or their inferior mental outfit. No Shanghai kidnaper would think of kidnaping a Chinese in foreign clothes, for the simple reason that he is not worth the candle. Who are the people wearing foreign clothes today in China? The college students, the clerks earning a hundred a month, the political busybodies who are always on the point of landing a job, the tangpu young men, the nouveaux riches, the nincompoops, the feebleminded... And then, of course, last but not least, we have Henry P'uyi, who has the incomparably bad taste to adopt a foreign name, foreign dress, and a pair of dark spectacles. That outfit of his alone will kill all his chances of coming back to the Dragon Throne, even if he has all the bayonets of the Mikado behind him. For you may tell any lies to the Chinese people, but you cannot convince them that the fellow who wears a foreign dress and dark spectacles is their"emperor."So long as he wears that foreign dress and so long as he calls himself Henry, Henry will be perfectly at home in the dockyards of Liverpool, but not on a Dragon Throne.

Now the philosophy between Chinese and Western dress is that the latter tries to reveal the human form, while the former tries to conceal it. But as the human body is essentially like the monkeys', usually the less of it revealed the better. Think of Gandhi in his loincloth! Only in a world of people blind in sense of beauty is the foreign dress tolerable. It is a platitude that the perfect human figure rarely exists. Let any one who doubts this go to Coney Island and see how beautiful real human forms are. But the Western dress is so designed that any man in the street can tell whether your waistline is thirty-two or thirty-eight. Why must one proclaim to the world that his waistline is thirty-two, and if it happens to be above normal, why can he not have the right to make it his private affair?

That is why I also believe in the foreign dress for young women of good figure between twenty and forty and for all children whose natural bodily rhythm has not yet been subjected to our uncivilized form of living. But to demand that all men and women reveal their figure to the eyes of the world is another story. While the graceful woman in foreign evening dress shines and charms in a way not even remotely dreamed of by the Oriental costume-makers, the average over-fed and over-slept lady of forty who finds herself in the golden horseshoe at an opera premiére is also one of the eyesores invented by the West. The Chinese dress is kinder to them. Like death, it levels the great and the small, the beautiful and the ugly. The Chinese dress is therefore more democratic.

So much for the esthetic considerations. Now for the reasons of hygiene and common sense. No sane-minded man can pretend that the collar, that survival of Cardinal Richelieu's and Sir Walter Raleigh's times, is conducive to health, and all thinking men in the West have repeatedly protested against it. While the Western female dress has achieved a large measure of comfort in this respect formerly denied to the fair sex, the male human neck is still considered by the Western educated public as so ugly and immoral and socially unpresentable that it must be concealed as much as his waistline must be revealed. That satanic device makes proper ventilation impossible in summer, proper protection against cold impossible in winter and proper thinking impossible at all times.

From the collar downwards, it is a story of continuous and unmitigated outrage of common sense. The clever foreigner who can invent Neon lights and Diesel engines has not enough common sense to see that the only part of his body which is free is his head. But why go into details—the tight-fitting underwear, which interferes with free ventilation, the vest which allows for no bending of the body, and the braces or the belt which allow for no natural difference in different states of nutrition? Of these, the least logical is the vest. Anybody who studies the natural forms of the naked human body knows that, except when in a perfectly straight position, the lines of the back and the front are never equal. Anyone who wears a stiff shirt front also knows by experience that it bulges every time the body bends forward. But the vest is designed on the assumption that these lines remain always equal, which compels one to keep in a perfectly straight position. As no one actually lives up to this standard, the consequence is that the end of the vest either protrudes or falls into creases pressing on the body at every movement. In the case of a male victim of obesity, the vest describes a protruding arc and invariably ends in the air, from which point the receding arc is taken up by the belt and the pants. Can anything invented by the human mind be more grotesque? Is it any wonder that a nudist movement has sprung up as a protest and a reaction against this grotesque bondage of the human body?

But if mankind were still in the quadruped stage, there could be some justification for the belt, which then could be adjusted as the saddle strap is adjusted on the horse. But, while mankind has adopted the erect position, his belt is designed on the assumption that he is still a quadruped, just as anatomy of the peritoneal muscles shows that these are designed for the quadruped position with all weight suspended from the backbone. The disastrous consequence of this erect position is that, while human mothers are liable to miscarriage and abortion, from which animals are exempt, the belt of the male dress also has a tendency to gravitate downwards. The only way to prevent this is to keep the belt so tight that it does not gravitate, but also with the result that it interferes with all natural intestinal movements.

I am quite convinced that when the Westerners have made more progress in the impersonal things, they will one day also come to devote more time to their personal things and exercise more common sense in the matter of dress. Western men are paying a severe penalty for their conservatism in this matter of dress and for their fear of innovation, while Western women long ago achieved simplicity and common sense in their dress. To speak not for the immediate decades but for the distant centuries, I am quite convinced that men will eventually evolve a dress for themselves that is logical and consistent with their biped position, as is already achieved in women's dress. Gradually all cumbersome belts and braces will be eliminated and men's dress will be so designed that it hangs naturally down from the shoulder in a sort of graceful and fitting form. There will be no meaningless padded shoulders and lapels, and in place of the present design, there will be a much more comfortable type of dress, more nearly resembling the house jacket. As I see it, the great difference then between men's dress and women's dress will be only that men wear pants while women wear skirts. So far as the upper part of the body is concerned, the same essential consideration of ease and comfort will prevail. Men's necks will be just as free as women's, the vest will correspondingly disappear and the jacket will be used exactly to the same extent as women's coats are used today. Most of the time men will go about without their jackets as women are doing today.

This involves, of course, a revolution in our present conception of the shirt. Instead of being something to be worn inside, it will be made of darker material and worn outside, from the lightest silk to the heaviest woollen material, according to the season, and cut accordingly for better appearance. And then they can slip the jacket on, whenever they want to, but more for consideration of the weather than for formality, for this future apparel will be practically correct and acceptable in any company. In order to destroy the insufferable belts and suspenders, there will be a kind of combination shirt and pants, to be slipped over the head as women's dresses are slipped over now, with certain adjustments, pretended or real, around the waist to bring out the figure better.

Even at present, a reform is possible for the elimination of the belt or suspenders while maintaining the present pattern of men's dress. The whole principle is: weight should be suspended from the shoulders and evenly distributed, and should not be clamped on to the vertical wall of the abdomen by sheer force of adhesion, friction and compression, and the human male waist should be liberated from its duty as a bottleneck, to the end that a system of loose-fitting undergarments may become possible. If we start on the road of progress without the vest, men can just have their shirts buttoned to the pants, as children's dresses are today. In time, then, as the shirt becomes outside apparel, it will be made of finer material, probably of the same color and quality as the pants, or matching them. Or if we start on dress reform with the vest as a necessary part, we should have a combination vest and pants, preserving their present form, but made of one piece, the back of the vest being reduced to two diagonal straps. Even at the present time, belts and suspenders can easily be eliminated by having six little appendages, four in front and two behind, sewed on to the inside of the vest, with buttonholes to fit into the buttons on the pants. As the vest comes outside the pants, there will be no visible difference from vests as they are worn at present. Once the innovations are started and men begin to think that their present dress designs are not coeternal with the universe, it will be possible to gradually modify and eliminate the vest itself, by having this combination garment so cut as to be better looking than an overall, but still going on the same principle.

It requires no imagination to see that for adjustment to varying climatic conditions, the Chinese dress is also the only logical mode. While the Westerner is compelled to wear underwear, one shirt, perhaps one vest and one coat, whether the weather temperature is below zero or above a hundred, the Chinese dress is infintely flexible. There is a story told of the fond Chinese mother who puts one gown on her boy when he sneezes once, puts on another when he sneezes twice, and puts on a third when he sneezes thrice. No Western mother can do that; she would be at her wit's end at the third sneeze. All she can do is to call for the doctor. I am led to believe that the only thing which saves the Chinese nation from extermination by tuberculosis and pneumonia is the cotton-padded gown.

十. 房屋和内部布置

　　"房屋"这个名词应该包括一切起居设备，或居屋的物质环境。因为人人知道择居之道，要点不在所见的内部什么样子，而在从这所屋子望出去的外景是什么样子，所着眼者实在于屋子的地位和四周的景物。我常看见上海的富翁，占着小小的一方地皮，中间有个一丈见方的小池，旁边有一座蚂蚁费三分钟即能够爬到顶上的假山，便自以为妙不可言，他不知道住在山腰茅屋中的穷人，竟可以拿山边湖上的全部景物作为自己的私产呢。这两者之间的优劣，简直是无从比拟。山中往往有地位极佳的房子，人在其中能将全部风景收到眼底，无论望到哪里，如遮着山尖的白云，飞过空中的鸟，山泉的淨，鸟喉的清越，种种景色，都等于自己所私有。这就是一个富翁，他的财产之多，远胜于住在城市中的百万富翁。城市中的人也未始不能看见偶尔在空中行过的云，但他绝不会实地去看看，即使看到了，也因这云没有别的景物为衬托，有什么好看的呢？这里的背景是完全不适宜的。

　　所以中国人对于房屋和花园的见解，都以屋子本身不过是整个环境中一个极小部分为中心观点，如一粒宝石必须用金银镶嵌之后，方能衬出它的灿烂光辉。所以一切人为的痕迹愈少愈妙，笔直的墙垣，应有倒挂的橱藤间节地遮蔽着。一所整方的房屋只合于工厂之用，因为只有工厂才以效用为第一个要件。如若作为住宅，便是大煞风景。依照陈继儒的简明说法，一所最合于中国理想的屋子应该如下：

> 门内有径，径欲曲；径转有屏，屏欲小；屏进有阶，阶欲平；阶畔有花，花欲鲜；花外有墙，墙欲低；墙内有松，松欲古；松底有石，石欲怪；石面有亭，亭欲朴；亭后有竹，竹欲疏；竹尽有室，室欲幽；室旁有路，路欲分；路合有桥，桥欲危；桥边有树，树欲高；树荫有草，草欲青；草上有渠，渠欲细；渠引有泉，泉欲瀑；泉去有山，山欲深；山下有屋，屋欲方；屋角有圃，圃欲宽；圃中有鹤，鹤欲舞；鹤报有客，客不俗；客至有酒，酒欲不却；酒行有醉，醉欲不归。

房屋必须有独立性方为住屋。李笠翁在他讨论生活艺术的《闲情偶寄》中，有好几处提到居室问题，在序文内曾畅论"自在"和"独立性"两点。我以为"自在"比"独立性"更重要。因为一个人不论有怎样宽大华丽的房屋，里边总有一间他所最喜爱，实在常处的房间，而且必是一间小而朴素，不甚整齐，和暖的房间。所以李笠翁说：

> 人之不能无屋，犹体之不能无衣。衣贵夏凉冬燠，房舍亦然。堂高数仞，榱题数尺，壮则壮矣，然宜于夏而不宜于冬。登贵人之堂，令人不寒而栗，虽势使之然，亦寥廓有以致之，我有重裘而彼难挟纩故也。及肩之墙，容膝之屋，俭则俭矣，然适于主而不适于宾。造寒士之庐，使人无忧而叹，虽气感之乎，亦境地有以迫之，此耐萧疏，而彼憎岑寂故也。吾愿

显者之居勿太高广。夫房舍与人，欲其相称。画山水者有诀云，丈山尺树，寸马豆人。使一丈之山，缀以二尺三尺之树，一寸之马，跨以似米似粟之人，称乎？不称乎？使显者之躯能如汤文之九尺十尺，则高数仞为宜；不则堂愈高而人愈觉其矮，地愈宽而体愈形其瘠。如何略小其堂而宽大其身之为得乎？……常见通侯贵戚，掷盈千累万之资，以治园圃，必先谕大匠曰：亭则法某人之制，榭则遵谁氏之规，勿使稍异。而操运斤之权者，至大厦告成，必骄语居功，谓其立户开窗，安廊置阁，事事皆仿名园，纤毫不谬。噫，陋矣！……

土木之事，最忌奢靡。匪特庶民之家当崇俭朴，即王公大人亦当以此为尚。盖居室之制，贵精不贵丽，贵新奇大雅，不贵纤巧烂漫。凡人止好富丽者，非好富丽，因其不能创异标新，舍富丽无所见长，只得以此塞责。譬如人有新衣二件，试令二人服之，一则雅素而新奇，一则辉煌而平易，观者之目，注在平易乎，在新奇乎？锦绣绮罗，谁不知贵，亦谁不见之；缟衣素裳，其制略新，则为众目所射，以其未尝睹也。

李笠翁在他所著的书中，讨论许多关于结构和布置上的要点。所涉及的物事有房屋、窗户、屏、灯、桌、椅、古玩、橱、床、箱、柜，等等。他极富创作思想，对每一件东西都有新颖的议论。他所创作的器具中，有许多种至今为人所乐用。最著名的是他在世时即已有人仿制出售芥子园信笺和窗户板壁的制法。他那部讨论生活艺术的书虽不很为人所知道，但初学画家所奉为圭臬的《芥子园画谱》极为著名。此外《李笠翁十种曲》也很著名。因为他是一个戏剧作家、音乐家、享乐家、服装设计家、美容专家，兼业余发明家，真可谓多才多艺。

他对于床的式样有极新颖的见解。据他说，每次迁入一所新屋时，所注意的第一件事情就是那张床。中国式的床大概都有高架可以挂帐子，其本身差不多等于一间小室。里面装置着帐榱床几和屉斗，以便安放书本茶壶鞋袜等零碎物件。李氏以为"床令生花"，就是床上并宜置几盆花草，他的方法是将一只特制的，阔约一尺，高仅二三寸的轻几，从帐顶悬下来。据他的意见，这只花几应该用彩纱包裹，并折成绉纹以像行云。这个几上便可以安放应时的盆花，即使偶缺，或焚龙涎香的炉，或佛手木瓜，以取其香。据他的意见：

> 若是，则身非身也，蝶也，飞宿眠食，尽在花间。人非人也，仙也，行起坐卧，无非乐境。予尝于梦酣睡足，将觉未觉之时，忽嗅腊梅之香，咽喉齿类，尽带幽芬，似从脏腑中出，不觉身轻欲举，谓此身必不复在人世间矣。既醒，语妻孥曰："我辈何人，遽有此乐，得无折尽平世之福乎？"妻孥曰："久赋常贫，未必不由于此。"此实事，非欺人语也。

李氏的发明中，在我看来，当以窗户的制法为最杰出。他曾发明"扇面窗"（湖上游艇所用）和"梅花窗"。中国人的习俗，扇面上都有书画，并有人癖嗜收集这种旧扇面，订成册页。扇面窗之制即取意于此。所以李氏以为游艇如安上扇面式的窗子，则艇中人从船窗观望两岸的景物，两岸的路人由船窗窥望艇中人的动作，都像在观看扇面画了。因为窗子之为物，其要点即在能任人从其中看得见外面的景物，正如我们所谓眼睛乃是灵魂的窗户。所以据李氏说起来，窗子的制法应以能在最有利的地位，望见最优美的景物为主。因而可以假借室外的风景，以补充室内自然成分的缺乏。他说：

坐于其中，两岸之湖光山色，寺观浮屠，云烟竹树，以及往来之樵人牧竖，醉翁游女，连人带马，尽入"便面"之中，作我天然图画。且又时时变幻，不为一定之形，非特舟行之际，摇一橹，变一象，撑一篙，换一景；即系缆时，风摇水动，亦刻刻异形。是一日之内，现出百千万幅佳山佳水……

予又尝作观山虚牖，名"尺幅窗"，又名"无心画"。姑妄言之，浮白轩中，后有小山一座，高不逾丈，宽止及寻，而其中则有舟崖碧水，茂林修竹；鸣禽响瀑，茅屋板桥，凡山居所有之物，无一不备。盖因善塑者肖予一像，神气宛然，又因予号笠翁，顾名思义，而为把钓之形。予思既执纶竿，必当坐之矶上，有石不可无水，有水不可无山，有山有水，不可无笠翁息钓归休之地，遂形此窟以居之。是此山原为像设，初无意于为窗也。

后见其物小而蕴大，有"须弥芥子"之义，尽日坐观，不忍阖牖，乃瞿然曰：是山也，而可以作画；是画也，而可以为窗；不过损予一日杖头钱，为装潢之具耳。遂命童子裁纸数幅，以为画之头尾，乃左右镶边。头尾贴于窗之上下，镶边贴于两旁，俨然堂画一幅，而但虚其中，非虚其中，欲以屋后之山代之也。坐而观之，则窗非窗也，画也，山非屋后之山，即画上之山也。不觉狂笑失声，妻孥群至，又复笑予所笑。而"无心画""尺幅窗"之制，从此始矣。

李氏对桌椅橱柜，也别有心裁。这里我只能提及一件他所发明的冬天所用的暖椅，凡是没有相当取暖设备的室中，这是一件很实用的器具。其制法是一张长椅，下面连着一个火柜。椅子的两旁各有一个高如矮桌的活动木架，可以随意旋转到椅子的正面，搁上一块板，当做桌子。火柜里有屉斗，以便置放炭盆。在这套桌椅上可以读书写字，坐卧随心。据李氏说，这暖椅每天只费炭四块，早晨加两块，下午再加两块，即可使坐者整天和暖舒服。"这椅子只须穿上两根杠子，便成一乘轿子，可供出门的代步。冷天坐着时，两足既不致受凉，而且可以随意在轿中吃喝。这椅子到了夏天，也可以改为凉椅。其法是将一只水缸安在椅背后，注满凉水，以取其凉意。

西方人已发明各种可以旋转的，可以折叠的，高矮大小可以调节的床椅和剃头椅，但是他们从没有想过创作可以拼拆的桌几和古玩架。这件东西在中国早已发明，并且制作极为精巧。可以拼拆的桌几名叫"燕几"，其制法的原则类于西方儿童所玩的积木，将一方方木块拼搭成种种物形。一幅七件的"燕几"，可以拼出正方长方或丁字形等的式样，多至四十余种。

还有一种名为蝶几。其中每一只几形状不是方的，而是三角形或菱形的，所以拼合起来，又可以拼成许多另外的式样。燕几大都供宴饮或抹牌之用，有时当中并留出一些空闲，以置放烛台。蝶几则既供饮宴抹牌之用，也可当做花盆架子，因为花盆架子本以式样不一为宜。这种蝶几每副共有十三件，可以拼成方形、长方形、菱形等，中间或留或不留空地。拼搭的方法并不一定，全看主妇的巧思去变化。

东西方主妇对于室内位置，大都欢喜时常变更式样，因此这类可以供她们欲望的需要。这种几桌所拼成的式样都是极为摩登式的，因为摩登器具都注意轮廓线简单化。而中国器具本来就是如此的。拼搭的艺术似乎就在轮廓线的简单化中求得各种不同的式样，我曾看见过一只古式花盆架，它的脚不是笔直而是半当中弯曲的。即以方桌和圆桌而言，做的时候即可分做成半圆形的两只，或分做成三角形的两只。如此拼起来时是一个圆桌或方桌，可供饮宴或抹牌之用。不用时，即可拆开来放在墙边，当做书架或花盆架。两只三角形的蝶几，倚墙并排摆在一处，看过去便好似从墙中凸出来的两座尖山。抹牌时所用的桌子，其大小都可以随人数的多寡而定。茶点宴饮时所用的桌子，可以随意拼成丁字形、马蹄形，或S形。如在较小的房间中，大家坐在这种式样的桌子上吃饭，岂不更为有趣？

中国江苏省常熟地方现在有一种照可以拼拆原则而制造的"匡几"。可以分拆的书架在西方也很普通，但常熟式的特点是在不用时可以依着大小的次序一个一个地套进去，而只成如衣箱大小一般的一个箱子。这书箱叠好时，很像一具极新式的书橱，但分开来时可以拼成许多个大小不一的书橱，其最小的长只尺余，可以置放在几上或枕边，拼叠的式样因此可以随时变更，以免多看了令人讨厌。

中国人对室内布置好像集中于两个观念：简单和空阔。凡是布置很讲究的房间，家具必不甚多，木料必是柚木，而打磨必极光亮，轮廓线必极简单，而大多必是圆角。柚木器具必须用手工打磨，其精工与否可判别价值的高下。室中一面靠墙处大概安一张半桌，上面放一只胆瓶。墙角边大概安着几个花盆架或古玩架，高矮不一，或安几只老树根所雕成的小矮凳。另一面墙边大概安一个书橱或古玩橱，式样必极曲折玲珑，极为摩登。墙上大概挂一两幅字画，字必雄劲，画取远淡空灵，而室中也须如这画一般空灵。中国的房屋中最特殊之点是用石板所铺成的院子，效用和西班牙式房屋的走廊相同，是平和幽静安宁的象征。

X. On House and Interiors

THE word"house"should include all the living conditions or the physical environment of one's house. For everyone knows it is more important in selecting a house to see what one looks out on from the house than what one sees in it. The location of the country and its surrounding landscape are the thing. I have seen rich men in Shanghai very proud of a tiny plot of land that they own, which includes a fish pond about ten feet across and an artificial hill that takes ants three minutes to crawl to the top, not knowing that many a poor man lives in a hut on a mountain side and owns the entire view of the hillside, the river and the lake as his private garden. There is absolutely no comparison between the two. There are houses situated in such beautiful scenery up in the mountains that there is no point whatsoever in fencing off a piece of land as one's own, because wherever he wanders, he owns the entire landscape, including the white clouds nestling against the hills, the birds flying in the sky and the natural symphony of falling cataracts and birds' song. That man is rich, rich beyond comparison with any millionaire living in a city. A man living in a city may see sailing clouds, too, but he seldom sees them actually, and when he does, the clouds are not set off against an outline of blue hills,   and then what is the point of seeing clouds? The background is all wrong.

The Chinese conception of house and garden is therefore determined by the central idea that the house itself is only a detail forming a part of the surrounding country, like a jewel in its setting, and harmonizing with it. For this reason, all signs of artificiality must be hidden as much as possible, and the rectilinear lines of the walls must be hidden or broken by overhanging branches. A perfectly square house, shaped like a magnified piece of brick, is justifiable in a factory building, because it is a factory building where efficiency is the first consideration. But a perfectly square house for a home to live in is an atrocity of the first order. The Chinese conception of an ideal home has been succinctly expressed by a writer in the following manner:

Inside the gate there is a footpath and the footpath must be winding. At the turning of the footpath there is an outdoor screen and the screen must be small. Behind the screen there is a terrace and the terrace must be level. On the banks of the terrace there are flowers and the flowers must be fresh. Beyond the flowers is a wall and the wall must be low. By the side of the wall, there is a pine tree and the pine tree must be old. At the foot of the pine tree there are rocks and the rocks must be quaint. Over the rocks there is a pavilion and the pavilion must be simple. Behind the pavilion are bamboos and the bamboos must be thin and sparse. At the end of the bamboos there is a house and the house must be secluded. By the side of the house there is a road and the road must branch off. At the point where several roads come together, there is a bridge and the bridge must be tantalizing to cross. At the end of the bridge there are trees and the trees must be tall. In the shade of the trees there is grass and the grass must be green. Above the grass plot there is a ditch and the ditch must be slender. At the top of the ditch there is a spring and the spring must gurgle. Above the spring there is a hill and the hill must be deep. Below the hill there is a hall and the hall must be square. At the corner of the hall there is a vegetable garden and the vegetable garden must be big. In the vegetable garden there is a stork and the stork must dance. The stork announces that there is a guest and the guest must not be vulgar. When the guest arrives there is wine and wine must not be declined. During the service of the wine, there is drunkenness and the drunken guest must not want to go home.

The charm of a house lies in its individuality. Li Liweng has several chapters on houses and interiors in his book on the Art of Living, and in the introductory remarks he emphasizes the two points of familiarity and individuality. Familiarity, I feel, is more important than individuality. For no matter how big and pretentious a house a man may have, there is always one particular room that he likes and really lives in, and that is invariably a small, unpretentious room disorderly and familiar and warm. So says Li:

A man cannot live without a house as his body cannot go about without clothing. And as it is true of clothing that it should be cool in summer and warm in winter, the same thing is true of a house. It is all very imposing to live in a hall twenty or thirty feet high with beams several feet across, but such a house is suitable for summer and not for winter. The reason why one shivers when he enters an official's mansion is because of its space. It is like wearing a fur coat too broad for girdling around the waist. On the other hand, a poor man's house with low walls and barely enough space to put one's knees in, while having the virtue of frugality, is suitable for the owner, but not suitable for entertaining guests. That is why we feel cramped and depressed without any reason when we enter a poor scholar's hut... I hope that the dwellings of officials will not be too high and big. For a house and the people living in it must harmonize as in a picture. Painters of landscape have a formula saying, "ten-feet mountains and one-foot trees; one-inch horses and bean-sized human beings." It would be inappropriate to draw trees of two or three feet on a hill of ten feet, or to draw a human being the size of a grain of rice or millet riding on a horse an inch tall. It would be all right for officials to live in halls twenty or thirty feet high, if their bodies were nine or ten feet. Otherwise the taller the building, the shorter the man appears, and the wider the space, the thinner the man seems. Would it not be much better to make his house a little smaller and his body a little stouter?...

I have seen high officials or relatives of officials who throw away thousands and ten thousands of dollars to build a garden and who begin by telling the architect, "For the pavilion, you copy the design of So-and-So, and for the covered terrace overlooking a pond, you follow the model of So-and-So, down to its last detail."When the mansion is completed, its owner will proudly tell people that every detail of the house, from its doors and windows to its corridors and towers, has been copied from some famous garden without the slightest deviation. Ah, what vulgarity! ...

Luxury and expensiveness are the things most to be avoided in architecture. This is so because not only the common people, but also the princes and high officials, should cherish the virtue of simplicity. For the important thing in a living house is not splendor, but refinement, not elaborate decorativeness, but novelty and elegance. People like to show off their rich splendor not because they love it, but because they are lacking in originality, and besides trying to show off, they are at a total loss to invent something else. That is why they have to put up with mere splendor. Ask two persons to put on two new dresses, one simple and elegant and original, and the other rich and decorative, but common. Will not the eye of spectators be directed to the original dress rather than to the common dress? Who doesn't know the value of silk and brocade and gauze, and who has not seen them? But a simple, plain dress with a novel design will attract the eyes of spectators because they have never seen it before.

There are points about house designs and interiors which Li Liweng goes into fully in his book. The subjects he deals with cover houses, windows, screens, lamps, tables, chairs, curios, cabinets, beds, trunks, and so on. Being an exceptionally original and inventive mind, he has something new to say on every topic, and some of his inventions have become a part of the Chinese tradition today. The most outstanding contributions are his letter paper, which were sold in his life time as"Chiehtseyüan letter paper,"and his windows and partition designs. Although his book on the Art of Living is not so generally well-known, he is always remembered in connection with the Chiehtseyüan Painting Patterns, the most generally used beginner's handbook of Chinese painting, and for his Ten Comedies, for he was a dramatist, musician, epicure, dress designer, beauty expert and amateur inventor all combined.

Li had new ideas about beds. He said that whenever he moved into a new house, the first thing he looked for and attended to was the bed. The Chinese bed has always been a curtained and framed affair, resembling a large cabinet or a small room in itself, with poles, shelves and drawers built around the pole, for placing books, tea-pots, shoes, stockings and odds and ends. Li conceived the idea that one should have flower stands as well in the bed. His method was to build a thin, tiny piece of wooden shelf, over a foot wide but only two or three inches deep, and have it fixed to the embroidered curtain. According to him, the wooden shelf should be wrapped up in embroidered silk to resemble a floating cloud, with certain irregularities. There he would put whatever flowers were in season, or burn"Dragon's Saliva" incense, or keep"Buddha's Fingers"or quince for their fragrance. Thus, he says,"My body is no longer a body, but a butterfly flitting about and sleeping and eating among flowers, and the man is no longer a man but a fairy, walking about and sitting and lying in a paradise. I have thus once in my sleep felt in a half-awake state the fragrance of plum flowers so that my throat and teeth and cheeks were permeated with this subtle fragrance, as if it came out from my chest. And I felt my body so light that I almost thought I was not living in a human world. After waking up I told my wife'Who are we to enjoy this happiness? Are we not thus"curtailing"the entire lot of happiness allowed to us ?' My wife replied,'Perhaps that's the reason why we always remain poor and low! The thing is true and not a lie.'"

Li's most outstanding contribution, I believe, is in his ideas about windows. He invented"fan windows"(for pleasure houseboats used on lakes), landscape windows, and plum-flower windows. The idea of having fan-shaped windows on the sides of a houseboat was connected with the Chinese habit of painting and writing on fans and collecting such fan paintings in albums. Li's idea was therefore that with the fan window on a boat as the frame, both the people inside the boat looking out on the scenery on the banks and the people walking on the banks looking at those having a wine or tea party in the boat would see the view like a picture on a Chinese fan. For the significance of the window lies in the fact that it is primarily a place for looking out on a view, as when we say that the eye is the"window"of the soul. It should be so designed as to look out on the best view and also to enable one to see the view in the most favorable manner, in this way introducing the element of nature into house interiors by"borrowing"from the outside landscape, as Li put it. Thus:

> *When a man is sitting in the boat, the light of the lake and the color of the hills, the temples, clouds, haze, bamboos, trees on the banks, as well as the woodcutters, shepherd boys, drunken old men and promenading ladies, will all be gathered within the framework of the fan and form a piece of natural painting. Moreover, it is a living and moving picture, changing all the time, not only when the boat is moving, giving us a new sight with every movement of the oar and a new view with every punting of the pole, but even when the boat is lying at anchor, when the wind moves and the water ripples, changing its form at every moment. Thus we are able to enjoy hundreds and thousands of beautiful paintings of hills and water in a day by means of this fan-shaped window...*

> *I have also made a window for looking out on hills, called landscape window, otherwise known also as"unintentional painting."I will tell how I came to make one. Behind my studio, the Studio of Frothy White (signifying"drinking"), there is a hill about ten feet high and seven feet wide only, decorated with a miniature scenery of red cliffs and blue water, thick forests and tall bamboos, singing birds and falling cataracts, thatched huts and wooden bridges, complete in all the things that we see in a mountain village. For at first a modeller of clay made a clay figure of myself with a wonderful expression, and furthermore, because my name Liweng meant"an old man with a bamboo hat,"also made me into a fisherman, holding a fishing pole and sitting on top of a rock. Then we thought since there was a rock, there must be also water, and since there*

was water, there must be also a hill, and since there were both hill and water, there must be a mountain retreat for the old man with a bamboo hat to retire and fish in his old age. That was how we gradually built up the entire scenery. It is clear therefore that the artificial hill grew out of a clay statue without any idea of making it serve the purpose of a window view. Later I saw that although the things were in miniature, their suggested universe was great, and it seemed to recall the Buddhist idea that a mustard seed and the Himalayas are equal in size, and therefore I sat there the whole day looking at it, and could not bear to close the window. And one day inspired I said to myself, "This hill can be made into a painting, and this painting can be made into a window. All it will cost me will be just one day's drink money to provide the 'mounting' for this painting." I therefore asked a boy servant to cut out several pieces of paper, and paste them above and below the window and at the sides, to serve as the mounting for a real picture. Thus the mounting was complete, and only the space usually occupied by the painting itself was left vacant, with the hill behind my house to take its place. Thus when one sits and looks at it, the window is no more a window but a piece of painting, and the hill is no longer the hill behind my house, but a hill in the painting. I could not help laughing out loud, and my wife and children, hearing my laughter, came to see it and joined in laughing at what I had been laughing at. This is the origin of the "unintentional painting," and the "landscape window."

In the matter of tables and chairs and cabinets, Li also had a number of novel ideas. I can only mention here his invention of a heated armchair for use in winter. It is quite a practicable and useful invention wherever the rooms are not properly heated. It is a long wooden settee on a raised wooden platform built into the settee itself. The platform is two or three feet deep, with upright wooden panels at the sides about the height of a low desk. The front of the settee is also provided with two wooden door panels, and as one goes up the platform, he closes the door, which together with the upright panels on the side forms a perfect support for a removable desk top. The sitter is thus encased behind the desk. The platform itself is provided with a drawer containing hot ashes and well burned and smokeless charcoal. The settee is so made that one can sit and work there, and also lie down when he is tired. Li claimed that the cost of thus providing a perfectly warm and comfortable place to work in was no more than four pieces of charcoal a day, two being added in the morning and two in the afternoon. He claimed further that, when traveling, two strong bamboo poles could be tied to its sides and the settee could be used like a regular sedan chair, with the certainty of avoiding cold feet, and the added advantage of keeping warm whatever food and drink was taken on the journey. For the summer, he also thought of a bench resembling a bathtub, with a porcelain tub specially ordered to fit into it. The tub could then be filled with cold water, reaching to the back of the seat and cooling it.

The Western world has invented rotating, collapsible, adjustable, reversible and convertible beds, sofas and barber chairs, but somehow it has not struck upon the idea of detachable and divisible tables and curio stands. This is a thing that has had a long development in China, showing considerable ingenuity. The principle of divisible tables, known as"yenchi,"originated with a game similar to building blocks for Western children, according to which a collection of blocks of wood forming a perfect square can be made into the most diverse symbolic figures of animals, human figures, utensils and furniture on a flat surface. A"yenchi" table of six pieces could be arranged to form one or several tables of different size, square or rectangular or T-shaped, or with the tops at various angles, making a total of forty ways of arrangement.

Another type, called tiehchi, or"butterfly tables,"differ from the yenchi in having triangular pieces and diagonal lines, and therefore the resulting shapes of pieces put together present a greater diversity of outline. Whereas the first type of yenchi tables was largely designed for dinner or card-tables of different size, sometimes with a vacant space in the center for candle stands, this second type is designed for dinner tables and card tables and flower and curio stands as well, because flower and curio stands require a greater diversity of arrangement. This butterfly table consists of thirteen pieces, and together they form square tables, rectangular tables, diamond shaped tables, with or without varying kinds of holes, the possibility of novel arrangements being limited only by the ingenuity of the housewife.

There is a great desire among housewives both of the East and West to vary their interior arrangements, and divisible flower stands or tea tables seem to provide the possibility for infinite variety. The resulting shapes of such tables look strangely modernistic, for modernistic furniture emphasizes the idea of simplicity of line, which is also characteristic of Chinese furniture. The art seems to lie in combining variety in line with simplicity. I have seen, for instance, an old Chinese redwood flower stand so made that its legs are not perfectly straight, but there is a slight turn in their middle. As for varying arrangements, the simplest way would be not to order a round or square table, but to have the round table consist of two semi-circular halves, and the square table consist of two triangular tops, forming a square with the common base of the triangles as its diagonal line. When not wanted for playing cards, such round or square tables can be broken up in two and placed against the wall with the base of the triangle or of the semi-circle against it, and used for flower pots or books. With the"butterfly tables,"one can then have, instead of a triangular table against the wall, one similar to it but with a twin apex to the triangle, like two peaks of a hill. Card tables can be made larger or smaller in size according to the company present. Tea tables can be made to look like two squares overlapping one another at one of their corners, or they may resemble a"T" shape or a"U" or an"S" shape. It might be quite interesting to have a small party sitting at dinner around a"U" or"S"shaped table in a small apartment.

There exists today a perfect copy of a detachable bookcase made of hardwood, found in Ch'angshu, Kiangsu. Sectional bookcases are common in the West, but the new feature about this detachable bookcase is that its sections are so designed that, when taken apart, they can be put one inside another, the whole occupying no more space than a large suitcase. As it stands, it looks like a strangely modernistic bookcase. But it is possible to vary it and modify it so that one can break it up and have two or three small bookcases, perhaps twelve, eighteen, or twenty-four inches long, to be put at the head of sofas and beds, instead of one big bookcase, standing forever at one point in the room.

The ideal of Chinese interiors seems to consist of the two ideas, simplicity and space. A well arranged room always has few pieces of furniture, and these are generally of mahogany with extremely polished surface and simple lines, usually curved around the ends. Mahogany is polished by hand, and the difference in polish, entailing enormous labor, accounts for a wide difference in price. A long board table without drawers generally stands against one wall, supporting one big"liver"vase. At another corner perhaps may be seen one, two or three mahogany flower or curio stands, of different height, and perhaps a few stools with gnarled roots as supports. A bookcase or curio cabinet stands on one side, with sections of various heights and levels, giving a strangely modernistic effect. And on the wall are just one or two scrolls, either of calligraphy, showing the sheer joy of brush movement, or of painting, with more empty space than brush strokes in it. And like the painting itself, the room should be k'ungling, or"empty and alive."The most distinctive feature of Chinese home design is the stone-paved courtyard, similar in effect to a Spanish cloister, and symbolizing peace, quiet and repose.

生活

的

艺术

The

Importance

of Living

第十章

享受大自然

一. 乐园已经丧失了吗

在这个行星上的万物之中，植物类根本谈不上对大自然有取什么态度的可能。所有的动物类也差不多全数没有取什么态度的可能。但其中竟有这么一个人类，会自有意识，并能意识到四周的环境，因而能够对它取一种态度，实在是一桩极奇怪的事情。人因为有智慧，便开始对宇宙发生疑问，开始对它的秘密探索，对它的意义开始寻求。他们对宇宙，同时有一种科学的和道德的态度。科学界人士注意寻求本人所生活的地球里外的化学合质，其四周空气的厚薄，辐射于上层空气的宇宙光线的多少和性质，山和石的组成，以及一般的支配、生命的定律。这种科学的兴趣和道德的态度固也有一种联系，但在它的本身，不过是单纯的求知欲和探索欲罢了。在另一方面，道德的态度便有许多差异。某些人是想和大自然融协和谐，某些人是想征服或统治和利用大自然，而某些人是高傲地贱视大自然。这个对自己的星球之高傲的贱视态度，乃是文明的一种奇特产物，尤其是某种宗教的奇特产物。这种态度起源于《失乐园》那个虚构的故事。所奇者是：这个故事不过是太古时代一种宗教传说的产物，现在竟会很普遍地被人信以为真。

对于这个故事是否真实，从来没有人发过疑问。总而言之，这个伊甸园是何等的美丽，而现在这个物质宇宙又是何等的丑恶。其实呢，自从夏娃亚当犯了罪之后，花树难道已不开花了吗？上帝难道因了一人犯罪，已诅咒苹果而禁止了它的结果吗？或他已决定将这花的颜色改为较灰暗而不像以前的鲜艳吗？百灵鸟、夜莺和鹦鸟难道已停止了它们的鸣叫吗？山顶难道已经没有了积雪，湖中已经没有了倒影吗？难道今日已经没有了日落时的红霞，没有了虹霓，没有了笼罩乡村的烟雾，没有了瀑布流泉和树荫吗？所以"乐园"已经"丧失"，我们现在是住在一个丑恶的宇宙中的神话，究竟是哪一个捏造的呢？我们真是上帝的忘恩负义的不肖儿女。

关于这个不肖的孩子，我们可以设一个寓言如下：从前有一个人，姓名姑且慢慢发表，他跑到上帝那里诉说，这个星球于他还不够好，要上帝给他一个珠玉为门的天堂。上帝先指着天空中的月亮，问他说："这不是一个很好的玩具吗？"他摇摇头说，他连看都不要看。上帝又指着远远的青山，问他说："这不是很美丽的景物吗？"他回说："太平淡无奇。"上帝又将兰花和三色花指给他看，叫他伸手摸摸花瓣是如何的软骨，并问他说："这颜色的配合岂不悦目吗？"他爽直回说："不。"上帝是无穷忍耐的，于是带他到水族动物池里，指着各种各色的热带鱼给他看，问他是不是觉得有趣。上帝又带他到一个树荫之下，用法力吹起一阵微风，问他是否觉得是一种享受？他回说："并不觉得。"上帝又带他到一处山边的湖畔，指出水中的微波，松林中的风过声，山石的幽静和湖光的反映给他看，但他依然回说，这些物事并不能提起他的兴致。至此，上帝以为这个他所手创的生物必是一个性情不很和善，而喜看较为刺激事物的人，所以就带他到洛基山的顶上，到美国西部的大峡谷让他看那些挂满钟乳、生满石笋的山洞，那些喷泉沙岗，那些沙漠中的仙人掌，到喜马拉雅山看雪景，到扬子江看三峡，到黄山看花岗石峰，到尼亚加拉看瀑布，再问他说，我岂不是已尽其可能将这个星球变为可以悦耳目，可以充肚腹的美丽世界吗？但是那个人依然向上帝吵着要一个珠玉为门的天堂，说这个星球在他还觉得是不够好。"你这个不知好歹、忘恩负义的畜生，"上帝斥他说，"如此的星球，你还觉得不够好吗？很好，我将要送你到地狱里去，让你看不到行云和花树，听不到流泉，将你幽囚到命终之日。"于是上帝立即送他去住在一座城市中的公寓里边。这个人的名字就是基督徒。

　　这个人的欲望显然很难满足。上帝是否能够另造一个使他满足的天堂？实在是一个疑问。即使造了出来，然而以他这种大富豪式心性，恐怕到了这个珠玉为门的天堂之后，不到两个星期，又会感到厌倦，而上帝也将感到束手无策，无法去满足这个不肖孩子的欲望了。现在我们大概都须承认现代的天文学，由于不断的探索整个可以看得到的宇宙，结果已使我们不能不承认这地球本身就是一个天堂。如若不然，则我们所梦想的天堂势必须占着空隙；既须占着空隙，则势必在苍穹里的星中，要不必在群星之间的空虚中。这天堂既然是在有月亮或没有月亮的星球中，那么我就想象不出这天堂怎样会比我们的地球更好。这天堂的月亮或许不止一个而有许多个，如粉红色的、紫色的、碧色的、绿色的、橙黄色的、水蓝色的、土耳其玉色的，此外或许还有更多的虹霓，但我颇疑惑看见两个月亮尚会讨厌的人，看见这许多月亮时，将更易于讨厌。难得看见雪景或虹霓尚会讨厌，常常看见更美丽的虹霓将更易于讨厌了。这天堂之中，或许将有六个季节而不是四个，将同样有春夏和日夜的交替，但我看不出这里边将有什么分别；如若一个人对地球上的春夏季节不感兴趣，又怎会对天堂中的春夏季节感到兴趣呢？我说这番话或许是极愚笨的，也许是极聪明的，但我总不能赞同佛教徒和基督徒以出世超凡思想所假设的虚无缥缈完全属于精神的天堂。以我个人而言，我宁愿住在这个地球，而不愿住在别个星球上。绝对没有一个人能说这个地球上的生活是单调乏味的。倘若一个人对于许多的气候和天空颜色的变化、许多随着月令而循环变换的鲜花依然不知满足，还不如赶紧自杀，而不必更徒然地去追寻一个或许只能使上帝满足而不能使人类满足的可能天堂了。

照着眼前可见的事实而言，大自然的景物声音气味和滋味，实在是和我们的看听闻吃器官具有一种神秘的和谐。这宇宙的景物声音和气味和我们的感受器官的和谐是如此完美这件事，使伏尔泰所讥笑的"宇宙目的论"有了一个绝好的论据。但我们不必一定都做"宇宙目的论"者。上帝或许会请我们去赴他的筵席，或许不请。中国人的态度是不问被邀请与否，总去赴席。菜肴既是这样的丰盛，而我们适又饥饿了，不吃也是呆子。尽管让哲学家去进行他们的形而上学的探讨，让他们去研究我们是否在被邀请之列，但聪明人必会在菜肴未冷之前，动手去吃，饥饿和好的常识常是并行的。

我们的地球实在是一个绝好的星球。第一，上面有日夜和早暮的彼此交替，热的白天接上一个风凉的夜里，人事甚忙的上午之前，必先来一个清爽的早晨。还有什么能比这些更好呢？第二，上面有本身都极完备的夏冬季节的交替，中间加插温和的春秋两季，以逐渐引进大寒和极热。还有什么能比这些更好呢？第三，上面有静而壮观的树，夏天给我们树荫，而冬天并不遮蔽掉暖人的太阳。还有什么比这个更好呢？第四，上面有各种不同的花果，按着月令循环交替。还有什么比这个更好呢？第五，上面有清朗的日子和云雾满天的日子彼此交替。还有什么比这个更好呢？第六，上面有春雨、夏雷、秋风、冬雪。还有什么比这些更好呢？第七，上面有孔雀、鹦鹉、鹩鸟、金丝雀等，或有着美丽的颜色，或有着清脆的鸣声。

还有什么比这些更好呢？第八，上面有动物园，里边有猢狲、虎、熊、骆驼、象、犀牛、鳄鱼、海狮、牛、马、狗、猫、狐狸、松鼠、土拨鼠，种类之多为人类意想所不能及。还有什么比这些更好呢？第九，上面有虹鱼、剑鱼、电鳝、鲸、柳条鱼、文蛤、鲍鱼、龙虾、淡水虾、甲鱼，种类之多也是人类意想不到的。还有什么比这些更好呢？第十，上面有伟大的红木树、喷发的火山、伟大的山洞、雄奇的山峰、起伏的山丘、幽静的湖沼、曲折的江河、有荫的堤岸。还有什么比这些更好呢？这张菜单，其花色简直是无穷尽的，可以合任何人的胃口。所以最聪明的法子就是：径自去享用这席菜肴，而不必憎嫌生活的单调。

I. Paradise Lost

IT is a curious thing that among the myriad creations on this planet, while the entire plant life is deprived from taking any attitude toward Nature and practically all animals can also have no"attitude"to speak of, there should be a creature called man who is both self-conscious and conscious of his surroundings and who can therefore take an attitude toward it. Man's intelligence begins to question the universe, to explore its secrets and to find out its meaning. There are both a scientific and a moral attitude toward the universe. The scientific man is interested in finding out the chemical composition of the inside and crust of the earth upon which he lives, the thickness of the atmosphere surrounding it, the quantity and nature of cosmic rays dashing about on the top layers of the atmosphere, the formation of its hills and rocks, and the law governing life in general. This scientific interest has a relationship to the moral attitude, but in itself it is a pure desire to know and to explore. The moral attitude, on the other hand, varies a great deal, being sometimes one of harmony with nature, sometimes one of conquest and subjugation, or one of control and utilization, and sometimes one of supercilious contempt. This last attitude of supercilious contempt toward our own planet is a very curious product of civilization and of certain religions in particular. It springs from the fiction of the "Lost Paradise,"which, strange to say, is pretty generally accepted as being true today, as a result of a primitive religious tradition.

It is amazing that no one ever questions the truth of the story of a lost Paradise. How beautiful, after all, was the Garden of Eden, and how ugly, after all, is the present physical universe? Have flowers ceased to bloom since Eve and Adam sinned? Has God cursed the apple tree and forbidden it to bear fruit because one man sinned, or has He decided that its blossoms should be made of duller or paler colors? Have orioles and nightingales and skylarks ceased to sing? Is there no snow upon the mountain tops and are there no reflections in the lakes? Are there no rosy sunsets today and no rainbows and no haze nestling over villages, and are there no falling cataracts and gurgling streams and shady trees? Who therefore invented the myth that the"Paradise"was" lost"and that today we are living in an ugly universe? We are indeed ungrateful spoiled children of God.

A parable has to be written of this spoiled child. Once upon a time there was a man whose name we will not yet mention. He came to God and complained that this planet was not good enough for him, and said he wanted a Heaven of Pearly Gates. And God first pointed out to the moon in the sky and asked him if it was not a good toy, and he shook his head. He said he didn't want to look at it. Then God pointed out to the blue hills in the distance and asked him if the lines were not beautiful, and he said they were common and ordinary. Next God showed him the petals of the orchid and the pansy, and asked him to put out his fingers and touch gently their velvety lining and asked if the color scheme was not exquisite, and the man said,"No."In His infinite patience, God took him to an aquarium, and showed him the gorgeous colors and shapes of Hawaiian fishes, and the man said he was not interested. God then took him under a shady tree and commanded a cool breeze to blow and asked him if he couldn't enjoy that, and the man replied again that he was not impressed. Next God took him to a mountain lake and showed him the light of the water, the sound of winds whistling through a pine forest, the serenity of the rocks and the beautiful reflections in the lake, and the man said that still he could not get excited over it. Thinking that this creature of His was not mild-tempered and wanted more exciting views, God took him then to the top of the Rocky Mountains, the Grand Canyon, and caves with stalactites and stalagmites, and geysers, and sand dunes, and the fairyfinger-shaped cactus plants on a desert, and the snow on the Himalayas, and the cliffs of the Yangtse Gorges, and the granite peaks of the Yellow Mountains, and the sweeping cataract of Niagara Falls, and asked him if He had not done everything possible to make this planet beautiful to delight his eyes and his ears and his stomach, and the man still clamored for a Heaven with Pearly Gates. "This planet," the man said, "is not good enough for me." "You presumptuous, ungrateful rat!" said God. "So this planet is not good enough for you. I will therefore send you to Hell where you shall not see the sailing clouds and the flowering trees, nor hear the gurgling brooks and live there forever till the end of your days." And God sent him to live in a city apartment. His name was Christian.

It is clear that this man is pretty difficult to satisfy. There is a question as to whether God can create a heaven to satisfy him. I am sure that with his millionaire complex, he will be pretty sick of the Pearly Gates during his second week in Heaven, and God will be at His wits' end to invent something else to please this spoiled child. Now it must be pretty generally accepted that modern astronomy, by exploring the entire visible universe, is forcing us to accept this earth itself as a very heaven, and the Heaven that we dream of must occupy some space, and occupying some space, it must be somewhere among the stars in the firmament, unless it is in the interstellar void. And since this Heaven is to be found in some star, with or without moons, my imagination rather fails to conceive of a better planet than our own. Of course there may be a dozen moons instead of one, colored pink, purple, Prussian blue, cabbage green, orange, lavender, aquamarine, and turquoise, and in addition there may be better and more frequent rainbows. But I suspect that a man who is not satisfied with one moon will also get tired of a dozen moons, and one who is not satisfied with an occasional snow scene or rainbow, will be equally tired of better and more frequent rainbows. There may be six seasons in a year instead of four, and there will be the same beautiful alternation of spring and summer and day and night, but I don't see how that will make a difference. If one doesn't enjoy spring and summer on earth, how can he enjoy spring and summer in Heaven? I must seem to be talking either like a great fool or an extremely wise man now, but certainly I don't share the Buddhist or Christian desire to escape from the senses and physical matter by assuming a heaven occupying no space and constructed out of sheer spirit. For myself, I would as soon live on this planet as on any other. Certainly no one can say that life on this planet is stale and monotonous. If a man cannot be satisfied with the variety of weather and the changing colors of the sky, the exquisite flavors of fruits appearing by rotation in the different seasons, and flowers blooming by rotation in the different months, that man had better commit suicide and not try to go on a futile chase after an impossible Heaven that may satisfy God himself and never satisfy man.

As the facts of the case really stand today, there is a perfect, and almost a mystic, co?rdination between the sights, sounds, smells and tastes of nature and our organs of seeing, hearing, smelling and eating. This co?rdination between the sights and sounds and smells of the universe and our own perceptive organs is so perfect that it forms a perfect argument for teleology, so much ridiculed by Voltaire. But we need not all be teleologists. God might have invited us to this feast, or he might not. The Chinese attitude is that we will join in the feast whether we are invited or not. It simply doesn't make sense not to taste the feast when the food looks so tempting and we have such an appetite. Let the philosophers carry on their metaphysical researches and try to find out whether we are among the invited guests, but the sensible man will eat up the food before it gets cold. Hunger always goes with good common sense.

Our planet is a very good planet. In the first place, there is the alternation of night and day, and morning and sunset, and a cool evening following upon a hot day, and a silent and clear dawn presaging a busy morning, and there is nothing better than that. In the second place, there is the alternation of summer and winter, perfect in themselves, but made still more perfect by being gradually ushered in by spring and autumn, and there is nothing better than that. In the third place, there are the silent and dignified trees, giving us shade in summer and not shutting out the warm sunshine in winter, and there is nothing better than that. In the fourth place, there are flowers blooming and fruits ripening by rotation in the different months, and there is nothing better than that. In the fifth place, there are cloudy and misty days alternating with clear and sunny days, and there is nothing better than that. In the sixth place, there are spring showers and summer thunderstorms and the dry crisp wind of autumn and the snow of winter, and there is nothing better than that. In the seventh place, there are peacocks and parrots and skylarks and canaries singing inimitable songs, and there is nothing better than that. In the eighth place, there is the zoo, with monkeys, tigers, bears, camels, elephants, rhinoceros, crocodiles, sea lions, cows, horses, dogs, cats, foxes, squirrels, woodchucks and more variety and ingenuity than we ever thought of, and there is nothing better than that. In the ninth place, there are rainbow fish, sword fish, electric eels, whales, minnows, clams, abalones, lobsters, shrimps, turtles and more variety and ingenuity than we ever thought of, and there is nothing better than that. In the tenth place, there are magnificent redwood trees, fire-spouting volcanoes, magnificent caves, majestic peaks, undulating hills, placid lakes, winding rivers and shady banks, and there is nothing better than that. The menu is practically endless to suit individual tastes, and the only sensible thing to do is to go and partake of the feast and not complain about the monotony of life.

二. 论宏大

　　大自然本身永远是一个疗养院，即使不能治愈别的病患，至少能治愈人类的"自大狂"症。人类应被安置于"适当的尺寸"中，并须永远被安置在用大自然做背景的地位上，这就是中国画家在所画的山水中总将人物画得极渺小的理由。在中国的《雪后观山》画幅中，那个观望山中雪景的人被画成粗看竟寻不到的尺寸，必须要仔细寻找方能觅到。这个人蹲身在一棵大松树下，在这十五英寸高的画面中，他身体的高度不过一英寸而已，而且全身不过寥寥数笔。又有一幅宋画，画着四个高士游于山野之间，举头观看头顶上如伞盖般的大树。一个人能偶尔觉得自己是十分渺小的，于他很有益处……所以许多中国人都以为游山玩水有一种化积效验，能使人清心净虑，扫除不少妄想。

　　人类往往易于忘却他实在是何等的渺小无能。一个人看见一座百层大厦时，往往会自负。治疗这种"自负症"的对症方法就是：将这所摩天大厦在想象中搬置到一座渺小的土丘上去，而习成一种分辨何者是伟大、何者不是伟大的更真见解。我们所以重海洋，是在它的广浩无边，重山岭是在它的高大绵延。黄山有许多高峰都是成千尺的整块花岗石从地面生成，连绵不绝长达半里多。这就是使中国画家的心灵受到感动的地方。它的幽静，它的不平伏的宏大和它那显然的永在，都可说是使中国人爱好画石的理由。一个人没有到过黄山绝不会相信世上有这么样的大石，十七世纪有一个"黄山画派"，即因爱好这种奇石而得名。

在另一方面，常和大自然的伟大为伍，当真可以使人的心境渐渐也成为伟大。我们自有一种把天然景色当做活动影片看的法子，而得到不亚于看活动影片的满足；自有一种把天边的乌云当做剧台后面的布景看，而得到不亚于看布景的满足；自有一种把山野丛林当做私人花园看，而得到不亚于游私人花园的满足；自有一种把奔腾澎湃的巨浪声音当做音乐听，而得到不亚于听音乐的满足；自有一种把山风当做冷气设备，而得到不亚于冷气设备的满足。我们随着天地之大而大，如中国一流的浪漫派才子刘伶所谓"大丈夫"的"以天地为庐"。

我生平所遇到的最好的景物是某晚在印度洋上所见。这景物的场面长有百里，高有三里。大自然表现了半小时的佳剧。有巨龙、雄狮等接连在天边行过——狮子昂首而摇，狮毛四面飘拂；巨龙婉转翻身，奋鳞舞爪——有穿着灰白色军服的兵士，戴有金色肩章的军官，排着队来往不绝，倏而合队，倏而退出。在这军队彼此追逐争战时，场面上的灯光忽而变换，白衣服的兵士忽而变为黄衣服，灰色衣服忽而变为紫衣服。至于背后的布景，则一忽儿已变为耀眼的金黄色。再过一刻，这大自然的"舞美师"渐渐将灯光低暗下去，紫衣服的兵士吞没了黄衣服的而渐渐变为深紫和灰色。在灯光完全熄灭之前的五分钟，又显现出一幅令人咋舌的惨怖黑暗景象。看这出生平所仅见的伟大的戏剧，我并没有花费分文。

　　这星球上面还有幽静的山，都是近乎治疗式的幽静。如幽静的峰、幽静的石、幽静的树，一切都是幽静而伟大的。凡是环抱形的山都是一所疗养院，人居其中即好似依偎在母亲的怀里。我虽不信基督教，但我确信伟大年久的树木和山居，实具有精神上的治疗功效，并不是治疗一块断骨或一方受着传染病的皮肤的场所，而是治疗一切俗念和灵魂病患的场所。

II. On Bigness

NATURE is itself always a sanatorium. If it can cure nothing else, it can cure man of megalomania. Man has to be"put in his place,"and he is always put in his place against nature's background. That is why Chinese paintings always paint human figures so small in a landscape. In a Chinese landscape called"Looking at a Mountain After Snow,"it is very difficult to find the human figure supposed to be looking at the mountain after snow. After a careful search, he will be discovered perching beneath a pine tree— his squatting body about an inch high in a painting fifteen high, and done in no more than a few rapid strokes. There is another Sung painting of four scholarly figures wandering in an autumn forest and raising their heads to look at the intertwining branches of majestic trees above them. It does one good to feel terribly small at times... That is why a mountain trip is supposed by the Chinese to have a cathartic effect, cleansing one's breast of a lot of foolish ambitions and unnecessary worries.

Man is liable to forget how small and often how futile he is. A man seeing a hundred-story building often gets conceited, and the best way to cure that insufferable conceit is to transport that skyscraper in one's imagination to a little contemptible hill and learn a truer sense of what may and what may not be called"enormous."What we like about the sea is its infiniteness, and what we like about the mountain is its enormity. There are peaks in Huangshan or the Yellow Mountains which are formed by single pieces of granite a thousand feet high from their visible base on the ground to their tops, and half a mile long. These are what inspire the Chinese artists, and their silence, their rugged enormity and their apparent eternity account partly for the Chinese love of rocks in pictures. It is hard to believe that there are such enormous rocks until one visits Huangshan, and there was a Huangshan School of painters in the seventeenth century, deriving their inspiration from these silent peaks of granite.

On the other hand, by association with nature's enormities, a man's heart may truly grow big also. There is a way of looking upon a landscape as a moving picture and being satisfied with nothing less big as a moving picture, a way of looking upon tropic clouds over the horizon as the backdrop of a stage and being satisfied with nothing less big as a backdrop, a way of looking upon the mountain forests as a private garden and being satisfied with nothing less as a private garden, a way of listening to the roaring waves as a concert and being satisfied with nothing less as a concert, and a way of looking upon the mountain breeze as an air-cooling system and being satisfied with nothing less as an air-cooling system. So do we become big, even as the earth and firmaments are big. Like the"Big Man"described by Yüan Tsi (A.D. 210-263), one of China's first romanticists, we"live in heaven and earth as our house. "

The best" spectacle"I ever saw took place one evening on the Indian Ocean. It was truly immense. The stage was a hundred miles wide and three miles high, and on it nature enacted a drama lasting half an hour, now with giant dragons, dinosaurs and lions moving across the sky—how the lions' heads swelled and their manes spread and how the dragons' backs bent and wriggled and curled!—now showing armies of white-clad and gray-uniformed soldiers and officers with golden epaulets, marching and counter-marching and united in combat and retreating again. As the battle and the chase were going on, the stage-lights changed, and the soldiers in white uniform burst out in orange and the soldiers in gray uniforms seemed to don purple, while the backdrop was a flaming iridescent gold. Then as Nature's stage technicians gradually dimmed the lights, the purple overcame and swallowed up the orange, and changed into deeper and deeper mauve and gray, presenting for the last five minutes a spectacle of unspeakable tragedy and black disaster before the lights went out. And I did not pay a single cent to watch the grandest show of my life.

There is, too, the silence of the mountains, and that silence is therapeutic–the silent peaks, the silent rocks, the silent trees, all silent and all majestic. Every good mountain with an enclosing gesture is a sanatorium. One feels good nestling like a baby on its breast. A disbeliever in Christian Science, I do believe in the spiritual, healing properties of grand, old trees and mountain resorts, not for curing a fractured shoulder-bone or an infected skin, but for curing the ambitions of the flesh and diseases of the soul.

三. 两个中国女子

享受大自然是一种艺术，因人的性情个性而异其趣，并且如别种艺术一般，极难于描写其中的技巧。其中一切都需出于自动，都需出于艺术天性的自动。所以在某一时候怎样去享受一树一石或一景，并无规则可定，因为没有景致是相同的。凡是懂这个道理的人，不必有人教他，即会知道怎样去享受自然。哈夫洛克·蔼理斯（Havelock Ellis）和凡·德·威尔德（Van der Velde，十九世纪末比利时杰出的设计家）说，夫妇在闺房静好之中，什么事可做，什么事不可做，什么是有趣的，什么是没趣的，绝不是可以用章程来规定的事情。这句话，实在是不朽之论。享受大自然中也同样如此。最好的探讨方法大概还是：从具有这种艺术天性的人们的生活中去研究爱好大自然。梦见一年以前所看到的一个景致，忽然想到一个地方去的愿望——这些都是突然而来的事情。凡有艺术天性的人，不论走到哪里，都会显出这个天性。凡真能享受大自然的作家，都会丢开他已定的纲要，而去自由地描写一场美丽的雪景或一场春天的晚景。新闻家和政治家的自传中，大都充满着过去经验的回忆，但是文学家的自传文中，应多谈一个快乐的夜里，或一次和几个朋友到一个山谷里去游玩的回忆。

在这一点上，我觉得拉迪亚德·吉卜林（Rudyard Kipling）和切斯·特顿（G.K.Chesterton）自传都是令人失望的。他们何以竟会将一生中的经历轻重倒置？真令人不解。他们所提到的，无非是人，人，人，而丝毫没有提到花鸟山丘和溪流。

中国文人的回忆文字和他们的信札在这一点上便不同了。信札中最重要的事情每每是告诉他的朋友一个晚上在湖上的经过，或在自传中记录他生平所认为快乐的一天和这天的经过。中国作家至少有很多个都喜爱记录夫妇闺房乐趣的回忆。其中冒辟疆所著的《影梅庵忆语》，沈三白的《浮生六记》和蒋坦的《秋灯琐忆》，更是极好的例子。冒沈二书是在夫人去世后所著，蒋书则是在夫人尚在的时候老年所著。我这里当先行引用《秋灯琐忆》中的几句话，书中主人公是作者的夫人秋芙，再引几段《浮生六记》中的话，书中主人公是作者的夫人芸娘。这两个女子虽不是极有学问的人或大诗家，但她们都有适当的性情。这并无关系，我们不必着眼于写出可传诸万世的好诗，只需学会怎样用诗句去记录一件有意义的事件、一次个人的心境，或用诗句来协助我们享受大自然。

甲　秋芙

秋芙每谓余云："人生百年，梦寐居半，愁病居半，襁褓垂老之日又居半。所仅存者，十一二年。况我辈蒲柳之质，犹未必百年者乎……"

秋月正佳，秋芙命雏鬟负琴，放舟两湖荷芰之间。时令自西溪归，及门，秋芙先出，因买瓜皮迹之。相遇于苏堤第二桥下，秋芙方鼓琴作《汉宫秋怨》曲，余为披襟而听。斯时四山沉烟，星月在水，玲琮杂鸣，不知天风声环佩声也。琴声未终，船唇已移近漪园南岸矣，因叩白云庵门，庵尼故相识也。坐次，采池中新莲，制羹以进，色香清冽，足沁肠腑，其视世味腥膻，何止薰莸之别。回船至段家桥，登岸，施竹簟于地，坐话良久。

闻城中尘嚣声，如蝇营营，殊聒人耳。其时星斗渐稀，潮气横白。听城头更鼓，已沉沉第四通矣，遂携琴刺船而去……

秋芙所种芭蕉，已叶大成阴，荫蔽帘幕；秋来风雨滴沥，枕上闻之，心与俱碎。一日，余戏题断句叶上云：

是谁多事种芭蕉？
早也潇潇，
晚也潇潇！
明日见叶上续书数行云：
是君心绪太无聊！
种了芭蕉，
又怨芭蕉！

字面柔媚，此秋芙戏笔也。然余于此，悟入正复不浅。

夜来闻风雨声，枕簟渐有凉意。秋芙方卸晚妆，余坐案旁，制《百花图记》未半，闻黄叶数声，吹堕窗下，秋芙顾镜吟曰：

昨日胜今日，今年老去年。

余怃然云："生年不满百，安能为他人拭涕？"辄为掷笔。夜深，秋芙思饮，瓦铫温暾，已无余火，欲呼小鬟，皆蒙头户间，为趾离召去久矣。余分案上灯置茶灶间，温莲子汤一瓯饮之。秋芙病肺十年，深秋咳嗽，必高枕始得熟睡。今年体力较强，拥髻相对，常至夜分，殆眠餐调摄之功欤？

595

余为秋芙制梅花画衣，香雪满身，望之如绿萼仙人，翩然尘世。每当春暮，翠袖凭栏，鬓边蝴蝶，犹栩栩然不知东风之既去也。

秋芙好棋，而不甚精。每夕必强余手谈，或至达旦。余戏举竹坨词云："簸钱斗草已都输，向持底今宵偿我？"秋芙故饰词云："君以我不能胜耶？请以所佩玉虎为赌。"下数十子，棋局渐输，秋芙纵膝上猫儿搅乱棋势。余笑云："子以玉奴自况欤？"秋芙嘿然，而银烛荧荧，已照见桃花上颊矣。自此更不复棋。

去年燕来较迟，帘外桃花，已零落殆半，夜深巢泥忽倾，堕雏于地，秋芙惧为猫儿所攫，急收取之，且为钉竹片于梁，以承其巢。今年燕子复来，故巢犹在，绕屋呢喃，殆犹忆去年护雏人耶？

虎跑泉上有木樨数株，偃伏石上。花时黄雪满阶，如游天香国中，足怡鼻观。余负花癖，与秋芙常煮茗其下。秋芙拗花簪鬓，额上发为树枝梢乱，余为蘸泉水掠之。临去折花数枝，插车背上，携入城闉，欲人知新秋消息也。

乙 芸

（按：在《浮生六记》中，一个不出名的文学家描写他夫妇的闺房中琐事的回忆。他俩都是富于艺术性的人，知道怎样尽量地及时行乐。文字极其自然，毫无虚饰。我颇觉得芸是中国文学中所记的女子中最为可爱的一个，他俩的一生很凄惨，但也很放荡，是心灵中所流露出来的真放荡。他俩以享受大自然为怡情悦性中必不可少的事件。以下三节描写他俩怎样度那快乐的牛郎织女相会节和中元节，以及怎样在苏州过夏。）

是年七夕，芸设香烛瓜果，同拜天孙于我取轩中。余镌"愿生生世世为夫妇"图章二方，余执朱文，芸执白文，以为往来书信之用。是夜月色颇佳，俯视河中，波光如练。轻罗小扇，并坐水窗，仰见一飞云过天，变态万状。芸曰："宇宙之大，同此一月，不知今日世间，亦有如我两人之情兴否？"余曰："纳凉玩月，到处有之。若品论云霞，或求之幽闺绣闼，慧心默证者固亦不少；若夫妇同观，所品论者恐不在此云霞耳。"未几，烛尽月沉，撤果归卧。

七月望，俗谓之鬼节。芸备小酌，拟邀月畅饮，夜忽阴云如晦。芸愀然曰："妾能与君白头偕老，月轮当出。"余亦索然。但见隔岸萤光明灭万点，梳织于柳堤蓼渚间。余与芸联句以遣闷怀，而两韵之后，逾联逾纵，想入非夷，随口乱道。芸已漱涎涕泪，笑倒余怀，不能成声矣。觉其鬓边茉莉浓香扑鼻，因拍其背，以他词解之曰："想古人以茉莉形色如珠，故供助妆压鬓，不知此花必沾油头粉面之气，其香更可爱，所供佛手，当退避三舍矣。"芸乃止笑曰："佛手乃香中君子，只在有意无意间；茉莉乃香中小人，故须借人之势，其香也如胁肩谄笑。"余曰："卿何远君子而近小人？"芸曰："我笑君子爱小人耳。"

正话间，漏已三滴，渐见风扫云开，一轮涌出，乃大喜，倚窗对酌。酒未三杯，忽闻桥下哄然一声，如有人堕。就窗细瞩，波明如镜，不见一物，唯闻河滩有只鸭急奔声。余知沧浪亭畔素有溺鬼，恐芸胆怯，未敢即言。芸曰："噫！此声也，胡为乎来哉？"不禁毛骨皆悚。急闭窗，携酒归房，一灯如豆，罗帐低垂，弓影杯蛇，惊神未定。剔灯入帐，芸已寒热大作，余亦继之，困顿两旬。真所谓乐极灾生，亦是白头不终之兆。

这书中可算是充满着美丽风雅，流露着对大自然的爱好。以下这段在苏州过夏的记录可见一斑：

迁仓米巷，余颜其卧楼曰"宾香阁"，盖以芸名而取如宾意也。院窄墙高，一无可取。后有厢楼，通藏书处，开窗对陆氏废园，但有荒凉之象。沧浪风景，时切芸怀。

有老妪居金母桥之东，埂巷之北，绕屋皆菜圃，编篱为门。门外有池约亩许，花光树影，错杂篱边。其地即元末张士诚王府废基也。屋西数武，瓦砾堆成土山。登其巅可远眺，地旷人稀，颇饶野趣。妪偶言及，芸神往不置。

越日至其地，屋仅二间，前后隔而为四，纸窗竹榻，颇有幽趣。老妪知余意，欣然出其卧室为赁，四壁糊以白纸，顿觉改观。于是禀知吾母，挈芸居焉。邻仅老夫妇二人，灌园为业，知余夫妇避暑于此，先来通殷勤，并钓池鱼，摘园蔬为馈。偿其价不受，芸做鞋报之，始谢而受。时方七月，绿树荫浓，水面风来，蝉鸣聒耳。邻老又为制鱼竿，与芸垂钓于柳荫深处。日落时，登土山观晚霞夕照，随意联吟，有"兽云吞落日，弓月弹流星"之句。少焉，月印池中，虫声四起，设竹榻于篱下。老妪报酒温饭熟，遂就月光对酌，微醺而饭。浴罢，则凉鞋蕉扇，或坐或卧，听邻老谈因果报应事。三鼓归卧，周体清凉，几不知身居城市矣。篱边倩邻老购菊，遍植之。九月花开，又与芸居十日。吾母亦欣然来观，持螯对菊，赏玩竟日。芸喜曰："他年当与君卜筑于此，买绕屋菜园十亩，课仆妪，植瓜蔬，以供薪水。君画我绣，以为诗酒之需。布衣菜饭，可乐终身，不必作远游计也。"余深然之。今即得有境地，而知己沦亡，可胜浩叹！

III. Two Chinese Ladies

THE enjoyment of Nature is an art, depending so much on one's mood and personality, and like all art, it is difficult to explain its technique. Everything must be spontaneous and rise spontaneously from an artistic temperament. It is therefore difficult to lay down rules for the enjoyment of this or that tree, this or that rock and this or that landscape in a particular moment, for no landscapes are exactly alike. He who understands will know how to enjoy Nature without being told. Havelock Ellis and Van der Velde are wise when they say that what is allowable and what is not allowable, or what is good and what is bad taste in the art of love between husband and wife in the intimacy of their bedroom, is not something that can be prescribed by rules. The same thing is true of the art of enjoying Nature. The best approach is probably by studying the lives of such people who have the artistic temperament in them. The feeling for Nature, one's dreams of a beautiful landscape seen a year ago, and one's sudden desire to visit a certain place—these things come in at the most unexpected moments. One who has the artistic temperament shows it wherever he goes, and writers who truly enjoy nature will go off in descriptions of a beautiful snow scene or a spring evening, forgetting entirely about the story or the plot. Autobiographies of journalists and statesmen are usually full of reminiscences of past events, while the autobiographies of literary men should mainly concern themselves with reminiscences of a happy night, or a visit with some of their friends to some valley. In this sense I find the autobiographies of Rudyard Kipling and G. K. Chesterton disappointing. Why are the important anecdotes of their lives regarded as so unimportant, and why are the unimportant anecdotes regarded as so important? Men, men, men, everywhere, and no mention of flowers and birds and hills and streams!

The reminiscences of Chinese literary men, and also their letters, differ in this respect. The important thing is to tell a friend in one's letter about a night on the lake, or to record in one's autobiography a perfectly happy day and how it was passed. In particular, Chinese writers, at least a number of them, have gone to the length of writing reminiscences about their married lives. Of these, Mao Pichiang's Reminiscences of My Concubine, Shen Sanpo's Six Chapters of a Floating Life, and Chiang T'an's Reminiscences Under the Lamp-Light are the best examples. The first two were written by the husbands after their wives' death, while the last was written in the author's old age during his wife's lifetime. We will begin with certain select passages from the Reminiscences Under the Lamp-Light with the author's wife Ch'iufu as the heroine, and follow it with selections from Six Chapters of a Floating Life, with Yün as the heroine. Both these women had the right temperament, although they were not particularly educated or good poets. It doesn't matter. No one should aim at writing immortal poetry; one should learn the writing of poems merely as a way to record a meaningful moment, a personal mood, or to help the enjoyment of Nature.

a. Ch'iufu

Ch'iufu often said to me, "A man's life lasts only a hundred years, and of this hundred sleep and dream occupy one half, days of illness and sorrow occupy one half, and the days of swaddling clothes and senile age again occupy one half. What we have got left is only a tenth or fifth part. Besides, we who are made of the stuff of willows can hardly expect to live a hundred years."

One day when the autumn moon was at its best, Ch'iufu asked a young maid to carry a ch'in and accompany her to a boating trip among the lotus flowers of the West Lake. I was then returning from the West River, and when I arrived and found that Ch'iufu had gone boating, I bought some melons and went after her. We met at the Second Bridge of the Su Tungp'o Embankment, when Ch'iufu was playing the sad ditty of "Autumn in Han Palace." Stopping to listen with my gown gathered in my hands, I listened to her music. At this moment, the hills all around were enveloped in the evening haze, and the reflections of the stars and the moon were seen in the water. Different musical sounds came to my ear so that I could not distinguish whether it was the sounds of the wind in the air, or the sounds of jingling jade. Before the song was completed, the bow of our boat had already touched the southern bank of the Garden of Swirling Waters. We then knocked at the gate of the White Cloud Convent, for we knew the nuns there. After sitting down for a while, the nuns served us with freshly picked lotus seeds prepared in soup. Their color and their fragrance were enough to cool one's intestines, a world different from the taste of meats and oily foods. Coming back, we landed at Tuan's Bridge, where we spread a bamboo matting on the ground and sat talking for a long time. The distant rumble of the city rather annoyed our ears like the humming of flies... Then the stars in the sky became fewer and fewer and the lake was blanketed with a stretch of white. We heard the drum on top of the city wall and realized that it was already the fourth watch [about 3 A. M.] and carried the ch'in and paddled the boat home.

The banana trees that Ch'iufu planted had already grown big leaves which cast their green shade across the screen. To have heard raindrops beating upon the leaves in autumn when lying inclined on a pillow was enough to break one's heart. So one day I playfully wrote three lines on one of the leaves:

What busybody planted this sapling?
Morning tapping,

Evening rapping!

Next day, I saw another three lines following them, which read:

It's you who're lonesome, fretting!
Banana getting,
Banana regretting!

The characters were delicately formed, and they came from Ch'iufu's playful pen. But I have learnt something from what she wrote.

One night we heard the noise of wind and rain, and the pillows and matting revealed the cooler spirit of autumn. Ch'iufu was just undressing for the night, and I was sitting by her side and had just gone through an album of hundred flowers with inscriptions that I was making. I heard the noise of several yellow leaves falling upon the floor from the window, and Ch'iufu sang the lines:

Yesterday was better than today;
And this year I'm older than the last.

I consoled her, saying,"One never lives a full hundred years. How can we have time to wipe the tears for others [the falling leaves]." And with a sigh I laid aside the painting brush. When the night was getting late, and Ch'iufu wanted to have something to drink, I found that the fire in the earthen stove had already died out, and the maid servants were all in dreamland, drooping their heads. I then took the oil lamp on the table and placed it under the little tea stove, and warmed up a cup of lotus seeds for her. Ch'iufu has been suffering from an affection in the lungs for ten years, and always coughs in late autumn and sleeps well only when upholstered on a high pillow. This year, she is feeling stronger, and we often sit face to face with each other deep into the night. Perhaps it is due to proper care and nourishment .

I made a dress with a plum-flower design for Ch'iufu, with fragrant snow all over her body, and at a distance she looked like a Plum Fairy standing alone in a world of mortal beings. In late spring, when her green sleeves were resting on the balcony, butterflies would flit about her temples, not knowing that the season of the Eastern Wind was already gone.

Last year, the swallows came back later than usual, and when they came, half of the peach blossoms outside the screen had already bloomed. One day, the clay from their nest fell down and a young swallow fell to the ground. Afraid that a wild cat might get it, Ch'iufu immediately took it up,   and made a bamboo support for its nest. This year the same swallows have returned and are chirping around the house. Do they perhaps remember the one who protected the young one last year?

Ch'iufu loves to play chess but is not very good at it. Every night, she would force me to play "the conversation of fingers" with her, sometimes till daybreak. I playfully quoted the line of Chu Chuchia, "At tossing coins and matching grass-blades you have both lost. I ask you with what are you going to pay me tonight?" "Are you so sure I cannot win?" she said, evading the question. "I will bet you this jade tiger." We then played and when twenty or thirty stones had been laid, and she was getting into a worse situation, she let the cat upon the chessboard to upset the game. "Are you regarding yourself as Yang Kueifei [who played the same trick upon Emperor T'ang Minghuang]?" I asked. She kept quiet, but the light of silver candles shone upon her peach-colored cheeks. After that, we did not play any more.

There are several cassia trees at the Hupao Spring, stretching low over some rocks. During blossom, its yellow flowers cover up the stone steps, its perfume making one feel like visiting the Kingdom of Divine Fragrance. I have a weakness for flowers and often boiled tea under them. Ch'iufu plucked the flowers and decorated her hair with them, but sometimes her hair would be caught or upset by the overhanging branches. I arranged it and smoothed it with the spring water. On our departure, we plucked a few twigs and brought them home, putting them on the back of our cart as we went through the city streets, that people might know the latest news of the new autumn.

b. Yün

In the Six Chapters of a Floating Life we have the reminiscences of an obscure Chinese painter about his married life with Yün. They were both simple artistic souls, trying to snatch every moment of happiness that came their way, and the story was told in a simple unaffected manner. Somehow Yün has seemed to me the most beautiful woman in Chinese literature. Theirs was a sad life, and yet it was one of the gayest, with a gaiety that came from the soul. It is interesting to see how the enjoyment of nature came in as a vital part of their spiritual experience. Below are three passages describing their enjoyment of the seventh of the seventh moon and the fifteenth of the seventh moon, both festivals, and of how they passed a summer inside the city of Soochow:

On the seventh night of the seventh moon of that year [1780] Yün prepared incense, candles and some melons and fruits, so that we might together worship the Grandson of Heaven in the Hall called "After My Heart." I had carved two seals with the inscription, "That we might remain husband and wife from incarnation to incarnation." I kept the seal with positive characters, while she kept the one with negative characters, to be used in our correspondence. That night, the moon was shining beautifully and when I looked down at the creek, the ripples shone like golden chains. We were wearing light silk dresses and sitting together with a small fan in our hands, before a window overlooking the creek. Looking up at the sky, we saw the clouds sailing through the heavens, changing at every moment into a myriad forms, and Yün said: "This moon is common to the whole universe. I wonder if there is another pair of lovers quite as passionate as ourselves looking at the same moon tonight?" And I said: "Oh, there are plenty of people who will be sitting in the cool evening and looking at the moon, and, perhaps also many women criticising or enjoying the clouds in their

chambers; but when a husband and wife are looking at the moon together, I hardly think
that the clouds will form the subject of their conversation."By and by, the candle-lights
went out, the moon sank in the sky, and we removed the fruits and went to bed.

The fifteenth of the seventh moon was All Souls' Day. Yün prepared a little
dinner, so that we could drink together with the moon as our company, but when night
came, the sky was suddenly overcast with dark clouds. Yün knitted her brow and said: "
If it be the wish of God that we two should live together until there are silver threads in
our hair, then the moon must come out again tonight."On my part I felt disheartened
also. As we looked across the creek, we saw will-o'-the-wisps flitting in crowds hither
and thither like ten thousand candle-lights, threading their way through the willows
and smartweeds. And then we began to compose a poem together, each saying two lines
at a time, the first completing the couplet which the other had begun, and the second
beginning another couplet for the other to finish, and after a few rhymes, the longer we
kept on, the more nonsensical it became, until it was a jumble of slapdash doggerel. By
this time, Yün was buried amidst tears and laughter and choking on my breast, while I
felt the fragrance of the jasmine in her hair assail my nostrils. I patted her on the shoulder
and said jokingly, "I thought that the jasmine was used for decoration in women's hair
because it was round like a pearl; I did not know that it is because its fragrance is so
much finer when it is mixed with the smell of women's hair and powder. When it smells
like that, even the citron cannot remotely compare with it."Then Yün stopped laughing
and said: "The citron is the gentleman among the different fragrant plants because its
fragrance is so slight that you can hardly detect it; on the other hand, the jasmine is a
common fellow because it borrows its fragrance partly from others. Therefore, the fragrance
of the jasmine is like that of a smiling sycophant." "Why, then,"I said, "do you keep
away from the gentleman and associate with the common fellow?"And Yün replied, "I
am amused by the gentleman that loves the common fellow."While we were thus
bandying words about, it was already midnight, and we saw the wind had blown away
the clouds in the sky and there appeared the full moon, round like a chariot wheel, and
we were greatly delighted. And so we began to drink by the side of the window, but before
we had tasted three cups, we heard suddenly the noise of a splash under the bridge, as if

some one had fallen into the water. We looked out through the window and saw there was not a thing, the water was as smooth as a mirror, except that we heard the noise of a duck scampering in the marshes. I knew that there was a ghost of some one drowned by the side of the Ts'anglang Pavilion, but knowing that Yün was very timid, dared not mention it to her. And Yün sighed and said: "Alas! whence cometh this noise?" and we shuddered all over. Quickly we shut the window and carried the wine pot back into the room. A lamp light was then burning as small as a pea, and the curtains moved in the dark, and we were shaking all over. We then put out the light and went inside the bed curtain, and Yün already had run up a high fever. Soon I had a high temperature myself, and our illness dragged on for about twenty days. True it is that when the cup of happiness overflows, disaster follows, as the saying goes, and this was also an omen that we should not be able to live together until old age.

The book is strewn literally with passages of such charm and beauty, showing an overflowing love of nature, but the following description of how they spent a summer must suffice:

After we had moved to Ts'angmi Alley, I called our bedroom the "Tower of Guests' Fragrance," with a reference to Yün's name, and to the story of Liang Hung and Meng Kuang who as husband and wife were always courteous to each other "like guests." We rather disliked the house because the walls were too high and the courtyard was too small. At the back, there was another house, leading to the library. Looking out of the window at the back, one could see the old garden of Mr. Lu, then in a dilapidated condition. Yün's thoughts still hovered about the beautiful scenery of the Ts'anglang Pavilion.

At this time there was an old peasant woman living on the east of Mother Gold's Bridge and the north of Kenghsiang. Her little cottage was surrounded on all sides by vegetable fields and had a wicker gate. Outside the gate, there was a pond about thirty yards across, surrounded by a wilderness of trees on all sides. ... A few paces to the west of the cottage, there was a mound filled with broken bricks, from the top of which one could command a view of the surrounding country, which was an open ground with a

*stretch of wild vegetation. Once the old woman happened to mention the place, and Yün
kept on thinking about it... So the next day I went there and found that the cottage
consisted only of two rooms, which could be partitioned into four. With paper windows
and bamboo beds, the house would make quite a delightfully cool place to stay in...*

*Our only neighbours were an old couple who raised vegetables for the market.
They knew that we were going to stay there for the summer, and came and called on
us, bringing us some fish from the pond and vegetables from their own fields. We offered
to pay for them, but as they wouldn't take any money, Yün made a pair of shoes for
them, which they were finally persuaded to accept. This was in July when the trees cast
a green shade over the place. The summer breeze blew over the water of the pond, and
cicadas filled the air with their singing the whole day. Our old neighbour also made a
fishing line for us, and we used to angle together under the shade. Late in the afternoons,
we would go up on the mound to look at the evening glow and compose lines of poetry,
when we felt so inclined. Two of the lines were:*

Beast-clouds swallow the sinking sun,
And the bow-moon shoots the falling stars.

*After a while, the moon cut her image in the water, insects began to cry all around,
and we placed a bamboo bed near the hedgerow to sit or lie upon. The old woman then
would inform us that wine had been warmed up and dinner prepared, and we would sit
down to have a little drink under the moon. After we had a bath, we would put on our
slippers and carry a fan, and lie or sit there, listening to old tales of retribution told by our
neighbour. When we came in to sleep about midnight, we felt our whole bodies nice and
cool, almost forgetting that we were living in a city.*

*There along the hedgerow, we asked the gardener to plant chrysanthemums. The
flowers bloomed in the ninth moon, and we continued to stay there for another ten days.
My mother was also quite delighted and came to see us there. So we ate crabs in the midst
of chrysanthemums and whiled away the whole day. Yün was quite enchanted with all
this and said: "Some day we must build a cottage here. We'll buy ten mow of ground,*

and around it we'll plant vegetables and melons for our food. You will paint and I will do embroidery, from which we could make enough money to buy wine and compose poems over dinners. Thus, clad in simple gowns and eating simple meals, we could live a very happy life together without going anywhere."I fully agreed with her. Now the place is still there while the one who knows my heart is dead. Alas, such is life!

四．论石与树

现在的事情，真使我莫名其妙。房屋都是造成方形的，整齐成列。道路也是笔直的，并且没树木。我们已不再看见曲径、老屋和花园中的井，城市中即使有两处私人花园，也不过是具体而微罢了。我们居然已做到将大自然推出我们生活之外的地步。我们住在没有屋顶的房子，房屋的尽处即算是屋顶，只要合于实用，便算了事，营造匠人也因看得讨厌而马虎完事。现在的房屋，简直像一个没有耐心的小孩用积木所搭成的房子，在没有加上屋面，尚未完成时，即已觉得讨厌而停工了。大自然的精神已经和现代的文明人脱离。我颇以为人类甚至已经企图把树木也文明化起来，只需看一看大道旁所植的树就知道：株数间隔何等整齐，还要把它们消一下毒，并且用剪子修整，使它们显出我们人类所认为美丽的形式。

我们现在种花，每每种成圆形，或星形，或字母形。如若当中有一株的枝叶偶尔横叉出齐整线之外，我们便视之如西点军校（West Point）学兵操练当中有一个学兵步伐错误一般可怕，赶紧要用剪子去剪它下来。凡尔赛所植的树，都是剪成圆锥形，一对一对极匀称地排列成圆形或长方形，如兵式操中的阵图一般。这就是人类的光荣和权力，如同训练兵丁一般去训练树木的能力。如若一对并植着的树高矮上略有参差，我们便觉得非剪齐不可，使它不至于扰乱我们的匀称感觉、人类的光荣和权力。

所以，当前的大问题就是：怎样去要回大自然、将大自然依旧引进人类的生活里边？这是一个极难于措置的问题。人们都是住在远离泥土的公寓中，即使他有着最好的艺术心性，将何从去着力呢？即使他有另租一间房屋的经济实力，但这里边怎样能够种植出一片草场，或开一口井，或种植一片竹园呢？一切的一切都是极端的错误，都是无从挽回的错误。除了摩天大厦，和夜间成排透露灯光的窗户之外，还有什么可以使人欣赏的东西呢？一个人越多看这种摩天大厦和夜间成排透露灯光的窗户，便会越自负人类文明的能力，而忘却人类本是何等渺小的生物。所以我只能认这个问题为无解决的可能，而搁在一旁。

所以，第一步我们须使每个人有很多的空地。不论什么借口，剥夺人类土地的文明总是不对的。假使将来产生一种文明，能使每个人都有一亩的田地，他才有下手的机会。他就可以有着自己所有的树，自己所有的石。他在选择地段的时节，必去选原有大树的地方。倘若果真没有大树，他必会赶紧去种植一些易于生长的树，如竹树、柳树之类。他不必再将鸟养在笼中，因为百鸟都会自己飞来。他必会听任青蛙留在近处，并且留些蝎子、蜘蛛。那时他的儿童才能在大自然中研究大自然，而不必从玻璃柜中去研究。儿童至少有机会去观察小鸡怎样从鸡蛋中孵出来，而对于两性问题不会再和那波士顿高等家庭中儿童一般一窍不通了。他们也有了机会可以看见蝎子和蜘蛛打架，他们的身上将时常很舒服地污秽了。

我在上文已经提过中国人的爱石心性，这就可以解释中国人都喜欢山水画的理由。但这解释还不过是基本的，尚不足以充分说明一般的爱石心理。基本的观念是石是伟大的、坚固的，暗示一种永久性。它们是幽静的、不能移动的，如大英雄一般的具着不屈不挠的精神。它们也是自立的，如隐士一般脱离尘世。它们也是长寿的，中国人对于长寿的东西都是喜爱的。最重要的是：从艺术观点看起来，它们就是魁伟雄奇，峥嵘古雅的模范。此外还有所谓"危"的感想，三百尺高的壁立巉岩总是奇景，即因它暗示着一个"危"字。

但应该讨论的地方还不止于此。一个人绝不能天天跑到山里去看石，所以必须把石头搬到家中。凡是花园里边的垒石和假山，布置总以"危"为尚，以期模仿天然山峰的峥嵘。这是西方人到中国游历时所不能领会了解的。但这不能怪西方人，因为大多数假山都是粗制滥造、俗不可耐，不能使人从中领略到真正的魁伟雄奇意味。用几块石头所叠成的假山，大都用水泥胶粘，而水泥的痕迹往往显露在外。真正合于艺术的假山，应该是像画中之山石一般。假山和画中山石所留于人心的艺术意味无疑地是相类而联系的，例如，宋朝的名画家米芾曾写了一部关于观石的书，另一宋朝作家曾写了一部石谱，书中详细描写几百种各处所产合于筑假山之用的石头。这些都显示宋代名画家时代，假山已经有了很高度的发展。

和这种山峰巨石的领略平行的，人类又发展了一种对园石的不同的领略，专注于颜色纹理面皱和结构，有时注意于击时所发出的声音。石愈小，愈是注意于结构和纹色。有许多人对集藏各种石砚和石章的癖好更增长了这一方面发展。这两种癖好被许多中国文士当做日常的功课。于是纹理细腻、颜色透明鲜艳成为最重要之点，再后，又有人癖好玉石所雕的鼻烟壶，情形也是如此。一颗上好的石章或一个好的鼻烟壶，往往可以值到六七百块钱。

要充分领略石头在室内和园内的用处，我们须先研究一下中国书法。因中国书法专在抽象的笔势和结构上用工夫，好的石块，一方面固然应该近乎雄奇不俗，但其结构更为重要。所谓结构并不是要它具着匀称的直线形、圆形或三角形，而应是天然的拙皱，老子在他所著的《道德经》中常称赞不雕之璞。我们千万不可粉饰天然，因为最好的艺术结晶也和好的诗文一般须像流水行云的自然，如中国评论家所谓不露斧凿之痕。这一点可以适用于艺术的任何一方面。我们所领略的是不规则当中的美丽，结构玲珑活泼当中的美丽，富家书房中常爱设用老树根所雕成的凳子，即是出于这种领略的观念。因此，中国花园中的假山大多是用未经斧凿的石块所叠成，有时是用丈余高的英石峰，有时是用河里或山洞里的石块，都是玲珑剔透，极尽拙皱之态的。有一位作家主张：如若石中的窟窿恰是圆形的，则应另外拿些小石子粘堆上去以减少其整圆的轮廓。上海和苏州附近花园中的假山大都是用从太湖底里所掘起的石块叠成的，石上都有水波的纹理，有时取到的石块如若还不够嵌空玲珑，则用斧凿修琢之后，依旧沉入水中，待过一两年后，再取出来应用，以便水波将斧凿之痕洗刷净尽。

对于树木的领略是较为易解的，并且当然是很普遍的。房屋的四周如若没有树木，便觉得光秃秃的如男女不穿衣服一般。树木和房屋之间的分别，只在房屋是造成的，而树木是生长的，凡是天然生长出来的东西总比人力造成的更为好看。为了实用上的便利，我们不能不将墙造成直的，将每层房屋造成平的。但在楼板这件事上，一所房屋中同层各房间的地板，其实并没有必须在同一水平线上的理由。不过我们已不可避免地偏向直线和方形，而这种直线和方形非用树木来调剂便不美观。此外在颜色设计上，我们不敢将房屋漆成绿色，但大自然敢将树木漆成绿色。

艺术上的智慧在于隐匿艺术。我们都是太好自显本领，在这一点上我不能不佩服清代的阮元。他于巡抚浙江的任上，在杭州西湖中造了一个小屿，即后人所称的"阮公屿"。这屿上并没有什么建筑，连亭子碑柱等都没有，他在这件创作上完全抹去了个人。现在这阮公屿依然峙立在西湖的水中，是约有百码方圆的一方平地，高出水面不过尺余，地上所有的不过是青葱飘拂的柳树。如在一个烟雾迷离的日子去远望这屿，你便能看到它好似从水中冉冉上升，杨柳的影子映在水中，冲破了湖面的单调，而使它增加了风韵。所以这阮公屿是和大自然完全和谐的。它不像那美国留学生回国后所造的灯塔式纪念塔般，令人看了触眼。这纪念塔是我每看见一次便眼痛一次的，我曾公开地许愿，我如若有一天做了强盗头而占据杭州，我的第一件行动便是用大炮将这个纪念塔轰去。

在数千百种的树木中，中国名士和诗人觉得有几种的结构和轮廓由于从书法家的观点上着着种种特别的美处，所以尤其宜于艺术家的欣赏。这就是说，虽然凡是树木都是好看的，但其中某某几种更是具着特别的姿势或风韵。所以他们特把这几种树木另提出来，而将它们联系于各种的指定感情。例如：橄榄树不如松树的峥嵘，杨柳虽柔媚但并不雄奇。有少数几种树木是常见于画幅和诗歌中的，其中最杰出的，如松树的雄伟，梅树的清奇，竹树的纤细令人生家屋之感，杨柳的柔媚令人如对婀娜的美女。

对松树的欣赏，或许可算最惹人注意和最具着诗的意义，它比别的树更能表征行为高尚的概念。因为树木也有最高尚和不高尚之别，也有雄奇和平淡之别，所以中国艺术家常称美松树的雄伟，如马修·阿诺德（Mathew Arnold）称美古希腊诗人荷马的伟大。在树木之中，想向杨柳去求雄伟，其徒然无效正如在诗人之中想向斯威本（Swinburne）去求雄奇。

美丽的种类种种不一：如柔和之美、优雅之美、雄伟之美、壮严之美、古怪之美、粗拙之美、力量之美、古色古香之美。松树就因为具着这种古色古香的性质，所以在树木中得到特别的位置。正如隐居的高士，宽袍大袖、扶着竹杖在山径中行走，而被认为是人类的最高理想一般。李笠翁因此曾说，坐在一个满植杨柳桃花的园中，而近旁没有松树，就等于坐在儿童女子之间，而旁边没有一个可以就教的老者一般。中国人也为了这个理由，于爱松之中尤爱松之老者，越老越好，因为它们更其雄伟。和松树并立的是柏树，也是以雄奇见称。它的树枝都是弯曲虬缠而向下的，向上的树枝象征少年和热望，而向下的树枝象征俯视年轻人的老者的佝偻姿势。

我曾说过，松的可爱处是在艺术上意义更深长，因为它代表幽静雄伟和出世，正和隐士的态度相类。这个可爱处常和玩石、在松下徘徊的老人联系在一起，如在中国画中所见的一般。当一个人立在松树下向上望时，心中会生出它是何等苍老、在宁静的独立中何等快乐的感想。老子说，石块无言，苍老的松树也无言，只是静静沉着地立在那里俯视世界，好似觉得已经阅历过多少人事沧桑，像有智慧的老人一般无所不晓，不过从不说话。这就是它神秘伟大的地方。

梅树的可爱处在于枝干的奇致、梅花的芬芳。诗人于欣赏树木时，常以松、竹、梅为寒冬三杰而称之为"岁寒三友"。因为竹和松是长青树，而梅在冬末春初时开花，所以梅树特别象征品质的高洁，一种寒冷高爽中的纯洁。它的香味是一种冷香，天气越冷，它越有精神。它也和兰花一样表征幽静中的风韵。宋代的隐居诗人林和靖曾以"梅妻鹤子"自傲。遗迹现在依旧在西湖的孤山，他墓旁还有一座鹤冢，每年诗人和名士去凭吊者很多。梅树的姿态和芬芳的可爱处，中国有一句古诗描写最好。那句诗是：

暗香浮动月黄昏

后来的诗人都认为这七个字已经尽了梅花的美处，更不能有所增减。

人的爱竹，爱的是干叶的纤弱，因此植于家中更多享受。它的美处是一种微笑般的美处，所给我们的乐处是一种温和的乐趣。竹以瘦细稀疏为妙，因此两三株和一片竹林同样可爱，不论在园中或画上。因为竹的可爱处在纤瘦，所以画在画上时只须两三枝即已足够，正如画梅花只须画一枝。纤瘦的竹枝最宜配怪石，所以画竹时，旁边总画上几块皱瘦玲珑的石头。

垂杨柳极易于生长，河边岸上也可以种植，这树象征女性的绝色美丽。张潮即因此认垂杨柳为世上四种最感人的物事之一，而说："柳令人感。"中国美人的细腰，中国的舞女穿着长袖宽袍于舞时都模拟着柳枝在风中回旋往复的姿势。因为柳树极易生长，中国有许多地方数里之中遍地是柳，当阵风吹过之时，便能激起所谓"柳浪"。此外黄莺和蝉都最喜欢栖于柳树，图画中画到杨柳时，每每画上几只黄莺和蝉以为点缀，所以"西湖十景"中，有一处的名称即是"柳浪闻莺"。

此外当然还有许多种可爱的树木，如梧桐树因树皮洁净，可以用小刀刻画诗词，而为人所爱。也有人喜爱盘绕在树根或山石上的巨藤，它们回环盘绕，和大树的直干适成一种对比。有时这种巨藤很像一条龙，于是称它为"卧龙"，横斜弯曲的老树枝干，也因为这个理由为人所爱。苏州太湖边的木渎地方有四棵老柏，其名为"清""奇""古""怪"。"清柏"的干很直，上面的枝叶四面铺张开来的如同伞形。"奇柏"横卧地上，树干有三个弯曲如Z形。"古柏"光皮秃顶，伸着半枯的树枝同人的手指一样。"怪柏"自根而上树干扭绞如同螺旋一般。

最重要的是爱树木，不单是爱树木本身，而连带爱着其他的天然物事如：石、云、鸟、虫和人。张潮曾说："艺花可以邀蝶，累石可以邀云，栽松可以邀风，……种蕉可以邀雨，植柳可以邀蝉。"人于爱树木之中连带爱着树上的鸟声；爱石之中连带爱着石旁的蟋蟀声。因为鸟必在树上，蟋蟀必在石旁方肯鸣叫。中国人喜爱善鸣的蛙、蟋蟀和蝉，更胜于爱猫、狗或别种家畜。动物之中，只有鹤的品格配得上松树和梅花。因为鹤也是隐逸的象征，一个高人看见一只鹤，甚至一只鹭，白而洁净，傲然独立于池中时，他便会期望自己也化成一只鹤。

郑板桥在写给他弟弟的信中，有一段论到不应该将鸟儿关在笼中，最能表现出人类怎样去和大自然融合而得到快乐（因为动物都是快乐的）的思想：

所云不得笼中养鸟，而余又未尝不爱鸟，但养之有道耳。欲养鸟，莫如多种树，使绕屋数百株，扶疏茂密，为鸟国鸟家。将旦时，睡梦初醒，尚辗转在被，听一片啁啾，如《云门》《咸池》之奏。及披衣而起，面漱口啜茗，见其扬翚振彩，倏往倏来，目不暇给，固非一笼一羽之乐而已。大率平生乐处，欲以天地为囿，江溪为池，各适其天，斯为大快，比之盆鱼笼鸟，其钜细仁忍何如也！

IV. On Rocks and Trees

I don't know what we are going to do now. We are building houses square
and are building them in a row, and we are having straight roads without
trees. There are no more crooked streets, no more old houses, no more wells
in one's garden, and whatever private garden there is in the city is usually a
caricature. We have quite successfully shut nature out from our lives, and
we are living in houses without roofs, the roofs being the neglected end of a
building, left in any old shape after the utilitarian purposes have been served
and the building contractor is a little tired and in a hurry to get through
his job. The average building looks like wooden blocks built by a peevish
or fickle child who is tired of the game before he finishes building, and
leaves them unfinished and uncrowned. The spirit of Nature has left the
modern civilized man, and it seems to me we are trying to civilize the trees
themselves. If we ever remember to plant them on boulevards, we usually
number them serially, disinfect them, cut them and trim them to assume a
shape that we humans consider beautiful.

We often plant flowers and lay them out on a plot so that they resemble
either a circle, or a star, or different letters of the alphabet, and we are
horrified when some of the flowers so planted get out of line, as we are
horrified when we see a West Point cadet march out of step, and we proceed

to cut them down with scissors. And at Versailles, we plant these conically cut trees in pairs and arrange them with perfect symmetry along a perfectly round circle or in perfectly rectilinear rows in army formation. Such is human glory and power and our ability to train and discipline the trees as we train and discipline uniformed soldiers. If one tree of a pair grows taller than the other, our hands itch to cut off its top so as not to let it disturb our sense of symmetry and human power and glory.

There exists, therefore, the great problem of recovering nature and bringing nature back to the home. This is an exasperating problem. What can one do with the best artistic temperament, when one lives in an apartment and away from the soil? How is one going to have a plot of grass or a well or a bamboo grove even if he is rich enough to rent a penthouse? Everything is wrong, utterly and irretrievably wrong. What has one got left to admire except tall skyscrapers and lighted windows in a row at night? Looking at these skyscrapers and these lighted windows in a row at night, one gets more and more conceited about the power of human civilization and forgets what puny little creatures human beings are. I am therefore forced to give up the problem as hopeless of solution.

We must begin, therefore, by giving man land and plenty of it. No matter what the excuse, a civilization that deprives man of land is wrong. But suppose in a future civilization every man is able to own an acre of land, then he has got something to start with. He can have trees, his own trees, and rocks, his own rocks. He will be careful to choose a site where there are already full-grown trees, and if there are not already full-grown trees, he will plant trees that grow fast enough for him, such as bamboos and willows. Then he will not have to keep birds in cages, for birds will come to him and he will see to it that there are frogs in the neighborhood, and preferably also some lizards and spiders. His children will then be able to study nature in Nature and not study nature in a glass case. At least his children will be able to watch how chickens hatch from their eggs and they need not be woefully ignorant about sex and reproduction as the children of"good"Boston families often are. And they will have the pleasure of watching a fight between lizards and spiders. And they will have the pleasure also of getting comfortably dirty.

The Chinese sentiment for rocks has already been explained, or hinted at, in a previous section. That explanation sufficiently accounts for the love of rocky peaks in Chinese landscape painting. This explanation is basic, but it does not sufficiently account for Chinese rock gardens and the love of rocks in general. The basic idea is that rocks are enormous, strong and suggest eternity. They are silent, ummovable and have strength of character like great heroes, and they are independent and detached from life like retired scholars. They are invariably old, and the Chinese love whatever is old. Above all, from the artistic point of view they have grandeur, majesty, ruggedness, and quaintness. There is the further sentiment of wei, which means"dangerous"but is really untranslatable. A tall cliff that rises abruptly three hundred feet above the ground is always fascinating to look at because of its suggestion of"danger."

But then it is necessary to go further. As one cannot visit the mountains every day, it is necessary to have rocks brought to the home. In the case of rock gardens and artificial rock grottoes, a subject which is difficult for Western travelers in China to understand and appreciate, the idea is still to retain a suggestion of the rugged,"dangerous"and majestic lines of rocky peaks. Western travelers are not to blame because most of the rockeries are done with atrocious taste, and fail to convey the suggestion of natural grandeur and majesty. Artificial grottoes built out of several pieces of rock are usually cemented together, and the cement shows. A really artistic rockery should have the composition and contrast of a painting. There is no question that the artistic appreciation of artificial rock sceneries and that of mountain rocks in landscape painting are closely associated, as we find the Sung painter, Mi Fu, was the author of a book on ink-stones, and there was a book Shihp'u on rocks by a Sung author, Tu Kuan, giving detailed descriptions of the quality of over a hundred kinds of rocks produced at different places and used for rockeries, showing that rockeries were already a highly developed art in the time of the great Sung painters.

Side by side with this appreciation of the grandeur of rocks on mountain peaks, there developed then a different appreciation of rocks in gardens, emphasizing their color, texture, surface, grains and sometimes the sounds they produced when struck. The smaller the stones, the more emphasis was laid on quality of texture and color of grains. The development along this direction was greatly helped by the hobby of collecting the finest ink-stones and seals, two things which the literary man in China daily associated with. Daintiness, texture, light or translucence and shades of color became then of the first importance, as also in the case of stone, jade and jadeite snuff bottles, which came later. A good stone seal or a good snuff bottle could cost six or seven hundred dollars.

For the fullest appreciation of all uses of stone in the house and gardens, however, one has to go back to Chinese calligraphy. For calligraphy is nothing but a study of rhythm and line and composition in the abstract. While really good pieces of rock should suggest majesty or detachment from life, it is even more important that the lines be correct. By line one does not mean a straight line, or a circle or a triangle, but the rugged lines of nature. Laotse,"The Old Boy,"always emphasized in his Taotehching the uncarved rock. Let us not tamper with Nature, for the best work of art, like the best poem or literary composition, is one which shows no sign of human effort, as natural as a winding river or a sailing cloud, or as the Chinese literary critics always say,"without ax and chisel marks."This applies to every field of art. The appreciation is of beauty in irregularity, in lines that suggest rhythm and movement and gesture. The appreciation of the gnarled roots of an oak tree, sometimes used as stools in a rich man's studio, is based on the same idea. Consequently most of the rockeries found in Chinese gardens are uncut rocks, which may be the fossilized bark of a tree ten or fifteen feet high standing vertically alone and unmovable like a great man, or of rocks found in lakes and caves, generally bearing perforations and having the utmost irregularity of outline. One writer suggested that if the perforations happen to be perfectly round, some little pebble should be inserted to break up the regularity of the circle. Rockeries near Shanghai and Soochow are mostly built of rocks from the Taihu Lake, bearing marks of former sea waves. Such rocks were dug out of the bottom of the lake, and sometimes when something was needed to correct their lines, they would be chiseled until they were perfect and let down into the water again for a year or so, so that the chisel marks might be obliterated by the movement of water.

The feeling for trees is easier to understand, and is, of course, universal. Houses without trees around them are naked, like men and women without clothing. The difference between trees and houses is that houses are built but trees grow, and anything which grows is always more beautiful to look at than anything which is built. There are considerations of practical convenience which force us to build our walls straight and our stories level, although in the matter of floors, there is absolutely no reason why the foors of different rooms in a house should not be on different levels. Nevertheless, there is an inevitable tendency to go in for straight lines and square shapes, and such straight lines and square shapes can be brought into pleasurable relief only by the company of trees. In the color scheme, too, we dare not paint our houses green. But nature dares and has painted the trees green.

The wisdom of art consists in concealing art. We are so anxious to show off. In this respect, I must pay my tribute to a great scholar, Yüan Yüan, who as governor had an islet built in the water of the West Lake, known today as Governor Yüan's Islet, and who refused to put a single human edifice on the place, not a pavilion, not a pillar, not even a monument. He completely obliterated himself as an architect. Today the Governor Yüan's Islet stands in the middle of the lake, a level piece of land about a hundred yards across, rising barely a foot above the water and planted all around with willow trees. And today as you stand looking at it on a misty day, the magic island seems to rise out of the water, and the willow trees cast their reflections in the water, breaking the monotony of the lake's surface and harmonizing with it. Therefore Governor Yüan's Islet is in perfect harmony with nature. It is not obtrusive to the eye, like the lighthouse-shaped monument next to it built by a student returned from America, which gives me inflammation of the eyelids every time I look at it. I have made a public promise that if one day I should emerge as a bandit general and capture Hangchow, my first official act would be to direct a cannon and blow that lighthouse-shaped thing to pieces.

Out of the myriad variety of trees, Chinese critics and poets have come to feel that there are a few which are particularly good for artistic enjoyment, due to their special lines and contours which are aesthetically beautiful from a calligrapher's point of view. The point is, that while all trees are beautiful, certain trees have a particular gesture or strength or gracefulness. These trees are therefore picked out from among the others and associated with definite sentiments. It is clear that an ordinary olive tree has no rugged manner, for which we go to the pine, and while a willow is graceful, it can never be said to be"majestic" or"inspiring."There are then a small number of trees which are more constantly painted in paintings and sung about in poems. Of these the most outstanding are the pine tree, enjoyed for its grand manner, the plum tree, enjoyed for its romantic manner, the bamboo tree, enjoyed for its delicacy of line and the suggestion of the home, and the willow tree, enjoyed for its gracefulness and its suggestion of slender women.

The enjoyment of the pine tree is probably most notable and of the greatest poetic significance. It typifies better than other trees the conception of nobility of manner. For there are trees noble and trees ignoble, trees distinguished for their grand manner and trees of the common manner. The Chinese artists therefore speak of the grand old manner of the pine tree, as Matthew Arnold spoke of the grand manner of Homer. It would be as hopeless to look for this grand manner in willows among the trees, as it is to look for the grand manner of poetry in Swinburne among the poets.There are so many kinds of beauty, beauty of tenderness, of gracefulness, of majesty, of austerity, of quaintness, of ruggedness, of sheer strength, and of a suggestion of the antique. It is this antique manner of the pine tree that gives the pine a special position among the trees, as it is the antique manner of a recluse scholar, clad in a loose-fitting gown, holding a bamboo cane and walking on a mountain path, that sets him off as the highest ideal among men. For this reason Li Liweng says that to sit in an orchard full of peach trees and flowers and willows without a pine nearby is like sitting in the company of young children and women without the presence of an austere master or old man, whom we can look up to. It is also for this reason that when Chinese admire pine trees, they go in for the old ones; the older the better, for then they become more majestic. Classed with the pine tree is the cypress which has the same manner, with twisting, encircling and ruggedly downward-pointing branches. While branches that stretch upwards toward heaven seem to symbolize youth and aspiration, downward-pointing branches seem to symbolize the posture of old men bending down toward youth.

I say the enjoyment of pine trees is artistically most significant, because it represents silence and majesty and detachment from life which are so similar to the manner of the recluse. This enjoyment is then associated with"stupid"rocks and with figures of old people loitering around underneath its shade, as we so often see in Chinese paintings. As one stands there beneath a pine tree, he looks up to it with a sense of its majesty and its old age, and its strange happiness in its own independence. Laotse says,"Nature does not talk,"nor does the old pine tree. There it stands silent and imperturbable; from its height it looks down upon us, thinking it has seen so many children grow up into maturity and so many middle-aged people pass on to old age. Likewise, old men, it understands everything, but it does not talk, and therein lie its mystery and grandeur.

The plum tree is enjoyed partly for its romantic manner in its branches, and partly for the fragrance in its flowers. It is curious that among the trees selected for our poetic enjoyment, the pine tree, the plum tree and the bamboo are associated with winter, being known as the"Three Friends of Winter,"for the bamboo tree and the pine tree are evergreens, while the plum tree blossoms at the end of winter and the beginning of spring. The plum tree, therefore, in particular, symbolizes purity of character, the purity that we find in the crisp, cold winter air. Its splendor is a cold splendor, and like the recluse, the cooler the atmosphere it finds itself in, the better it prospers. Like the orchid flower, it typifies the idea of charm in seclusion. A Sung poet and recluse, Lin Hoching, declared that he had married plum trees as his wives, and had a stork for his son. Today the site of his seclusion on the Kushan in the middle of the West Lake is an object of pilgrimage for poets and scholars, and below his tomb is the tomb of the stork, his"son."Now the appreciation of the plum tree, of its type of fragrance and its outline, is best expressed by this poet in his famous line of seven words:

An hsiang fou tung yin heng sheh

"Its dim fragrance floats around, its shadow leaning across."It is admitted by all poets that the essence of the beauty of the plum is expressed in those seven words and cannot be improved upon.

The bamboo tree is loved for its delicacy of trunk and leaves, and being more delicate, it is more enjoyed in the intimacy of a scholar's home. Its beauty is more a kind of smiling beauty and the happiness it gives us is mild and temperate. Bamboos are best enjoyed when they are thin and slender and sparse, and for this reason two or three trees are as good as a whole bamboo grove, either in life or in painting. The appreciation of its slender outlines makes it possible to paint just two or three twigs of bamboo in a picture, as it is also possible to paint a single twig of plum flowers. Somehow its slender lines go very well with the rugged lines of rocks, and hence one finds always one or two rocks painted along with a few bamboos. Such rocks are invariably painted as having the beauty of Slenderness.

The willow grows easily anywhere and often on a bank. It is the feminine tree par excellence. That is why Chang Ch'ao counts the willow among the four things in the universe which touch man's heart most profoundly, and why he says the willow tree makes a man sentimental. Chinese ladies of slender waist are said to have"willow waists,"and Chinese female dancers, with their long sleeves and their flowing robe, try to simulate the movement of willow branches swaying and bowing in the wind. As willows grow most easily, there are places in China where willows are planted for miles around and then when a wind blows over them, the effect of the combination is spoken of as"willow-waves,"or liulang. Furthermore, as orioles love to perch on their hanging branches, they are associated in pictures or in life with the presence of orioles, or with cicadas which also love to rest there. One of the ten scenic spots of the West Lake is therefore called Liulang Wen Ying, or"Listening to Orioles among Willow-Waves."

There are of course other trees, and a good number of them admired for other reasons, like the wut'ung (Sterculia platanifolia), admired for the cleanliness of its bark and the possibility of carving poems on its smooth surface with a knife. There is also great love of gigantic old creepers, two or three inches across at their roots, encircling old trees or rocks. Their encircling and undulating movement contrasts pleasurably with the straight trunks of erect trees. Sometimes a particularly good creeper suggests a sleeping dragon and is given that name. Old trees that have zigzag and more or less sloping trunks are also greatly loved and valued for this reason. At Mutu, a point on the Taihu Lake near Soochow, there are four such cypress trees which have been given the four respective names"Pure,""Rare,""An tique"and"Quaint.""Pure"goes up by a long, straight trunk, spreading out a foliage on top resembling an umbrella;"Rare"crouches on the ground and rolls along in three zigzag bands like the letter"Z";"Antique"is bald and bare at the top and broad and stumpy, with its straggling limbs half dried up and resembling a man's fingers; and"Quaint's"trunk twists around in spiral formation all the way up to its highest branches.

Above all, the enjoyment of trees is not only in and for themselves, but in association with other elements of nature, such as rocks, clouds, birds, insects and human beings. Chang Ch'ao says that"planting flowers serves to invite butterflies, piling up rocks serves to invite clouds, planting pine trees serves to invite the wind,... planting banana trees serves to invite the rain, and planting willow trees serves to invite the cicada."One enjoys the sounds of birds along with the trees, and enjoys the sounds of crickets along with the rocks, for birds sing where trees are, and crickets sing where rocks are found. The Chinese enjoyment of croaking frogs, chirping crickets and intoning cicadas is immeasurably greater than their love of cats and dogs and other animal pets. Among all the animals, the only one which belongs in the same category with pine trees and plum trees is the stork, because he, too, is the symbol of the recluse. As one sees a stork, or even a heron, standing motionless in the marshes of some secluded pond, dignified, elegant and white and pure, the scholar wishes that he were a stork himself.

The final picture of man harmonizing with nature and happy because the animals are happy is best expressed by Cheng Panch'iao (1693-1765) in his letter to his younger brother, pointing out his disapproval of keeping birds in cages:

In regard to what I said about not keeping birds in cages, I wish to add that it isn't that I don't love birds, but there is a proper way of loving them. The best way of keeping birds is to plant hundreds of trees around the house, and let them find in their green shade a bird kingdom and bird homes. So then, at dawn, when we have waked up from sleep and are still tossing about in bed, we hear a chorus of chirping songs like a celestial symphony. When we have got up and put on our gowns and are washing our faces or gargling our mouths or sipping the morning tea, we see their gorgeous plumes flitting to and fro, and before we have time to look at one, our eyes are attracted by another—

an enjoyment that is not to be compared with looking at a single bird in a single cage. Generally the enjoyment of life should come from a view regarding the universe as a park and the rivers and lakes as a pond, so that all beings can live according to their nature, and great indeed is such happiness! How does this compare in kindness and cruelty and in the magnitude of enjoyment with the enjoyment of a bird in a cage, or of a fish in a jar!

五．论花和折枝花

　　现在的人对于花和插花的爱好这件事，似乎都出以不经意。其实呢，要享受花草也和享受树木一般，须先下一番选择工夫，分别品格的高低，而配以天然的季节和景物。就拿香味这一端讲起来，香味很烈的如茉莉，较文静的如紫丁香，最文静细致的如兰花。中国人认为花的香味越文静的品格越高。再拿颜色来讲，深浅也种种不一。有许多浓艳如少妇，有许多淡雅如闺中的处女，有许多似乎是专供大众欣赏的，而另有些幽香自怡，不媚凡俗。有许多以鲜艳见长，有许多则以淡雅显高。最重要的是：凡是花木都和开花的季节和景物有连带特性。例如我们提到玫瑰时，便自然想到风清日和的春天；提到荷花时，便想到风凉夏早的池边；提到桂花时，便想到秋高气爽的中秋月圆时节；提到菊花时，便想到深秋对菊持螯吃蟹时的景物；提到梅花时，便想到冬日的瑞雪；联想到水仙花，便会使我们想到新年的快乐景色。每一种花似乎都和开花时的环境完全融洽和谐，使人易于记忆什么花代表什么季节，如同冬青树代表圣诞节一般。

　　兰花、菊花和莲花也如松竹一般，具着某种特别的品质而特别为人所重视，在中国的文学中视之为高人的象征。其中兰花更因为具着一种特式的美丽而为人所敬爱。中国诗人于花中最爱梅花，这一点上文中已经有所说明，称之为"花魁"，因为梅花开于新年，正是一年之先。但是各人的意见当然也有不同的，所以有许多人则尊牡丹为"花王"，尤其是在唐代。另一方面说起来，牡丹以浓艳见长，所以象征富贵，而梅花以清瘦见长，所以象征隐逸清苦。因此，前者是物质的，而后者是精神的。中国有一位文人极推崇牡丹，原因是当唐朝武则天临朝的时代，她一时忽发狂兴，诏谕苑中百花必须冬月的某天一齐开放，百花都不敢不按时开放，唯有牡丹独违圣旨，比规定的时刻迟了数小时方始开花。因此触了武则天的怒，而下诏将苑中几百盆牡丹一起从京都（西安）贬到洛阳去。从此牡丹失去了恩宠，但其种未绝，以后盛于洛阳。我以为中国人不很重视玫瑰花的理由，大概是玫瑰和牡丹同其浓艳，而不及牡丹富丽堂皇，所以被抑在下的。据中国旧书的说法，牡丹共有九十种，各有一个诗意的特别名称。

兰花和牡丹的品格截然不同，是幽雅的象征，因为兰花是常生于幽谷的。文人称它具有"孤芳独赏"的美德，它从不取媚于人，也不愿移居城市之中，即使移植了，灌溉看顾也须特别当心，否则立刻枯死。所以中国书中常称深闺的美女和隐居山僻不求名利的高人为"空谷幽兰"。兰花的香味是如此文静，它不求取悦于人，但能领略的人就知道它的香味何等高洁啊！这使它成为不求斗于世的高人和真正友谊的象征。有一本古书上说："久居芝兰之室，则不闻其香。"就因为这人的鼻已充满了花香了。依李笠翁的说法，兰花不宜遍置各处，而只宜限于一室，方能于进出之时欣赏其幽香。美国的兰花形式较大，颜色较为富丽，但似乎没有这种文静的香味。我的家乡福建是中国有名的"建兰"产地。建兰的花瓣较小，长只一寸，颜色淡绿，种在紫砂盆中，异常好看。最著名的一种名叫"郑孟良"，颜色和水差不多，浸在水中竟可花水不分。牡丹都以产地为名，而兰花都以从前种它的高人为名，如美国花草之以种者的名字为名一般，例如李司马、黄八哥之类。

无疑，兰花的难于种植和异常文静的香味，使它得到高贵的身价。各种花木中以兰花最为娇嫩，稍不经心，便会枯死。所以爱艺兰者都是亲手灌溉整理，不肯假手于仆役，我曾看见过爱兰花者之专心护视不亚于人之爱护其父母。奇花异卉也如稀有的金石古玩一般，在占有上很易引起同好者的妒忌。例如向人索取枝芽而被拒绝者，每每会变成极端的仇恨。中国某种笔记中，曾载某人向他的朋友索取一种奇的枝芽未能如愿，即下手偷窃，因此被控获罪。对于这种情形，沈复在他所著的《浮生六记》中有一段极好的描写：

> 花以兰为最，取其幽香韵致也，而瓣品之稍堪入谱者不可多得。兰坡临终时，赠余荷瓣素心春兰一盆，皆肩平心阔，茎细瓣净，可以入谱者。余珍如拱璧。值余幕游于外，芸能亲为灌溉，花叶颇茂。不二年，一日忽萎死。起根视之，皆白如玉，且兰芽勃然。初不可解，以为无福消受，浩叹而已。事后始悉有人欲分不允，故用滚汤灌杀也。从此誓不植兰。

正如梅花是诗人林和靖的爱物，莲花是儒家周莲溪的爱物一般，菊花是诗人陶渊明的爱物。菊花开于深秋，所以也具冷香色之誉。菊花之冷色和牡丹的浓艳，极容易分辨。菊花的种类甚多，据我所知，宋代名士范成大是赐以菊花各种美名的始创者。种类的繁多似乎是菊的特色，其花形和颜色的种类多到不胜数计。白和黄色的是花的正宗，紫和红色的为花的变体，所以品格即次。白色和黄色的菊花有银盏、银玲、金玲、玉盆、玉玲、玉绣球等美称，也有用古代美人的名字如杨贵妃和西施之类的。花的形式有时如时髦女人的鬓发，有时如少女头上一绺一绺的长发。花的香味也各有不同，以含有麝香味或龙脑香味者为最上。

湖莲自成一种，且据我看来，是花中之最美者。消夏而没有莲花，实不能称为美满。如若屋旁没有种荷的池子，则可以将它种在大缸中，不过这种方法缺少了一片连绵、花叶交映、露滴花开、芳香裹里的佳景（美国的水莲和中国的荷花不同）。宋代名士周莲溪著文解释他爱莲的理由，并说莲花是"出淤泥而不染"，所以可比之为贤人，这完全是儒家的口气。再从实用方面讲起来，这花从顶到根，没有一样是废物。莲根即藕，是绝佳的水果；荷叶可用以包食物；花可供人赏玩；子即莲子，尤其是食物中的仙品，可以新鲜时生吃，或晒干后煮了吃。

海棠花的式样和苹果花很有些相像，也是诗人所爱的花之一。这花虽盛产于杜甫的故乡四川，但他的诗中恰一字不曾提过。这件事很奇怪，猜测之说很多，其中以杜甫的母亲名海棠，所以他避讳一说最为近情。

我以为兰花之外，香味最佳者是桂花和水仙花。这水仙盛产于我的故乡漳州，从前曾大批贩运至美国，但后来因美国国务院说这种过于芬芳的花或有滋生细菌的可能，因而突然禁止进口。水仙的茎和根部都洁净如翠玉，况且种在盆中只用水和石子而不用泥，极为清洁，在这情形之下，何以能滋生细菌？所以这种禁令，实令人莫测高深。杜鹃花虽极美丽，但人都称之为凄凉的花。因为据说从前一个人走遍天下去找寻被后母所逐出的哥哥，但终究未能寻到，死后化为杜鹃终日泣血，而这杜鹃花就是从杜鹃的血泪中所生出来的。

折花插瓶一事，其郑重也和品第花的本身差不多。这种艺术远在十一世纪中即有普遍的发展，十九世纪《浮生六记》的作者沈复在"闲情记趣"中曾论到插花的艺术，插花适当，可以使之美如图画。

> 唯每年篱东菊绽，积兴成癖，喜摘插瓶，不爱盆玩。非盆玩不足观，以家无园圃，不能自植；货于市者，俱丛杂无致，故不取耳。其插花朵，数宜单，不宜双。每瓶取一种，不取二色。瓶口取阔大，不取窄小，阔大者舒展。不拘自五七花三四十花，必于瓶口中一丛怒起，以不散漫，不挤轧，不靠瓶口为妙；所谓"起把宜紧"也。或亭亭玉立，或飞舞横斜。花取参差，间以花蕊，以免飞钹耍盘之病。叶取不乱，梗取不强。用针宜藏，针长宁断之，毋令针针露梗；所谓"瓶口宜清"也。视桌之大小，一桌三瓶至七瓶而止，多则眉目不分，即同市井之菊屏矣。几之高低，自三四寸至二尺五六寸而止，必须参差高下，互相照应，以气势联络为止。若中高两低，后高前低，成排对列，又犯俗所为"锦灰堆"矣。或密或疏，或进或出，全在会心者得画意乃可。

> 若盆碗盘洗，用漂青、松香、榆皮面和油，先熬以稻灰，收成胶，以铜片按钉向上，将膏火化，粘铜片和盘盆碗洗中。俟冷，将花用铁丝扎把，插于钉上，宜偏斜取势，不可居中，更宜枝疏叶清，不可拥挤；然后加水，用碗沙小许掩铜片，使观者疑丛花生于碗底方妙。若以木本花果插瓶，剪裁之法（不能色色自见，倩人举折者每不合意），必先执在手中，横斜以观其势，反侧取其态，相定之后，剪去杂枝，以疏瘦古怪为佳。再

思其梗如何入瓶，或折或曲，斜入瓶口，方免背叶侧花之患，若一枝到手，先拘其梗之直者插瓶中，势必枝乱梗强，花侧叶背，既难取态，更无韵致矣。折梗打曲之法，锯其梗之半而嵌以砖石，则直者曲矣。如患梗倒，敲一两钉以菀之。即枫叶竹枝，乱草荆棘，均堪入选，或绿竹一竿，配以枸杞数粒；几茎细草，伴以荆棘两枝；苟位置得宜，另有世外之趣。

V. On Flowers and Flower Arrangements

THERE seems to be a certain randomness about the enjoyment of flowers and flower arrangements, as we know it today. The enjoyment of flowers, like the enjoyment of trees, must begin with the selection of certain noble varieties, with a sense of grading of their relative standing, and with the association of definite sentiments and surroundings with definite flowers. To begin with, there is the matter of fragrance, from the strong and obvious, like that of jasmine, to the delicate, like that of lilac, and finally to the most refined and subtle kind, like that of the Chinese orchid. The more subtle and less easily perceivable its fragrance, the more noble the flower may be regarded. Then there is the matter of color and appearance and charm, which again varies a great deal. Some are like buxom lassies and others are like slender, poetic, quiet ladies. Some seem to pander their charms to the crowd, and others are happy in their own fragrant being and seem contented merely to dream their hours away. Some go in for a dash of color, while others have a milder and more restrained taste. Above all, flowers are always associated with the outward surroundings and seasons of their bloom. The rose is naturally associated in our minds with a bright sunny spring day; the lotus is naturally associated with a cool summer morning on a pond; the cassia is naturally associated with the harvest moon and mid-autumn festivities; the chrysanthemum is associated with the eating of crabs in late autumn; the plum is naturally associated with snow and together with the narcissus it forms a definite part of our enjoyment of the New Year. Each seems perfect in its own natural surroundings, and it is the easiest of all things for lovers of flowers to make them stand in our mind for definite pictures of the different seasons, as the holly stands for Christmas.

Like the pine tree and the bamboo, the orchid, the chrysanthemum and the lotus are selected for certain definite qualities and stand in Chinese literature as symbols for the gentleman, the orchid more particularly for an exotic beauty. The plum flower is probably most beloved by Chinese poets among all flowers, and has been partly dealt with already in the previous section. It is said to be the"first"among the flowers, because it comes with the New Year and therefore stands first in the procession of flowers in the course of the year. Opinions differ, of course, and the peony has been traditionally regarded as the"king of flowers,"particularly in the T'ang Dynasty. On the other hand, the peony being rich in its colors and its petals, is rather regarded as the symbol of the rich and happy man, whereas the plum flower is the poet's flower, and symbol of the quiet, poor scholar, and therefore the latter is spiritual as the former is materialistic. One scholar voiced his sympathy for the peony only because of the fact that, when Empress Wu of the T'ang Dynasty commanded one day, in one of her megalomaniac whims, that all the flowers in the imperial garden should bloom on a certain day in mid-winter, just because she wanted it, the peony was the only one that dared to offend her Imperial Majesty by blooming a few hours late, and consequently, all the thousands of pots of peony flowers were banished by imperial decree from Sian, the capital, to Loyang. Although falling out of imperial favor, the cult of the peony was still maintained and Loyang became a center for peony flowers. I think the reason that the Chinese do not place more importance on the rose is because its color and shape belong in the same class with the peony, but have been overshadowed by the latter's gorgeousness.

According to early Chinese sources, there were ninety varieties of the peony distinguished, and each was given a most poetic name.

Unlike the peony, the orchid stands as the symbol of secluded charm because it is often found in a deserted shady valley. It is said to have the virtue of"enjoying its own lonely charm,"not caring whether people look at it or not, and extremely unwilling to be moved into the city. If it consents to be moved, it must be cultivated on its own terms, or it dies. Hence we often speak of a beautiful secluded maiden, or a great scholar living away in the mountains with contempt for power and fame, as"a secluded orchid in a deserted valley."Its fragrance is so subtle that it doesn't seem to make a particular effort to please anybody, but when people do appreciate it, how divine is its fragrance! This makes it a symbol for the gentleman not caring to cater to the public, and also for true friendship, because an ancient book says,"After entering and remaining in a house with orchids for a long time, one ceases to feel the fragrance,"when he himself is permeated with it. Li Liweng advised that the best way to enjoy orchids was not to place them in all rooms, but only in one room and then to enjoy the fragrance when passing out and in. American orchids do not seem to have this subtle fragrance, but on the other hand, are bigger and more gorgeous in shape and color. In my native city and province, we are supposed to have the best orchids in China, known as"Fukien orchids."The flower is pale green with spots of purple and is of a very much smaller size, the petals being slightly over an inch long. The best and most highly valued variety, the Ch'en Mengliang, has such a color that it is barely visible when immersed in water, being of the same color as the water itself. Unlike the peony, whose varieties are known after their place of origin, the different famous varieties of the orchid are known, like many American flower varieties, after their owners, as"General P'u, ""Quartermaster Shun,""Judge Li,""Eighth Brother Huang,""Ch'en Mengliang, ""Hsü Chingch'u. "

There is no question that the extreme difficulty of cultivating orchids and the flower's extreme delicacy of health contributed to the idea of its nobility of character. Among all the flowers, the orchid is the one that most easily withers or rots away with the slightest mishandling. Hence an orchid-lover always attends to it with his personal care and does not leave it to the servants, and I have seen people caring for their orchids like their own parents. An extremely valuable plant aroused as much jealousy as a particularly good piece of bronze or vase, and hatred from a friend's refusal to give away its new offshoots could be extremely bitter. Chinese notebooks record the case of a scholar who was refused new offshoots from a plant and was sentenced to jail for stealing it. This sentiment is well expressed by Shen Fu in Six Chapters of a Floating Life in the following manner:

The orchid was prized most among all the flowers because of its subdued fragrance and graceful charm, but it was difficult to obtain really good classic varieties. When Lanp'o died, he presented me with a pot of spring orchids, whose flowers had lotus-shaped petals; the centre of the flowers was broad and white, the petals were very neat and even at the "shoulders," and the stems were very slender. This type was classical, and I prized it like a piece of old jade. When I was working away from home, Yün used to take care of it personally and it grew beautifully. After two years, it died suddenly one day. I dug up its roots and found that they were white like marble, while nothing was wrong with the sprouts, either. At first, I could not understand this, but ascribed it with a sigh merely to my own bad luck, which might be unworthy to raise such flowers. Later on, I found out that some one had asked for some of the flowers from the same pot, had been refused, and had therefore killed it by pouring boiling water over it. Thenceforth I swore I would never grow orchids again.

The chrysanthemum is the flower of the poet T'ao Yüanming, as the plum flower was the flower of the poet Lin Hoching, and the lotus was the flower of the Confucian doctrinaire, Chou Liench'i. Blooming in late autumn, it shares the idea of "cold fragrance" and "cold splendor." The contrast between the cold splendor of the chrysanthemum and the gorgeous splendor, say, of the peony is easily seen and understood. Hundreds of varieties exist, and so far as I know, a great Sung scholar, Fan Ch'engta, started the fashion of recording its different varieties with the most beautiful names. Variety seems to be the very essence of the chrysanthemum flower, both variety of shape and of color. The white and the yellow are regarded as the"orthodox"colors of the flower, while purple and red are regarded as deviations and therefore given a low grading. The colors of white and yellow gave rise to the names of the varieties like"Silver Bowl,""Silver Bells,""Golden Bells,""Jade Basin,""Jade Bells,""Jade Embroidered Ball."Some were given the names of famous beauties, like"Yang Kueifei"and"Hsishih."Sometimes their shapes resemble a lady's close-cropped hair and sometimes their quills resemble flowing locks. Some varieties have more fragrance than others, and the best are supposed to have the fragrance of musk, or of an incense called"Dragon's Brains."

The lotus or water lily is in a class by itself and seems to me personally the most beautiful of all flowers, when we consider the flower, including its stem and its leaves floating on the water, as a whole. It is impossible to enjoy summer without having lotus flowers around, and if one does not have a house near a pond, he can grow them in big earthen jars. In this case, however, we miss much of the beauty of a half a mile's stretch of lotus flowers, their perfume pervading the air, and their white and red tipped blossoms contrasting with their broad green leaves with water running on them like liquid pearls. (The American water lilies are different from the lotus.) The Sung scholar Chou wrote an essay explaining why he loved the lotus and pointing out that the lotus, like the gentleman, grew out of dirty water but was not contaminated by it. He was talking like a regular Confucian doctrinaire. From the utilitarian point of view, every part of the flower is utilized. The lotus root is used to make a cooling drink, its leaves are used for wrapping fruits or food to be steamed, its flowers are enjoyed for their shape and fragrance, and finally the lotus seed is regarded as the food of the fairies, either eaten raw, fresh from the pod, or dried and sugared.

The hait'ang pyrus, resembling apple-blossoms, enjoys as great a popularity among poets as any other flower, although Tu Fu failed to make a single mention of this flower which grew in his native province, Szechuen. Various explanations have been offered, but the most plausible one was that the hait'ang was his mother's name and he had avoided it out of deference to his mother. There are only two flowers for whose fragrance I am willing to forego the orchid, and they are the cassia and the narcissus. The last is also a special product of my native city, Changchow, and its import into the United States in the form of cultivated roots at one time ran to hundreds of thousands of dollars, until the Department of Agriculture saw fit to deprive the American people of this flower with a heavenly fragrance, in order to protect them from possible germs. The notion that the white roots of the narcissus, as clean as a fairy itself, and intended to be planted not in mud but in a glass or china basin of water supported with pebbles, and prepared with the utmost care,   could contain germs is most fantastic. The azalea is supposed to be a tragic flower, in spite of its smiling beauty, because it was supposed to spring from the tears of blood of the cuckoo, who was formerly a boy in search of his lost brother persecuted out of home by a stepmother.

Quite as important as the selection and grading of the flowers themselves is their arrangement in vases. This was an art that could be traced back at least as far as the eleventh century. The author of Six Chapters of a Floating Life, written at the beginning of the nineteenth century, gives a description of the art of arranging flowers to resemble a picture with good composition in his chapter on"The Little Pleasures of Life":

The chrysanthemum, however, was my passion in the autumn of every year. I loved to arrange these flowers in vases, but not to raise them in pots, not because I did not want to have them that way, but because I had no garden in my home and could not take care of them myself. Those I bought at the market were not properly trained and not to my liking. When arranging chrysanthemum flowers in vases, one should take an odd, not even number, and each vase should have flowers of only one colour. The mouth of the vase should be broad so that the flowers can lie easily together. Whether there be half a dozen flowers or even thirty or forty of them in a vase, they should be so arranged as to come up together straight from the mouth of the vase, neither overcrowded, nor too much spread out, nor leaning against the mouth of the vase. This is called"keeping the handle firm."Sometimes they can stand gracefully erect, and sometimes spread out in different directions. In order to avoid a bare monotonous effect, they should be mixed with some flower buds and arranged in a kind of studied disorderliness. The leaves should not be too thick and the stems should not be too stiff. In using pins to hold the stems up, one should break the long pins off, rather than expose them. This is called"keeping the mouth of the vase clear."Place from three to seven vases on a table, depending on the size of the latter, for if there were too many of them, they would be overcrowded, looking like chrysanthemum screens at the market. The stands for the vases should be of different height, from three or four inches to two and a half feet, so that the different vases at different heights would balance one another and belong intimately to one another as in a picture with unity of composition. To put one vase high in the centre with two low at the sides, or to put a low one in front and a tall one behind, or to arrange them in symmetrical pairs, would be to create what is vulgarly called"a heap of gorgeous refuse." Proper spacing and arrangement must depend on the individual's understanding of pictorial composition.

In the case of flower bowls or open dishes, the method of making a support for the flowers is to mix refined resin with elm bark, flour and oil, and heat the mixture with hot hay ashes until it becomes a kind of glue, and with it glue some nails upside down on to a piece of copper. This copper plate can then be heated up and glued on to the bottom of the bowl or dish. When it is cold, tie the flowers in groups by means of wire and stick them on those nails. The flowers should be allowed to incline sideways and not shoot up from the centre; it is also important that the stems and leaves should not come too closely together. After this is done, put some water in the bowl and cover up the copper support with some clean sand, so that the flowers will seem to grow directly from the bottom of the bowl.

When picking branches from flower trees for decoration in vases, it is important to know how to trim them before putting them in the vase, for one cannot always go and pick them oneself, and those picked by others are often unsatisfactory. Hold the branch in your hand and turn it back and forth in different ways in order to see how it lies most expressively. After one has made up one's mind about it, lop off the superfluous branches, with the idea of making the twig look thin and sparse and quaintly beautiful. Next think how the stem is going to lie in the vase and with what kind of bend, so that when it is put there, the leaves and flowers can be shown to the best advantage. If one just takes any old branch in hand, chooses a straight section and puts it in the vase, the consequence will be that the stem will be too stiff, the branches will be too close together and the flowers and leaves will be turned in the wrong direction, devoid of all charm and expression. To make a straight twig crooked, cut a mark half way across the stem and insert a little piece of broken brick or stone at the joint; the straight branch will then become a bent one. In case the stem is too weak, put one or two pins to strengthen it. By means of this method, even maple leaves and bamboo twigs or even ordinary blades of grass and thistles will look very well for decoration. Put a twig of green bamboo side by side with a few berries of Chinese matrimony vine or arrange some fine blades of grass together with some branches of thistle. They will look quite poetic, if the arrangement is correct.

六．袁中郎的瓶花

关于折花插瓶一事，十六世纪中的作家袁中郎在他的著作中讨论得最为透彻。他所著的《瓶史》极为日本人所爱好，因此日本有所谓"袁派"插花家。他在这书的小引中说："夫山水花木者，名之所不在，奔竞之所不至也，天下之人，栖止于嚣崖利薮，目眯尘沙，心疲计算，欲有之而有所不暇，故幽人韵士得以乘间而踞为一日之有。"他又说明瓶花之乐不得"狃以为常"，它不过是居于城市者的"暂时快心事"，而不可"忘山水之大乐"。

他在《瓶史》中提及书房中插花为饰时所应该留意之处，并说胡乱插供，不如无花。最后则论及插花的各种铜瓶和瓷瓶，花瓶可以分两类：凡富贵之家有汉代大铜瓶和大厅堂者，应供高大的花卉；寻常的韵士则应用小瓶，供小枝。但在选择上须下功夫。花中唯牡丹和莲花必须用大瓶插供。

对于插花一事，他说：

插花不可太繁，亦不可太瘦，多不过二种三种。高低疏密，如画苑布置方妙。置瓶忌两对，忌一律，忌成行列，忌以绳束缚；夫花之所谓整齐者，正以参差不论，意态天然。如子瞻之文，随意断续；青莲之诗不拘对偶，此真整齐也。若夫枝叶相当，红白相配，此省曹墀下树，墓门华表也，恶得为整齐哉？

室中天然几一，藤床一，几宜阔厚，宜细滑，凡本地边栏漆桌描金螺钿床，乃彩花瓶架之类，皆置不用。

又对于浴花和浇花一事，他所说的话极能道出花的性情和精神：

夫花有喜怒寤寐，晓夕浴花者，得其候，乃为膏雨。澹云薄日，夕阳佳月，花之晓也。狂号连雨，烈焰浓寒，花之夕也。唇檀拱目，媚体藏风，花之喜也。晕酣神敛，烟色迷离，花之愁也。欹枝困槛，如不胜风，花之梦也。嫣然流盼，光华溢目，花之醒也。晓则空亭大厦，昏则曲房奥室，愁则屏气危坐，喜则欢呼调笑，梦则垂帘下帷，醒则分膏理泽，所以悦其性情，时其起居也。浴晓者上也，浴寐者次也，浴喜者下也。若夫浴夕浴愁，直花刑耳，又何取焉。

浴之之法，用泉甘而清者，细微浇注，如微雨解醒，清露润甲，不可以手触花，及指尖折剔，亦不可付之庸奴猥婢。浴梅宜隐士，浴海棠宜韵客，浴牡丹芍药宜靓妆少女，浴榴宜艳色婢，浴木樨宜清慧儿，浴莲宜娇媚妾，浴菊宜好古而奇者，浴腊梅宜清瘦僧。然寒花性不耐浴，当以轻绡护之。

据袁氏的说法，凡插瓶的花，某种须和着插，如婢之配主。因为中国自古以来，大人家的主妇必有一个终身服侍的侍婢，所以就产生了美丽的主妇如有艳婢在旁为配，便更加出色的观念。主婢都宜娇美，但何者是属于主妇式的美，何者是属于侍婢式的美，连我自己亦说不出。主婢如若不相配称，其触目难看等于披屋和正屋的不相称。将这个观念引用到花上，袁氏以为在瓶花的配侍上，梅花宜以山茶为婢，海棠宜以苹婆丁香为婢，牡丹宜以玫瑰木香为婢，芍药宜以罂粟蜀葵为婢，石榴宜以紫薇大红千叶木槿为婢，莲花宜以玉簪为婢，木樨宜以芙蓉为婢，菊花宜以秋海棠为婢，蜡梅宜以水仙为婢。婢也各自具着她自己的姿态，种类不同，正和她们的主妇一般。她们的名称虽是婢，但当中并没有轻视的意思。她们都被比做历史上有名的侍婢，如：水仙神骨清绝，是织女的梁玉清；山茶玫瑰明艳，是石崇的翔风和羊家的净琬；山矾洁而逸，是鱼玄玑的绿翘；丁香瘦、玉簪寒、秋海棠娇然有酸态，是郑康成（郑玄，为汉大儒，曾注经书）的侍儿。

他以为一个人不论对于什么艺术，即小如下棋，也须癖好成痴，方能有所成就，对于花的爱好也是如此：

余观世上语言无味面目可憎之人，皆无癖之人耳。若真有所癖，将沉湎酣溺，性命死生以之，何暇及钱奴宦贾之事。古之负花癖者，闻人谈一异花，虽深谷峻岭，不惮蹑屫而从之。至于浓寒盛暑，皮肤皴鳞，汗垢为泥，皆所不知。一花将萼，则移枕携襆睡卧其下，以观花之由微至盛至落至萎地而后去。或千株万本以穷其变，或单枝数房以极其趣，或嗅叶而知花之大小，或见根而辨色之红白，是之谓真爱花，是之谓真好事也。

又对于赏花一事，他说：

茗赏者上也，谈赏者次也，酒赏者下也，若夫内酒越茶及一切庸秽凡俗之语，此花神之深恶痛斥者，宁闭口枯坐，勿遭花恼可也，夫赏花有地有时，不得其时而漫然命客，皆为唐突。寒花宜初雪，宜雪霁，宜新月，宜暖房，温花宜晴日，宜轻塞，宜华堂。暑花宜暑月，宜雨后，宜快风，宜佳木荫，宜竹下，宜水阁。凉花宜爽月，宜夕阳，宜空阶，宜苔径，宜古藤巉石边。若不论风日，不择佳地，神气散缓，了不相属，此与妓舍酒馆中花何异哉？

最后他列举十四种花快意，和二十二种花折辱如下：

花快意

明窗　净室　古鼎　宋砚　松涛　溪声　主人好事能诗　门僧解烹茶　苏州人送酒　座客工画　花卉盛开　快心友临门　手抄艺花书　夜深炉鸣妻妾校花故实

花折辱

主人频拜客　俗子阑入　蟠枝　庸僧谈禅　窗下狗斗　莲子胡同歌童弋阳腔　丑女折戴　论升迁　强作怜爱　应酬诗债未了

花盛开债未偿　家人催算账　检《韵府》押字　破书狼藉　福建牙人吴中赝画　鼠矢　蜗涎　僮仆偃蹇　令初行酒尽　与酒馆为邻　案上有黄金白雪　中原紫气等诗

VI. The "Vase Flowers" of Yüan Zhonglang

PROBABLY the best treatise on the arrangement of flowers was written by Yüan Chunglang, one of my favorite authors in other respects, living at the end of the sixteenth century. His book on the arrangement of flowers in vases, called P'ingshih, is highly valued in Japan, and there is known to be a"Yüan School"of flower arrangement. He began in his preface by noting that since hills and water and flowers and bamboos luckily lay outside the scope of the strugglers for fame and power, and furthermore, since such people were so busy with their engrossing pursuits and therefore had no time for the enjoyment of hills and water and flowers and bamboos, the retiring scholar was enabled to snatch this opportunity and monopolize the enjoyment of the latter for himself. He explained, however, that the enjoyment of vase flowers should never be regarded as normal, but at best only as a temporary substitute for people living in cities, and their enjoyment should not cause one to forget the greater happiness of enjoying the hills and lakes themselves.

Proceeding from a consideration that one should be careful in admitting flowers for decoration in his studio, and that it would be better to have no flowers at all than to have promiscuous varieties admitted, he went on to describe the various types of bronze and porcelain vases to be used. Two types are distinguished. Those who are rich and possess antique bronze vessels of the Han Dynasty and have big halls should have big flowers and tall branches standing in huge vases. On the other hand, scholars should have smaller branches of flowers to go with smaller vases, which should also be carefully selected. The only exceptions allowed are the peony and the lotus, which being big flowers should be placed in big vases.

In putting flowers in vases:

One should avoid having them too profuse, or too meager. At most, two or three varieties may be put in a vase, and their relative height and arrangement should aim at the composition of a good picture. In placing flower vases, one should avoid having them in pairs, or uniform, or in a straight row. One should also avoid binding the flowers with string. For the neatness of flowers lies exactly in their irregularity and naturalness of manner, like the prose of Su Tungp'o, which flows on or stops as it pleases, and like the poems of Li Po, which do not necessarily go in couplets. This is true neatness. How can it be called neatness when the branches and leaves merely match each other and red is mixed with white? The latter resemble the trees in the courtyard of minor provincial officials or the stone gateways leading to a tomb.

The room in which flowers are placed should contain a simple table and a cane couch. The table must be broad and thick and should be of fine wood and have a smooth surface. All lacquered tables with decorated margins, golden-painted couches and stands with colored floral designs should be eliminated.

With regard to the "bathing" of flowers, or watering them, the author shows a loving insight into the moods and sentiments of the flowers themselves:

For flowers have their moods of happiness and sorrow and their time of sleep. If one bathes flowers in their morning and evening, at the proper time, the water is like good rain to them. A day with light clouds and a mild sun and the sunset and beautiful moon are morning to the flowers. A big storm, a pouring rain, a scorching sun and bitter cold are evening to the flowers. When their stands bask in the sunlight and their delicate bodies are protected from wind, that is the happy mood of the flowers. When they seem drunk or quiet and tired and when the day is misty, that is the sorrowful mood of the flowers. When their branches incline and rest sideways as if unable to hold themselves erect, that is when the flowers are dreaming in their sleep. When they seem to smile and look about, with a shining light in their eyes, that is when the flowers have waked up from their sleep. In their"morning"they should be placed in an empty pavilion or a big house; in their"evening,"they should be placed in a small room or a secluded chamber; when they are sad, they should sit quietly with abated breath, and when they are happy, they should smile and shout and tease each other; during their sleep, they should let down the curtain, and when they have waked up, they should attend to their toilet. All this is done to please their nature and regulate their times of getting up and going to bed. To bathe flowers in their"morning"is the best; to bathe them when they are asleep, is second; and to bathe them when they are happy is the last. As for bathing them during their"evening,"or during their sorrow, it really seems more like a way of punishing the flowers.

The way of bathing flowers is to use fresh and sweet water from a spring and pour it down gently in small quantities, like a small shower awakening a drunken man, or like the gentle dew itself permeating their body. One should avoid touching the flowers with his hands, or picking them with the tips of fingers, and the work cannot be entrusted to stupid manservants or dirty maids. Plum flowers should be bathed by recluse scholars, the hait'ang by charming guests, the peony by beautifully dressed young girls, the pomegranate by beautiful slave girls, the cassia by intelligent children, the lotus by fascinating concubines, the chrysanthemum by remarkable persons who love the ancients, and the winter plum by a slender monk. On the other hand, flowers blooming in the cold season should not be bathed, but should be protected by thin silk gauze.

According to Yüan, certain flowers go with certain other flowers as their minors or"maids"in a vase. As personal maids who attended to a lady for life were an institution in old China, there developed the notion that beautiful

ladies looked perfect when they had pretty maids by their side as their
necessary adjuncts. Both ladies and maids should be beautiful, but there is a
je ne sais quoi which stamps one type of beauty as belonging to a maid rather
than to a mistress. Maids who were out of harmony with their mistresses
were like stables that did not match a manor house. Carrying the notion over
to flowers, Yuan found that, for their "maids" in the vase, the plum flower
should have camelias, the hait'ang should have apple blossoms and lilacs,
the peony should have cinnamon roses, the Paeonia albiflora should have
poppies and Szechuen sunflowers, the pomegranate should have crape myrtle
and Hisbiscus syriacus, the lotus should have white day lilies, the cassia
should have Hisbiscus mutabilis, the chrysanthemum should have "autumn
hait'ang," and the winter plum should have narcissus. Each maid is exquisite
in its own way, and they differ in their voluptuous or elegant charms like their
mistresses. Not that any slight was intended upon these flower maids, for
they were comparable to the famous maids of history, the narcissus ethereal
down to her bones like Liang Yüch'ing, the maid of the Spinster in heaven,
the camelia and the rose fresh and youthful like the maids Hsiangfeng and
Chingwan of the Shih and Yang families (of Chin Dynasty), the shanfan
flower clean and "romantic" like the maid servant of the tragic nun-poetess,
Yü Hsüanch'i, while the lilac was slender, the white day lily was cool and
the "autumn hait'ang" was coy, but savored a little of pedantry like the maid
of Cheng K'angch'eng (scholar of the Han Dynasty and profuse commentator
on Confucian classics).

Holding to his central idea that anyone who achieves notable results in
any line, even in such matters as playing chess, must love it to a point of craze,
Yüan develops the same idea with regard to the love of flowers as a hobby:

I have found that all the people in the world who are dull in their conversation
and hateful to look at in their faces are those who have no hobbies... When the ancient
people who had a weakness for flowers heard there was a remarkable variety, they would

*travel across high mountain passes and deep ravines in search of them, unconscious
of bodily fatigue, bitter cold or scorching heat, and their peeling skins, and completely
oblivious of their bodies soiled with mud. When a flower was about to bud, they would
move their beds and pillows to sleep under them, watching how the flowers passed from
infancy to maturity and finally dropped off and died. Or they would plant thousands in
their orchards to study how they varied, or have just a few in their rooms to exhaust their
interest. Some would be able to tell the size of flowers from smelling their leaves, and
some were able to tell from the roots the color of their flowers. These were the people who
were true lovers of flowers and who had a true weakness for them.*

In regard to the"enjoyment"(or shang) of flowers, it is pointed out that:

*Enjoying them with tea is the best, enjoying them with conversation second, and
enjoying them with wine the third. As for all forms of noisy behaviour and common
vulgar prattle, they are an insult to the spirits of flowers. One should rather sit dumb like
a fool than offend them. There is a proper place and time for the enjoyment of flowers
and to enjoy them without regard to the proper circumstances would be a sacrilege. The
flowers in the cold season should be enjoyed at the beginning of snow, or after the sky
has cleared after a snowfall, or during crescent moon, or in a warm room. The flowers in
the warm [spring] season should be enjoyed on a clear day, or on a slightly chilly day,
in a beautiful hall. The flowers of summer should be enjoyed after rain, in a refreshing
breeze, in the shade of nice trees, beneath bamboos, or on a water terrace. The flowers of
the cool [autumn] season should be enjoyed under a cool moon, at sunset, on the brink
of a stone hall pavement, on a mossy garden path, or in the neighborhood of rugged rocks
surrounded by ancient creepers. If one looks at flowers without regard to wind and sun
and place, or when one's thoughts are wandering and bear no relation to the flowers,
what difference is it from seeing flowers in singsong houses and wine taverns?*

Finally, Yuan lists the following fourteen conditions as"pleasing"to the
flowers, and"twenty-two"conditions as being disgraceful or humiliating to

them:
Conditions that please the flowers:

A clear window   Guests in the room are
A clean room   exquisite
Antique tripods   Many flowers in bloom
Sung ink-stones   A carefree friend has arrived
"Pine waves"and river sounds   Copying books on flower cultivation
The owner loving hobbies and poetryKettle sings deep at night
Visiting monk understands tea   Wife and concubines editing
A native of Chichow arrives with wine. stories of flowers
Conditions humiliating to the flowers :
The owner constantly seeing
guests
A stupid servant putting in extra
branches, upsetting the
arrangement
Common monks talking zen
Dogs fighting before the window
Singing boys of Lientes Alley
Yiyang [Kiangsi] tunes
Ugly women plucking flowers and
decorating their hair with them
Discussing people's official promotion
and demotion
False expressions of love
Poems written for courtesy
Flowers in full bloom before one
has paid his debts
The family asking for accounts
Writing poems by consulting

rhyming dictionaries
Books in bad condition lying about
Fukien agents
Kiangsu spurious paintings
Faeces of mice and rats
Trailing marks of slime left by snails
Servants lying about
Wine runs out after one begins to play
wine games
Being neighbor to wine-shops
A piece of writing with phrases like
the "purple morning air"[common
in imperial eulogies] on the desk

七．张潮的警句

　　我们已知道享受大自然不单是限于艺术和图画。显现于我们面前的大自然是整个的，包括一切声音、颜色、式样、精神和气氛。人则以了解生活的艺术家的资格去选择大自然的精神，而使它和自己的精神融合起来，这是一切中国诗文作家所共持的态度。不过其中以十七世纪中叶的诗人张潮在他所著《幽梦影》一书中说得最透彻。这书是一种文人的格言，中国古代类似的著作很多，但都不如这书。这种格言都是取材于旧谚，正如安特森的故事之取材于英国古代的童语，和休勃的歌曲之取材民间俚曲。这部书极为中国文人所爱读，有许多人曾于读后加上自己的评注，庄谐都有。但我因为篇幅关系，只能译引其中论及享受大自然的一部分。此外，我因他对于人生问题所说的话是如此澈彻警惕，所以也译引了一些在后面：

　　论何者为宜

　　花不可以无蝶，山不可以无泉，石不可以无苔，水不可以无藻，乔木不可以无藤萝，人不可以无癖。

　　赏花宜对佳人，醉月宜对韵人，映雪宜对高人。

艺花可以邀蝶，累石可以邀云，栽松可以邀风，贮水可以邀萍，筑台可以邀月，种蕉可以邀雨，植柳可以邀蝉。

楼上看山，城头看雪，灯前看月，舟中看霞，月下看美人，另是一番情境。

梅边之石宜古，松下之石宜拙，竹旁之石宜瘦，盆内之石宜巧。

有青山方有绿水，水唯借色于山；有美酒便有佳诗，诗亦乞灵于酒。

镜不幸而遇嫫母，砚不幸而遇俗子，剑不幸而遇庸将，皆无可奈何之事。

论花与美人

花不可见其落，月不可见其沉，美人不可见其夭。

种花须见其开，待月须见其满，著书须见其成，美人须见其畅适，方有实际；否则皆为虚设。

看晓妆宜于傅粉之后。

貌有丑而可观者，有虽不丑而不足观者；文有不通而可爱者，有虽通而极可厌者。此未易与浅人道也。

以爱花之心爱美人，则领略自饶别趣；以爱美人之心爱花，则护惜倍有深情。

美人之胜于花者，解语也；花之胜于美人者，生香也。二者不可得兼，舍生香而解语者也。

养花胆瓶，其式之高低大小须与花相称；而色之浅深浓淡，又须与花相反。

凡花色之娇媚者多不甚香，瓣之千层者多不结实。甚矣全才之难也！兼之者，其唯莲乎。

梅令人高，兰令人幽，菊令人野，莲令人淡，春海棠令人艳，牡丹令人豪，蕉与竹令人韵，秋海棠令人媚，松令人逸，桐令人清，柳令人感。

所谓美人者，以花为貌，以鸟为声，以月为神，以柳为态，以玉为骨，以冰雪为肤，以秋水为姿，以诗词为心，吾无间然矣。

天下无书则已，有则必当读；无酒则已，有则必当饮；无名山则已，有则必当游；无花月则已，有则必当赏玩；无才子佳人则已，有则必当爱慕怜惜。

媸颜陋质，不与镜为仇，亦以镜为无知之死物耳；使镜而有知，必遭扑破矣。

买得一本好花，犹且爱怜而护惜之；矧其为"解语花"乎！

若无诗酒，则山水为具文；若无佳丽，则花月皆虚设。才子而美姿容，佳人而工著作，断不能永年者。匪独为造物之所忌，盖此种原不独为一时之宝，乃古今万世之宝，故不欲久留人世以娶亵耳。

论山水

物之能感人者：在天莫如月，在乐莫如琴，在动物莫如鸟，在植物莫如柳。

为月忧云，为书忧蠹，为花忧风雨，为才子佳人忧命薄，真是菩萨心肠。

昔人云："若无花月美人，不愿生此世界。"予益一语云："若无翰墨棋酒，不必定作人身。"

山之光，水之声，月之色，花之香，文人之韵致，美人之姿态，皆无可名状，无可执著；真足以摄召魂梦，颠倒情思。

因雪想高士；因花想美人；因酒想侠客；因月想好友；因山水想得意

诗文。

有地上之山水，有画上之山水，有梦中之山水，有胸中之山水。地上者妙在丘壑深邃；画上者妙在笔墨淋漓；梦中者妙在景象变幻；胸中者妙在位置自如。

游历之山水，不必过求其妙，若因之卜居，则不可不求其妙。

笋为蔬中尤物；荔枝为果中尤物；蟹为水族中尤物；酒为饮食中尤物；月为天文中尤物；西湖为山水中尤物；词曲为文字中尤物。

游玩山水，亦复有缘；苟机缘未至，则虽近在数十里之内亦无暇到也。

镜中之影，著色人物也；月下之影，写意人物也；镜中之影，钩边画也；月下之影，没骨画也。月中山河之影，天文中地理也；水中星月之象，地理中天文也。

论春秋

春者，天之本怀；秋者，天之别调。

古人以冬为"三余"，予谓当以夏为"三余"：晨起者夜之余；夜坐者昼之余；午睡者应酬人事之余。古人诗曰："我爱夏日长。"洵不诬也。

律己宜带秋气，处世宜带春气。

诗文之体，得秋气为佳，词曲之体，得春气为佳。

论　声

春听鸟声，夏听蝉声，秋听虫声，冬听雪声；白昼听棋声，月下听箫声，山中听松声，水际听欸乃声，方不虚此生耳。若恶少斥辱，悍妻诟谇，真不若耳聋也。

闻鹅声如在白门；闻橹声如在三吴；闻滩声如在浙江；闻羸马项下铃铎声，如在长安道上。

凡声皆宜远听；唯琴声则远近皆宜。

松下听琴，月下听箫，涧边听瀑布，山中听梵呗，觉耳中别有不同。

水之为声有四：有瀑布声，有流泉声，有滩声，有沟浍声。风之为声有三：有松涛声，有秋草声，有波浪声。雨之为声有二：有梧叶荷叶上声，有承檐溜竹筒中声。

论　雨

雨之为物，能令昼短，能令夜长。

春雨如恩诏；夏雨如赦书；秋雨如挽歌。

春雨宜读书；夏雨宜弈棋；秋雨宜检藏；冬雨宜饮酒。

吾欲致书雨师：春雨宜始于上元节后，至清明十日前之内，及谷雨节中；夏雨宜于每月上弦之前及下弦之后；秋雨宜于孟秋季秋之上下二旬；至若三冬，正可不必雨也。

论风月

新月恨其易沉，缺月恨其迟上。

月下听禅，旨趣益远；月下说剑，肝胆益真；月下论诗，风致益幽；月下对美人，情意益笃。

玩月之法：皎洁则宜仰观，朦胧则宜俯视。

春风如酒；夏风如茗；秋风如烟；冬风如姜芥。

论闲与友

天下有一人知己，可以不恨。

能闲世人之所忙者，方能忙世人之所闲。

人莫乐于闲，非无所事事之谓也。闲则能读书，闲则能游名胜，闲则能交益友，闲则能饮酒，闲则能著书。天下之乐，孰大于是？

云映日而成霞，泉挂岩而成瀑。所托者异，而名亦因之。此友道之所以可贵也。

上元须酌豪友；端午须酌丽友，七夕须酌韵友，中秋须酌淡友；重九须酌逸友。

对渊博友，如读异书；对风雅友，如读名人诗文；对谨饬友，如读圣贤经传；对滑稽友，如阅传奇小说。

一介之士，必有密友。密友不必定是刎颈之交。大率虽千百里之遥，皆可相信，而不为浮言所动；闻有谤之者，即多方为之辨析而后已；事之宜行宜止者，代为筹划决断；或事当利害关头，有所需而后济者，即不必与闻，亦不虑其负我与否，竟为力承其事，此皆所谓密友也。

求知己于朋友易；求知己于妻妾难；求知己于君臣则尤难之难。

发前人未发之论，方是奇书；言妻子难言之情，乃为密友。

邻居须得良朋始佳。若田夫樵子，仅能辨五谷而测晴雨，久且数，未免生厌矣。而友之中又当以能诗为第一，能谈次之，能画次之，能歌又次之，解觞政者又次之。

论书与读书

少年读书，如隙中窥月；中年读书，如庭中望月；老年读书，如台上玩月。皆以阅历之浅深，为所得之浅深耳。

能读无字之书，方可惊人妙句；能会难通之解，方可参最上禅机。

古今至文，皆血泪所成。

《水浒传》是一部怒书，《西厢记》是一部悟书，《金瓶梅》是一部哀书。

文章是案头之山水，山水是地上之文章。

读书最乐，若读史书，则喜少怒多，究之怒处亦乐处也。

读经宜冬，其神专也；读史宜夏，其时久也；读诸子宜秋，其致别也；读诸集宜春，其机畅也。

文人讲武事，大都纸上谈兵；武将论文章，半属道听途说。

善读书者，无之而非书：山水亦书也，棋酒亦书也，花月亦书也。善游山水者，无之而非山水：书史亦山水也，诗酒亦山水也，花月亦山水也。

昔人欲以十年读书，十年游山，十年检藏。予谓检藏尽可不必十年，只二三载足矣。若读书与游山，虽或相倍蓰，恐亦不足以偿所愿也。必也如黄九烟前辈之所云："人生必三百岁"而后可乎？

古人云："诗必穷而后工。"盖穷则语多感慨，易于见长耳。若富贵中人，既不可忧贫叹贱，所谈者不过风云月露而已，诗安得佳？苟思所变，计唯有出游一法。即以所见之山川风土，物产人情，或当疮痍兵燹之余，或值旱涝灾侵之后，无一不可寓之诗中，借他人之穷愁，以供我之咏叹，则诗亦不必待穷而后工也。

论一般生活

"情"之一字，所以维持世界；"才"之一字，所以粉饰乾坤。

宁为小人之所骂，毋为君子之所鄙；宁为盲主司之所摈弃，毋为诸名宿之所不知。

人须求可入诗，物须求可入画。

景有言之极幽，而实萧索者，烟雨也；境有言之极雅，而实难堪者，贫病也；声有言之极韵，而实粗鄙者，卖花声也。

躬耕吾所不能，学灌园而已矣；樵薪吾所不能，学薙草而已矣。

一恨书囊易蛀；二恨夏夜有蚊；三恨月台易漏；四恨菊叶多焦；五恨松多大蚁；六恨竹多落叶；七恨桂荷易谢；八恨薜萝藏虺；九恨架花生刺；十恨河豚多毒。

窗内人于窗纸上作字，吾于窗外观之，极佳。

当为花中之萱草，毋为鸟中之杜鹃。

值太平世，生湖山郡，官长廉静，家道优裕，娶妇贤淑，生子聪慧，人生如此，可云全福。

胸藏丘壑，城市不异山林；兴寄烟霞，阎浮有如蓬岛。

清宵独坐，邀月言愁；良夜孤眠，呼蛩语恨。

居城市中，当以画幅当山水，以盆景当苑囿，以书籍当友朋。

延名师，训子弟，入名山，习举业，丐名士，代捉刀，三者都无是处。

方外不必戒酒，但须戒俗；红裙不必通文，但须得趣。

厌催租之败意，丞宜早完粮；喜老衲之谈禅，难免常常布施。

万事可忘，难忘者名心一段；千般易淡，未淡者美酒三杯。

酒可以当茶；茶不可以当酒；诗可以当文，文不可以当诗；曲可以当词，词不可以当曲；月可以当灯，灯不可以当月；笔可以当口，口不可以当笔；婢可以当奴，奴不可以当婢。

胸中小不平，可以酒消之；世间大不平，非剑不能消也。忙人园亭，宜与住宅相连；闲人园亭，不妨与住宅相远。

有山林隐逸之乐而不知享者：渔樵也，农圃也，缁黄也；有园亭姬妾之乐而不能享不善享者：富商也，大僚也。

痛可忍，而痒不可忍；苦可耐，而酸不可耐。

闲人之砚，固欲其佳；而忙人之砚，尤不可不佳；娱情之妾，固欲其美；而广嗣之妾，亦不可不美。

鹤令人逸；马令人俊；兰令人幽；松令人古。

······

美味以大嚼尽之，奇境以粗游了之，深情以浅语传之，良辰以酒食度之，富贵以骄奢处之，俱失造化本怀。

VII. The Epigrams of Zhang Ch'ao

WE have seen that the enjoyment of nature does not lie merely in art and painting. Nature enters into our life as a whole. It is all sound and color and shape and mood and atmosphere, and man as the perceiving artist of life begins to select the proper moods of nature and harmonize them with his own. This is the attitude of all Chinese writers of poetry or prose, but I think its best expression is found in the epigrams of Chang Ch'ao (mid-seventeenth century), in his book Yumengying (or Sweet Dream Shadows). This is a book of literary maxims, of which there are many collections, but none comparable to those written by Chang Ch'ao himself. Such literary maxims stand in relation to popular proverbs as the fairy tales of Anderson stand in relation to old English fairy tales, or as Schubert's art songs stand in relation to folk melodies. His book has been so beloved that a group of Chinese scholars have added comments of their own to each of his maxims, in a most delightful chatty vein. I am compelled, however, to translate only some of the best of his maxims about the enjoyment of Nature. A few of his maxims on human life are so good and form such a vital part of the whole that I shall include some of them at the end.

On What Is Proper

It is absolutely necessary that flowers should have butterflies, hills should have springs, rocks should have moss, water should have water-cress, tall trees should have entwining creepers, and human beings should have hobbies.

One should enjoy flowers in the company of beauties, get drunk under the moon in the company of charming friends, and enjoy the light of snow in the company of highminded scholars.

Planting flowers serves to invite butterflies, piling up rocks serves to invite the clouds, planting pine trees serves to invite the wind, keeping a reservoir of water serves to invite duckweed, building a terrace serves to invite the moon, planting banana trees serves to invite the rain, and planting willow trees serves to invite the cicada.

One always gets a different feeling when looking at hills from the top of a tower, looking at snow from a city wall, looking at the flowers in the lamp-light, looking at colored clouds in a boat, and looking at beautiful women under the moonlight.

Rocks lying near a plum tree should look"antique,"those beneath a pine tree should look"stupid,"those by the side of bamboo trees should look"slender,"and those inside a flower basin should be exquisite.

Blue waters come from green hills, for the water borrows its color from the hills; good poems come from flavory wine, for poetry begs its inspiration from the wine.

When the mirror meets with an ugly woman, when a rare ink-stone finds a vulgar owner, and when a good sword is in the hands of a common general, there is utterly nothing to be done about it.

On Flowers and Women

One should not see flowers wither, see the moon decline below the horizon, or see beautiful women die in their youth.

One should see flowers when they are in bloom, after planting the flowers; should see the moon when it is full, after waiting for the moon;

should see a book completed, after starting to write it; and should see beautiful women when they are gay and happy. Otherwise our purpose is defeated.

One should look at beautiful ladies in the morning toilet after they have powdered themselves.

There are faces that are ugly but stand looking at, and other faces that do not stand looking at, although not ugly; there are writings which are lovable although ungrammatical, and there are other writings which are extremely grammatical, but are disgusting. This is something that I cannot explain to superficial persons.

If one loves flowers with the same heart that he loves beauties, he feels a special charm in them; if one loves beautiful women with the same heart that he loves flowers, he feels a special tenderness and protective affection.

Beautiful women are better than flowers because they understand human language, and flowers are better than beautiful women because they give off fragrance; but if one cannot have both at the same time, he should forsake the fragrant ones and take the talking ones.

In putting flowers in liver-colored vases, one should arrange them so that the size and height of the vase match with those of the flowers, while the shade and depth of its color should contrast with them.

Most of the flowers that are seductive and beautiful are not fragrant, and flowers that have layers upon layers of petals mostly are ill-formed. Alas, rare is a perfect personality! Only the lotus combines both.

The plum flower makes a man feel highminded, the orchid makes a man feel secluded, the chrysanthemum makes a man simple-hearted, the lotus makes a man contented, the spring hait'ang makes a man passionate, the peony makes a man chivalrous, the bamboo and the banana tree make a man charming, the autumn hait'ang makes a man graceful, the pine tree makes a man feel like a recluse, the wut'ung (sterculia platanifolia) makes a man clean-hearted, and the willow makes a man sentimental.

If a beauty should have the face of a flower, the voice of a bird, the soul of the moon, the expression of a willow, the charm of an autumn lake, bones of jade and skin of snow, and a heart of poetry, I should be perfectly satisfied. [I should say so!–Tr.]

If there are no books in this world, then nothing need be said, but since there are books, they must be read; if there is no wine, then nothing need be said, but since there is wine, it must be drunk; if there are no famous hills, then nothing need be said, but since there are, they must be visited; if there are no flowers and no moon, then nothing need be said, but since there are, they must be enjoyed and"played"; if there are no talented men and beautiful women, then nothing need be said, but since there are, they must be loved and protected.

The reason why a looking-glass doesn't become the enemy of ugly-looking women is because it has no feeling; if it had, it certainly would have been smashed to pieces.

One feels tender toward even a good potted flower that he has just bought; how much more should he be tender toward a"talking flower! "

Without wine and poetry, hills and water would exist for no purpose; without the company of beautiful ladies, flowers and the moon would be wasted. Talented men who are at the same time handsome, and beautiful ladies who at the same time can write, can never live a long life. This is not only because the gods are jealous of them, but because this type of person is not only the treasure of one generation, but the treasure of all ages, so that the Creator doesn't want to leave them in this world too long, for fear of sacrilege.

On Hills and Water

Of all the things in the universe, those that touch man most profoundly are: the moon in heaven, the ch'in in music, the cuckoo among animals, and the willow tree among plants.

To worry with the moon about clouds, to worry with books about moths, to worry with flowers about storms, and to worry with talented men and beautiful women about a harsh fate is to have the heart of a Buddha.

One dies without regret if there is one in the whole world a"bosom friend,"or one who"knows his heart."

An ancient writer said that if there were no flowers and moon and beautiful women, he would not want to be born in this world, and I might add, if there were no pen and ink and chess and wine, there was no purpose in being born a man.

The light of hills, the sound of water, the color of the moon, the fragrance of flowers, the charm of literary men, and the expression of beautiful women are all illusive and indescribable. They make one lose sleep dreaming about them and lose appetite thinking about them.

The snow reminds one of a highminded scholar; the flower reminds one of beautiful ladies; wine reminds one of good swordsmen; the moon reminds one of good friends; and hills and water remind one of good verse and good prose that the author himself is pleased with.

There are landscapes on earth, landscapes in painting, landscapes in dreams, and landscapes in one's breast. The beauty of landscapes on earth lies in depth and irregularity of outline; the beauty of landscapes in painting lies in the freedom and luxuriousness of the brush and ink; the beauty of landscapes in dreams lies in their strangely changing views; and the beauty of landscapes in one's breast lies in the fact that everything is in its proper place.

For places that we pass by during our travel, we need not be fastidious in our artistic demands, but for places where we are going to settle down for life we must be fastidious in such demands.

The bamboo shoot is a phenomenon among the vegetables; the lich'i is a phenomenon among fruits; the crab is a phenomenon among aquatic animals; wine is a phenomenon among our foods and drinks; the moon is a phenomenon in the firmament; the West Lake is a phenomenon among hills and waters; and the Sung lyrics (ts'e) and Yüan dramatic poems (ch'ü) are phenomena in literature.

In order to see famous hills and rivers, one must have also predestined luck; unless the appointed time has come, one has no time to see them even though they are situated within a dozen miles.

The images in a looking-glass are portraits in color, but the images [shadows] under a moonlight are pen sketches. The images in a looking-glass are paintings with solid outlines, but the images under a moonlight are"paintings without bones."The images of hills and waters in the moon are geography in heaven, and the images of stars and the moon in water are astronomy on earth.

On Spring and Autumn

Spring is the natural frame of mind of heaven; autumn is one of its changing moods.

The ancient people regarded winter as the"extra"[or resting period] of the other three seasons, but I think we should regard summer as the season of"three extras": getting up at a summer dawn is the extra of the night; sitting at a summer night is the extra of the day; and an afternoon nap is the extra of social intercourse. Indeed,"I love the long summer days,"as an ancient poet says.

One should discipline oneself in the spirit of autumn, and deal with others in the spirit of spring.

Good prose and"T'ang poems"should have the spirit of autumn; good Sung lyrics and Yuan dramatic poems should have the spirit of spring.

On Sounds

One should listen to the sounds of birds in spring, to the sounds of cicadas in summer, to the sounds of insects in autumn and the sounds of snowfall in winter; he should listen to the sounds of playing chess in the daytime, the sounds of flute under the moonlight, the sounds of pine trees in the mountains, and the sounds of ripples on the waterside. Then he shall not have lived in vain. But when a young loafer starts a racket in the street or when one's wife is scolding, one might just as well be deaf.

Hearing the sound of geese makes one feel like in Nanking; hearing the sound of oars makes one feel like in Soochow, Ch'angchow and Huchow; hearing the sound of waves on the beach makes one feel like being in Chekiang; and hearing the sound of bells beneath the necks of thin horses makes one feel like being on the road to Sian.

All sounds should be listened to at a distance; only the sounds of the ch'in can be listened to both at a distance and nearby.

There is a special flavor about one's ears when listening to ch'in music under pine trees, listening to a flute in the moonlight, listening to a waterfall by a brook, and listening to Buddhist chants in the mountains.

There are four kinds of sounds of water: the sounds of cataracts, of gushing springs, of rapids, and of gullets. There are three kinds of sounds of wind: the sounds of"pine waves,"of autumn leaves, and of storm upon the water. There are two kinds of sounds of rain: the sounds of raindrops upon the leaves of wu'tung and lotus, and the sounds of rain water coming down from the eaves into bamboo pails.

On Rain

This thing called rain can make the days seem short and the nights seem long.

A spring rain is like an imperial edict conferring an honor; a summer rain is like a writ of pardon for a condemned criminal; an autumn rain is like a dirge.

A rainy day in spring is suitable for reading; a rainy day in summer is suitable for playing chess; a rainy day in autumn is suitable for going over things in the trunks or in the attic; and a rainy day in winter is suitable for drinking.

I would write a letter to the God of Rain and tell him that rain in spring should come after the fifteenth of the first moon [when the Lantern Festival is over], and continue till ten days before ch'ingming [the third day of the third moon, at which time the peach-trees begin to blossom], and come also at kuyü [time for planting rice]; that summer rain should come in the first and last ten days of every month [so as not to interfere with our enjoyment of the moon]; that autumn rain should come in the first and last ten days of the seventh and the ninth moon [leaving the eighth moon, or mid-autumn, entirely dry for enjoyment of the harvest moon]; and that as for the three months of winter, no rain is called for at all.

On the Moon, Wind and Water

One is exasperated at the crescent moon for declining so early, and exasperated at the waning moon in its third quarter for coming up so late.

To listen to a Buddhist lesson under the moon makes one's mental mood more detached; to discuss swordmanship under the moon makes one's courage more inspired; to discuss poetry under the moon makes one's personal flavor more charming in seclusion; and to look at beautiful women under the moon makes one's passion deeper.

The method of "playing" the moon is to look up at it from a low place when it is clear and bright, and to look down at it from a height when it is hazy and unclear.

The spring wind is like wine; the summer wind is like tea; the autumn wind is like smoke; and the winter wind is like ginger.

On Leisure and Friendship

Only those who take leisurely what the people of the world are busy about can be busy about what the people of the world take leisurely.

There is nothing that man enjoys more than leisure, and this does not mean that one simply does nothing during that time. Leisure enables one to read, to travel to famous places, to form beneficial friendships, to drink wine, and to write books. What greater pleasures can there be in the world than these?

When a cloud reflects the sun, it becomes a colored cloud (hsia), and when a spring gullet flows over a cliff, it becomes a waterfall. By a different association, it is given a new name. That is why friendship is so valuable.

When celebrating the Lantern Festival on the fifteenth of the first moon, one should drink with nonchalant friends; when celebrating the Dragon Boat Festival on the fifth of the fifth moon, one should drink with handsome friends; when celebrating the annual reunion of the Cowherd and the Spinning Maid in Heaven on the seventh day of the seventh moon, one should drink with friends who have charm; when looking at the harvest moon, at the Mid-Autumn Festival, one should drink with quiet or mild-tempered friends; when going up to high mountains on the ninth day of the ninth moon, one should drink with romantic friends.

To talk with learned friends is like reading a rare book; to talk with poetic friends is like reading the poems and prose of distinguished writers; to talk with friends who are careful and proper in their conduct is like reading the classics of the sages; and to talk with witty friends is like reading a novel of romance.

Every quiet scholar is bound to have some bosom friends. By" bosom friends"I do not mean necessarily those who have sworn a life-and-death friendship with us. Generally bosom friends are those who, although separated by hundreds or thousands of miles, still have implicit faith in us and refuse to believe rumors against us; those who on hearing a rumor, try every means to explain it away; those who in given moments advise us as to what to do and what not to do; and those who at the critical hour come to our help, and, sometimes without our knowing, undertake of their own accord to settle a financial account, or make a decision, without for a moment questioning whether by doing so they are not making themselves open to criticism of perhaps injuring our interests.

It is easier to find bosom friends ("those who know our hearts") among friends than among one's wife and concubines, and it is still more difficult to find a bosom friend in the relationship between ruler and ministers.

A"remarkable book"is one which says things that have never been said before, and a"bosom friend"is one who unburdens to us his family secrets.

Living in the country is only enjoyable when one has got good friends with him. One soon gets tired of the peasants and woodcutters who know only how to distinguish the different kinds of grains and to forecast the weather. Again, among the different kinds of friends, those who can write poems are the best, those who can talk or hold a conversation come second, those who can paint come next, those who can sing come fourth, and those who understand wine games come last.

On Books and Reading

Reading books in one's youth is like looking at the moon through a crevice; reading books in middle age is like looking at the moon in one's courtyard; and reading books in old age is like looking at the moon on an open terrace. This is because the depth of benefits of reading varies in proportion to the depth of one's own experience.

Only one who can read books without words [i .e., the book of life itself] can say strikingly beautiful things; and only one who understands truth difficult to explain by words can grasp the highest Buddhist wisdom.

All immortal literature of the ancients and the moderns was written with blood and tears.

All Men Are Brothers (Shuihu) is a book of anger, The Monkey Epic(Hsiyuchi) is a book of spiritual awakening, and Gold-Vase Plum(Chinp'ingmei)[a pornographic novel]is a book of sorrow.

Literature is landscape on the desk, and a landscape is literature on the earth.

Reading is the greatest of all joys, but there is more anger than joy in reading history. But after all there is pleasure in such anger.

One should read the classics in winter, because then one's mind is more concentrated; read history in summer, because one has more time; read the ancient philosophers in autumn, because they have such charming ideas; and read the collected works of later authors in spring, because then Nature is coming back to life.

When literary men talk about military affairs, it is mostly military science in the studio[literally," discussing soldiers on paper"]; and when military generals discuss literature, it is mostly rumors picked up on hearsay.

A man who knows how to read finds everything becomes a book wherever he goes: hills and waters are also books, and so are chess and wine, and so are the moon and flowers. A good traveler finds that everything becomes a landscape wherever he goes: books and history are landscapes, and so are wine and poetry, and so are the moon and flowers.

An ancient writer said that he would like to have ten years devoted to reading, ten years devoted to travel and ten years devoted to preservation and arrangement of what he had got. I think that preservation should not take ten years and two or three years should be enough. As for reading and travel, I do not think even twice or five times the period suggested would be enough to satisfy my desires. To do so one would have to live three hundred years, as Huang Chiuyen says.

The ancient people said that"poetry becomes good only after one becomes poor or unsuccessful,"for the reason that an unsuccessful man usually has a lot of things to say, and it is thus easy to show himself to advantage. How can the poetry of the rich and successful people be good when they neither sigh over their poverty nor complain about their being unpromoted, and when all they write about are the wind, the clouds, the moon and the dew? The only way for such a person to write poetry is to travel, so that all he sees on his way, the hills and rivers and people's customs and ways of life, and perhaps the sufferings of people during war or famine, may all go into his poems. Thus borrowing from the sorrows of other people, for the purpose of his own songs and sighs, one can write good poetry without waiting to be poor or unsuccessful.

On Living in General

Passion holds up the bottom of the universe and genius paints up its roof.

Better be insulted by common people than be despised by gentlemen; better be flunked by an official examiner than be unknown to a famous scholar.

A man should so live as to be like a poem, and a thing should so look as to be like a picture.

There are scenes which sound very exquisite, but are really sad and forlorn, as for instance a scene of mist and rain; there are situations which sound very poetic, but are really hard to bear, as for instance sickness and poverty; and there are sounds which seem charming when mentioned, but are really vulgar, as for instance the voices of girls selling flowers.

I cannot be a farmer myself, and all I can do is to water the garden; I cannot be a woodcutter myself, and all I can do is to pull out the weeds.

My regrets, or things that exasperate me, are ten: (1) that book bags are easily eaten by moths, (2) that summer nights are spoiled by mosquitos, (3) that a moon terrace easily leaks, (4) that the leaves of chrysanthemums often wither, (5) that pine trees are full of big ants, (6) that bamboo leaves fall in great quantities upon the ground, (7) that the cassia and lotus flowers easily wither, (8) that the pilo plant often conceals snakes, (9) that flowers on a trellis have thorns, and (10) that porcupines are often poisonous to eat.

It is extremely pretty to stand outside a window and see someone writing characters on the window paper from the inside.

One should be the hsüan [Hemerocallis flava, a plant called"Forget-sorrow"] among the flowers, and not be the cuckoo [reputed to shed tears of blood which grow up into azaleas] among the birds.

To be born in times of peace in a district with hills and lakes when the magistrate is just and upright, and to live in a family of comfortable means, marry an understanding wife and have intelligent sons—this is what I call a perfect life.

To have hills and valleys in one's breast enables one to live in a city as in a mountain wood, and to be devoted to clouds transforms the Southern Continent into a fairy isle.

To sit alone on a quiet night—to invite the moon and tell her one's sorrow—to keep alone on a good night—and to call the insects and tell them one's regrets.

One living in a city should regard paintings as his landscape, miniature sceneries in a pot as his garden, and books as his friends.

To ask a famous scholar to teach one's children, to go into a famous mountain and learn the art of writing examination essays, and to ask a famous writer to be his literary ghost–all these three things are utterly wrong.

A monk need not abstain from wine, he needs only abstain from vulgarity; a red petticoat need not understand literature, she need only understand what is artistically interesting.

If one is annoyed by the coming of tax-gatherers, he should pay the land taxes early; if one enjoys talking Buddhism with monks, he cannot help making contributions to temples from time to time.

It is easy to forget everything except this one thought of fame; it is easy to grow indifferent to everything except three cups of wine.

Wine can take the place of tea, but tea cannot take the place of wine; poems can take the place of prose, but prose cannot take the place of poems; Yüan dramatic poems can take the place of Sung lyrics, but Sung lyrics cannot take the place of Yüan dramatic poems; the moon can take the place of lamps, but lamps cannot take the place of the moon; the pen can take the place of the mouth, but the mouth cannot take the place of the pen; a maid servant can take the place of a man servant, but a man servant cannot take the place of a maid.

A little injustice in the breast can be drowned by wine; but a great injustice in the world can be drowned only by the sword.

A busy man's private garden must be situated next to his house; while a man of leisure may have his private garden separated from his house at a distance.

There are people who have the pleasures of a mountain recluse lying before them and don't know how to enjoy them–fishermen, woodcutters, farmers, gardeners and monks; there are people who have the pleasures of gardens, pavilions and concubines before them and don't know how to enjoy them–rich merchants and high officials.

It is easy to stand a pain, but difficult to stand an itch; it is easy to bear the bitter taste, but difficult to bear the sour taste.

It is true that the ink-stone of a man of leisure should be exquisite, but a busy man's ink-stone should equally be exquisite; it is true that a concubine for pleasure should be pretty but a concubine for the continuation of the family line should also be pretty.

The stork gives a man the romantic manner, the horse gives a man the heroic manner, the orchid gives a man the recluse's manner, and the pine gives a man the grand manner of the ancients.

......

It is against the will of God to eat delicate food hastily, to pass gorgeous views hurriedly, to express deep sentiments superficially, to pass a beautiful day steeped in food and drinks, and to enjoy your wealth steeped in luxuries.

生活

的

艺术

The

Importance

of Living

第十一章

旅行的享受

一. 论游览

旅行在从前是行乐之一，但现在已变成一种实业。旅行在现代确已比在一百年前便利了不少。政府和所设的旅行机关，已尽力下了一番功夫以提倡旅行，结果是现代的人大概都比前几代的人多旅行了一些。不过旅行到了现代，似乎已是一种没落的艺术。我们如要了解何以谓之旅行，必须先能辨别其实不能算是旅行的各种虚假旅行。

第一种虚假旅行，即旅行以求心胸的必进。这种心胸的必进，现在似乎已行之过度，我很疑惑一个人的心胸，是不是能够这般容易改进。无论如何，俱乐部和演讲会对此的成绩都未见得良好。但我们既然这样专心于改进我们的心胸，则我们至少须在闲暇的日子，让我们的心胸放一天假，休息一下子。这种对旅行的不正确的概念，产生了现代导游者的组织，这是我所认为无事忙者令人最难忍受的讨厌东西。当我们走过一个广场或铜像时，他们硬叫我们去听他讲述生于一七九二年四月二十三日、死于一八五二年十二月二日等。我曾看见过女修道士带着一群学校儿童去参观一所公墓，当她们立在一块墓碑前面时，一个修道士就拿出一本书来，讲给儿童听，死者的生死月日，结婚的年月，他太太的姓名，和其他许多不知所云的事实。我敢断定这种废话，必已使儿童完全丧失了这次旅行的兴趣。成年人在导游的指引之下，也教成了这样的儿童，有许多比较好学不倦的人，竟还会拿着铅笔和日记簿速记下来。中国人在有许多名胜的地方旅行时，也受到同样的麻烦。不过中国的导游不是职业人员，而只是些水果小贩、驴夫和农家的童子，性情略比职业导游活泼，但所讲的话不像职业导游那么准确。某一天，我到苏州去浏览虎丘山，回来时，脑筋中竟充满了自相矛盾的史实和年代，因为据引导我的贩橘童子告诉我，高悬在剑池四十尺之上的那座石桥，就是古美人西施的晨妆处（实则西施的梳妆台远在十里之外）。其实这童子只不过想向我兜卖一些橘子，但因此使我知道民间传说怎样会渐渐地远离事实而变为荒诞不经。

第二种虚假的旅行，即为了谈话资料而旅行，以便事后可以夸说。我曾在杭州名泉和名茶的产地虎跑看见过旅行者将自己持杯饮茶时的姿势摄入照片。拿一张在虎跑品茶的照片给朋友看，当然是一件很风雅的事情，所怕的就是他重视照片而忘却了茶味。这种事情很易使人的心胸受到束缚，尤其是自带照相机的人，如我们在巴黎或伦敦的游览中所见者，他们的时间和注意力已完全消耗于拍摄照片之中，以致无暇去细看各种景物了。这种照片固然可供他们在空闲的时候慢慢地阅看，但如此照片，世界各处哪里买不到，又何必巴巴地费了许多事特地自己跑去拍摄呢？这类历史的名胜，渐渐成为夸说资料，而不是游览资料。一个人所到的地方越多，他所记忆的也越丰富，因而可以夸说的也越多。这种寻求学问的驱策使人在旅行时不能不于一日中，求能看到最可能多的名胜地。他手里拿着一张游览地点程序表，到过一处，即用铅笔划去一个名字。我疑心这类旅行家在假期中，也是讲究效能的。

这种愚拙的旅行，当然产生了第三种虚伪旅行家：定了游览程序的旅行家。他们在事先早已算定将在维也纳或布达佩斯耽搁多少时候。他们在起程之前，都先预定下游览的程序，临时如上课一般切实遵时而行。他们好似在家时，在旅行时也是受月份牌和时钟的指挥。

我主张真正的旅行动机，应完全和这些相反。第一，旅行的真正动机应为旅行以求忘其身之所在，或较为诗意的说法——旅行以求忘却一切。凡是一个人，不论阶级比他高者对他的感想怎样，在自己的家中总是唯我独尊的，同时须受种种俗尚、规则、习惯和责任的束缚。一个银行家总不能做到叫别人当他是一个寻常人，而忘却自己是一个银行家。因此在我看来，旅行的真正理由实是在于变换所处的社会，使他人拿他当一个寻常人看待。介绍信于一个人做商业旅行时是一件有用之物，但商业旅行在本质上是不能置于旅行之列的。一个人倘在旅行时带着介绍信，他便难于期望恢复他自由人类的本来面目，也难于期望显出他于人造的地位之外的人类天然地位。我们应知道一个人到了一处陌生地方时，除了受朋友的招待和介绍到同等阶级的社会去周旋的舒适外，还有比这更好的——由一个童子领着到深山丛林里自由游览的享受。他有机会去享受在餐馆里做手势点一道熏鸡，或向一个东京警察做手势问道的乐趣。得过这种旅行经验的人，至少在回到家后，可以不必如平时一样一味依赖他的车夫和贴身侍者了。

一个真正的旅行家必是一个流浪者，经历着流浪者的快乐、诱惑和探险意念。旅行必须流浪式，否则不成其为旅行。旅行的要点于无责任、无定时、无来往信札、无嗫嚅好问的邻人、无来客、无目的地。一个好的旅行家绝不知道他往哪里去，更好的甚至不知从何处而来，他甚至忘却了自己的姓名。屠隆曾在他所著的《冥廖子游》中很透彻地阐明这一点——这游记我译引在下文里边。他在某处陌生的地方并无一个朋友，但恰如某女尼所说："无所特善视者，尽善视普世人也。"没有特别的朋友，就是人尽可友，他普爱世人，所以处身于其中，领略他们的可爱处和他们的习俗。这种好处是坐着游览汽车看古迹的旅行家所无从领略的，因为他们只有在旅馆里边和从本国同来的游伴谈天的机会。最可笑的是有许多美国旅行家到巴黎之后，必认定到同游者都去吃的餐馆中去吃饭，好似借此可以一见同船来的人，并可以吃到和在家时所吃一样的烘饼。英国人到了上海之后必住到英国人所开设的旅馆里边去，在早餐时照常吃着火腿煎蛋和涂着橘皮酱的面包，闲时在小饮室里坐坐，遇到有人邀他坐一次人力车时，必很羞缩地拒绝。他们当然是极讲究卫生的，但又何必到上海去呢？如此的旅行家，绝没有和当地的人士在精神上融合的机会，因此也就丧失了一种旅行中最大的益处。

　　流浪精神使人能在旅行中和大自然更加接近。所以这一类旅行家每喜欢到阒无人迹的山中去，以便可以幽然享受和大自然融合之乐。所以这些旅行家在预备出行时，绝不会到百货公司去费许多时刻选购一套红色或蓝色的游泳衣，买唇膏尚可容许，因为旅行家大概都是崇奉唇骚者，喜欢色色自然，而一个女人如若没有了好唇膏，便会不自然的。但这终究为了他们乃是到人所共赴的避暑地方或海滨去，而在这种地方是完全得不到和大自然发生更深关系的益处的。往往有人到了一处名泉欣然自语："这可真是幽然独处了。"但是在旅馆吃过晚饭在起居室内拿起一张报纸随便看看时，即看见上面载着某甲夫人曾在星期一到过这地方。次日早晨他去"独"步时，又遇到隔夜方到的某乙全家。星期四的晚上，他又很快乐地知道某丙夫妇也将要到这幽静的山谷中度夏。接着就是某甲夫人请某乙全家吃茶点，某乙请某丙夫妇打牌。并能听见某丙夫人喊着说："奇啊，这不是好像依旧在纽约吗？"

　　我以为除此以外，另有一种旅行，不为看什么事物，也不为看什么人的旅行，所看的不过是松鼠、麝鼠、土拨鼠、云和树。我有一位美国女友曾告诉我，有一次，她怎样被几个中国朋友邀到杭州附近的某山去看"虚无一物"。据说，那一天早晨雾气很浓。当她们上山时，雾气越加浓厚，甚至可以听得见露珠滴在草上的声音。这时除了浓雾之外，不见一物，她很失望。"但你必须上去，因为顶上有奇景可见呢。"她的中国朋友劝她。于是她跟着向上走去。不久，只看见远处一块被云所包围的怪石，别人都视为好景。"那里是什么？"她问。"这就是倒植莲花。"她的朋友回答。她很为懊恼，就想回身。"但是顶上还有更奇的景致哩。"她的朋友又劝。这时她的衣服已半潮，但她已放弃反抗，所以依旧跟着别人上去。最后，她们已达山顶，四围只见一片云雾和天边隐约可见的山峰。"但这里实在没有什么可看啊。"她责问。"对了，我们特为上来看虚无一物的。"她的中国朋友回答。

观看景物和观看虚无，有极大的区别。有许多特去观看景物的，其实并没有看到什么景物，但有许多去观看虚无的能看到许多事物。我每听到一位作家到外国去"搜集新著的资料"时，总在暗暗好笑，难道他的本乡本国中，其人情和风俗上已没有了可供他采集的资料吗？难道他的论文资料竟已穷尽吗？纺织区难道是太缺乏浪漫性吗？格恩赛岛太沉寂，不足为一部杰出小说的背景吗？所以我们须回到"旅行在于看得见物事的能力之哲学问题"，这就可使到远处去旅行和下午在田间闲步之间失去它们的区别。

依金圣叹之说，两者是相同的。旅行者所必需的行具就是如他在著名的戏曲《西厢记》的评语中所说："胸中的一副别才。眉下的一副别眼。"其要点在于此人是否有易觉的心和能见之眼。倘若他没有这两种能力，即使跑到山里去，也是白费时间和金钱。在另一方面，倘若他有这两种能力，则不必到山里去，即坐在家里远望，或步行田间去观察一片行云、一只狗、一道竹篱或一棵树，也能同样享受到旅行的快乐。我现在译引一段金氏所论真正旅行艺术的说辞：

吾读世间游记，而知世真无善游人也。夫善游之人也者，其于天下之一切海山方兵，洞天福地，固不辞千里万里，而必一至以尽探其奇也。然其胸中之一副别才，眉下之一双别眼，则方且不必直至海山方兵，洞天福地，而后乃今始曰："我且探其奇也。"夫昨之日而至一洞天，凡罄若干日之足力目力心力，而既毕其事矣；明之日，又将至一福地，又将罄若干日之足力目力心力，而于以从事。彼从旁之人不能心知其故，则不免曰："连日之游快哉！始毕一洞天，乃又造一福地。"殊不知先生且正不然。其离前之洞天，而未到后之福地，中间不多，虽所隔止于三二十里，又少而或止于八、七、六、五、四、三、二里；又少而或止于一里半里，此先生则于一里半里之中间，其胸中之所谓一副别才，眉下之一双别眼，即何尝不以待洞天福地之法而待之哉？

今夫以造化之大本领、大聪明、大气力而忽然结撰而成一洞天、一福地，是真骇目惊心之事，不必人道也。然吾每每谛视天地之间随分一鸟、一盆、一花、一草，乃至鸟之一毛、鱼之一鳞、花之一瓣、草之一叶，则初未有不费彼造化者之大本领、大聪明、大气力，而后结撰而得成名者也。谚云："狮子搏象用全力，搏兔亦全力。"彼造化者则真然矣。生洞天福地用全力，生随分之一鸟、一鱼、一盆、一花、一草，以至一毛、一鳞、一瓣、一叶，殆无不用尽全力。由是言之，然则世间之所谓骇目惊心之事，固不必定至于洞天福地而后有此，亦为信然也。

抑即所谓洞天福地也者，亦尝计其云：如之何结撰也哉？庄生有言：

"指马之百体非马，而马系前者，立其百体而谓之马也。"此于大泽，百材皆度；观乎大山，水石同坛。夫人诚知百材万木，杂然同坛之为大泽大山，而其于游也，斯庶几矣。其层峦绝，则积石而成，是穹窿也；其飞流悬瀑，则积泉而成，是灌输也。果石石而察之，殆初无异于一拳者也；试泉泉而寻之，殆初无异于细流者也。且不直此也，老氏之言曰："三十辐共一毂，当其无，有车之用；埏埴以为器，当其无，有器之用；凿户牖以为室，当其无，有室之用。"然则一一洞天福中间，所有回看为峰，延看为岭，仰看为壁，俯看为溪，以至正者坪，侧者坡，跨者梁，夹者洞，虽其奇奇妙妙至于不可方物，而吾有以知其奇之所以奇，妙之所以妙，则固必在于所谓当其无之处也矣。盖当其无，则是无峰、无岭、无壁、无溪、无坪坡梁洞之地也。然而当其无斯，则真吾胸中一副别才之所翱翔，眉下一双别眼之所排荡也。

夫吾胸中有其别才，眉下有其别眼，而皆必于当其无处，而后翱翔，而后排荡，然则我真胡为必至于洞天福地？正如顷所云，离于前未到于后之中间，三十二里，即少止于一里半里，此亦何地不有所谓当其无之处耶？一略彴小桥、一槎枒独树、一水、一村、一篱、一犬，吾翱翔焉，吾排荡焉。此其于洞天福地之奇奇妙妙，诚未能知为在彼，而为在此也？

且人亦都不必胸中之真有别才，眉下之真有别眼也。必曰，先有别才而后翱翔，先有别眼而后排荡，则是善游之人，必至旷世而不得一遇也。如圣叹意者，天下亦何别才别眼之与，有但肯翱翔焉，斯即别才矣；果能排荡焉，斯即别眼矣。米老之相石也曰："要秀、要皱、要透、要瘦。"今此一里半里之一水、一村、一篱、一犬则皆极秀、极透、极皱、极瘦者也，我亦定不以如米老之相石故耳。诚亲见其秀处、皱处、透处、瘦处乃在于此，斯虽欲不于是焉翱翔，不于是焉排荡，亦岂可得哉？且彼洞天福地之为峰、为岭、为壁、为溪、为坪坡梁涧，是亦岂能多有其奇奇妙妙者乎？亦都不过能秀、能皱、能透、能瘦焉耳。由斯一言，然则必至于洞天福地而后游，此其不游之处，盖以多多矣。且必至于洞天福地而后游，此其洞天福地，亦终于不游已也。何也？彼不能知一篱、一犬之奇妙者，必彼所见之洞天福也，皆适得其不奇不妙者也。

嶯山云："千载以来，独有宣圣是第一善游人。其次则数王羲之。"或有征其说者，嶯山云："宣圣吾深感其'食不厌精，脍不厌细'之二言。王羲之吾见若干帖，所有字画，皆非献之所能窥也。"圣叹曰："先生此言，疑杀天下人去也。"又嶯山每语圣叹云："王羲之若闲居家中，必就庭花逐枝逐朵细数其须。门生执巾侍立其侧，常至终日都无一语。"圣叹问此故事出于何书？嶯山云："吾知之。"盖嶯山奇之特如此，惜乎天下之人，不遇嶯山一倾倒其风流也。

I. On Going about and Seeing Things

TRAVEL used to be a pleasure, now it has become an industry. No doubt there are greater facilities for traveling today than a hundred years ago, and governments with their official travel bureaus have exploited the tourist trade, with the result that the modern man travels on the whole much more than his grandfather. Nevertheless travel seems to have become a lost art. In order to understand the art of travel, one should first of all beware of the different types of false travel, which is no travel at all.

The first kind of false travel is travel to improve one's mind. This matter of improving one's mind has undoubtedly been overdone. I doubt very much whether one's mind can be so easily improved. Anyway there is very little evidence of it at clubs and lectures. But if we are usually so serious as to be bent upon improving our minds, we should at least during a vacation let the mind lie fallow, and give it a holiday. This false idea of travel has given rise to the institution of tourist guides, the most intolerable chattering kind of interfering busybodies that I can imagine. One cannot pass a square or a bronze statue without his attention being called to the fact that So-and-So was born on April 23, 1792, and died on December 2, 1852. I have seen convent sisters escorting school children to a cemetery, and as the group stopped before a tombstone, reading to them from a book the dates of the deceased, the age at which he married, the name and surname of his wife, and such learned nonsense, which I am sure spoiled the pleasure of the entire trip for the children. The grown-ups themselves are turned into a group of school children being vociferously lectured to by the guide, and in the case of travelers of the more studious type,   also taking notes very assiduously like good school children. Chinese tourists suffer like American tourists at Radio City, with the difference that Chinese guides are not professional, but are fruit-sellers, donkey drivers and peasant boys, whose information is less correct, if their personalities are more lively. Visiting Huch'iu Hill at Soochow one day, I came back with a terrible confusion of historical dates and sequence, for the awe-inspiring bridge suspended forty feet over the Sword Pond, with two round holes in the stone slabs of the bridge through which a sword had flown up as a dragon, became, according to my orange-selling boy, the place where the ancient beauty Hsishih attended to her morning toilet! (Hsishih's"dressing table"was actually about ten miles away from the place.) All he wanted was to sell me some oranges. But then I had a chance of seeing how folklore was changed and modified and"metamorphosed."

The second kind of false travel is travel for conversation, in order that one may talk about it afterwards. I have seen visitors at Hup'ao of Hangchow, a place famous for its tea and spring water, having their picture taken in the act of lifting tea cups to their lips. To be sure, it is a highly poetic sentiment to show friends a picture of themselves drinking tea at Hup'ao. The danger is that one spends less thought on the actual taste of the tea than on the photograph itself. This sort of thing can become an obsession, especially with travelers provided with cameras, as we so often see on sight-seeing buses in Paris and London. The tourists are so busy with their cameras that they have no time to look at the places themselves. Of course they have the privilege of looking at them in the pictures afterwards when they go home, but it is obvious that pictures of Trafalgar Square or the Champs Elysees can be bought in New York or in Peiping. As these historical places become places to be talked about afterwards instead of places to be looked at, it is natural that the more places one visits, the richer the memory will be, and the more places there will be to talk about. This urge for learning and scholarship therefore impels the tourist to cover as many points as possible in a day. He has in his hand a program of places to visit, and as he comes to a place, he checks it off with a pencil on the program. I suspect such travelers are trying to be efficient even on their holiday.

This sort of foolish travel necessarily produces the third type of false travelers, who travel by schedule, knowing beforehand exactly how many hours they are going to spend in Vienna or Budapest. Before such a traveler departs, he makes a perfect schedule for himself and religiously adheres to it. Bound by the clock and run by the calendar as he is at home, he is still bound by the clock and run by the calendar while abroad.

In place of these false types of travel, I propose that the true motives of travel are, or should be, otherwise. In the first place, the true motive of travel should be travel to become lost and unknown. More poetically, we may describe it as travel to forget. Everyone is quite respectable in his home town, no matter what the higher social circles think of him. He is tied by a set of conventions, rules, habits and duties. A banker finds it difficult to be treated just as an ordinary human being at home and to forget that he is a banker, and it seems to me, the real excuse for travel is that he shall be able to find himself in a community in which he is just an ordinary human being. Letters of introduction are all very well for people on business trips, but business trips are by definition outside the category of pure travel. A man stands a poorer chance of discovering himself as a human being if he brings along with him fetters of introduction, and of finding out exactly how God made him as a human being, apart from the artificial accidents of social standing. Against the comforts of being well-received by friends in a foreign country and guided efficiently through the social strata of one's own class, there is the greater excitement of a boy scout in a forest, left to his own devices. He has the chance to prove for himself that he can order a fried chicken with the language of fingers alone, or find his way about town by communicating with a Tokyo policeman. At least, such a traveler can come home with a less tenderfootish dependence upon his chauffeur and butler.

A true traveler is always a vagabond, with the joys, temptations and sense of adventure of the vagabond. Either travel is"vagabonding"or it is no travel at all. The essence of travel is to have no duties, no fixed hours, no mail, no inquisitive neighbors, no receiving delegations, and no destination. A good traveler is one who does not know where he is going to, and a perfect traveler does not know where he came from. He does not even know his own name and surname. This point has been emphasized by T'u Lung in his idealized sketch of the Travels of Mingliaotse—which I have translated in the next section. Probably he hasn't got a single friend in a strange land, but as a Chinese nun expressed it,"Not to care for anybody in particular is to care for mankind in general."Having no particular friend is having everybody as one's friend. Loving mankind in general, he mixes with them and goes about observing the charms of people and their customs. This kind of benefit is entirely missed by those travelers on the sight-seeing buses, who stay in the hotel, converse with their fellow passengers from the home country, and in the case of many American travelers in Paris, make a point of eating at the favorite rendezvous of American tourists, where they can be sure of seeing all their fellow passengers who came over on the same ship all over again, and can eat American doughnuts which taste exactly as they taste at home. English travelers in Shanghai make sure that they put up at an English hotel where they can have their bacon and eggs and toast with marmalade at breakfast, and hang about the cocktail lounge and fight shy of any inducement to get them to take a rickshaw ride. They are terribly hygienic, to be sure, but why go to Shanghai at all? Such travelers never allow themselves the time and leisure for entering into the spirit of the people and thus forfeit one of the greatest benefits of traveling.

The spirit of vagabondage makes it possible for people taking a vacation to get closer to Nature. Travelers of this kind will therefore insist on going to the summer resorts where there are the fewest people and one can have some sort of real solitude and communion with Nature. Travelers of this sort, therefore, do not in their preparation for journeys go into a department store and take a lot of time to select a pink or a blue bathing suit. Lipstick is still allowable because a vacationist, being a follower of Jean Jacques Rousseau, wants to be natural, and no lady can be quite natural without a good lipstick. But that is due to the fact that one goes to the summer resorts and beaches where everybody goes, and the entire benefit of a closer association with Nature is lost or forgotten. One goes to a famous spring and says to himself, "Now I am entirely by myself," but after supper at the hotel, he takes up a paper in the lounge and discovers that Mrs. B—came up to the place on Monday. Next morning on his "solitary" walk, he encounters the entire family of the Dudleys, who arrived by train the night before. On Thursday night he finds out to his great delight that Mrs. S—and her husband are also having a vacation in this wonderful secluded valley. Mrs. S—then invites the Dudleys to tea, and the Dudleys invite Mr. and Mrs. S—to a bridge party, and you can hear Mrs. S—exclaim, "Isn't it wonderful? It is just like being in New York, isn't it?"

I may suggest that there is a different kind of travel, travel to see nothing and to see nobody, but the squirrels and muskrats and woodchucks and clouds and trees. A friend of mine, an American lady, described for me how she went with some Chinese friends to a hill in the neighborhood of Hangchow, in order to see nothing. It was a misty day in the morning, and as they went up, the mist became heavier and heavier. One could hear the soft beat of drops of moisture on the leaves of grass. There was nothing to be seen but fog. The American lady was discouraged. "But you must come along; there's a wonderful sight on top," insisted her Chinese friends. She went up with them and after a while saw an ugly rock in the distance enveloped by the clouds, which had been heralded as a great sight."What is there?"she asked."That is the Inverted Lotus,"her friends replied. Somewhat mortified, she was ready to go down."But there is a still more wonderful sight on top,"they said. Her dress was already half damp with the moisture, but she had given up the fight already and went on with them. Finally they reached the summit. All about them was an expanse of mists and fogs, with the outline of distant hills barely visible on the horizon."But there is nothing to see here,"my American friend protested."That is exactly the point. We come up to see nothing,"her Chinese friends replied.

There is all the difference between seeing things and seeing nothing. Many travelers who see things really see nothing, and many who see nothing see a great deal. I am always amused at hearing of an author going to a foreign country to"get material for his new book,"as if he had exhausted all there was to see in humanity in his own town or country and as if the theme could ever be exhausted."Thrums" must be unromantic and the Island of Guernsey too dull to build a great novel upon! We come therefore to the philosophy of travel as consisting of the capacity to see things, which abolishes the distinction between travel to a distant country and going about the fields of an afternoon.

The two become the same thing, as Chin Shengt'an insisted. The most necessary outfit a traveler has to carry along with him is"a special talent in his breast and a special vision below his eyebrows,"as the Chinese dramatic critic expressed it in his famous running comment on the drama Western Chamber. The point is whether one has got the heart to feel and the eyes to see. If he hasn't, his visits to the mountains are a pure waste of time and money; on the other hand, if he has got"a special talent in his breast and a special vision below his eyebrows,"he can get the greatest joy of travel even without going to the mountains, by staying at home and watching and going about the field to watch a sailing cloud, or a dog, or a hedge, or a lonely tree. I translate below Chin's dissertation on the true art of travel:

I have read the travel sketches of people and realize that very few people understand the art of travelling. Surely the man who knows how to travel will not be frightened by a long journey to see all the sights of the land and sea and explore all their grandeur and mystery. But a certain talent in his breast and a certain vision below his eyebrows tells him that it is not necessary to go to all the famous beauty spots of land and sea in order to explore nature's wonder and mystery. One day he goes to a stone cave after using up a great deal of the energy of his legs, his eyes and his mind, and when he has done that, the next day he goes again to another blessed spot and uses up some more of the energy of his legs, his eyes and his mind. People who do not understand him will say, "What a wonderful time you have been having, visiting places these days! Just after seeing one stone cave, you have gone ahead to visit another blessed spot." They have entirely missed the point. For there is a distance between the two places he visited perhaps twenty or thirty li, or perhaps, eight, seven, six, five, four, three, two li, or perhaps even one li or just half a li. With the special talent in his breast and the special vision beneath his pair of eyebrows, has he not looked at that little distance of a li or half a li in the same way he looked at the stone cave and the blessed spot?

It is true that there is something which terrifies the eye and surprises the soul to find that Mother Nature with her great skill and wisdom and energy has suddenly produced a thing like a stone cave or a blessed spot. But I have often stared casually at little things of this universe, a bird, a fish, a flower, or a small plant, and even at a bird's feather, a fish's scale, a flower petal and a blade of grass, and realized how Mother Nature has also created it with all her great skill and wisdom and energy. As it is said that the lion uses the same energy to attack an elephant as to attack a wild rabbit, so does Mother Nature truly do the same thing. She uses all her energy in producing a stone cave or a blessed spot, but she also uses all her energy in producing a bird, a fish, a flower, a blade of grass, or even a feather, a scale, a petal, a leaf. Therefore, it is not alone the stone cave or the blessed spot that terrifies the eye and surprises the soul in this world.

Furthermore, have we ever thought how the stone cave and the blessed spot were produced? Chuangtse has wisely said, "To comprehend the different organs of the horse is not to comprehend the horse itself. What we call the horse exists before its different organs." To take another analogy, we see forests growing around the great lakes and timber and rocks spread all over the great mountains. It gives the traveler joy to know that the great forests and timber and rocks are assembled together to form the great lakes and great mountains. But the towering peaks are formed by little rocks and the falling cataracts are formed and nourished by little springs of water. If we examine them one by one, we see that the stones are no bigger than the palm of one's hand, and the springs are no bigger than little rivulets. Laotse has said: "Thirty spokes are grouped around the hub of a wheel, and when they lose their own individuality, we have a functioning cart. We knead clay into a vessel and when the clay loses its own existence, we have a usable utensil. We make a hole in the wall to make windows and doors, and when the windows and doors lose their own existence, we have a house to live in." And so when we view a stone cave or a blessed spot and see the vertically uprising peaks, horizontally stretching mountain passes, those that go up and form a precipice, those that go down and form a river, those that are level and form a precipice, those that go down and form a river, those that are level and form a plateau, those that are inclined and form a hillside, those that stretch across and become bridges, and those that come together and become ravines, we realize that however incomparably manifold they are in their greatness and mystery, this mystery and grandeur arises when the parts lose their individual existence. For when they lose their own existence, there are no passes, no precipices, no rivers and no plateaus, hillsides, bridges and ravines. But it is exactly in their non-existence that the special talent in our breast and the special vision below our eyebrows wander and float at ease. And since this special talent in our breast and this special vision below our eyebrows can wander and float at ease only when these things are non-existent, why then must we insist on going to the stone cave and to the blessed spot?

If, therefore with the special talent in my breast and the special vision below my eyebrows, I can still wander and float about at ease only when these things lose their individual existence, isn't it then unnecessary that I visit the stone cave and the blessed spot? For as I have just said, in the distance of twenty or thirty li, or even one or half li, are there not also everywhere things that lose their existence? A crooked little bridge—a shaggy lonely tree—a glimpse of water—a village—a hedge—a dog—how do I know that the mystery and the grandeur of the stone cave and the blessed spot where I may wander and float about at ease are not even here ?...

Besides, it is not necessary that we have a special talent in our breast and a special vision below our eyebrows: should one require a special talent in order to float about and require a special vision in order to wander at ease, we might not find a single person in the world who understands the art of traveling. According to Shengt'an [the writer himself] , there are no special talents and no special visions: to be willing to float about is already to have the special talent, and to be able to wander about at ease is already to have a special vision. The criteria of Old Mi [Mi Fu] for judging rocks are hsiou, tsou, t'ou and sou [delicacy, undulation, clarity and slenderness] . Now a little patch of water, a village, a bridge, a tree, a hedge or a dog at the distance of one li or half a li has each its great delicacy, great undulation, great clarity and great slenderness. If we fail to see this, it is because we do not understand how to look at them as Old Mi looked at the rocks. If we only see their delicacy, their undulation, their clarity and their slenderness, we cannot help but wander and float about at ease among them. What is there in the grandeur and mystery of peaks and mountain passes and precipices and rivers and the plateaux, the slopes, the bridges and the ravines in the stone cave and the blessed spot outside their being delicate, undulated, clear and slender? Those who insist on visiting the stone caves and the blessed spots therefore have left much that they have not visited; in fact, they have not visited any place at all. For those who fail to see the mystery and grandeur of a single hedge or a dog have seen only what is not grand and what is not mysterious in the stone caves and the blessed spots.

Toushan [Chin's friend] said, "The one who understands best the art of travel in all history is Confucius, and the second, Wang Hsichih [acknowledged master of Chinese calligraphy] ."On being asked to explain, Toushan said,"I know this of Confucius from the two sentences that for him,'rice could not be white enough, and mincemeat could not be chopped fine enough', and I know this of Wang from seeing examples of his calligraphy. There are things in it which his son Hsienchih could not even understand. ""What you have just said is devastating to the entire mankind," said I. Toushan once told me,"When Wang Hsichih was at home, he often counted the pistils of every flower on every branch in his yard, and he would be thus occupied the whole day without saying a word, while his students stood by with towels at his side. ""Where did you find the authority for this statement?"asked Shengt'an."I found it in my own heart,"replied Toushan. Such a marvelous person is Toushan. Alas, that the world has not discovered Toushan and admired his romantic imagination!

二.《冥寥子游》

甲　出游之由

冥寥子为吏，困世法，与人吐匿情之谭，行不典之礼。何谓"匿情之谭"？主宾长揖，寒暄而外，不敢多设一语。平生无斯须之旧，一见握手，动称肺腑，掉臂去之，转盼胡越。面颂盛德，则夷也；不旋踵而背语，蹯也。燕坐之间，实辨有口，乃托简重；身有秽行，谬为清言。惧里言漏实，庄语触忌，则一切置之，而别为浮游不根之谭，甚而假优伶之讴歌以乱之，即耳目口鼻，悉非我有，嗔喜笑骂，总属不真。俗已如此，虽欲力矫之不能。何谓"不典之礼"？宾客酬应，无论尊贵，虽其平交，终日磬折俯首。何仇于天而目与之远，何亲于地而目与之近。贵人才一启口，诺声如雷，一举手而我头已抢地矣。彼此相诣，绝不欲见，而下马投刺，徒终日仆仆。夫往来通情，非举行故事也。先王制礼，固如是乎？褒衣束带，缚如槛猿，虱肤，痒甚而不可扪。跬步闲行，辄恐逾官守。马上以目注鼻，视不越尺寸，视越尺寸，人即从旁侦之。溺下至不可忍，而无故莫敢驻足。其大者"三尺"在前，清议在后。寒暑撼其外，得失煎其中，岂唯绳墨之失哉！虽有豪杰快士，通脱自喜，不涉此途则已，一涉此途，不得不俯而就其笼络。冥寥子将纵心广意而游于潹之乡矣。

或曰："吾闻之，道士处静不枯，处动不喧，居尘出尘，无缚无解；俄而柳生其肘，有鸟巢于其顶，此亦冥静沉寥之极也。供爨下之役，拾地上之残，此亦卑琐秽贱之极也，而至人皆冥之。子厌仕路之踬踣，而乐奇游之清旷，无乃心为境杀乎？"

冥寥子曰："得道之人，入水不濡，入火不焦，触实若虚，蹈虚若实。靡入不适，靡境不冥，则其固然。余乃好道，非得道者也。得道者，把柄在我，虚空粉碎。投之嚣喧秽贱，若浊水青莲，淤而不染，故可无择乎？所之余，则安能若柳之从风，风宁则宁，风摇则摇；若沙之在水，水清则清，水浊则浊。余尝终日清静，以晷刻失之，终岁清静，以一日失之。欲听其所之，而在境不乱，不可得也。使天子可以修道，则巢许何以箕颖？使国王可以修道，则释迦何以雪山？使列侯可以修道，则子房何以谢病？使庶官可以修道，则通明何以挂冠？余将广心纵意而于潹之乡矣。"

或曰："愿闻子游。"

冥寥子曰："夫游者，所以开耳目，舒神气，穷九州，览八荒，采真访道，庶几至人。啖云芝，逢石髓，御风骑气，冷然而飘，眇不知其何之，然后归而掩关面壁，了大事矣。余非得道者，宅神以内，养德以澹，游气以虚，敢不力诸，然而未也。宅神以内，忽而驰于外；养德以澹，忽而移于浓；游气以虚，忽而著于意。其中不宁，则稍假外镇之；其心无以自得，则或取境娱之。故余之游迹奇矣。"

乙　旅行之法

"挟一烟霞之友与俱，各一瓢一衲，百钱自随。不取盈，而取令百钱常满，以备非常。两人乞食，无问城郭村落，朱门白屋，仙观僧庐戒所。乞以食不以酒，以蔬不以肉。其乞辞以孙不以哀，界则去之，其不界者亦去之，要以苟免饥而已。有疑物色者，晦而自免去；有见凌者，屈体忍之。有不得已，无所从乞，即以所携百钱用其一二，遇便即补足焉，非甚不得已，不用也。

"行不择所之，居不择所止。其行甚缓，日或十里，或二十里，或三十、四十、五十里而止。不取多，多恐其罢也。行或遇山川之间，青泉白石，水禽山鸟，可爱玩，即不及往，选沙汀磐石之上，或坐而眺焉。邂逅樵人渔夫，村氓野老，不通姓氏，不作寒暄，而约略谈田野之趣。移晷乃去，别而不关情也。

"大寒大暑，必投栖焉而不行，惧寒暑之气侵人也。行必让路，津必让渡。江湖风涛，则止不渡，或半渡而风涛作，则凝神定气，委命达生，曰：'苟渡而溺，天也。'即恐，宁免乎？如其不免，则游止矣；幸而获免，游如初。

遭恶少年于道，或误触之，少年行其无礼，则孙辞谢之。谢之而不免，则游止矣；幸而获免，游如初。

有疾病，则投所止而调焉，其同行者稍为求药，而己则处之泰如。内视反听，无怖死。如是则重病必轻，轻病立愈。如其大运行尽，则游止矣；幸而获免，游如初。踪迹所至，逻者疑焉，而以细人见禽，或以情脱，或以知免。如其不免，则游止矣；幸而获免，游如初。行而托宿石庵茅舍，无论也，托宿而不及，即寺门岩阿，穷檐之外，大树之下，可以偃息。或山鬼伺之，虎狼窥之，奈何？山鬼无能为苦，虎狼无术以制之，不有命在天乎？以'四大'委之，而神气了不为动。卒填其喙，数也，则游止矣。幸而获免，游如初。"

丙 高山之顶

"其游以五岳四渎，洞天福地为主，而以散在九州之名山大川佐之，亦止及九州所辖，人迹所到而已。其在赤县神州之外，若须弥昆仑及海上之十州三岛，身无羽翼，恐不能及也。所遇亦止江湖之士，山泽之臞而已。若扶桑青童，旸谷神王，桐柏、小有、王母、云林诸真，身无仙骨，恐不能觐也。

"其登五岳也，竦立罡风之上，游览四海之外，万峰如螺，万水如带，万木如荠。星河摩于巾领，白云出于怀袖，鹔鹴举手可拾，日月掠双鬓而过之。即啸语亦不敢纵，非唯惊山灵，殆恐咫尺通乎帝座矣。上界晴灏，万里无纤翳，下方雷雨晦冥而不知，唯闻霹雳声细于儿啼。斯时也，目光眩瞀，魂气跃跃出圹垠，即欲乘长风而去，何之乎？或西日欲匿，东月初吐，烟霞晃射，紫翠倏弈，峰峦远近，乍浓乍淡。又或五夜闻钟声，大殿门不关，虎啸有风，飒飒去，披衣起视，则兔魄斜堕，残雪在半岭，烟光溟濛，前山不甚了了。于斯时，清冷逼人，心意欲绝。又或岳帝端居，群灵来朝，幢节参差，铃管萧萧，殿角云气，幕霞绡，恍惚可睹，似近而遥。快哉！灵人之音，何彼冷风之断之也？

"五岳而外，名山复不少矣，若四明、天台、金华、括苍、金庭、天姥、武夷、匡庐、峨嵋、终南、中条、五台、太和、罗浮、会稽、茅山、九华、林屋诸洞天福地，称仙灵之窟宅，神仙之奥区者，莫可殚数。芒履竹杖，纵不能遍历，随其力之所能到而邀焉。饮神瀵之水，问仙鼠之名，啖胡麻之饭，餐柏上之露。或绝壁危峰陡插天表，人不能到，则以索自垣而登。或石梁中断，玉扉忽开，奋而阑入，无恐谽砑，縠之洞，深黑而不见底，仅通一线，仰逗天光，以火自爇而入焉无恐，以寻高流羽士，肉芝瑶草，及仙人之遗蜕处。

"游于大川，若洞庭、云梦、瞿塘、巫峡、具区、彭蠡、扬子、钱塘，空阔浩森，鱼龙神怪之所出没。微风不动，空如镜也；神龙不怒，抱珠卧也。水光接天，明月下照，龙女江妃，试轻绡，蹑文履，张羽盖，吹洞箫而去，凌波径渡，良久而灭，胡其冷爽也。恶风击之，洪涛隐起，鸥夷贾怒，天吴助之。大地若磨焉，寓县若簸焉，恍乎张龙公挟九子，擘青天而飞去，胡其险壮也。又秀媚靓妆，莫如虎林之西湖。杨柳夹岸，桃花临水，则丽华、贵嫔之开晓镜也。菱叶吐华，芙渠濯濯，朝光澄鲜，芳香袭人，则宜主、合德之出浴也。天清日朗，风物明媚，朱阁朝临，兰桡夕泛，则杨家妃子之笑也。烟雨如黛，群山黯淡，奇绝变幻，亦大可喜，则吴王西子之颦也。"

丁　回到尘世

冥寥子散步西泠六桥，已而深入天竺灵鹫，礼古先生。罢而出，访丁野鹤于烟霞石屋之间。入潮音落迦，则冥寥子之家山也，观音大士道场在焉。采莲花而观大海，岂不胜哉！

意兴既远，汗漫而行万里，足下耳目，偶惬其性，或旬日居之，终朝趺坐，以炼三宝。《道德》五千言，其窍与妙乎？玉清金笥，其忘与觅乎？扶桑玉书，其不问邻乎？阴符二篇，其机在目乎？太上指其观心，古佛操其定慧。因禅定以求参同，则兀如非枯也。

仙灵之宫，真如之寺，金身妙相，焜耀如月。烛既明矣，香既清矣，羽人衲子，分蒲团而坐，啜茗进果，翻经阅藏。小倦则相与调息，入定，久之而起，则月在藤萝，萧籁阒然，沙弥以头触地，童子据药炉而眠，于斯时，虽有尘心，何由而入也？

若在旷野，矮墙茅屋，酸风吹扉，淡日照林，牛羊归乎长坂，饥鸟噪于平田，老翁敝衣乱发，而曝短桑之下，老妇以瓦盆贮水而进麦。当其情境凄绝，亦萧瑟有致哉！若道人之游，以此为厌薄，则不如无游也。

若入通都大邑，人烟辐辏，车马填委，冥寥子行歌而观之：若集百货，若屠沽者，若倚门而讴者，若列肆而卜者，若聚讼者，若戏鱼龙角抵者，若樗蒲蹴踘者，冥寥子无不寓目焉。兴到，入酒肆，沽浊醪，焚枯鱼生菜，两人对饮；微醒，长吟采芝之曲，徘徊四顾，意豁如也。惊诧市人，何物道者，披蓝缕萧然，而风韵乃尔乎？众共疑之。盖仙人云，须臾，径去不见。

高门大第，王公贵人，置酒为高会，金盈座，玉盘进醴，堂上乐作，歌声遏云，老隶守门，拄杖在手。道人闯入乞食焉。双眸炯碧，意度轩轩，而高唱曰："诸君且勿喧，听道人歌花上露。"

> 花上露，
> 何盈盈，
> 不畏冷风至，
> 但畏朝阳生。
> 江水既东注，
> 天河复西倾；
> 铜台化丘陇，
> 田父纷来畊。
> 三公不如一日醉，
> 万金难买千秋名。
> 请君为欢调凤笙！

> 花上露，
> 浓于酒，
> 清晓光如珠，

如珠惜不久。

高坟郁累累，

白杨起风吼；

狐狸走在前，

狝猴啼其后。

流香渠上红粉残，

祈年宫里苍苔厚。

请君为欢早回首！

　　歌罢，若有一客怒曰："道者何为？吾辈饮方欢，而渠乃来败人意。"亟以胡饼遣之。道人则受胡饼趋出。一客谓其从者曰："急追还道者。"前一客曰："饮方欢，恨渠来溷人。以胡饼逐之善矣，何故追还？"后一客曰："仆察道者有异，欲令还而熟视之。"前一客曰："乞儿也！何异之有彼？渠意所需，一残羹冷炙而足。"又一客曰："味初歌词，小不类乞者。"

　　座上若有一红绡歌姬离席曰："以儿所见，此道者，天上谪神仙也。儿察其眉宇清淑，音吐俊亮。谬为乞儿状，而举止实微露其都雅。歌辞深秀，乃金台宫中语，固非人间下里之音，况吐乞儿口哉！神仙好晦迹而游人间，乞追之，勿失。"

　　最后一客曰："何关渠事，亦饮酒耳，试令追还道者，固无奇矣。"

　　红绡者不服，曰："儿固与诸公无缘。"

　　又若有一青绡者复离席曰："诸公等以此为赌墅可乎？试令返道者，果有异，则言有异者胜；返之而无奇，则言无奇者胜。"诸公大哄曰："善。"令从者追之，则化为乌有先生矣。从者返命，前一客曰："吾固知其不可测也。"红绡者愀然曰："是甫出门而即乌有耶，惜哉，失一异人！"

　　冥寥子曳杖逍遥而出郭门。连经十数大城，皆不入。至一处，见峰峦背郭，楼阁玲珑，琳宫梵宇，参差掩映，下临清池。时方春日韶秀，鸟鸣嘉树，百卉敷荣，城中士女，新妆服。雕车绣鞍，竞出行春。或荫茂树而飞觥，或就芳草而布席，或登朱楼，或棹青雀，或并辔而寻芳，或连袂而踏歌。冥寥子乐之，为之踟蹰良久。

　　俄而有一书生，肤清神爽，翩翩而来。长揖冥寥子曰："道者亦出行春乎？仆有少酒在前溪小阁樱桃之下，朋侪不乏，而欲邀道者助少趣，能从我去否？"

　　冥寥子欣然便行，至其处，若见六七书生，皆少年俊雅。先一书生笑谓诸君曰："吾辈在此行春，无杂客，适见此道差不俗，今日之尊罍，欲与道者共之。诸君以为何如？"咸应曰："善。"

　　于是以次就坐，道者坐末席。酒酣畅洽，谈议横生，臧否人物，扬挖风雅。有称怀春之诗者，有咏禾黍之篇者，有谈廊庙之筹策者，有及山林之远韵者，辨博纷纶，各极其至，道人在座，饮啖而已。先书生虽在剧谈中，顾独数目道人，曰："道者安得独无言？"道人曰："公等清言妙理，听之欣赏而不能尽解，又何能出一辞？"

　　少选，诸君尽起行陌上，折花攀柳。时多妖丽，藤芜芍药，往往目成。而道人独行入山径，良久而出。诸君曰："道者独行入山何为？"曰："贫道适以双柑斗酒，往听黄鹂声耳。"一书生曰："道者安得作许语，差不俗。庸知非黄冠中之都水、贺监耶？"道人深自谦抑。

　　诸君复还就坐，一人曰："今日之游，不可无作。"一人应曰："良是。"

　　有一人则先成一诗曰：

疏烟醉杨柳，

微雨沐桃花；

不畏清尊尽，

前溪是酒家。

一人曰：

厨冷分山翠，

楼空入水烟；

青阳君不醉，

风雨送残年。

……

道人曰："诸公诗各佳甚。"一人曰："道人能赏我辈之诗，必善此技，某等愿闻。"道人起立，谦让再三，诸君固请不辍，道人不得已……乃吟曰：

沿溪踏沙行，

水绿霞红处；

仙犬忽惊人，

吠入桃花去。

诸君大惊，起拜曰："咄咄，道者作天仙之语，我辈固知非常人也。"于是竞问道人姓名，但笑而不答。问者不已，道人曰："诸公何用知道人名，云水野人，邂逅一笑，即见呼以'云水野人'可矣。"诸君既心异道人，于是力欲挽入城郭，道人笑曰："贫道浪游至此，四海为家，诸公谬爱，即追随入城，无所不可。"

　　遂相携入城，以次更宿诸君家。自是或登高堂，或入曲房，或文字之饮，或歌舞之场，道人无不往者。城中传闻有一"云水野人"，好事者争相致之，道人悉赴。人与之饮酒，即饮酒；与之谈诗文，即谈诗文；挈之出游，即出游；询以姓名，则笑而不答。其谈诗文，剖析今古，规合体裁颇核；或称先王，间及世务，兼善诙谐。人愈益喜之。而尤习于养生家言。

　　偶观歌舞，近靡曼，或调之以察其意，道人欣然，似类有标韵者。至主人灭烛留髡，燕笑媟狎，即正容危坐，人莫能窥。夜尝少卧，借主人一蒲团，结跏趺其上，倦则即其上假寐而已。人以此益异焉。

　　居月余，一日忽告去。诸君苦留之，不可得。各出金钱布帛诸物相赠，作诗送行。临别，诸公皆来会，惆怅握手，有泣下者。冥寥子至郭门，第备足百钱，悉出诸公所赠诸物，散给贫者而去。诸公闻之益叹息，莫测所以。

　　戊　出游的哲学

　　冥寥子行，出一山路，深窅峭险，乔木千章，藤萝交荫，仰视不见天日。人烟杳然，樵牧尽绝。但闻四旁鸟啼猿啸，阴风肃肃而恐人。冥寥子与其友行许久。忽见一老翁，庞眉秀颊，目有绿筋，发垂两肩，抱膝而坐大石之上，冥寥子前揖之。老翁为起，注目良久，不交一言。冥寥子长跽进曰："此深山无人处，安得有跫然者？翁殆得道异人也。弟子生平好道，中岁无闻，石火膏油，心切悲叹，愿垂慈旨以开迷。"老翁佯为弗闻。固请之，乃稍教以虚静无为之旨。无何别去，目送久之而灭。山深境绝处，安得无若而翁者耶？

　　又或随其所到，有故人在焉——畴昔以诗文交者，以道德交者，以经济交者，以心相知者，以气相期者，思一见之，则不复匿姓名，径造其家。

故人见肃，见冥寥子衣冠稍异，怪问之。答曰："余业谢人间事，通明季真吾师也。"曰："婚嫁毕乎？""未也。以俟其毕，如河之清？向子平去则不返，余犹将指家山，聊以适我性尔。"于是款之清斋，追往道故数十年之前，俯仰一笑，俱属梦境。友人乃低回慨叹且羡："冥寥子其无累之人耶。夫贵势高张，荣华渗溂，人之所易溺也。白首班行，龙钟盘跚，犹恋其物而不肯舍。一旦去之，攒眉向人。业问车马而迟行，出国门而回首。既返田舍，不屑屑焉艺种理麻豆面，日夜间长安之耗，而遗书当路故人焉。胸中数往数来，直至属纩乃已。有大拜命下之日，即其属纩之辰，有目瞑数时，而朝使使后至者，大可笑也。子何修而早自脱屣若此？"

冥寥子曰："余闲中观焉，殆有所伤而悟也。余观于天：日月星汉，何冗而早夜西驰？今日之日，一去即失；虽有明日，非今日矣。今年之年，一去即失；虽有明年，非今年矣。天日自幻，吾日自短，三万六千朝而外，吾不得而有也；天年自多，吾年自少，百岁而外，吾不得而有也。又况其所谓'百'者，所谓'三万六千'者，人生常不得满。而其间风雨忧愁，尘劳奔走之日常多；良时嘉会，风月美好，胸怀宽闲，精神和畅，琴歌酒德，乐而婆娑者，知能几何？

"日月之行，疾于弹丸。当其辚辘而欲堕西岩，虽有拔山扛鼎之力，不能挽之而东；虽有苏、张之口，不能说之而东；虽有樗里晏婴之知，不能转之而东；虽有触虹、蹈海之精诚，不能感之而东。古今谈此事以为长恨。

"余观于地：高岸为谷，深谷为陵，江湖汤汤，日夜东下而不止。方平先生曰：'余自接待以来，已三见沧海为桑田矣。'

"余观于万物：生老病死，为阴阳所摩，如膏之在鼎，火下熬之，不斯须而乾尽；如烛在风中，摇摇然泪枯烬落，顷刻而灭；如断梗之在大海，前浪推之，后浪叠之，泛泛去之而莫知所栖泊。又况七情见戕，声色见伐，忧喜太极，思忧过劳，命无百年之固，而气作千秋之期，身坐膏火之中，而心营天地之外，及其血气告衰，神明不守，安得不速坏乎？

"王侯将相，甲第如云，击钟而食，动以千指。平旦开门，宾客拥入；日昃张宴，粉黛成行。道人过之，呵声雷鸣，而不敢窥；后数十年又过之，则蔓草瓦砾，被以霜露，风凄日冷，不见片瓦，儿童放牛牧豕之场，乃畴昔燕乐鼓舞处也。方其鼎盛豪华，谐谑欢笑时，宁知遂有今日。大荣衰歇，何其一瞬也！岂止金谷铜台，披香太液，经百千年而后沦没哉？暇日出郭，登丘陇，郁郁累累，燕韩耶？晋魏耶？王侯耶？厮养耶？英雄耶？子耶？黄壤茫茫，是乌可知？吾想其生时耽荣好利，竞气争名，规其所难图，而猎其所无益；忧劳经营，畴不其然，一朝长寝，万虑俱毕。

"余尝宿于官舍，送往迎来，不知其更几主宰也。余尝阅乎朝籍，去故登新，不知其更几名也。余尝出关门，临津渡，陟高冈，眺原野，舟车络驿，山川莽苍，不知其送人几许也。叹息沉吟，或继以涕泗，则吾念灰矣。"

友人曰："晏子有言：'古而无死，则爽鸠氏之乐也。'齐景公流涕悲伤，识者讥其不达。今吾子见光景之驶疾，知代谢之无常，而感慨系之，至于沉痛，得毋屈达人之

识乎？"

冥寥子曰："不然。代谢故伤，伤乃悟也。齐景公恨荣华之难久，而欲据而有之，以极生人之乐，我则感富贵之无常，而欲推而远之，以了性命之期，趋不同也。"

曰："于今者遂已得道乎？"

冥寥子曰："余好道，非得道者也。"

曰："子好道，而游者何？"

冥寥子曰："夫游，岂道哉！余厌仕路踶踌，人事烦嚣，而聊以自放者
也。欲了大事，须俟闭关。"

曰："子一瓢一衲，行歌乞食，有以自娱乎？"

冥寥子曰："余闻之师，盖有少趣在澹。烹羊宰牛，水陆毕陈，其始亦
甚甘也。及其厌饱膨膨脝，滋觉甚若，不如青苏白饭，气清体平，习而安
之，殊有余味。妖姬娈童，尽态极妍，挝鼓吹笙，满堂鼎沸，其始亦甚乐
也。及其兴尽意败，转生悲凉，不如焚香摊书，兀兀晏坐，气韵萧疏，久
而益远。某虽尝滥进贤冠，家无负郭，橐无阿堵，止有图书数卷，载之以
西，波臣惧为某累，一举而捐之水滨。此身之外，遂无长物，境寂而累遣，
体逸而心闲，其趣讵不长哉？一衲一瓢，任其所之，居不择处，与不择物。
来不问主，去不留名。在冷不嫌，入嚣不涸。故我之游，亦学道也。"

其人乃欣然而喜曰："聆子之言，如服清凉散，不自知，其烦热之去
体也。"

……

顷之，一少年来，载手而骂冥寥子曰："道人乞食，得食则去，饶舌
何为？是妖人也。吾且闻之官。"攘臂欲殴冥寥子，冥寥子笑而不答。或
劝之，乃解。

于是冥寥子行歌而去，夜宿逆旅，威有妇人，冶容艳态，而窥于门，
须臾渐迫，微辞见调。冥寥子私念："此非妖也耶？"端坐不应。妇人曰：
"吾仙人也，愍子勤心好道，故来度子。且与子宿缘，幸无见疑。吾将与
子共游于蓬莱度索之间矣。"冥寥子又念："昔闻成子学道荆山，试而不遇，
卒为邪鬼所惑，失其左目，遂不得道而绝。《真诰》以为犹是成子用志不
专，颇有邪心故也。夫鬼狐惑人，伤生殒命，固也，不可近。即圣贤见

试，不遇，亦非所以专精而凝神也。端坐如初。妇人瞥然不见。为鬼狐，为魔试，皆不可知矣。

冥寥子游三年，足迹几遍天下。目之所见，耳之所闻，身之所接，物态非常，情境靡一，无非炼心之助。虽浪迹亦不为无补哉！

于是归而茸一茆四明山中，终身不出。

II. "The Travels of Mingliaotse"

a . The Reason for the Flight

MINGLIAOTSE was once an official, and he was tired of the ways of the world, of having to say things against his heart and to perform ceremonies against good form. What is"to say things against ones heart? "A host and his visitor make a low bow to each other, and after a few casual remarks about the weather, dare not make another comment. People we have met for the first time shake hands with us and insist they are our bosom friends, but after they have parted from us, we are totally indifferent to each other. When we praise a person, we compare him to the saint Poyi, and as soon as he has turned away, we talk behind his back and compare him to the thief Cheh. And when we are sitting at ease enjoying a conversation, we try to preserve a curt dignity, although we have so much we should like to say to each other; and we gabble about noble ideals, while we have immoral conduct. Being afraid that to unbosom our heart would betray the truth and to tell the truth would hurt, we lay these thoughts aside and let the conversation drift aimlessly on trivial topics. Sometimes we even play the actor and sigh or shout to cover up our thoughts, so that our ears, our eyes, our mouth and our nose are no more our own, and our anger, our joy, our laughter and our condemnations are no longer genuine. Such is the established convention of society, and there is no way of rectifying it. And what is"to perform ceremonies against good form?"In dealings with our fellowmen, no matter of what rank, we bow and kowtow the whole day, although they are our old friends. We dissociate ourselves without reason from some, as if they were our mortal enemies, and equally without reason try to get close to others, although they have no real affinity to us. Hardly has a nobleman opened his mouth when we answer,"Yes, sir!"with a roar, and yet he need only raise an arm, and our heads may be chopped off. We see two people calling on each other, and although they hate to see each other's faces, they spend their days busily dismounting from their horses and leaving their cards. Now calling on a friend to inquire after his welfare should not be merely an empty form. Did the ancient kings who established these ceremonies mean it to be this way? We put on our gowns and girdles, feeling like caged monkeys, so that even when a louse biles our body and our skin itches, we cannot scratch it. And when we are walking at leisure in the streets, we are afraid of disobeying the law. Immediately our eyes look at our nose and we dare not look beyond a short distance, and if we look beyond a short distance, other people will look at us and try to detect what we are doing. When we want to ease ourselves and the feeling is intense, we hardly dare to stop without some excuse. The higher officials are ever mindful of the sword in front and other people's criticism behind. The cold and hot seasons disturb their bodies, and the desire of possession and the fear of loss trouble their hearts. Thus they suffer greater loss than comes from the mere fear of being incorrect. Even the noblest and most chivalrous spirits, who have a sense of wise disenchantment and are pleased with their own being, fall into this trap once they have become officials. So, wishing to emancipate his heart and liberate his will, Mingliaotse sets forth to travel in the Country of the Nonchalant.

Some one may say:"I have heard that the follower of Tao lives in quiet and does not feel lonely, and lives in a crowd and does not feel the noise. He lives in the world and yet is out of it, is without bondage and without need for emancipation, and soon a willow tree grows from his left armpit and a bird makes its nest on the top of his head. This is the height of the culture of quietism and emancipation. To be a servant in the kitchen, or to pick up the waste on the ground, is one of the lowest of professions, and yet the saint is not disturbed by it. Are you not making your spirit the servant of your body, when you are afraid of the restrictions of the official life, and desire to travel to unusual places?"

And Mingliaotse replies:"He who has attained the Tao can go into water without becoming wet, jump into fire without being burned, walk upon reality as if it were a void and travel on a void as if it were reality. He can be at home wherever he is and be alone in whatever surroundings. That is natural with him. But I am not one who has attained the Tao, I am merely a lover of Tao. One who has attained the Tao is master of himself, and the universe is dissolved for him. Throw him in the company of the noisy and the dirty, and he will be like a lotus flower growing from muddy water, touched by it, yet unstained. Therefore he does not have to choose where to go. I am yet unqualified for this, for I am like a willow tree following the wind—when the wind is quiet, then I am quiet, and when the wind moves, I move, too. I am like sand in the water—which is clean or muddy as the water is. I have often achieved purity and quiet for a whole day and then lost it in a moment, and have sometimes achieved purity and quiet for a year and then lost it in a day. It has not been possible for me to let everything alone and not be disturbed by material surroundings. If an emperor could follow the Tao, why did Ch'ao Fu and Hsü Yu have to go to the Chi Hill and the Ying River? If a prince could follow the Tao, why did Sakyamuni have to go to the Himalayas? If a duke could follow the Tao, why did Chang Liang have to ask for sick leave? And if a minor official could follow the Tao, why did T'ao Yüanming have to resign from his office? I am going to emancipate my heart and release my spirit and travel in the Country of the Nonchalant."

"Let me hear about your travels,"the friend says, and Mingliaotse replies:

"One who travels does so in order to open his ears and eyes and relax his spirit. He explores the Nine States and travels over the Eight Barbarian Countries, in the hope that he may gather the Divine Essence and meet great Taoists, and that he may eat of the plant of eternal life and find the marrow of rocks. Riding upon the wind and sailing upon ether, he goes coolly whithersoever the wind may carry him. After these wanderings, he comes back, shuts himself up and sits looking at the blank wall, and in this way ends his life. I am not one who has attained the Tao. I would like to house my spirit within my body, to nourish my virtue by mildness, and to travel in ether by becoming a void. But I cannot do it yet. I tried to house my spirit within my body, but suddenly it disappeared outside; I tried to nourish my virtue by mildness, but suddenly it shifted to intensity of feeling; and I tried to wander, in ether by keeping in the void, but suddenly there sprang up in me a desire. And so, being unable to find peace within myself, I made use of the external surroundings to calm my spirit, and being unable to find delight within my heart, I borrowed a landscape to please it. Therefore strange were my travels. "

b . The Way of Traveling

"I go forth with a friend who loves the mountain haze and each of us carries a gourd and wears a cassock, taking with us a hundred cash. We do not want more, but try always to keep it at a hundred to meet emergencies. And we two go begging through cities and through hamlets, at vermilion gates and at white mansions, before Taoist temples and monks' huts. We are careful of what we beg for, asking for rice and not for wine, and for vegetables and not for meat. The tone of our begging is humble, but not tragic. If people give, we then leave them, and if people don't give, we also leave them; the whole object being merely to forestall hunger. If some people are rude, we take it with a bow. Sometimes when there is no place at which to beg and we cannot do otherwise, we spend one or two from the hundred cash we carry along with us, and make it up whenever possible. But we do not spend any cash unless we are actually forced to it.

"We travel without a destination and stop over wherever we find ourselves, and we go very slowly, perhaps ten li a day, perhaps twenty, or perhaps thirty, forty or fifty. We do not try to do too much, lest we feel tired. And when we come to mountains and streams, and are enchanted with the springs and white rocks and water fowls and mountain birds, we choose a spot on a river islet and sit on a rock, looking at the distance. And when we meet woodcutters or fishermen or villagers or rustic old men, we do not ask for their names and surnames, nor give ours, nor talk about the weather, but chat briefly on the charms of the country life. After a while we part company with them without regret.

"In times of great cold or great heat, we have to seek shelter, lest we be affected by the weather. On the road, we stand aside and let other people pass, and at a ferry, we wait to let other people get into the boat first. But if there is a storm we do not try to cross the water, or if a storm comes up when we are already halfway across, we calm our spirit, and leaving it to fate with an understanding of life, we say,'If we should be drowned while crossing, it is Heaven's will. Can we escape by worrying over it?'If we cannot escape, there our journey ends. If fortunately we escape, then we go on as before. If we meet some rough young fellow on the way or accidentally bump into him, and if the young man is rude, we politely apologize to him. If after the apology we still cannot escape a fight, then there our journey ends. But if we escape it, then we go on as before. If one of us falls ill, we stop to attend to the illness, and the other tries to beg a little for some medicine, but he himself takes it calmly. He looks within himself and is not afraid of death. And so a severe illness is changed into a light illness, and a light illness is immediately cured. If it is willed that our days are numbered, then there our journey ends. But if we escape it, then we go on as before. It is natural that during our wanderings, we might arouse the suspicion of detectives or guards and might be arrested as spies. We then try to escape either by our cunning or by our sincerity. If we cannot escape, then there our journey ends. But if we escape, then we go on as before. Of course we stop over for the night at a mat-shed hut or a stone lodge, but if it is impossible to find such a place, we stop for the night lying outside a temple gate, or beneath a rock cave, or outside people's house walls or beneath tall trees. Perhaps the mountain spirits and tigers or wolves may be looking upon us, and what are we to do? The mountain spirits can do us no harm, but we are unable to defend ourselves against the tigers or wolves. But haven't we a fate controlled up in heaven? We therefore leave it to the laws of the universe, and we do not even change the color of our face. If we are eaten up, that is our fate, and there our journey ends. But if we escape it, then we go on as before...

c . At Austere Heights

"As for my destinations, I visit chiefly the Five Sacred Mountains and the Four Sacred Waters and generally the sacred places on mountain tops, and secondarily include also the famous mountains and rivers of the Nine States. But I limit myself to only those parts within the jurisdiction of the Nine States and where human beings have set foot. As for those regions lying outside the Celestial Empire, like the Himalayas and the Ten Small Islands and Three Big Islands of the China Sea, I do not think I should be able to go there, not being provided with a pair of wings. I expect also to meet only traveling scholars of the lakes and rivers or retired men of the mountains; as for the various immortals, I do not think I shall be able to come across them, not being provided with an immortal body myself.

"When I go up to the Five Sacred Mountains, there I stand high above the celestial winds and look beyond the Four Seas, and the myriad mountain peaks appear like little snails, the myriad rivers seem like winding girdles, and the myriad trees appear like kale. The Milky Way seems to graze by my collar, white clouds pass through my sleeves, the eagles of the air seem within my arm's reach, and the sun and the moon brush my temples and pass by. And there I have to speak in a low voice, not only for fear I may frighten away the spirit of the mountain, but also lest it be overheard by God on His throne. Above us there is the pure firmament, without a single speck of dust in that vast expanse of space, while below us rain and thunder and stormy darkness take place without our knowledge, and the rumbling of the thunder is heard only as the gurgling of a baby. At this moment, my eyesight is dazzled by the light, and my spirit seems to fly out of the limitations of space, and I feel as if I were riding upon the far-journeying winds, but do not know where to go. Or when the western sun is about to hide itself and the eastern moon is bursting out from below the horizon, there the light of the clouds shines forth in all directions and the purple and the blue scintillate in the sky and the distant and the near-by peaks change from a deeper into a lighter hue in a short moment. Or again in the middle of the night, I hear the sound of the temple bells and the roar of a tiger, followed by a gust of rustling wind, and the door of the main temple hall being open, I slip on my gown and get up, and lo! there the Spirit of the Rabbit is reclining, and some remains of the last snowfall still cover up the upper slopes, the light of the night lies like a mass of undefined white, and the distant mountains present a hardly visible outline. At such a moment, I feel my body steeped in the cool air and all desires of the flesh have melted away. Or perhaps I see the God of the Sacred Mountain sitting in state, giving audience to the inferior spirits. There is a profusion of banners and canopies, and the air is filled with the music of the flute and the bells, and the palace roofs are clothed in a mantle of clouds and scarfs of haze, seeming to have and yet not to have a visible outline, and giving the illusion of now being so near and now being so far away. Ah, thrice happy is it to hear the music of the gods, and why is it suddenly interrupted by a gust of cold wind?

"Besides these Five Sacred Mountains, there are a number of other famous mountains, like Szeming, T'ient'ai, Chinhua, Kuats'ang, Chint'ing, T'ienmu, Wuyi, Lushan, Omei, Chungnan, Chungt'iao, Wut'ai, T'aiho, Lofu, Kweich'i, Maoshan, Chiuhua, and Linwu, and such sacred places without number, which have been called the dwellings of the fairies or the abodes of the spirits. I go forth in sandals, carrying a bamboo cane, and though I may not be able to visit them all, I wander about as far as my energy permits. I drink the water of the Gods' Faeces, inquire after the name of the Fairy Mouse, chew the rice of sesame and drink the dew of pine trees. When I come to a steep peak or overhanging precipice which rises abruptly into the sky, never scaled by man, I tie myself to a rope and climb up to the top. Coming to a broken stone bridge, or an old gate suddenly discovered open, I walk into it without fear; or coming to a rocky cave so dark that one cannot see its bottom, with but a single ray of light coming in through a crevice in its roof, I light a torch and go in by myself without fear, in the hope of finding some highminded Taoists, or immortal plants, or perhaps the bodily remains of some Taoists who have gone up to heaven.

"I visit also the famous rivers and lakes, like the Tungt'ing, the Yünmeng, the Chüt'ang, the Wuhsia, the Chüch'ü, the P'engli, the Yangtse, and the Ch'ient'ang. Such deep expanses of water are the abodes of fish, dragons, and the water spirits. When the air is calm and the water smooth like a mirror, we know that then the Divine Dragon is peacefully asleep, holding a pearl in its breast. When the lights of the water merge with the color of the sky under a clear moon, we know then it is the Princess of the Dragon King and the Mistress of the River coming out in a canopied procession, flute in hand and clad in their new scarfs of light gauze, treading in embroidered shoes upon the rippling waves. This procession continues for some time and then disappears. Ah, how cool it is then! Or a furious wind lashes upon the water and gigantic waves rise, and we know then it is the spirit of Ch'ihyi in anger, assisted by T'ienwu. Then the great earth churns about like a mill and our earthly abode shakes and rolls like a sifter, and we seem to see Old Dragon Chang breaking his way into heaven, carrying his nine sons in his arms. Ah, how magnificent it is then! Or if we go in for the gentle beauty of well-dressed women, there is no better place than the West Lake of Hangchow, where the willows line the banks and the peach flowers look at their own image in the water, and we know then it is the Imperial Consort Lihua opening her vanity box in the morning. When the water-caltrops are in bloom and the lotus flowers look fresh and gay on a bright morning and the place is filled with a subtle fragrance, we know then that the beauties Yichu and Hoteh are coming out of their bath. When the sky is clear and the sun is shining and the things of the place have a bright charm, and people are leaning upon their balconies in the vermilion towers in the morning, or boating on the lake with painted oars in the evening, we know then it is the Queen Yang Kweifei in her smiling mood. When a mist and rain hang over the lake and the many hills are enveloped in gray, changing into the most unexpected colors, it is also a source of great delight, for we know then it is Hsishih, Queen of the Wu kingdom, knitting her eyebrows.

d . Back to Humanity

Then Mingliaotse walks leisurely past the Six Bridges of Hsiling and goes up to T'ienchu and Lingch'iao, where after visiting some ancient scholars, he comes out and looks for Wild Stork Ting in some stone cave amidst the clouds. Then there is Ch'aoyin (Pootoo) which is the monastic home of Mingliaotse, and where the temple in honor of the Goddess of Mercy is situated. Mingliaotse goes there to pick lotus flowers and look at the great sea. Ah, is this not a great delight!

And so wandering farther and farther, happy of heart. Mingliaotse proceeds leisurely, covering a distance of ten thousand li on foot. And when he is pleased with what he sees or hears, he stays at a place for ten days.

[At a temple] he sits still with crossed legs to master the Three Precious Spirits. The five thousand words of Taotehching—is not the philosophy subtle and fine? The Golden Casket of Taoist books—is it already lost or still to be found? The Jade Book of Fusang—shall he not ask his neighbors concerning its whereabouts? The Two Books of Yinfu—does their secret lie right before his eyes? The Supreme Ruler guides his perceptive mind, and the Ancient Buddha directs his spiritual wisdom. And so trying to understand the law of the changing universe, he is not lonely during his contemplation.

In the temple of Buddha,   there is the gracious appearance of his golden body, irradiating a glorious halo. The candles have been lighted and the incense smoke fills the air with a light fragrance, and there the

Taoists or monks are seated in order on their straw cushions, drinking tea and eating fruit and perusing the classics. After a while, when they are tired, they control their respiration and enter the stage of quietude. After a long time, they get up, and see the moon shining from behind the wisterias, while the universe lies hushed in silence. An acolyte is kowtowing with his head against the ground, and a boy servant is taking a nap near the stove where the medicine of fairies is being stewed. At this moment how can an earthly thought enter our minds, even if it is there?

When out in the open country, he sees low walls enclosing mud huts thatched with straw. A piercing wind is blowing through the door and a mild sun is shining upon the forests. The cattle and sheep are returning home to the hillside, and hungry birds are making a noisy sound in the fields on the plain. An old farmer in ragged clothes and disheveled hair is sunning himself beneath a small mulberry tree, and an old woman is holding an earthen vessel filled with water and serving a wheat meal. When the landscape and the mood of the moment are so sad, one feels too that it is as beautiful as a picture. If a Taoist on travel should regard such views as too ordinary, he might just as well not travel at all.

On entering a big city, where crowds jostle and the traffic of carts and horses fills the streets, Mingliaotse goes along singing and observing the people—storekeepers, butchers, minstrel singers, fortune-tellers, people occupied in a dispute, jugglers, animal trainers, gamblers and sportsmen. Mingliaotse looks at them all. And when the spirit moves him, he enters a wine shop and orders some strong wine with dried fish and green vegetables, and the two men drink across a table. Thus getting warmed up, they sing the ditty Gathering the Immortal Plant, and look about, supremely satisfied with themselves. The people of the street wonder at the sight of these two ragged souls carrying themselves with such an air of charm and happiness, and suspect that they are perhaps fairies incarnated. After a short while, they suddenly disappear altogether.

In the great mansions behind tall gates, dukes and princes or officials of high ranks are having a wine feast. Food is being served on jade plates, and beauties are sitting around the table. An orchestra is playing in the hall and the sound of song pierces the clouds. An old servant with a cane in hand is watching the door. Mingliaotse goes right in to beg for food. With his bright, wide-open eyes and a dignified air, he shouts to the company, "Stop all this noise and listen to a Taoist singing the song of 'Dew Drops on Flowers'":

Dew drops on flowers,
Oh, how gay!
Fear not the cutting wind,
But dread the coming day!
Eastward flows the River,
Westward the Milky Way.
Farmers till the field where
Once the Bronze Tower lay.
Better to have got
A day with this all-precious pot
Than future names remembered not.
Oh, make merry while ye may!

Dew drops on flowers,
Oh, how bright!
So long they last, they shine
Like pearls in morning light,
Where grave mounds dot the wilds,

And winds whine through the night;
Foxes' howls and screech-owls
On poplars ghastly white.
See where red leaves blow,
Down on the Fragrant Gullet flow,
And mosses over Ch'inien Palace grow.
Oh, make merry while ye might!

After Mingliaotse has finished his song, one of the guests seems to be angry and says:"Who is this Taoist to spoil our enjoyment in the midst of our wine feast? Give him a piece of sesame cake and send him away!"Mingliaotse receives the cake and leaves. Then another guest speaks to his attendant, saying,"Quick! Ask that Taoist to come back!""But we were just enjoying our wine,"says the former,"and he came to spoil our pleasure. That was why I sent him away with a piece of sesame cake. It's just fine. Why do you want him to come back?""It seems to me,"replies the latter,"there's something unusual about this Taoist, and I want to ask him to return to have a good look at him.""Why, he is only a beggar!"replies the former."What is unusual about him? All he wants is a cold dish of leftovers."Then another guest joins in and says,"It doesn't seem from the song he sang that he is just a beggar. "

At this moment, a sing-song girl in red gauze rises up from her seat and says,"According to my humble opinion, this Taoist is a fallen fairy from heaven. His eyes and forehead are delicately formed and his voice is strong and clear. He is disguised as a beggar, but somet1hing in his behavior betrays his noble breeding. The song he sang is graceful and deep in meaning, more like the song of fairies in heaven than the song of men on earth. What beggar could have sung such a song! He is a fairy traveling among mortals in disguise. Please ask him to come back, for we must not lose him.""What has all this to do with him?"says the last one."All he wants is perhaps a drink of wine.

You ask him to come back and we'll find out that he is a common fellow after all. "

The girl in red gauze is still unconvinced and remarks, "Well, all I can say is, we haven't the luck to meet with immortals."

Then another girl in green gauze rises up from her seat and says, "Will the gentlemen make a bet with me? Ask the Taoist to return, and if he is an extraordinary person then those who say he is extraordinary win the bet, and if we find that he is a common fellow, then those who say he is a common fellow win.""Good!"shout the gentlemen together. They then send a servant to go after Mingliaotse, but he has completely disappeared, and the servant returns with the news."I knew that he was no common fellow!"says the former."Alas, we have just lost an immortal!"says the girl in red gauze."Why, he just went out of the door and has completely disappeared! "

Mingliaotse then proceeds with his cane and leisurely passes out of the outer city gate. He passes by a dozen big cities without entering one of them, until he comes to a place where he sees a city wall nestling against a mountain range. There are fine, tall towers and spacious, magnificent temples, whose roofs overlap one another in irregular formations, overlooking a clear pond below. It is a beautiful spring day, birds are singing on splendid trees and all the flowers are in their full glory. The men and women of the city, clad in their new clothes and riding in carved carriages or sitting on embroidered saddles, have come out of the city to"pace the spring."Some are drinking in the shade of tall trees, and some have spread a mat on the fragrant grass, and others have climbed up to a high vermilion tower, or are rowing on"green sparrow"boats; again others are riding shoulder to shoulder to visit the flowers, or are walking hand in hand and singing folk songs. Mingliaotse feels extremely happy and hangs about for a long time.

By and by a scholar with a clean face and nice complexion appears, coming along gracefully in his long gown. Making a low bow to Mingliaotse, he says, "Do Taoists come out to pace the spring, too? I have a few friends having a picnic over there, beneath the cherry trees in front of the little tower across the river. It is a jolly company and I shall be much pleased if you can join us. Can you come along? "

Mingliaotse gladly follows the young man, and when he arrives, he sees six or seven scholars, all handsome and young. The first young man introduces him to the company with a smile. "My friends, this is just a spring party among ourselves. I just met this Taoist gentleman on the road and saw he was not at all vulgar, and I therefore propose that we share our cups with him. What do you think?" "Good!" they all reply.

So then they all take their seats in order and Mingliaotse sits at the end of the table. When sufficient wine has been served and everybody is tipsy and happy, the conversation waxes more and more brilliant, and they pass witty remarks about the different people and the gentry. Some declaim poems celebrating the spring, some sing the song of gathering flowers, some discuss the policies of the court, and some tell of the secluded charm of hills and woods. There is then an exciting conversation going on, with each trying to outdo the others, while the Taoist merely occupies himself by chewing his rice. The first young man looks several times at Mingliaotse amidst his busy conversation, and says, "We must hear something from this Taoist teacher, too." And Mingliaotse replies, "Why, I am just enjoying the many fine and wise things you gentlemen have been saying, and have not been able to understand them all. How can I contribute anything to your conversation?"

After a while, the company rise up to take a walk in the rice fields, some plucking flowers and others pulling willow branches on the way. The place is full of beauties, and everywhere one's eyes turn, one sees beautiful peonies and mimu. But Mingliaotse wanders alone into a hill path, and comes out again after a long while. "Why did you go alone?"the gentleman ask."I was going with two oranges and a gallon of wine to listen to the orioles,"replies Mingliaotse."Why, this is a most extraordinary man, by the way he talks,"says one scholar, and Mingliaotse replies by a courteous remark concerning his unworthiness. So then the company sit down again, and one man says, "It won't do to go home from such an outing without writing some poems,"and another expresses his approval.

Soon one person has completed his poem first, which reads:

The willows drunk with circling haze,
And rain-bathed peach flowers brightly gleam.
Fear not thy fragrant cup be empty;
A tavern lies across the stream.

Another finishes his poem, which reads:

My kitchen shares the mountain green;
My tower is sprinkled wet with spray.
If ye drink not when spring is here,
Soon comes the windy, wintry day.

After three other persons have contributed their quatrains, Mingliaotse is invited to make his contribution. He rises to his feet, and after some expressions of his diffidence and the friends' insistence, he sings:

I tread along the sandy bank,
Where clouds are golden, water clear;
The startled fairy hounds go barking—

Into the peach grove disappear.

Amazed by this poem, the company rise from their seats and make a low bow to Mingliaotse. "Tut! Tut! To hear such celestial words from a monk! We knew that you were an extraordinary person. And they all come round to ask for his name and surname, but Mingliaotse only smiles without giving reply. As they still insist, Mingliaotse says, "What do you want to know my name for? I am merely a rustic person wandering among clouds and waters, and we've met with a smile. You can just call me 'The Rustic Fellow of Clouds and Waters.'" This intrigues the company still more, and they express their desire to invite him to go into the city with them. "I am merely a poor monk enjoying a vagabond's travel, and all the world is my home," replies Mingliaotse with a smile. "But since you are so kind, I will come along. "

They then go back to the city together, and Mingliaotse stays at their homes by turn. During the days that follow, he finds himself now in a rich man's hall, and now in a well-hidden small studio, now enjoying a literary wine dinner and now watching performances of dance and song, and Mingliaotse goes to all the places to which he is invited. The people of the city hear of the Rustic Fellow of Clouds and Waters, and the socially active people shower him with invitations, and he visits them all. When people give him drink, he drinks; when people discuss poetry and literature, he discusses poetry and literature with them; when people take him out for an excursion, he goes with them; but when they ask for his name and surname, he merely smiles without reply. In his discussion of poetry and literature, he makes very apt remarks about the ancient and modern writers and gives a penetrating analysis of their styles and forms. Sometimes he discusses the political order of the ancient kings and makes passing comments on current affairs and enchants the people still more by his witty remarks.

Especially well-versed is he in the teaching of Taoism regarding"nourishing the spirit."Sometimes, when he is watching dancing or singing which borders on the bawdy and people make ribald jokes to find out his attitude toward these things, he seems to be enjoying himself, like the romantic scholars. But when it comes to extinguishing the candle and the host asks him to stay with some girl entertainer, and when the party becomes really rowdy, he sits upright with an austere appearance, and nobody can make anything of him. When he takes a nap during the night, he asks for a straw cushion from the host and sits with crossed legs on it, and merely dozes off when he is tired. For this reason, admiration and wonder grow about him.

After more than a month's stay, he takes leave suddenly one day, against the persistent entreaty of the people. His friends give him money and cloth for presents, and write poems of farewell to him. At the farewell party, the gentlemen all come to give him a send-off; sadly they hold his hands and some shed tears. Mingliaotse arrives at the outer city gate, and after having reserved a hundred cash for himself, he distributes the gentlemen's presents to the poor and goes away. When his friends hear of this, they sigh and marvel still more, knowing not what to make of him.

e . Philosophy of the Flight

Mingliaotse then follows a mountain path, and finds himself in deep, rugged mountains. Thousands of old trees, with creepers growing on them, spread their deep shade so that one walking underneath cannot see the sky. There is not a trace of human habitation, and not even a woodcutter or a cowherd is in sight. He hears only the cries of birds and monkeys around him, and a gust of infernally cold wind makes him shiver. Mingliaotse proceeds with his friend for a long while, when they suddenly see an old man with a majestic forehead and a delicate face and green veins showing on his eyeballs. His hair falls down to the shoulders, and he is sitting on a rock, hugging his knees. Mingliaotse goes forward and makes a bow. The old man rises to his feet and looks at him steadliy for a long time, but does not say a word. Going down on his knees, Mingliaotse speaks to him, "Is Father an extraordinary person who has attained the Tao? How otherwise can I find the sound of footsteps in this deep mountain solitude? Your disciple has always loved the Tao and in his middle age has not yet found it. I feel sad at the vanity of this life which rapidly burns out like a flash from a flint, or like oil in a pan. Will you please take pity on me and disperse my ignorance?" The old man pretends not to hear him. But after Mingliaotse insists upon his request, he merely teaches him a few words about being carefree and quiet and the idea of inaction, and after a little while, goes his way. Mingliaotse's eyes follow him for a long while until he altogether disappears. How does one explain the existence of such an old man in this deep mountain solitude?

Then wandering farther on, he chances to come upon an old friend of his. Sometimes when he thinks of those people with whom he had formed a friendship on the basis of love of prose and poetry, or of respect for each other's character, or of business relations, or of personal intimacy and mutual understanding of one another's hearts, or of a mutual confidence in one another's future, he begins to desire to see them again. Then he goes straight to the home of his friend, without concealing his indentity. The friend bows to greet him, and seeing that Mingliaotse is clad in such a strange dress, is surprised and asks him some questions. "I have already retired from the world, and Chichen of T'ungming is my master," explains Mingliaotse. "Are your sons and daughters all married?" the friend asks. "No, not yet," replies Mingliaotse. "When they are all married, then I shall be free of all cares, like the clearing up of the water of the Yellow River. Tsep'ing went away and never returned home, but I am still looking forward to returning to the hills of my homeland, in order to live in harmony with my original nature."The host then gives him vegetarian food, and they begin to talk of the days twenty or thirty years ago, and surveying the past with a laugh, feel that everything has passed like a dream. The friend then bows his head and sighs, expressing his envy of the carefree life that Mingliaotse is leading.

"Are you not indeed a carefree man!" his friend says."Now wealth and power and the glories of this world are things in which people easily get drowned. I sometimes see an old man with white hair on his head marching slowly with a stoop in an official procession, still clinging to these things and unwilling to let them go. If one day he quits office, he looks about with knitted eyebrows. Inquiring if the carriage is ready, he is still slow to depart, and passing out of the city gate of the capital, he still looks back. When back at his farm, he still disdains to occupy himself with planting rice and hemp and beans, and morning and night he will be asking for news from the capital. Or he will still be writing letters to his friends at the court, and such thoughts flit back and forth in his breast ceaselessly until he draws his last breath. Sometimes an imperial order for his recall to office arrives at the moment when he is breathing his last, and sometimes the official messenger arrives with the news just a few hours after he closes his eyes. Isn't this ridiculous? How have you trained yourself that you are able to emancipate yourself from such things in good time?"

"I have looked at life in my leisure,"replies Mingliaotse."It seems that I have come to an awakening through a sense of life's tragedy. I have looked at the skies, and wondered how the sun and the moon and the stars and the Milky Way rush westward day and night like busy people. Today passes and never returns, and although tomorrow comes, it is no longer today. This year passes and never returns, and although there is next year, it is no longer this year. And so the days of Nature are steadily lengthened or unrolled, while the days of my life are daily shortened, and outside the thirty-six thousand mornings, time does not belong to me. The years of Nature are steadily unrolled, while the years of my life become steadily shortened, and beyond a hundred, they do not belong to me. Furthermore, the so-called'hundred'and the so-called'thirty-six thousand'in life are not always as we wish them to be, and among the days and years, most are passed in bad weather and sorrow and worry and running about. How many moments are there when the day is beautiful and the company enjoyable, when the moon and the wind are good and our heart is at ease and our spirit happy, when there is music and song and wine and we can enjoy ourselves and while away the hours!

"The sun and the moon pursue their courses, as fast as the bullet, and when their wheels are about to go behind the Western Precipices, the arms of the strongest man on earth cannot hold them and make them travel eastward, even the eloquence of Su Ch'in and Chang Yi cannot persuade them to travel eastward, even the wit and the strategy of Ch'ulitse and Yen Ying cannot change their minds and make them travel eastward, even the sincerity of Chingwei who knocked herself against the Rainbow and was transformed into a bird, trying to fill the sea of her regrets with pebbles, cannot touch their hearts and make them travel eastward. Writers throughout the ages who discussed this point have always held it as a matter of eternal regret.

"And I have looked at the earth, where high banks have been leveled into valleys and deep valleys have been heaved up into mountains, and the water of the rivers and lakes flows night and day eternally eastward into the sea. Fang p'ing has said,'Since I took over my duty, I have seen the sea three times changed into a mulberry field, and vice versa.'

"Again I have looked at the living things of this world, how they are born and grow old and fall ill and die, being ground thus in the mill of yin and yang, like oil in a frying pan which, being heated by fire from below, dries up in a short while; or like a candle which flickers in the wind and soon goes out, its tears being dried up and its soot fallen to the table; or like a boat cut adrift on a big sea, washed forward by successive waves and floating it knows not whither. Besides, the seven desires of man continue to burn him up and the pleasures of the flesh pare him down; he is sometimes too much disappointed and sometimes too much elated, and usually too much worried. Without more than a hundred years at his disposal, he plans to live for a thousand years, and while sitting like oil on fire, his ambitions stretch beyond the universe. Why wonder, therefore, that his being quickly deteriorates when old age comes along and his vital energy is used up and his spirit wanders away from its abode?

"I have seen princes and dukes and generals and premiers whose crowded roofs form a skyline like the clouds. When the dinner bell sounds, a thousand people are seen eating together, and when their gates are opened in the morning, crowds of visitors rush in. Day and night they give feasts and their halls are full of painted women. When a monk passes by, they shout at him with a thundering voice and he dares not even to look at the house. But after twenty or thirty years, the monk passes by again, and he sees a stretch of wild grass and broken tiles covered by dew and frost, and a cold sun is shining upon the place and a moaning wind passes by, with not a roof left standing. What was once the scene of song and dance and merrymaking is today the pasture ground of a few cowherd boys. Did they ever realize, when they were at the height of their prosperity and laughing and merrymaking, that this day would come? And why did the great glories of this world pass away in the twinkling of an eye? Was it alone, like the Garden of Chinku, the Tower of T'ungt'ai, the Hall of P'ihsiang and the Pool of T'aiyi that gradually became ruins after the passage of hundreds or thousands of years? On my leisure days,

I have passed out of the city and gone up on hills, where I saw a stretch of grave mounds. Do these belong to Yen or Han or Chin or Wei? Or were these people princes and dukes, or were they pages and servants? Or were they heroes or were they fools? How can I know from this stretch of yellow soil? I thought how they, when they were living, clung to glory and wealth, vied with one another in their ambitions and struggled for fame, how they planned what they could never achieve and acquired what they could never use. Which one of them did not worry and plan and strive? One morning their eyes closed for the eternal sleep, and they left all their worries behind.

"I have stopped over at the residences of officials and wondered how many had taken another's place as the host of the house. I have looked at the records of the personnel at court and wondered how many times old names had been struck off and new names put in. I have been at mountain passes and at ferries, and have gone up a high hill to look down upon the plain, and I have seen continuous processions of boats and carriages and wondered how many travelers they have carried away. And so I sighed in silence, and sometimes my tears dropped and my heart's desires were turned cold like ashes."

"I have heard it said by Yentse," replies his friend, "that Sangch'iu was happy over the fact that there was no death, and that the King Ching of Ch'i shed tears and was grieved on account of death, and wise people criticized him for not understanding life. Are you not also lacking in the wisdom of those who understand life, when you feel sad and even shed tears because of the swift passing of time and the instability of life?"

"No," replies Mingliaotse. "I have felt sad from the sense of the instability of life, and I have come to an awakening from this feeling of sadness. King Ching of Ch'i feared that his power and glory were temporary and wanted to enjoy them forever and exhaust the happiness of human life. On the contrary, I feel the instability of wealth and power and wish to keep them at a distance, in order to run my normal course of life. Our aims are different. "

"Have you already attained the Tao, then?"

"No, I have not attained the Tao. I am only one who loves it,"replies Mingliaotse.

"Why do you wander about, if you love the Tao?"

"Oh, no, do not confuse my wanderings with the Tao,"replies Mingliaotse."I was merely tired of the restrictions of the official life and the bother of worldly affairs, and I am traveling merely to free myself from them. As for winding up the 'great business of life,'I shall have to wait until I return and shut myself up. "

"Are you happy, going about with a gourd and a cassock and begging and singing for your living?"

"I have heard from my teacher,"replies Mingliaotse, "that the art of attaining happiness consists in keeping your pleasures mild. When people come to a feast where lambs and cows are killed and all the delicacies that come from land and sea are laid out on the table, they all enjoy them at first, but when they come to the point of satiety, they begin to feel a sense of repulsion. Much better are meals of plain rice and green vegetables, which are mild and simple and good for one's health and which will be found to have a lasting flavor, after one gets used to them. People also enjoy themselves at first in parties where there are beautiful women and boys, where some beat the drum and others play the sheng and a lot of things are going on in the hall. But after the mood is past, one gets, on the contrary, a sense of sadness. It is much better to light incense and open your book and sit quietly and leisurely, maintaining a calm of spirit, and the charm deepens as you go along. Although I was at one time of my life an official, I had no property or wealth outside a few books. At first I traveled with these books, but fearing that they might be the cause of envy on the part of the water spirits, I threw them into the water. And now I haven't got a thing outside this body. Does not then the charm of life remain for me long and lasting, when my burdens are gone, my surroundings are quiet, my body is free and my heart is leisurely? With a cassock and a gourd, I go wherever I like, stay wherever I choose and take whatever I get. Staying at a place, I do not inquire after its owner, and going away, I do not leave my name. I do not feel ill at ease when I am left in the cold, and do not become contaminated when I am in noisy company. Therefore, the purpose of my wanderings is also to learn the Tao. "

Having heard this, his friend says with a happy smile," Your words make me feel like having taken a dose of cooling medicine. The disturbing fever has left my body without my knowing it."

[Here follows a discussion on the identity of the Three Religions and the existence of God and Buddha and fairies and ghosts .]

After a while a young man comes along and, pointing his finger at Mingliaotse, shouts,"Get away from here, you beggar! A monk ought to go away quietly when he receives his food. If you keep on babbling nonsense, I must regard you as a sorcerer and prosecute you at court."The young man rolls up his sleeve as if to strike at Mingliaotse, but the latter merely smiles without making a reply. The quarrel is settled by some passer-by.

Mingliaotse goes away singing his songs. At night he stops at an inn, and there is a very well-dressed woman peeping in at the door. Gradually she approaches and begins slyly to tease him. Mingliaotse thinks to himself that she must be an evil spirit, and remains sitting alone."I am a fairy,"says the woman,"and I have come to save you because I know that you have been trying very hard to learn the Tao. Besides, I had an appointment with you in the previous incarnation. Please don't doubt me. I will accompany you to the Enchanted Land."Mingliaotse remembers that when Lü

Ch'engtse was learning the Tao at Chingshan, he was once thus deceived by a temptress and finally enslaved by an evil spirit. He lost his left eye, and died without being able to attain the Tao. Even the Classics regard the failure of Lü Ch'engtse as due to his lack of complete control of his mind and to the existence of evil desires. It is natural that, when ghosts and fox spirits tempt people, they destroy their life, and they should therefore be avoided. But even if sages and saints should make a mistake and be thus deceived, it would be a wrong way to control their mind and preserve their spirit. So he sits austerely as before and the woman all of a sudden disappears. Who can know whether she was a ghost or a fox spirit or a temptress?

Thus for three years Mingliaotse continues in his travels, wandering almost over the entire world. Everything that he sees with his eyes or hears with his ears, or touches with his body, and all the different situations and meetings, are thus used for the benefit of training his mind. And so such vagabond travel is not entirely without its benefits.

He then returns and builds himself a hut in the hills of Szeming and never leaves it again.

生活　The

的　Importance

艺术　of Living

第十二章

文化的享受

/ *Chapter Twelve* **The Enjoyment of Culture** /

一．智识上的鉴别力

　　教育和文化的目标，只在于发展智识上的鉴别力和良好的行为。一个理想的受教育者，不一定要学富五车，而只须明于鉴别善恶；能够辨别何者是可爱，何者是可憎的，即在智识上能鉴别。最令人难受的，莫过于遇着一个胸中满装着历史的事实人物，并且对苏俄或捷克的时事极为熟悉，但见解和态度完全错误的人。我曾遇见过这一类人，他们在谈话时，无论什么题目，总有一些材料要发表出来，但是他们的见地完全是可笑可怜的。他们的学问是广博的，但毫无鉴别能力。博学不过是将许多学问或事实填塞进去，而鉴别力是美术的判别问题，中国人于评论一个文人时，必拿他的学行和识见分开来讲。对于历史家尤其应该如此区别。一个满腹学问的人，或许很易于写成一部历史，但所说的话或竟是毫无主见与识别的。而在论人和论事时，或竟是只知依入门户，并无卓识。这种人就是所谓缺乏智识上的鉴别力。强记事实是一件极容易的事情。历史上一个指定时代中的事实，我们极易强记，但分别轻重和是非是一件极难的事情，有恃于一个人的见解力了。

　　所以一个真有学问的人，其实就是一个善于辨别是非者。这就是所谓鉴别力，而有了鉴别力，雅韵即会随之而生。一个人如若想有鉴别力，他必须先有见事明敏的能力，独立的判断力，不为一切社会的、政治的、文学的、艺术的或学院式诱惑所威胁或眩惑。一个人在成人时代中，四周当然必有无数各种各式的诱惑，如：名利诱惑、爱国诱惑、政治诱惑、宗教诱惑，惑人的诗人、惑人的艺术家、惑人的独裁者与惑人的心理学家。当一个心理分析家告诉我们，幼年时代的脏腑效能的种种不同运用切实有关一个人日后生活中的志向、挑衅心和责任心，或便秘症引起暴躁的性情时，凡有识力者对之只可付诸一笑。当一个人错误时，他就是错误的，不必因震于他的大名或震于他的高深学问，而对他有所畏惧。

因此识和胆是相关联的，中国人每以胆识并列，而据我们所知，胆力或独立的判别力，实在是人类中一种稀有的美德。凡是后来有所成就的思想家和作家，大多在青年时即显露出智力上的胆力。这种人绝不肯盲捧一个名震一时的诗人，他如真心钦佩一个诗人时，必会说出他钦佩的理由。这就是依赖着他的内心判别而来的，这就是我们所谓文学上的鉴别力。他也绝不肯盲捧一个风行一时的画派，这就是艺术上的鉴别力。他也绝不肯盲从一个流行的哲理或一个时髦的学说，不论他们有着何等大名做后盾。他除了内心信服之外，绝不肯昧昧然信服一个作家。如若那个作家能使他信服，那个作家就是不错的；但如若那作家不能使他信服，则那个作家是错误的，而他自己是对的，这就是智识上的鉴别力。这种智力上的胆力和独立的判断力，无疑必需一己的内心中先具着一种稚气的、天真的自信心。但一己的内心所能依赖的也只有这一点，所以一个学生一旦放弃他个人判断的权利时，便顿然易于被一切人生的诱惑动摇了。

孔子好像已经觉得学而不思比思而不学更不好，所以说："学而不思则罔，思而不学则殆。"他必因看见弟子之中这种学而不思的人太多了，所以提出这种警告。这个警告其实也是现代的学校所极为需要的。我们都知道现在一般的教育和一般的学校制度，都偏于割舍了鉴别力以求学问，视强记事实即为教育的本身目标，好像富于学问即会使人成为一名高士。但是学校中为什么要贬视思想？为什么要歪曲学制，而将愉快的求学企图变成了机械式严定尺寸划一的和被动的强记事实？我们为什么要把智识置于思想之前？我们为什么愿意称呼一个仅是读足了心理学、中古历史、伦理学和宗教学学分的大学毕业生为学成之士？这种学分和文凭何以会取代了教育的真正目标？何以会使学生们心目中也认为如此？

理由很简单：我们用这个制度，因为我们是在将民众整批地教育，如在工厂里边一般。而一涉工厂的范围，则一切都须依着呆板的机械式制度去行事了。为了保护学校的名誉和将产物标准化，学校要发给文凭，以为证明。为了须发文凭，便不能没有次第；为了须分次第，便不能没有记分；为了须记分，便不能没有大小考试了。这全部的程序，成为一个整个的合于逻辑的必然事件，而使人无从避免。但机械式大小考试，为害之大，远过于我们所能想见。因为它立刻使人注重强记事实而忽略了鉴别力的发展。我本人曾当过教师，很知道出历史题目确比一般的泛常普通智识题目较为容易，印批分数时也较为省力。

而危险在于这种制度一经订立之后，我们即易于忘却我们已渐渐或将要脱离教育的真正理想目标，即我所说的智识上鉴别力的发展。所以孔子说："多见而识之，知之次也。"这句话，仍有牢记的价值。世上实在无所谓必修科目，无必读之书，甚至莎士比亚剧本也是如此。学校好似已采用一种愚笨的概念，以为只须从历史或地理中采集若干有限的资料，便足以供一个学者所必需。我曾受过相当的教育，但我至今弄不清楚西班牙首都叫什么名字，并且有一个时期还以为哈瓦那是一个邻近古巴的海岛呢。必修课程的规定，其危险在于它义涵一个人只要读完这个课程，便已在事实上知晓了一个学者所应知晓的事情。所以一个毕业生离校之后，即不再企图更事学问或再读一些书，因为已经学完了一切应该知道的学问了。这也无怪其然，因为这是一个合于逻辑的结果。

我们须放弃一个人的智识有法子考验或测量的概念。庄子说得好："吾生也有涯，而知也无涯。"寻求学识，终不过是像去发现一个新大陆，或如阿纳托尔·法朗士（Anatole France）所说："一个心灵的探险行为。"我们如用一种坦白的、好奇的、富于冒险性的心胸去维持这个探索精神，则这种寻求行为便永远是一种快乐而不是痛苦了，我们应该舍弃那种规定的、划一的、被动的强记事实方法，而将这种积极滋长的个人快乐定为理想目标。文凭和学分一旦废除或仅仅值其所实值，学问的寻求即能趋于积极。因为那时做学生的至少要自问为什么而读书了。这句问话，在目下是无须他来答复的，因为现在每个学生都知道他为了要升入二年级，所以在一年级读书；为了要升入三年级，所以在二年级读书。这种外加的意念其实都应该丢弃，因为寻求知识完全是自己的事情，而和旁人不相干。现在的学生，有许多是为了大学的注册主任的关系而读书，有许多是为了他们的父母或教师或未来的太太而读书，以便取悦于耗费了许多金钱培植他们的父母，或以便取悦于看待他们很好很热心的教师，或以便将来可以多赚些钱去养他们的家口。我以为这类思想都是属于不道德的。寻求智识完全是自己的事情，而和旁人无干，只有如此，教育方能成为一种快乐并趋于积极。

I. Good Taste in Knowledge

THE aim of education or culture is merely the development of good taste in knowledge and good form in conduct. The cultured man or the ideal educated man is not necessarily one who is well-read or learned, but one who likes and dislikes the right things. To know what to love and what to hate is to have taste in knowledge. ... I have met such persons, and found that there was no topic that might come up in the course of the conversation concerning which they did not have some facts or figures to produce, but whose points of view were deplorable. Such persons have erudition, but no discernment, or taste. Erudition is a mere matter of cramming of facts or information, while taste or discernment is a matter of artistic judgment. In speaking of a scholar, the Chinese generally distinguish between a man's scholarship, conduct, and taste or discernment. This is particularly so with regard to historians; a book of history may be written with the most fastidious scholarship, yet be totally lacking in insight or discernment, and in the judgment or interpretation of persons and events in history, the author may show no originality or depth of understanding. Such a person, we say, has no taste in knowledge. To be well-informed, or to accumulate facts and details, is the easiest of all things. There are many facts in a given historical period that can be easily crammed into our mind, but discernment in the selection of significant facts is a vastly more difficult thing and depends upon one's point of view.

An educated man, therefore, is one who has the right loves and hatreds. This we call taste, and with taste comes charm. Now to have taste or discernment requires a capacity for thinking things through to the bottom, an independence of judgment, and an unwillingness to be bulldozed by any form of humbug, social, political, literary, artistic, or academic. There is no doubt that we are surrounded in our adult life with a wealth of humbugs: fame humbugs, wealth humbugs, patriotic humbugs, political humbugs, religious humbugs and humbug poets, humbug artists, humbug dictators and humbug psychologists. When a psychoanalyst tells us that the performing of the functions of the bowels during childhood has a definite connection with ambition and aggressiveness and sense of duty in one's later life, or that constipation leads to stinginess of character, all that a man with taste can do is to feel amused. When a man is wrong, he is wrong, and there is no need for one to be impressed and overawed by a great name or by the number of books that he has read and we haven't.

Taste then is closely associated with courage, as the Chinese always associate shih with tan, and courage or independence of judgment, as we know, is such a rare virtue among mankind. We see this intellectual courage or independence during the childhood of all thinkers and writers who in later life amount to anything. Such a person refuses to like a certain poet even if he has the greatest vogue during his time; then when he truly likes a poet, he is able to say why he likes him, and it is an appeal to his inner judgment. This is what we call taste in literature. He also refuses to give his approval to the current school of painting, if it jars upon his artistic instinct. This is taste in art. He also refuses to be impressed by a philosophic vogue or a fashionable theory, even though it were backed by the greatest name. He is unwilling to be convinced by any author until he is convinced at heart; if the author convinces him, then the author is right, but if the author cannot convince him, then he is right and the author wrong. This is taste in knowledge. No doubt such intellectual courage or independence of judgment requires a certain childish, na?ve confidence in oneself, but this self is the only thing that one can cling to, and the moment a student gives up his right of personal judgment, he is in for accepting all the humbugs of life.

Confucius seemed to have felt that scholarship without thinking was more dangerous than thinking unbacked by scholarship; he said,"Thinking without learning makes one flighty, and learning without thinking is a disaster."He must have seen enough students of the latter type in his days for him to utter this warning, a warning very much needed in the modern schools. It is well known that modern education and the modern school system in general tend to encourage scholarship at the expense of discernment and look upon the cramming of information as an end in itself, as if a great amount of scholarship could already make an educated man. But why is thought discouraged at school? Why has the educational system twisted and distorted the pleasant pursuit of knowledge into a mechanical, measured, uniform and passive cramming of information? Why do we place more importance on knowledge than on thought? How do we come to call a college graduate an educated man simply because he has made up the necessary units or weekhours of psychology, medieval history, logic, and"religion"? Why are there school marks and diplomas, and how did it come about that the mark and the diploma have, in the student's mind, come to take the place of the true aim of education?

The reason is simple. We have this system because we are educating people in masses, as if in a factory, and anything which happens inside a factory must go by a dead and mechanical system. In order to protect its name and standardize its products, a school must certify them with diplomas. With diplomas, then, comes the necessity of grading, and with the necessity of grading come school marks, and in order to have school marks, there must be recitations, examinations, and tests. The whole thing forms an entirely logical sequence and there is no escape from it. But the consequences of having mechanical examinations and tests are more fatal than we imagine. For it immediately throws the emphasis on memorization of facts rather than on the development of taste or judgment. I have been a teacher myself and know that it is easier to make a set of questions on historical dates than on vague opinions on vague questions. It is also easier to mark the papers.

The danger is that after having instituted this system, we are liable to forget that we have already wavered, or are apt to waver from the true ideal of education, which as I say is the development of good taste in knowledge. It is still useful to remember what Confucius said:"That scholarship which consists in the memorization of facts does not qualify one to be a teacher."There are no such things as compulsory subjects, no books, even Shakespeare's, that one must read. The school seems to proceed on the foolish idea that we can delimit a minimum stock of learning in history or geography which we can consider the absolute requisite of an educated man. I am pretty well educated, although I am in utter confusion about the capital of Spain, and at one time thought that Havana was the name of an island next to Cuba. The danger of prescribing a course of compulsory studies is that it implies that a man who has gone through the prescribed course ipso facto knows all there is to know for an educated man. It is therefore entirely logical that a graduate ceases to learn anything or to read books after he leaves school, because he has already learned all there is to know.

We must give up the idea that a man's knowledge can be tested or measured in any form whatsoever. Chuangtse has well said,"Alas, my life is limited, while knowledge is limitless!"The pursuit of knowledge is, after all, only like the exploration of a new continent, or"an adventure of the soul,"as Anatole France says, and it will remain a pleasure, instead of becoming a torture, if the spirit of exploration with an open, questioning, curious and adventurous mind is maintained. Instead of the measured, uniform and passive cramming of information, we have to place this ideal of a positive, growing individual pleasure. Once the diploma and the marks are abolished, or treated for what they are worth, the pursuit of knowledge becomes positive, for the student is at least forced to ask himself why he studies at all. At present, the question is already answered for the student, for there is no question in his mind that he studies as a freshman in order to become a sophomore, and studies as a sophomore in order to become a junior. All such extraneous considerations should be brushed aside, for the acquisition of knowledge is nobody else's business but one's own. At present, all students study for the registrar, and many of the good students study for their parents or teachers or their future wives, that they may not seem ungrateful to their parents who are spending so much money for their support at college, or because they wish to appear nice to a teacher who is nice and conscientious to them, or that they may go out of school and earn a higher salary to feed their families. I suggest that all such thoughts are immoral. The pursuit of knowledge should remain nobody else's business but one's own, and only then can education become a pleasure and become positive.

二．以艺术为游戏和个性

艺术是创造，也是消遣。这两个概念中，我以为以艺术为消遣，或以艺术为人类精神的一种游戏，是更为重要的。我虽然最喜欢各式不朽的创作，不论它是图书、建筑或文学，但我相信只有在许多一般的人民都喜欢以艺术为消遣，而不一定希望有不朽的成就时，真正艺术精神方能成为普遍面弥漫于社会之中。这正如学校中的学生，重在要他们多数能随便玩玩网球或足球，而不必定求他们能产生少数几个能加入全国竞赛的锦标运动员或球员。儿童或成人，也重在能创作一些物事以为消遣，而不必定求其能产生一个罗丹（Rodin，十九世纪之法国大雕刻家）。我宁愿学校中教授儿童做些塑泥手工，宁愿一切银行总理和经济专家能自制圣诞贺卡，无论这个思想是如何的可笑，而以为这样实在较胜于少数几个艺术家为了职业关系而从事这些工作。换句话说，我赞成一切的业余主义。我喜欢业余哲学家、业余诗人、业余植物学家和业余航空家。我觉得在晚间听听一个朋友随便弹奏一两种乐器，乐趣不亚于去听一次第一流的职业音乐会。一个人在自己的房里看一个朋友随便试演几套魔术，乐趣更胜于到剧院去看一次台上所表演的职业魔术。父母看自己的子女表演业余式戏剧所得的乐趣，更胜于到剧场去看一次莎士比亚戏剧。我们知道这些都是出于自动的，而真正艺术精神只有在自动中方有。这也就是我重视中国画为高士的一种消遣，而不限是一个职业艺术家的作品的理由。只有在游戏精神能够维持时，艺术方不至于沦为商业化。

游戏的特性，在于游戏都是出于无理由的，也绝不能有理由。游戏本身就是理由。这个见地，有天演历史为证明。美丽是一种"生存竞争说"所无从解释的东西；世界上甚至有对生物具着毁灭性的美丽方式：例如鹿的过于发育的美角。达尔文发觉他的"自然选择说"实在无从解释植物和动物中的美丽分子，所以不能不另定一个性的选择为附加原则。我们如若不能承认艺术实只是一种体力和心力的泛滥，自由而不受羁绊，只为自己而存在，则无从了解艺术和它的要素。"为艺术而从事艺术"的口号常受旁人的贬责，但我以为这不是一个可容政治家参加议论的问题，不过是一个关于一切艺术创作的心理起源的无可争论的事实。希特勒贬斥许多种现代艺术为不道德，但我认为那种替希特勒作画真像，放到新艺术博物院去取媚这个炙手可热的统治者的画家，乃正是不道德之中最不道德的人。这不是艺术，简直是卖淫。商业式艺术不过是妨碍艺术创作的精神，而政治式艺术竟毁灭了它。因为艺术的灵魂是自由。现代独裁者拟想产生一种政治式艺术，实在是做绝不可能的企图。他们似乎还没有觉得艺术不能借刺刀强迫而产生，正如我们不能用金钱向妓女买到真正的爱情。

如要了解艺术的要素，我们必须从艺术是力的泛滥的物体基础去研究。这就是所谓艺术或创作的冲动。艺术家每喜欢用"灵感"这个名词，即表示本人也不知道这行动是从哪里来的。这其实不过是一种内心鼓动关系，如科学家去做一种发现真理时的行动，或探险家去做一次发现一个新海岛时的冲动。这里边并无理由可说。我们在今日有了生物学知识的协助，渐能知道我们思想生活的整个组织是受着血液中"荷尔蒙"（Hormones）增减的支配，对各项器官和控制这种器官的神经系所起作用的调节。动怒和惧怕，不过某种液汁的分泌关系。天才本身，在我看来，也不过是腺分泌过量供给的结果。中国某无名小说作家虽然并没有"荷尔蒙"的知识，居然能臆测到一切活动的起源，以为是我们体内的虫的缘故。通奸是由于虫在那里咬大肠，因而鼓动个人泄欲。志愿、挑衅心和爱名位，也是由于某一种虫在那里作怪，使人片刻不得安逸，直到他的志愿达到了目的才罢休。著作一本书，例如一本小说，也是由于某一种虫在那里鼓动和迫促那作者无理由地去创作。"荷尔蒙"和虫这两个名词中，我宁取虫，因为它好像更为生动。

虫的供给过量，或只是常量，一个人便将被迫去做一些创作。因为这时他是自己也做不了主的。当一个小孩的体力供给过量时，他便会将寻常的跨步改做跳跃。当一个人的体力供给过量时，他即将跨步改为跳舞，不过是一种低效能的跨步；所谓低效能者，是从实用主义者耗费力量的见解而言，而并不从美术的见地而言。跳舞者并不径直走向目的地，而是迂回地兜着大圈子走过去。一个人在跳舞时绝不会顾到爱国的，所以命令一个人遵照着资本家或法西斯主义或普罗主义的预定方式跳舞，简直就是毁灭游戏的精神，以及使跳舞的神圣效能减低。如若一个共产主义者企图去达到一种政治目的，或企图去做一个忠实的同志，他只可跨步而不当跳舞……难道人类在和一切别种动物比较之下，还嫌他们的工作不够量，所以连这些些的空闲去从事游戏和艺术，也须受那个怪物（即国家的权力）的干涉吗？

这种对于艺术只为游戏的真性质的了解，或许可以有助于澄清艺术和道德的关系。所谓美者不过是合式而已。世上有合式的行为，如同有合式的画或桥一般。艺术的范围并不仅限于图画、音乐和跳舞，因为无论什么东西，都有合式的。赛跑中有运动员的合式；一个自幼至长，更自长至老的人，在每个时代中都有相配的行为而具着行为上的合式；一次布置周密，指导有方，因而获得最后胜利的总统竞选活动，也自具着其进行上的合式；即小如一个人的笑和咳嗽，也有合式和不合式之别。如中国旧官僚习气即属于合式。凡属人类的活动，都各有它的表显方式，所以要想将艺术的表显限制于音乐跳舞和图画这几个小范围内，是不可能的。

所以艺术有了这样较广泛的解释之后，行为上的合式和艺术上的优美个性便有了密切关系，并成为同样的重要。我们的身体动作上可以具有一种逾常的美点，如一首音韵和谐的诗的节调上逾常的美点一般。一个人一有那种过量的力量供给，便会在一切行动中显出飘逸和潇洒，并顾到合式。飘逸和潇洒是从体力充足的感觉而产生，他感觉到能把一个行动做到超过仅仅看得过的地位而做得非常合式。在较为抽象的范围中，我们能在一切做得好的动作中看到这种美点。做一次优美动作或简洁动作的冲动，本来就是一个美术的冲动，甚至如一件谋杀行为，或一件阴谋行为，只要在动作上做得简洁，看去也是美的。就是在人生的一切小节上，也可能有飘逸潇洒和胜任的姿势。凡是我们所谓的礼貌，都属于这一类。一次行得适宜恰当的问候，我们称之为优美惬人意的问候；反过来说，一次行得不好的问候，便谓之拙劣讨人嫌的问候了。

中国人说话和一切人生动作上的礼貌的发展，在晋代末叶（三四世纪）达到最高点。这就是"清谈"最流行的时代。这时女子的服装尤其讲究，男子中则有许多个以美貌出名。这时并盛行留"美髯"和穿着宽大的长袍。这种长袍的裁制很特别，能使一个人缩手到衣里去搔身体上任何部分的痒处。当时一切举动都是出之以潇洒的。拂帚，即拿几绺马鬃扎在一根柄上以供驱除蝇蚋之用，成为谈天时一种重要的道具。这种闲谈在文学中至今尚称之为"帚谈"。这帚的用处，是在随谈随拂，以助谈思。扇子也是谈天时一种优美的道具，可以在谈时忽开忽折，或微微地摇动着，正如一个美国老妇在谈天时，将她的眼镜忽而除下忽而又戴上的神情一般，都是悦目。在实用上讲起来，拂帚和扇子与英国人的单面眼镜差不多，但它们都是谈天时的道具，如手杖之为闲步时的道具。我所亲见的各种西方礼貌中，最悦目的当为普鲁士绅士在室内向女客并足行鞠躬礼时和德国少女叉腿向人行礼时的姿势。我觉得这两种姿势都美丽无比，可惜现在都已经被淘汰了。

中国人所行的礼貌，种类很多，一举手一投足中的姿势都经过研究教导。从前满洲人的"打千"，姿势是极为悦目的：她走进房中时，把一只手垂直在身体的前面，然后用优美的姿势，把一只膝屈一下子，如若房中的人不止一个，她可以在屈膝的当儿，将身体向四周旋转一下，对在座的众人，打一个总千。下棋的高手在落子时，姿势也极好看：他用两指拈起一粒棋子，用很优美的姿势，轻轻地推上棋盘。富于礼貌的满洲人，他们发怒时的姿势也极美丽：他穿着装有"马蹄袖"的袍子，这马蹄袖平时都是翻转着里子向外的，他在表示极不高兴时，就将两手一垂，将翻起的袖子往下一甩，走出房去，这就是所谓"拂袖而去"。

文雅的满洲官员，说话时的音调极为悦耳，有着美妙的节奏和有高有低的音韵。他说话时很慢，一个字一个字地吐出来。说话中夹着许多诗文中的成语，以表示学问的渊博。做官人的笑和痰嗽，姿势确实悦人耳目的：他们在痰嗽时，大都出之以三个音节；第一第二是往里一吸，打扫喉咙，到第三节，方把痰从一声咳嗽之中吐将出来。只要他的姿势做得极美化，我倒并不以他把痰吐在地上为嫌，因为我从小即生长于这种微菌之中，并没有觉得受到什么影响。他的笑也是极富有音韵而美化的：起首时略带一些矜持，轻笑两声，然后纵声一笑。他如已有白须，那就更为好看。

笑术更是中国优伶所必须苦练的，为演剧中重要动作之一。观众看见剧中人笑得美妙时，大都报以彩声。笑术不是一件容易的事情，因为笑的种类甚多：如快乐时的笑，看见别人中圈套时的笑，蔑视的笑，其中最难于模拟的则是一个人受到挫败时的苦笑。中国的剧场观众，最注意伶人的各种小动作，称之为"台步"或"做工"。伶人的举手、投足、扭颈、转身、拂袖、掀髯，都有一定的尺寸，须经过严格的训练。所以中国人将各种戏文分为两类：一类是唱工戏，另一类就是做工戏。所谓做工者，即指一切手足的动作和表情。中国伶人在表示不赞同的摇头，表示疑忌的掀眉，和表示满意的掀髯中，都有一定的姿势。

艺术和德行，只是一件艺术作品的一个特有之点，仍在那个艺术家的个性表现时方发生关系。一个具有伟大个性的艺术家产生伟大艺术；一个具有卑琐个性的艺术家产生卑琐的艺术；一个多情的艺术家产生多情的艺术；一个逸乐的艺术家产生逸乐的艺术；一个温柔的艺术家产生温柔的艺术；一个细巧的艺术家产生细巧的艺术。这就是艺术和德行的关系的总括。所以德行并不是一件可以从外面灌输进去的东西。它只是艺术家的灵魂的自然表现，而必须发于内心。它不是属于一个选择问题，而是一件不可逃避的事实。心肠卑鄙的画家绝不能产生伟大的画作，而心胸伟大的画家也绝不会产生卑鄙的画作，就是有性命的出入时，他也是不屈和不肯苟从的。

中国人对于艺术的品，或称人品、品格的见解是极有兴趣的。其中也包含品第高下的意义，如我们次第画家或诗人为"第一品"或"第二品"。

又我们尝试茶的滋味，称之为"品茶"。各种人在他们的各种动作中都表现了所谓的品，例如一个赌徒，如他在赌时的脾气很坏，即谓之赌品不好；一个酒徒如在醉后的行为很坏，即谓之酒品不好。棋手也有棋品高下之别。中国一部最早的评诗著作，书名印为《诗品》。该书的内容即是品评诗人的高下。此外还有评画的著作，书名即是《画品》。

所以，因了这个"品"的思想，一般人都深信一个艺术家的优劣完全系于其人格的高低。这人格是属于德行的，也是属于艺术的。它意在注重人类了解心、高尚心、出世、不俗、不卑鄙、不琐屑的观念。在这种意义上，它类似英文中所谓Manner（风格）或Style（派头）。一个任性的或不肯墨守成规的艺术家必显出他的任性或不肯墨守成规的风格；一个风雅的人必自然显出他的风雅风格；一个伟大的艺术家绝不肯俯就成规。在这个意义上，个性或风格实即是艺术的灵魂。中国人都默信一个画家本身的道德和美术的个性是伟大的，否则他绝不能成为伟大的画家。中国人评骘书画时，最高的标准不在于作者的技巧是否纯熟，而只在于作者是否有高尚的性格。技巧纯熟的作品往往会是风格很低的。在英文中我们即谓之缺乏"特性"。

因此，这一来我们达到了一切艺术的中心问题。中国大军事家兼政治家曾国藩在他的家书中曾说过，书法的两种重要原则为：形和神。并说当时的名书法家何绍基很赞同他的说法和钦佩他的卓见。一切艺术既然都属于有形之物，其中当然有一个机械的问题，即技巧问题，凡是艺术家都应精通的。不过因为艺术也是属于精神的，所以在一切形式的创作中，最重要的因素即是个人的表现。在艺术作品中，最富有意义的部分即是技巧以外的个性。在文字著作中的，唯一最重要的东西即是作者所特有的笔法和感情，如他所表现于爱憎之中的。这种个性或个人的表现常有被技巧所掩没的危险，而一切初学者不论是书画或演剧，最大的难关即在难于任着己意做去。其中的理由当然是初学者每每被范型或技巧束缚而不敢逾越，但不论哪一种形式，如缺乏这种个人的因素，便不能合式。凡是合式的物事或动作，必有一种飘逸的神态，所以悦目的就在于这个神态。不论它是一个锦标高尔夫球员甩动球棍的神态，或是一个人一帆风顺功成名就时的神态，或是一个美式足球运动员抱着足球在场中飞奔的神态，里边必须有一种真性的流露，这个真性必不可被技巧所毁损，而必须在技巧之中自由而愉快地充沛着。一列火车循着弧线转弯时，一只快艇张着满帆饱风向前飞驶时，都有一种极悦目的神态。一只燕子飞翔时，一只鹰攫身扑取别的动物时，一匹赛场中的马"很合式"地冲进底线时，也都有着这种悦目的神态。

我们所定的资格是：一切艺术必须有它的个性，而所谓的个性无非就是作品中所显露的作者的性灵，中国人称之为心胸。一件作品如若缺少这个个性，便成了死的东西。这个缺点是不论怎样高明的技巧都不能弥补的。如若缺乏个性，美丽的本身也将成为平凡无奇了。有许多希望成为好莱坞电影明星的女子都没有能够了解这一点，而只知拼命地模仿玛琳·黛德丽（Marlene Dietrich，主演《蓝天使》，好莱坞著名影星）或珍·哈露（Jean Harlow，二十世纪美国女电影演员），因此使物色人才的导演觉得非常失望。平庸的美貌女子很多很多，但鲜艳活泼的千中难得其一。她们为什么不去模仿玛丽·德瑞斯勒（Marie Dressler，好莱坞著名女演员）的身段和神情？一切的艺术都是相通的，以性灵的流露这一原则为根据，不论是在电影的表现中，或是在书画中，或是在文学著作中。其实从玛丽·德瑞斯勒和莱昂纳尔·巴里摩尔（Lionel Barrymore，第四届奥斯卡影帝）的表演中，即能意会出写作的秘诀。养成这个个性的可爱乃是一切艺术的重要基础，因为不论一位艺术家做一些什么东西，他的性灵总是能在他的作品中显露出来。

个性的培植是道德的，也是美术的，当中需要学问和雅韵。雅韵近乎风味，或许是一个艺术家生而已有的。但要能欣赏一件作品，非有学问不可。这个情形在书画之中极为显明。我们从一幅字中，即能看出作者是否曾见过魏拓，倘若他真的见过，这学问就使他的作品具着一种古气。但除此之外，他也须将自己的个性加进去。至于个性的强弱当然是高低不一的。如他是属于一种细致富于情感的心胸，他于作品的风格上必现出细致和富于情感；如他是喜爱雄豪的，他的风格也必是趋于雄豪的。因此，在书画中，尤其是在书中，我们可以从而看到各式各样的美点。在这种完美的作品中，个性已和技巧融合于一起，不能再加以分析。这美点可以是属于古怪或任性之类，可以是属于粗豪之类，可以是属于雄壮之类，可以是属于自由的性灵之类，可以是属于大胆不循俗例之类，可以是属于浪漫的风韵之类，可以是属于拘泥之类，可以是属于柔媚之类，可以是属于庄严之类，可以是属于简单笨拙之类，可以是属于齐整之类，可以是属于敏捷之类，有时甚至可以是属于故意的鬼怪之类。世上只有一种美点是不可能的，因为它根本不存在，这就是忙劳生活的美点。

II. Art as Play and Personality

ART is both creation and recreation. Of the two ideas, I think art as recreation or as sheer play of the human spirit is more important. Much as I appreciate all forms of immortal creative work, whether in painting, architecture or literature, I think the spirit of true art can become more general and permeate society only when a lot of people are enjoying art as a pastime, without any hope of achieving immortality. As it is more important that all college students should play tennis or football with indifferent skill than that a college should produce a few champion athletes or football players for the national contests, so it is also more important that all children and all grown-ups should be able to create something of their own as their pastime than that the nation should produce a Rodin. I would rather have all school children taught to model clay and all bank presidents and economic experts able to make their own Christmas cards, however ridiculous the attempt may be, than to have only a few artists who work at art as a profession. That is to say, I am for amateurism in all fields. I like amateur philosophers, amateur poets, amateur photographers, amateur magicians, amateur architects who build their own houses, amateur musicians, amateur botanists and amateur aviators. I get as much pleasure out of listening to a friend playing a sonatina of an evening in an indifferent manner as out of listening to a first-class professional concert. And everyone enjoys an amateur parlor magician, who is one of his friends, more than he enjoys a professional magician on the stage, and every parent enjoys the amateur dramatics of his own children much more heartily than he enjoys a Shakespearean play. We know that it is spontaneous, and in spontaneity alone lies the true spirit of art. That is why I regard it as so important that in China painting is essentially the pastime of a scholar and not of a professional artist. It is only when the spirit of play is kept that art can escape being commercialized.

Now it is characteristic of play that one plays without reason and there must be no reason for it. Play is its own good reason. This view is borne out by the history of evolution. Beauty is something that cannot be accounted for by the struggle for existence, and there are forms of beauty that are destructive even to the animal, like the over-developed horns of a deer. Darwin saw that he could never account for the beauties of plant and animal life by natural selection, and he had to introduce the great secondary principle of sexual selection. We fail to understand art and the essence of art if we do not recognize it as merely an overflow of physical and mental energy, free and unhampered and existing for its own sake. This is the much decried formula of "art for art's sake." I regard this not as a question upon which the politicians have the right to say anything, but merely as an incontrovertible fact regarding the psychological origin of all artistic creation. Hitler has denounced many forms of modern art as immoral, but I consider that those painters who paint portraits of Hitler, to be shown at the new Art Museum in order to please the powerful ruler, are the most immoral of all. That is not art, but prostitution. If commercial art often injures the spirit of artistic creation, political art is sure to kill it. For freedom is the very soul of art. Modern dictators are attempting the impossible when they try to produce a political art. They don't seem to realize that you cannot produce art by the force of the bayonet any more than you can buy real love from a prostitute.

In order to understand the essence of art at all, we have to go back to the physical basis of art as an overflow of energy. This is known as an artistic or creative impulse. The use of the very word"inspiration"shows that the artist himself hardly knows where the impulse comes from. It is merely a matter of inner urge, like the scientist's impulse for the discovery of truth, or the explorer's impulse for discovering a new island. There is no accounting for it. We are beginning to see today, with the help of biological knowledge, that the whole organization of our mental life is regulated by the increase or decrease and distribution of hormones in the blood, acting on the various organs and the nervous system controlling these organs. Even anger or fear is merely a matter of the supply of adrenalin. Genius itself, it seems to me, is but an over-supply of glandular secretions. An obscure Chinese novelist, without the modern knowledge of hormones, made a correct guess about the origin of all activity as due to"worms"in our body. Adultery is a matter of worms gnawing our intestines and impelling the man to satisfy his desire. Ambition and aggressiveness and love of fame or power are also due to certain other worms giving the person no rest until he has achieved the object of his ambition. The writing of a book, say a novel, is again due to a species of worms which impel and urge the author to create for no reason whatsoever. Between hormones and worms, I prefer to believe in the latter. The term is more vivid.

Given an over-supply or even a normal supply of worms, a man is bound to create something or other, because he cannot help himself. When a child has an over-supply of energy, his normal walking is transformed into hopping or skipping. When a man has an over-supply of energy, his walking is transformed into prancing or dancing. So, then, dancing is nothing but inefficient walking, inefficient in the sense that there is a waste of energy from an utilitarian, not an aesthetic, point of view. Instead of going straight to a point, which is the quickest road, a dancer waltzes and goes in a circle. No one really tries to be patriotic when he is dancing, and to command a man to dance according to the capitalist or fascist or proletarian ideology is to destroy the spirit of play and glorious inefficiency in dancing... As if man in civilization didn't work too much already, in comparison with every other species and variety of the animal kingdom, so that even the little leisure he has, the little time for play and art, must too be invaded by the claims of that monster, the State!

This understanding of the true nature of art as consisting in mere play may help to clarify the problem of the relationship between art and morality. Beauty is merely good form, and there is good form in conduct as well as in good painting or a beautiful bridge. Art is very much broader than painting and music and dancing, because there is good form in everything. There is good form in an athlete at a race; there is good form in a man leading a beautiful life from childhood and youth to maturity and old age, each appropriate in its own time; there is good form in a presidential campaign well directed, well maneuvered and leading gradually to a finale of victory, and there is good form, too, in one's laughter or spitting, as so carefully practised by the old Mandarins in China. Every human activity has a form and expression, and all forms of expressions lie within the definition of art. It is therefore impossible to relegate the art of expression to the few fields of music and dancing and painting.

With this broader interpretation of art, therefore, good form in conduct and good personality in art are closely related and are equally important. There can be a luxury in our bodily movements, as in the movement of a symphonic poem. Given that over-supply of energy, there is an ease and gracefulness and attendance to form in whatever we do. Now ease and gracefulness come from a feeling of physical competence, a feeling of ability to do a thing more than well—to do it beautifully. In the more abstract realms, we see this beauty in anybody doing a nice job. The impulse to do a nice job or a neat job is essentially an aesthetic impulse... In the more concrete details of our life, there is, or there can be, ease and gracefulness and competence, too. All the things we call"the amenities of life"belong in this category. Paying a compliment well and appropriately is called a beautiful compliment, and on the other hand, paying a compliment with bad taste is called an awkward compliment.

The development of the amenities of speech and life and personal habits reached a high point at the end of the Chin Dynasty (third and fourth centuries, A.D.) in China. That was the time when"leisurely conversations"were in vogue. The greatest sophistication was seen in women's dress, and there were a great number of men noted for their handsomeness. There was a fashion for growing"beautiful beards,"and men learned to wabble about clad in extremely loose gowns. The dress was so designed that there was no part of one's body unreachable in case one wanted to scratch an itch. Everything was gracefully done. The chu, a bundle of hair from the horse's tail tied together around a handle for driving away mosquitoes or flies, became an important accessory of conversation, and today such leisurely conversations are still known in literary works as chut'an or"chu conversations." The idea was that one was to hold the chu in his hand and wave it gracefully about in the air during conversation. The fan came in also as a beautiful adjunct to conversation, the conversationalist opening, waving and closing it, as an American old man would take off his spectacles and put them on again during a speech, and was just as beautiful to look at. In point of utility, the chu or the fan was only slightly more useful than an Englishman's monocle, but they were all parts of the style of conversation, as a cane is a part of the style of walking. Among the most beautiful amenities of life I have seen in the West are the clicking of heels of Prussian gentlemen bowing to a lady in a parlor and the curtsying of German girls, with one leg crossed behind the other. That I consider a supremely beautiful gesture, and it is a pity that this custom has gone out of vogue.

Many are the social amenities practised in China. The gestures of one's fingers, hands and arms are carefully cultivated. The method of greeting among the Manchus, known as tach'ien, is also a beautiful thing to look at. The person comes into the room, and letting one arm fall straight down at the side, he bends one of his legs and makes a graceful dip. In case there is company sitting around the room, he makes a graceful turn around the axis of his unbent leg while in that position, thus making a general greeting to the entire company. One should also watch a cultivated chess-player put his stones on the chessboard. Holding one of those tiny white or black stones carefully balanced on his forefinger, he gently pushes it from behind by an outward movement of his thumb and an inward movement of the forefinger, and lands it beautifully on the board. A cultured Mandarin made extremely beautiful gestures when he was angry. He wore a gown with the sleeves tucked up at the lower ends showing the silk lining, known as"horse-hoof sleeves,"and when he was greatly displeased, he would brandish his right arm or both arms downwards and with an audible jerk bring the tucked-up"horse-hoof"down, and gracefully wobble out of the room. This is known as fohsiu, or to"brush one's sleeves and leave. "

The speech of a cultured Mandarin official is also a beautiful thing to hear. His words come out with a beautiful cadence, and the musical tones of the Peking accent have a graceful musical rise and fall. His syllables are pronounced gracefully and slowly, and in the case of real scholars, his language is set up with jewels from the Chinese literary language. And then one should hear how the Mandarin laughs or spits. It is positively delightful. The spitting is done generally in three musical beats, the first two being sounds of drawing in and clearing the throat in preparation for the final beat of spitting out, which is executed with a quick forcefulness: staccato after legato. I really don't mind the germs thus let out into the air, if the spitting is aesthetically done, for I have survived the germs without any appreciable effect on my health. His laughter is an equally regulated and artistically rhythmical affair, slightly artificial and stylized, and finishing off in an increasing generous volume, pleasantly softened by a white beard when there is one.

Such laughter is a carefully cultivated art with an actor as part of his technique of acting, and theater-goers always enjoy and applaud a perfectly executed laugh. This is of course a very difficult thing, because there are so many kinds of laughter: the laughter of happiness, the laughter at some one falling into one's trap, the laughter of sneer or contempt, and most difficult of all, the laughter of despair, of a man caught and defeated by the force of over-whelming circumstances. Chinese theater-goers watch for these things and for the hand gestures and steps of an actor, the latter being known as t'aipu, or "stage steps." Every movement of the arm, every inclination of the head, every twist of the neck, every bend of the back, every waving movement of the flowing sleeve, and of course every step of the foot, is a carefully practiced gesture. The Chinese classify acting into the two classes of singing and acting, and there are plays with emphasis on singing, and other dramas with emphasis on acting. By "acting" is meant these gestures of the body, the hands and the face, as much as the more general acting of emotions and expressions. Chinese actors have to learn how to shake their heads in disapproval, how to lift their eyebrows in suspicion, and how to gently stroke their beard in peace and satisfaction...

Art has a relationship to morals only insofar as the peculiar quality of a work of art is an expression of the artist's personality. An artist with a grand personality produces grand art; an artist with a trivial personality produces trivial art; a sentimental artist produces sentimental art, a voluptuous artist produces voluptuous art, a tender artist produces tender art, and an artist of delicacy produces delicate art. There we have the relationship between art and morality in a nutshell. Morality, therefore, is not a thing that can be superimposed from the outside. ... It must grow from the inside as the natural expression of the artist's soul. And it is not a question of choice, but an inescapable fact. The mean-hearted artist cannot produce a great painting, and a big-hearted artist cannot produce a mean picture, even if his life were at stake.

The Chinese notion of p'in in art is extremely interesting, sometimes spoken of as jenp'in ("personality of the man") or p'inkeh ("personality of character"). There is also an idea of grading, as we speak of artists or poets of"the first p'in" or"second p'in,"and we also speak of tasting or sampling good tea as"to p'in tea."There are then a whole category of expressions in connection with the personality of a person as shown in a particular action. In the first place, a bad gambler, or a gambler who shows bad temper or bad taste, is said to have a bad tup'in or a bad"gambling personality."A drinker who is apt to behave disgracefully after a hard drink is said to have a bad chiup'in or bad"wine personality."A good or bad chess-player is said to have a good or bad ch'ip'in or"chess personality."The earliest Chinese book of poetic criticism is known as Shihp'in(Personalities of Poetry), with a grading of the different poets, and of course there are books of art criticism known as huap'in or"Personalities in Painting. "

Connected with this idea of p'in, therefore, there is the generally accepted belief that an artist's work is strictly determined by his personality. This"personality" is both moral and artistic. It tends to emphasize the notion of human understanding, high-mindedness, detachment from life, absence of pettiness or triviality or vulgarity. In this sense it is akin to English"manner"or"style."A wayward or unconventional artist will show a wayward or unconventional style and a person of charm will naturally show charm and delicacy in his style, and a great artist with good taste will not stoop to"mannerisms."In this sense, personality is the very soul of art. The Chinese have always accepted implicitly the belief that no painter can be great unless his own moral and aesthetic personality is great, and in judging calligraphy and painting, the highest criterion is not whether the artist shows good technique but whether he has or has not a high personality. A work, showing perfect technique, may nevertheless show a"low" personality, and then, as we would say in English, that work lacks"character. "

We have come thus to the central problem of all art. Tseng Kuofan said in one of his family letters that the only two living principles of art in calligraphy are form and expression, and that one of the greatest calligraphists of the time, Ho Shaochi, approved of his formula and appreciated his insight. Since all art is concrete, there is always a mechanical problem, the problem of technique, which has to be mastered, but as art is also spirit, the vital element in all forms of creation is the personal expression. It is the artist's individuality, over against his mere technique, that is the only significant thing in a work of art. In writing, the only important thing in a book is the author's personal style and feeling, as shown in his judgment and likes and dislikes. There is a constant danger of this personality or personal expression being submerged by the technique, and the greatest difficulty of all beginners, whether in painting or writing or acting, is to let oneself go. The reason is, of course, that the beginner is scared by the form or technique. But no form without this personal element can be good form at all. All good form has a swing, and it is the swing that is beautiful to look at, whether it is the swing of a champion golf-player's club, or of a man rocketing to success, or of a football player carrying the ball down the field. There must be a flow of expression, and that power of expression must not be hampered by the technique, but must be able to move freely and happily in it. There is that swing—so beautiful to look at—in a train going around a curve, or a yacht going at full speed with straight sails. There is that swing in the flight of a swallow, or of a hawk dashing down on its prey, or of a champion horse racing to the finish"in good form,"as we say.

We require that all art must have character, and character is nothing but what the work of art suggests or reveals concerning the artist's personality or soul or heart or, as the Chinese put it,"breast."Without that character or personality, a work of art is dead, and no amount of virtuosity or mere perfection of technique can save it from lifelessness or lack of vitality. Without that highly individual thing called personality, beauty itself becomes banal. So many girls aspiring to be Hollywood stars do not know this, and content themselves with imitating Marlene Dietrich or Jean Harlow, thus exasperating a movie director looking for talent. There are so many banally pretty faces, and so little fresh, individual beauty. Why don't they study the acting of Marie Dressier? All art is one and based on the same principle of expression or personality, whether it is acting in a movie picture or painting or literary authorship. Really, by looking at the acting of Marie Dressier or Lionel Barrymore, one can learn the secret of style in writing. To cultivate the charm of that personality is the important basis for all art, for no matter what an artist does, his character shows in his work.

The cultivation of personality is both moral and aesthetic, and it requires both scholarship and refinement. Refinement is something nearer to taste and may be just born in an artist, but the highest pleasure of looking at a book of art is felt only when it is supported by scholarship. This is particularly clear in painting and calligraphy. One can tell from a piece of calligraphy whether the writer has or has not seen a great number of Wei rubbings. If he has, this scholarship gives him a certain antique manner, but in addition to that, he must put into it his own soul or personality, which varies of course. If he is a delicate and sentimental soul, he will show a delicate and sentimental style, but if he loves strength or massive power, he will also adopt a style that goes in for strength and massive power. Thus in painting and calligraphy, particularly the latter, we are able to see a whole category of aesthetic qualities or different types of beauty, and no one will be able to separate the beauty of the finished product and the beauty of the artist's own soul. There may be beauty of whimsicality and waywardness, beauty of rugged strength, beauty of massive power, beauty of spiritual freedom, beauty of courage and dash, beauty of romantic charm, beauty of restraint, beauty of soft gracefulness, beauty of austerity, beauty of simplicity and"stupidity,"beauty of mere regularity, beauty of swiftness, and sometimes even beauty of affected ugliness. There is only one form of beauty that is impossible because it does not exist, and that is the beauty of strenuousness or of the strenuous life.

三．读书的艺术

读书是文明生活中人所共认的一种乐趣，极为无福享受此种乐趣的人所羡慕。我们如把一生爱读书的人和一生不知读书的人比较一下，便能了解这一点。凡是没有读书癖好的人，就时间和空间而言，简直是等于幽囚在周遭的环境里边。他的一生完全落于日常例行公事的圈禁中。他只有和少数几个朋友或熟人接触谈天的机会，他只能看见眼前的景物，他没有逃出这所牢狱的法子。但在他拿起一本书时，他已立刻走进了另一个世界。如若所拿的又是一部好书，他便已得到了一个和一位最善谈者接触的机会。这位善谈者引领他走进另外一个国界，或另外一个时代，或向他倾吐自己胸中的不平，或和他讨论一个他从来不知道的生活问题。一本古书使读者在心灵上和长眠已久的古人如相互面对，当他读下去时，便会想象到这位古作家是怎样的形态和怎样的一种人，孟子和大史家司马迁都表示这个意见。一个人在每天二十四小时中，能有两小时的工夫撇开一切俗世烦扰而走到另一个世界去游览一番，这种幸福自然是被无形牢狱所拘囚的人们所极羡慕的。这种环境的变更，在心理的效果上，其实等于出门旅行。

但读书的益处还不只这一些。读者常会被携带到一个思考和熟虑的世界里边去。即使是一篇描写事实的文章，躬亲其事和从书中读到事情的经过，其间也有很大的不同。因为这种事实一经描写到书中便成为一幅景物，而读者便成为一个脱身是非真正的旁观者了。所以真正有益的读书，便是能引领我们进到这个沉思境界的读书，而不是单单去知道一些事实经过的读书。人们往往耗费许多时间去读新闻报纸，我以为这不能算是读书。因为一般的新闻报纸读者，他们的目的只是要得知一些毫无深层价值的事实经过罢了。

据我的意见，宋朝苏东坡的好友诗人黄山谷所说的话实在是一个读书目标的最佳共式。他说："三日不读书，便觉语言无味，面目可憎。"他的意思当然是人如读书即会有风韵，富风味。这就是读书的唯一目标。唯有抱着这个目标去读书，方可称为知道读书之术。一个人并不是为了要使心智进步而读书，因为读书之时如怀着这个念头，读书的一切乐趣便完全丧失了。犯这一类毛病的人必在自己的心中说，我必须读莎士比亚，我必须读索福克勒斯（Sophocles），我必须读艾略特博士（Dr. Eliot）的全部著作，以便可以成为一个有学问的人。我以为这个人永远不会成为有学问者。他必在某天的晚上出于勉强地去读莎士比亚的《哈姆雷特》（Hamlet），放下书时，将好像是从一个噩梦中苏醒一般。其实呢，他除了可说一声已经读过这本书之外，并未得到什么益处。凡是以出于勉强的态度去读书的人，都是些不懂读书艺术的人。这类抱着求知目标而读书，其实等于一个参议员在发表意见之前的阅读旧案和报告书。这是在搜寻公事上的资料，而不得谓之读书。

因此，必须是意在为培植面目的可爱和语言的有味而读书，照着黄山谷的说法，方可算做真正的读书。这个所谓"面目可爱"，显然须做异于体美的解释。黄山谷所谓"面目可憎"者，并不是相貌的丑恶。所以世有可憎的美面，也有可爱的丑面。我的本国朋友中，有一位头尖如炸弹形一般，但这个人终是悦目的。西方的作家中，我从肖像中看来，相貌最可爱者当属切斯特顿，他的胡须、眼镜、丛眉、眉间的皱纹，团聚在一起是多么怪异可爱啊！这个形容使人觉得他的前脑中充满着何等丰富的活泼思想，好像随时从他异常尖锐的双目中爆发出来。这就是黄山谷所谓"可爱的面目"，不是由花粉胭脂所妆成的面目，而是由思想力所华饰的面目。至于怎样可以"语言有味"，全在他的书是怎样的读法。一个读者如能从书中得到它的味道，他便会在谈吐中显露出来。他的谈吐如有味，则他的著作中也自然会富有滋味。

因此，我以为味道乃是读书的关键，而这个味道也必然因此是各有所嗜的，如人对于食物一般。最合卫生的吃食方法终是择其所嗜而吃，方能保证其必然消化。读书也和吃食相同。在我是美味的，也许在别人是毒药。一个教师绝不能强迫他的学生去读他们所不爱好的读物；而做父母的，也不能强迫子女吃他们不喜欢吃的东西。一个读者如对于一种读物并无胃口，则他所浪费在读书的时间完全是虚耗的，正如袁中郎所说："若不惬意，放置之俟他人。"

所以世上并无一个人所必须读的书，因为我们的智力兴趣是如同树木一般的生长，如同河水一般的流向前去的，只要有汁液，树木必会生长；只要泉源不涸，河水必会长流；当它碰到石壁时，它自会转弯；当它流到一片可爱的低谷时，它必会暂时停留一下子；当它流到一个深的山池时，它必会觉得满足而停在那里；当它流过急湍时，它必会迅速前行。如此，它无须用力，也无须预定目标，自能必然有一天流到海中。世上并没有人人必读的书，但有必须在某一时间，必须在某一地点，必须在某种环境之中，必须在某一时代方可以读的书。我颇以为读书也和婚姻相同，是由姻缘或命运所决定。世上即使有人人必读的书如《圣经》，但读它必应有一定的时期。当一个人的思想和经验尚没有达到可读一本名著的相当时期时，他即使勉强去读，也必觉得其味甚劣。孔子说，五十读易。他的意思就是说，四十五岁时还不能读。一个人没有到识力成熟的时候，绝不能领略《论语》中孔子话语中淡淡的滋味和他已成熟的智慧。

再者，一个人在不同的时候读同一部书，可以得到不同的滋味。例如我们在和一位作家谈过一次后或看见过他的面目后，再去读他的著作，必会觉到更多的领略。又如在和一位作家反目之后，再去读他的著作，也会得到另一种滋味。一个人在四十岁时读《易经》所得的滋味，必和在五十岁人生阅历更丰富时读它所得的滋味不同。所以将一本书重读一遍，也是有益的并也可以从而得到新的乐趣。我在学校时教师命读《Westward Ho》（《向西去啊》，英国查尔斯·金斯利著）和《The History of Henry Esmond》两书，那时我已能领略《Westward Ho》的滋味，但对于《The History of Henry Esmond》觉得很是乏味，直到后来回想到的时候，方觉得它也是很有滋味的，不过当时未能为我领略罢了。

所以读书是一件涉及两方面的事情：一在作者，一在读者。作者固然对读者做了不少的贡献，但读者也能借着自己的悟性和经验，从书中悟会出同量的收获。宋代某大儒在提到《论语》时说，读《论语》的人很多很多。有些人读了之后，一无所得，有些人对其中某一两句略感兴趣，但有些人则会在读了之后，手舞足蹈起来。

我以为一个人能发现他所爱好的作家，实在是他的智力进展里边一件最重要的事情。世上原有所谓性情相近这件事，所以一个人必须从古今中外的作家去找寻和自己的性情相近的人。一个人唯有借着这个方法，才能从读书中获得益处。他必须不受拘束地去找寻自己的先生。一个人所最喜爱的作家是谁？这句问话，没有人能回答，即在本人也未必能答出来。这好似一见钟情，一个读者不能由旁人指点着去爱好这个或那个作家。但他一旦遇到所爱好的作家时，他的天性必会立刻使他知道的。这类忽然寻到所爱好的作家的例子甚多。世上常有古今异代相距千百年的学者，因思想和感觉的相同，竟会在书页上会面时完全融洽和谐，如面对着自己的肖像一般。在中国语文中，我们称这种精神的融洽为"灵魂的转世"。例如苏东坡乃是庄周或陶渊明转世，袁中郎乃是苏东坡转世之类。苏东坡曾说，当他初次读庄子时，觉得他幼时的思想和见地正和这书中所论者完全相同。当袁中郎于某夜偶然抽到一本诗集而发现一位同时代的不出名作家徐文长时，会不知不觉地从床上跳起来，叫起他的朋友，两人共读共叫，甚至童仆都被惊醒。乔治·艾略特（George Eliot）描摹他的第一次读卢梭，称之为"一次触电"。尼采（Nietzsche）于初读叔本华（Schopenhauer）时也有同样的感觉。但叔本华是一位乖戾的先生，而尼采是一个暴躁的学生，无怪后来这学生就背叛他的先生了。

只有这种读书法，这种自己去找寻所喜爱的作家，方是对读者有益的。这犹如一个男人和一个女子一见钟情，一切必都美满。他会觉得她的身材高矮正合度，相貌恰到好处，头发的颜色正深浅合度，说话的声音恰高低合度，谈吐和思想也都一切合度。这青年不必经教师的教导，自会去爱她。读书也是如此，他自会觉得某一个作家恰称自己的爱好。他会觉得这作家的笔法、心胸、见地、思态都是合式的，于是他对这作家的著作能字字领略，句句理会。并因为两人之间有一种精神上的融洽，所以一切都能融会贯通。他已中了那作家的魔术，他也愿意中这魔术。不久之后，他的音容笑貌也会变得和那作家的音容笑貌一模一样了。如此，他实已沉浸在深切爱好那作家之中，而能从这类书籍里边得到滋养他灵魂的资料。不过数年之后，这魔法会渐渐退去，他对这个爱人会渐渐觉得有些厌倦。于是他会去找寻新的文字爱人，等到他有过三四个这类爱人，把他们的作品完全吞吸之后，他自己便也成为一位作家了。世上有许多读者从来不会和作家相爱，这正如世上有许多男女虽到处调情，但始终不会和某一个人发生切近的关系，他们能读一切的作品，结果终是毫无所得。

如此的读书艺术的概念，显然把读书为一种责任或义务的概念压了下去。在中国，我们常听到勉人"苦读"的话头。从前有一个勤苦的读书人在夜里读书时，每以锥刺股，使他不致睡去。还有一个读书人在夜里读书时，命一个女婢在旁边以便在他睡去时惊醒他，这种读法太没意思了。一个人在读书的时候，正当那古代的聪明作家对他说话时而忽然睡去，他应当立刻上床去安睡。用锥刺股或用婢叫醒，无论做到什么程度，绝不能使他得到什么益处，这种人已完全丧失了读书快乐的感觉。凡是有所成就的读书人绝不懂什么叫做"勤研"或"苦读"，他们只知道爱好一本书，而不知其然地读下去。

这个问题解决之后，读书的时间和地点问题也同时得到了答案，即读书用不着相当的地点和时间。一个人想读书时，随时随地可读。一个人倘懂得读书的享受，不论在学校里边或学校外边都可以读，即在学校里边也不致妨碍他的兴趣。曾国藩在家书中答复他的弟弟想到京师读书以求深造时说：

苟能发奋自立，则家塾可读书；即旷野之地，热闹之场，亦可读书；负薪牧承皆可读书。苟不能发奋自立，则家塾不宜读书；即清净之乡，神仙之境皆不能读书。

有些人在要读书时常常想起许多借口。刚要开始读时，他会憎厌房里太冷，或椅子太硬，或亮光太烈，而说不能读，还有些作家每每憎厌蚊子太多或纸张太劣，或街上太闹，而说无从写作。宋代大儒欧阳修自承最佳的写作时候乃是"三上"：即枕上、马上和厕上。清代学者顾千里当夏天时，常"裸而读经"即以此得名。反之，一个人如若不愿意读书，则一年四季之中自有不能读书的理由：

春天不是读书天，夏日炎炎正好眠，秋去冬来真迅速，收拾书包过新年。

那么究竟怎样才算是真正的读书艺术呢？简单的答语就是：随手拿过一本书，想读时，便读一下子。如想真正得到享受，读书必须出于完全自动。一个人尽可以拿一本《离骚》或一本欧玛尔·海亚姆（Omar Khayyam）的《鲁拜集》，一手挽着爱人，同到河边去读。如若那时天空中有美丽的云霞，他尽可以放下手中的书，抬头赏玩。也可以一面看，一面读，中间吸一斗烟，或喝一杯茶，更可以增添他的乐趣。或如在冬天的雪夜，一个人坐在火炉的旁边，炉上壶水轻沸，手边放着烟袋烟斗，他尽可以搬过十余本关于哲学、经济、诗文、传记的书籍堆在身边的椅子上，以闲适的态度，随手拿过一本来翻阅。如觉得合意时，便可读下去，否则便可换一本。金圣叹以为在雪夜里关紧了门读一本禁书乃是人生至乐之一。陈眉公描写读书之时说，古人都称书籍画幅为"柔篇"，所以最适宜的阅读方式就是须出于写意，这种心境使人养成随事忍耐的性情。所以他又说，真正善于读书的人，对于书中的错字绝不计较，正如善于旅行的人对于上山时一段崎岖不平的路径，或如出门观看雪景的人对于一座破桥，或如隐居乡间的人对于乡下的粗人，或如一心赏花的人对于味道不好的酒一般，都是不加计较的。

中国最伟大的女词人李清照的自传中，有一段极尽描写读书之乐之能事。她和她的丈夫在领到国子监的膏火银后，常跑到庙集去，在旧书和古玩摊上翻阅残书简篇和金石铭文，遇到爱好的，即买下来。归途之中，必再买些水果，回到家后一面切果，一面赏玩新买来的碑拓，或一面品茶，一面校对各版的异同。她在所著《金石录》后跋中，有一段自述说：

> 余性偶强记，每饭罢，坐归来堂，烹一茶，指堆积书史言：某事在某书某卷，第几页，第几行，以中否角胜负，为饮茶先后。中即举杯大笑，至茶倾覆怀中，反不得饮而起。甘心老是乡矣！故虽处忧患困穷，而志不屈。收书既成……于是几案罗列，枕席枕籍会意心谋，目往神授，乐在声色犬马之上。

这一段自述文，是她在老年时丈夫已经故世后所写的。这时正当金人进扰中原，华北遍地烽烟，她也无日不在流离逃难之中。

III. The Art of Reading

READING or the enjoyment of books has always been regarded among the charms of a cultured life and is respected and envied by those who rarely give themselves that privilege. This is easy to understand when we compare the difference between the life of a man who does no reading and that of a man who does. The man who has not the habit of reading is imprisoned in his immediate world, in respect to time and space. His life falls into a set routine; he is limited to contact and conversation with a few friends and acquaintances, and he sees only what happens in his immediate neighborhood. From this prison there is no escape. But the moment he takes up a book, he immediately enters a different world, and if it is a good book, he is immediately put in touch with one of the best talkers of the world. This talker leads him on and carries him into a different country or a different age, or unburdens to him some of his personal regrets, or discusses with him some special line or aspect of life that the reader knows nothing about. An ancient author puts him in communion with a dead spirit of long ago, and as he reads along, he begins to imagine what that ancient author looked like and what type of person he was. Both Mencius and Ssema Ch'ien, China's greatest historian, have expressed the same idea. Now to be able to live two hours out of twelve in a different world and take one's thoughts off the claims of the immediate present is, of course, a privilege to be envied by people shut up in their bodily prison. Such a change of environment is really similar to travel in its psychological effect.

But there is more to it than this. The reader is always carried away into a world of thought and reflection. Even if it is a book about physical events, there is a difference between seeing such events in person or living through them, and reading about them in books, for then the events always assume the quality of a spectacle and the reader becomes a detached spectator. The best reading is therefore that which leads us into this contemplative mood, and not that which is merely occupied with the report of events. The tremendous amount of time spent on newspapers I regard as not reading at all, for the average readers of papers are mainly concerned with getting reports about events and happenings without contemplative value.

The best formula for the object of reading, in my opinion, was stated by Huang Shanku, a Sung poet and friend of Su Tungp'o. He said, "A scholar who hasn't read anything for three days feels that his talk has no flavor (becomes insipid), and his own face becomes hateful to look at (in the mirror)." What he means, of course, is that reading gives a man a certain charm and flavor, which is the entire object of reading, and only reading with this object can be called an art. One doesn't read to "improve one's mind," because when one begins to think of improving his mind, all the pleasure of reading is gone. He is the type of person who says to himself: "I must read Shakespeare, and I must read Sophocles, and I must read the entire Five Foot Shelf of Dr. Eliot, so I can become an educated man." I'm sure that man will never become educated. He will force himself one evening to read Shakespeare's Hamlet and come away, as if from a bad dream, with no greater benefit than that he is able to say that he has "read" Hamlet. Anyone who reads a book with a sense of obligation does not understand the art of reading. This type of reading with a business purpose is in no way different from a senator's reading up of files and reports before he makes a speech. It is asking for business advice and information, and not reading at all.

Reading for the cultivation of personal charm of appearance and flavor in speech is then, according to Huang, the only admissible kind of reading. This charm of appearance must evidently be interpreted as something other than physical beauty. What Huang means by"hateful to look at"is not physical ugliness. There are ugly faces that have a fascinating charm and beautiful faces that are insipid to look at. I have among my Chinese friends one whose head is shaped like a bomb and yet who is nevertheless always a pleasure to see. The most beautiful face among Western authors, so far as I have seen them in pictures, was that of G. K. Chesterton. There was such a diabolical conglomeration of mustache, glasses, fairly bushy eyebrows and knitted lines where the eyebrows met! One felt there were a vast number of ideas playing about inside that forehead, ready at any time to burst out from those quizzically penetrating eyes. That is what Huang would call a beautiful face, a face not made up by powder and rouge, but by the sheer force of thinking. As for flavor of speech, it all depends on one's way of reading. Whether one has"flavor"or not in his talk, depends on his method of reading. If a reader gets the flavor of books, he will show that flavor in his conversations, and if he has flavor in his conversations, he cannot help also having a flavor in his writing.

Hence I consider flavor or taste as the key to all reading. It necessarily follows that taste is selective and individual, like the taste for food. The most hygienic way of eating is, after all, eating what one likes, for then one is sure of his digestion. In reading as in eating, what is one man's meat may be another's poison. A teacher cannot force his pupils to like what he likes in reading, and a parent cannot expect his children to have the same tastes as himself. And if the reader has no taste for what he reads, all the time is wasted. As Yüan Chunglang says,"You can leave the books that you don't like alone, and let other people read them."

There can be, therefore, no books that one absolutely must read. For our intellectual interests grow like a tree or flow like a river. So long as there is proper sap, the tree will grow anyhow, and so long as there is fresh current from the spring, the water will flow. When water strikes a granite cliff, it just goes around it; when it finds itself in a pleasant low valley, it stops and meanders there a while; when it finds itself in a deep mountain pond, it is content to stay there; when it finds itself traveling over rapids, it hurries forward. Thus, without any effort or determined aim, it is sure of reaching the sea some day. There are no books in this world that everybody must read, but only books that a person must read at a certain time in a given place under given circumstances and at a given period of his life. I rather think that reading, like matrimony, is determined by fate or yinyüan. Even if there is a certain book that every one must read, like the Bible, there is a time for it. When one's thoughts and experience have not reached a certain point for reading a masterpiece, the masterpiece will leave only a bad flavor on his palate. Confucius said,"When one is fifty, one may read the Book of Changes,"which means that one should not read it at forty-five. The extremely mild flavor of Confucius' own sayings in the Analects and his mature wisdom cannot be appreciated until one becomes mature himself.

Furthermore, the same reader reading the same book at different periods, gets a different flavor out of it. For instance, we enjoy a book more after we have had a personal talk with the author himself, or even after having seen a picture of his face, and one gets again a different flavor sometimes after one has broken off friendship with the author. A person gets a kind of flavor from reading the Book of Changes at forty, and gets another kind of flavor reading it at fifty, after he has seen more changes in life. Therefore, all good books can be read with profit and renewed pleasure a second time. I was made to read Westward Ho! and Henry Esmond in my college days, but while I was capable of appreciating Westward Ho! in my 'teens, the real flavor of Henry Esmond escaped me entirely until I reflected about it later on, and suspected there was vastly more charm in that book than I had then been capable of appreciating.

Reading, therefore, is an act consisting of two sides, the author and the reader. The net gain comes as much from the reader's contribution through his own insight and experience as from the author's own. In speaking about the Confucian Analects, the Sung Confucianist Ch'eng Yich'uan said, "There are readers and readers. Some read the Analects and feel that nothing has happened, some are pleased with one or two lines in it, and some begin to wave their hands and dance on their legs unconsciously."

I regard the discovery of one's favorite author as the most critical event in one's intellectual development. There is such a thing as the affinity of spirits, and among the authors of ancient and modern times, one must try to find an author whose spirit is akin with his own. Only in this way can one get any real good out of reading. One has to be independent and search out his masters. Who is one's favorite author, no one can tell, probably not even the man himself. It is like love at first sight. The reader cannot be told to love this one or that one, but when he has found the author he loves, he knows it himself by a kind of instinct. We have such famous cases of discoveries of authors. Scholars seem to have lived in different ages, separated by centuries, and yet their modes of thinking and feeling were so akin that their coming together across the pages of a book was like a person finding his own image. In Chinese phraseology, we speak of these kindred spirits as re-incarnations of the same soul, as Su Tungp'o was said to be a re-incarnation of Chuangtse or T'ao Yüanming, and Yüan Chunglang was said to be the re-incarnation of Su Tungp'o. Su Tungp'o said that when he first read Chuangtse, he felt as if all the time since his childhood he had been thinking the same things and taking the same views himself. When Yüan Chunglang discovered one night Hsü Wench'ang, a contemporary unknown to him, in a small book of poems, he jumped out of bed and shouted to his friend, and his friend began to read it and shout in turn, and then they both read and shouted again until their servant was completely puzzled. George Eliot described her first reading of Rousseau as an electric shock. Nietzsche felt the same thing about Schopenhauer, but Schopenhauer was a peevish master and Nietzsche was a violent-tempered pupil, and it was natural that the pupil later rebelled against the teacher.

It is only this kind of reading, this discovery of one's favorite author, that will do one any good at all. Like a man falling in love with his sweetheart at first sight, everything is right. She is of the right height, has the right face, the right color of hair, the right quality of voice and the right way of speaking and smiling. This author is not something that a young man need be told about by his teacher. The author is just right for him; his style, his taste, his point of view, his mode of thinking, are all right. And then the reader proceeds to devour every word and every line that the author writes, and because there is a spiritual affinity, he absorbs and readily digests everything. The author has cast a spell over him, and he is glad to be under the spell, and in time his own voice and manner and way of smiling and way of talking become like the author's own. Thus he truly steeps himself in his literary lover and derives from these books sustenance for his soul. After a few years, the spell is over and he grows a little tired of this lover and seeks for new literary lovers, and after he has had three or four lovers and completely eaten them up, he emerges as an author himself. There are many readers who never fall in love, like many young men and women who flirt around and are incapable of forming a deep attachment to a particular person. They can read any and all authors, and they never amount to anything.

Such a conception of the art of reading completely precludes the idea of reading as a duty or as an obligation. In China, one often encourages students to "study bitterly." There was a famous scholar who studied bitterly and who stuck an awl in his calf when he fell asleep while studying at night. There was another scholar who had a maid stand by his side as he was studying at night, to wake him up every time he fell asleep. This was nonsensical. If one has a book lying before him and falls asleep while some wise ancient author is talking to him, he should just go to bed. No amount of sticking an awl in his calf or of shaking him up by a maid will do him any good. Such a man has lost all sense of the pleasure of reading. Scholars who are worth anything at all never know what is called"a hard grind"or what"bitter study"means. They merely love books and read on because they cannot help themselves.

With this question solved, the question of time and place for reading is also provided with an answer. There is no proper time and place for reading. When the mood for reading comes, one can read anywhere. If one knows the enjoyment of reading, he will read in school or out of school, and in spite of all schools. He can study even in the best schools. Tseng Kuofan, in one of his family letters concerning the expressed desire of one of his younger brothers to come to the capital and study at a better school, replied that:"If one has the desire to study, he can study at a country school, or even on a desert or in busy streets, and even as a woodcutter or a swineherd. But if one has no desire to study, then not only is the country school not proper for study, but even a quiet country home or a fairy island is not a proper place for study."There are people who adopt a self-important posture at the desk when they are about to do some reading, and then complain they are unable to read because the room is too cold, or the chair is too hard, or the light is too strong. And there are writers who complain that they cannot write because there are too many mosquitos, or the writing paper is too shiny, or the noise from the street is too great. The great Sung scholar, Ouyang Hsiu, confessed to"three on's"for doing his best writing: on the pillow, on horseback and on the toilet. Another famous Ch'ing scholar, Ku Ch'ienli, was known for his habit of"reading Confucian classics naked"in summer. On the other hand, there is a good reason for not doing any reading in any of the seasons of the year, if one does not like reading:

To study in spring is treason;
And summer is sleep's best reason;

If winter hurries the fall,

Then stop till next spring season.

What, then, is the true art of reading? The simple answer is to just take up a book and read when the mood comes. To be thoroughly enjoyed, reading must be entirely spontaneous. One takes a limp volume of Lisao, or of Omar Khayyam, and goes away hand in hand with his love to read on a river bank. If there are good clouds over one's head, let them read the clouds and forget the books, or read the books and the clouds at the same time. Between times, a good pipe or a good cup of tea makes it still more perfect. Or perhaps on a snowy night, when one is sitting before the fireside, and there is a kettle singing on the hearth and a good pouch of tobacco at the side, one gathers ten or a dozen books on philosophy, economics, poetry, biography and piles them up on the couch, and then leisurely turns over a few of them and gently lights on the one which strikes his fancy at the moment. Chin Shengt'an regards reading a banned book behind closed doors on a snowy night as one of the greatest pleasures of life. The mood for reading is perfectly described by Ch'en Chiju (Meikung):" The ancient people called books and paintings'limp volumes'and'soft volumes'; therefore the best style of reading a book or opening an album is the leisurely style."In this mood, one develops patience for everything. As the same author says,"The real master tolerates misprints when reading history, as a good traveller tolerates bad roads when climbing a mountain, one going to watch a snow scene tolerates a flimsy bridge, one choosing to live in the country tolerates vulgar people, and one bent on looking at flowers tolerates bad wine."

The best description of the pleasure of reading I found in the autobiography of China's greatest poetess, Li Ch'ingchao (Yi-an, 1081-1141). She and her husband would go to the temple, where secondhand books and rubbings from stone inscriptions were sold, on the day he got his monthly stipend as a student at the Imperial Academy. Then they would buy some fruit on the way back, and coming home, they began to pare the fruit and examine the newly bought rubbings together, or drink tea and compare the variants in different editions. As described in her autobiographical sketch known as Postscript to Chinshihlu (a book on bronze and stone inscriptions) :

I have a power for memory, and sitting quietly after supper in the Homecoming Hall, we would boil a pot of tea and, pointing to the piles of books on the shelves, make a guess as to on what line of what page in what volume of a certain book a passage occurred and see who was right, the one making the correct guess having the privilege of drinking his cup of tea first. When a guess was correct, we would lift the cup high and break out into a loud laughter, so much so that sometimes the tea was spilled on our dress and we were not able to drink. We were then content to live and grow old in such a world! Therefore we held our heads high, although we were living in poverty and sorrow... In time our collection grew bigger and bigger and the books and art objects were piled up on tables and desks and beds, and we enjoyed them with our eyes and our minds and planned and discussed over them, tasting a happiness above those enjoying dogs and horses and music and dance...

This sketch was written in her old age after her husband had died, when she was a lonely old woman fleeing from place to place.

四．写作的艺术

写作的艺术，其范围的广泛，远过于写作的技巧。实在说起来，凡是期望成为作家的初学者，都应该叫他们先把写作的技巧完全撇开，暂时不必顾及这些小节，专在心灵上用功夫，发展出一种真实的文学个性，去做他的写作基础。这个方法应该对他很有益处。基础已经打好，真实的文学个性已经培养成功时，笔法自然而然会产生，一切技巧也自然而然地跟着纯熟。只要他的立意精警，文法上略有不妥之处也是不妨的，这种小小的错误，自有那出版者的编校员会替他改正的。反之，一个初学者如若忽略了文学个性的培植，则无论怎样去研究文法和文章，也是不能成为作家的。布封（Buffon）说得好："笔法即作者。"笔法并不是一个方式，也不是一个写作方法中的制度或饰件，其实不过是读者对于作者的心胸特性，深刻或浅泛的，有见识或无见识和其他各种特质，如：机智、幽默、讥嘲、体会、柔婉、敏锐、了解力、仁慈的乖戾或乖戾的仁慈、冷酷、实际的常识和对于一切物事的一般态度所得的一种印象罢了。可知世上绝不能有教人学会"幽默技巧"的袖珍指南，或"乖戾的仁慈三小时速通法"，或"常识速成十五法"，或"感觉敏锐速成十一法"。

我们须超过写作艺术的表面而更进一步。我们在做到这一步时，便会觉得写作艺术这个问题其实包括整个文学思想、见地、情感和读写的问题。当我在中国做恢复性灵和提倡更活泼简易的散文体的文学运动时，不得不写下许多篇文章，发展我对一般的文学的见地，尤其是对于写作的见地。我可以试写出一组关于文学的警语，而以"雪茄烟灰"为题。

甲　技巧和个性

做文法教师的论文学，实等于木匠谈论美术。评论家专从写作技巧上分析文章，这其实等于一个工程师用测量仪丈量泰山的高度和结构。

世上无所谓写作的技巧。我心目中所认为有价值的中国作家，也都是这般说法。

写作技巧之于文学，正如教条之于教派——都是属于性情琐屑者的顾及小节。

初学者往往被技巧之论所眩惑——小说的技巧、剧本的技巧、音乐的技巧、演剧的技巧。他不知道写作的技巧和作家的家世并没有关系，演剧的技巧和名艺人的家世并没有关系。他不知道世上有所谓个性，这个性其实就是一切艺术和文学成就的基础。

乙　文学的欣赏

当一个人读了许多本名著，而觉得其中某作家叙事灵活生动，某作家细腻有致，某作家文意畅达，某作家笔致楚楚动人，某作家味如醇酒佳酿时，他应坦白地承认爱好他们、欣赏他们，只要他的欣赏是出乎本心的。读过这许多的作品后，他便有了一个相当的经验基础，即能辨识何者是温文，何者是醇熟，何者是力量，何者是雄壮，何者是光彩，何者是辛辣，何者是细腻，何者是风韵。在他尝过这许多种滋味之后，他不必借指南的帮助，也能知道何者是优美的文学了。

一个念文学的学生第一件事情就是：学习辨别各种不同的滋味。其中最优美的是温文和醇熟，但也是最难于学到的。温文和平淡，其间相差极微。

一个写作者，如若思想浅薄，缺乏创造性，则大概将从简单的文体入手，终至于奄无生气。只有新鲜的鱼可以清炖，如若已宿，便须加酱油、胡椒和芥末——越多越好。

优美的作家正如杨贵妃的姐姐一般，可以不假脂粉，素面朝天。宫中别的美人便少不了这两件东西，这就是英文作家中极少敢于用简单文体的理由。

丙　文体和思想

作品的优劣，全看它的风韵和滋味如何，是否有风韵和滋味。所谓风韵并无规则可言，发自一篇作品，正如烟气发自烟斗，云气发自山头，并不自知它的去向。最优美的文体就是如苏东坡的文体一般近于"行云流水"。

文体是文字、思想和个性的混合物，但有许多文体是完全单靠着文字而成的。

清澈的思想用不明朗的文字表现者，事实上很少。不清澈的思想而表现极明白者倒很多。如此的文体，实可称为明白的不明朗。

用不明朗的文字表现清澈的思想，乃是终身不娶者的文体。因为他永远无须向他的妻子做任何解释，如：伊曼纽尔·康德（Immanuel Kant）之类。萨缪尔·巴特勒（Samuel Butler，英国作家，译有《荷马史诗》）有时也是这样的古怪。

一个人的文体常被他的"文学爱人"藻饰。他在思想上和表现方式上，每每会渐渐地近似这位爱人。初学者只有借这个方法，才能培植出他的文体。等到阅世较深之后，他自会从中发现自己，而创成自己的文体。

一个人如若对某作家向来是憎恶的，则阅读这作家的作品必不能得到丝毫的助益。我颇希望学校中的教师能记住这句话。

一个人的品性，一部分是天生的，他的文体也是如此。还有一部分则完全是由感染而来的。

一个人如没有自己所喜爱的作家，即等于一个飘荡的灵魂。他始终是一个不成胎的卵子，不结子的雄蕊。所喜爱的作家或文学爱人，就是他灵魂的花粉。

世上有合于各色各种脾胃的作家，但一个人必须花些工夫，方能寻到。

一本书犹如一个人的生活，或一个城市的画像。有许多读者只看到纽约或巴黎的画像，而并没有看见纽约或巴黎的本身。聪明的读者则既读书，也亲阅生活的本身。宇宙即是一本大书，生活即是一所大的学校。一个善读者必拿那作家从里面翻到外面，如叫花子将他的衣服翻转来捉虱子一般。

有些作家能如叫花子积满了虱子的衣服一般，很有趣地不断地挑拨他们的读者，痒也是世间一件趣事。

初学者最好应从读表示反对意见的作品入手。如此，他绝不致误为骗子所欺蒙。读过表示反对意见的作品后，他即已有了准备，而可以去读表示正面意见的作品，富于评断力的心胸即是如此发展出来的。

作家都有他所爱用的字眼，每一个字都有它的生命史和个性。这生命史和个性是普通的字典所不载的，除非是如《袖珍牛津字典》（Concise of Pocket Oxford Dictionary）一类的字典。好的字典和《袖珍牛津字典》，都是颇堪一读的。

世上有两个文字矿：一是老矿，一是新矿。老矿在书中，新矿在普通人的语言中。次等的艺术家都从老矿去掘取材料，唯有高等的艺术家会从新矿中去掘取材料。老矿的产物都已经过溶解，但新矿的产物不然。

王中（公元二十七年至一百年）将"专家"和"学者"加以区别，也将"作家"和"思想家"加以区别。我以为当一个专家的学识宽博后，他即成为学者，一个作家的智慧深切后，他即成为思想家。

学者在写作中，大都借材于别的学者。他所引用的旧典成语越多，越像一位学者。一个思想家于写作时，则都借材于自己肚中的概念，越是一个伟大的思想家，越会依赖于自己的肚腹。

一个学者像一只吐出所吃的食物以饲小鸟的老鹰，一个思想家则像一条蚕，所吐的不是桑叶而是丝。

一个人的观念在写作之前，都有一个"怀孕"时期，也像胚胎在母腹中必有一个怀孕时期一般。当一个人所喜爱的作家已在他的心灵中将火星燃着，开始发动了一个活的观念流泉时，这就是所谓"怀孕"。当一个人在他的观念还没有经过怀孕的时期，即急于将它写出付印时，这就是错认肚腹泻泄时的疼痛为怀孕足月时的阵痛。当一个人出卖他的良心而做违心之论时，这就是堕胎，那胚胎落地即死。当一个作者觉得他的头脑中有如电阵一般的搅扰，觉得非将他的观念发泄出来不能安逸，乃将它们写在纸上而觉如释重负时，这就是文学的产生。因此，一个作家对于他的文学作品，自会有一种如母亲对于子女一般的慈爱感情，因此，自己的作品必是较好的，犹如一个女子在为人之妻后必是更可爱的。

作家的笑正好如鞋匠的锥，越用越锐利，到后来竟可以尖如缝衣之针，但他观念的范围必日渐广博，犹如一个人的登山观景，爬得越高，所望见者越远。

一个作家因为憎恶一个人，而拟握笔写一篇极力攻击他的文章，但一方面并没有看到那个人的好处，这个作家便没有写作这篇攻击文章的资格。

丁　自我发挥派

十六世纪末叶，袁氏三弟兄所创的"性灵派"或称"公安派"（袁氏三弟兄为公安县人），即是自我发挥的学派。"性"即个人的"性情"，"灵"即个人的"心灵"。

写作不过是发挥一己的性情，或表演一己的心灵。所谓"神通"，就是这心灵的流动，实际上确是由于血液内"荷尔蒙"的泛滥所致。

我们在读一本古书或阅一幅古画时，其实不过是在观看那作家的心灵流动。有时这心力之流如若干涸，或精神如若颓唐时，即是最高手的书画家也会缺乏精神和活泼的。

这"神通"是在早晨，当一个人于好梦沉酣中自然醒觉时来到。此后，他喝过一杯早茶，阅读一张报纸，而没有看到什么烦心的消息，慢慢走到书室里边，坐在一张明窗前的写字台边，窗外风日晴和，在这种时候，他必能写出优美的文章、优美的诗、优美的书札，必能作出优美的画，并题优美的款字在上面。

这所谓"自我"或"个性"，乃是一束肢体肌肉、神经、理智、情感、学养、悟力、经验偏见所组成。它一部分是天成的，而一部分是养成的；一部分是生而就有的，而一部分是培植出来的。一个人的性情是在出世之时，甚至在出世之前即已成为固定的。有些是天生硬心肠和卑鄙的；有些是天生坦白磊落、尚侠慷慨的；也有些是天生柔弱胆怯、多愁多虑的。这些都深隐于骨髓之中，因此，即使是最良好的教师和最聪明的父母，也没有法子变更一个人的个性。另有许多品质，则是出世之后由教育和经验而得到的。但因为一个人的思想观念和印象乃是在不同的生活时代，从种种不一的源泉和各种不同的影响潮流中所得到的，因此他的观念、偏见和见地有时会极端自相矛盾。一个人爱狗而恶猫，但也有人爱猫而恶狗。所以人类个性型式的研究，乃是一切科学中最为复杂的科学。

"自我发挥派"叫我们在写作中只可表达我们自己的思想和感觉，出乎本意的爱好，出乎本意的憎恶，出乎本意的恐惧和出乎本意的癖嗜，在表现这些时，不可隐恶而扬善，不可畏惧外界的嘲笑，也不可畏惧有背于古圣或时贤的。

"自我发挥派"的作家对一篇文章专喜爱其中个性最流露的一节，专喜爱一节中个性最流露的一句，专喜爱一句中个性最流露的一个表现语词。他在描写或叙述一幅景物、一个情感或一件事实时，只就自己所目击的景物，自己所感觉的情感，自己所了解的事实而加以描写或叙述。凡符合这条定例者，都是真文学，不符合者，即不是真文学。

《红楼梦》中的女子林黛玉，即是一个"自我发挥派"。她曾说："若果有了奇句，连平仄虚实不对，却使得的。"

"自我发挥派"因为专喜爱发乎本心的感觉，所以自然蔑视文体上的藻饰，因此这派人士在写作中专重天真和温文，他们尊奉孟子"言以达志"的说法。

文学的美处，不过是达意罢了。

这一派的弊病，在于学者不慎即会流于平淡（袁中郎），或流于怪僻（金圣叹），或过于离经叛道（李卓吾）。因此后来的儒家都非常憎恶这个学派。但以事实而论，中国的思想和文学全靠他们这班自出心裁的作家出力，方不至于完全灭绝。在以后的数十年中，他们必会得到其应有的地位。

中国正统派文学的目标明明在于表现古圣的心胸，而不是表现作者自己的心胸，所以完全是死的。"性灵派"文学的目标是在于表现作者自己的心胸，而不是古圣的心胸，所以是活的。

这派学者都有一种自尊心和独立心，这使他们不至于逾越本分而以危言耸人的听闻，如若孔孟的说话偶然和他们的见地相合，良心上可以赞同，他们不会矫情而持异说。但是，如若良心上不能赞同时，他们便不肯将孔孟随便放过去。他们是不为金钱所动、不为威武所屈的。

发乎本心的文学，不过是对于宇宙和人生的一种好奇心。凡是目力明确，不为外物所惑的人，都能时常保持这个好奇心。所以他不必歪曲事实以求景物能视若新奇。别人觉得这派学者的观念和见地十分新奇，即因他们都是看惯了矫揉造作的景物。

凡是有弱点的作家，必会亲近性灵派。这派中的作家都反对模仿古人或今人的，反对一切文学技巧的定例，袁氏弟兄相信让手和口自然做去，自能得合式的结果。李笠翁相信文章之要在于韵趣。袁子才相信做文章无所谓技巧。北宋作家黄山谷相信文章的章句都是偶然而得的，正如木中被虫所蚀的洞一般。

戊　家常的文体

用家常文体的作家是以真诚的态度说话。他把他的弱点完全显露出来，所以他是从无防人之心的。

作家和读者之间的关系，不应像师生的关系，而应像厮熟朋友的关系，只有如此，方能渐渐生出热情。

凡在写作中不敢用"我"字的人绝不能成为一个好作家。

我喜爱说谎者更胜于喜爱说实话者。我喜爱不谨慎的说谎者更胜于喜爱谨慎的说谎者。他的不谨慎，表示他的深爱读者。

　　我深信一个不谨慎的蠢人，而不敢相信一个律师。这不谨慎的蠢人是一个国家中最好的外交家，能得到人民的信仰。

　　我心目中所认为最好的杂志是一个半月刊，但不必真正出书，只需每两星期一次召集许多人，群聚在一间小室之中，让他们去随意谈天，每次以两小时为度，读者即是旁听的人。这就等于一次绝好的夜谈。完毕之后，读者即可去睡觉，在明天早晨起身去办公时，不论他是一个银行职员，或一个会计，或一个学校教师到校去张贴布告，他必会觉得隔夜的滋味还留在齿颊之间。

　　各地方的菜馆大小不一，有些是高厅大厦，金碧辉煌，可设盛宴；有些是专供小饮。我所最喜欢的是同着两三个知己朋友到这种小馆子里去小饮，而极不愿意赴要人或富翁的盛宴。我们在这小馆子里边又吃又喝，随便谈天，互相嘲谑，甚至杯翻酒泼，这种快乐是盛宴上的座客所享不到的，也是梦想不到的。

　　世界上有富翁的花园和大厦，但山中也有不少的小筑。这种小筑有些虽也布置得很精雅，但氛围终和朱红色大门、绿色窗户、仆婢排立的富家大厦截然不同。当一个人走进这种小筑时，他没听见忠狗的吠声，他没看见足恭谄笑的侍者和阍人讨厌的面孔。在离开那里，走出大门的时候，他没看见门外矗立两旁的一对"不洁的石狮子"。十七世纪某中国作家有一段绝好这种境地的描写，这好似周、程、张、朱正在伏羲殿内互相揖让，就座之时，苏东坡和东方朔忽然赤足半裸地也走了进来，拍着手互相嘲笑作乐。旁观的人或许愕然惊怪，但这些高士不过互相目视，做会心的微笑罢了。

　　己　什么是美

　　所谓文学的美和一切物事的美，大都有赖于变换和动作，并且以生活为基础。凡是活的东西都有变换和活动，而凡是有变换和活动的东西自然也有美。当我们看到山岩深谷和溪流具着远胜于运河的奇峭之美，而并不是经由建筑家用计算方法造成时，试问我们对于文学和写作怎样可以定出规例来？星辰是天之文，名山大河是地之文；风吹云变，我们就得到一个锦缎的花纹图案；霜降叶落，我们就得到了秋天之色。那些星辰在穹苍中循着它们的轨道而运行时，何曾想到地球上会有人在那里欣赏它们？然而我们终在无意之间发现了天狗星和牛郎星。地球的外壳在收缩引张之际推起了高高的山，陷下了深深的海，其实地球又何曾出于有意地创造出那五座名岳，为我们崇拜的目的？然而太华和昆仑终已矗立于地面，高下起伏，绵延千里，玉女和仙童立在危岩之上，显然是供我们欣赏的。这些就是大艺术造化家自由随意的挥洒。当天上的云行过山头而遇到强劲的山风时，它何曾想到有意露出裙边巾角以供我们赏玩？然而它们自然会整理，有时如鱼鳞，有时如锦缎，有时如赛跑的狗，如怒吼的狮子，如纵跳的凤凰，如踞跃的麒麟，都像是文学的杰作。当秋天的树木受到风霜雨露的摧残，正致力于减少它们的呼吸以保全它们的本力时，它们还会有这空闲去涂粉涂脂，以供古道行人的欣赏吗？然而他们终是那么冷洁幽寂，远胜于王维、米芾的书画。

所以凡是宇宙中活的东西都有着文学的美。枯藤的美胜于王羲之的字，悬崖的壮严胜于张猛龙的碑铭。所以我们知道"文"或文学的美是天成的。凡是尽其天性的，都有"文"或美的轮廓为其外饰，所以"文"或轮廓形式的美是内生的，而不是外来的。马的蹄是为适于奔跑而造，老虎的爪是为适于扑攫而造，鹤的腿是为适于涉水塘而造，熊的掌则是为适于在冰上爬行而造，这马、虎、鹤、熊，自己又何曾想到它们的形式的美呢？它们所做的事情无非是为生活而运用其效能，并取着最宜于他行动的姿势。但是从我们的观点说起来，则看到马蹄、虎爪、鹤腿、熊掌，都有一种惊人的美，或是雄壮有力的美，或是细巧有劲的美，或是骨骼清奇的美，或是关节粗拙的美。此外则象爪如"隶书"，狮鬃如"飞白"，争斗时的蛇屈曲扭绕如"草书"，飞龙如"篆书"，牛腿如"八分"，鹿如"小楷"。它们的美都生自姿势和活动，它们的体形都是身体效能的结果。这也就是写作之美的秘诀。"式"之所需，不能强加阻抑；"式"所不需，便当立刻停止。因此，一篇文学名作正如大自然本身的一个伸展，在无式之中成就佳式。美格和美点能自然而生，因为所谓的"式"，乃是动作的美，而不是定形的美。凡是活动的东西都有一个"式"，所以也就有美、力和文，或形式和轮廓的美。

IV. The Art of Writing

THE art of writing is very much broader than the art of writing itself, or of the writing technique. In fact, it would be helpful to a beginner who aspires to be a writer first to dispel in him any overconcern with the technique of writing, and tell him to stop trifling with such superficial matters and get down to the depths of his soul, to the end of developing a genuine literary personality as the foundation of all authorship. When the foundation is properly laid and a genuine literary personality is cultivated, style follows as a natural consequence and the little points of technique will take care of themselves. It really does not matter if he is a little confused about points of rhetoric and grammar, provided he can turn out good stuff. There are always professional readers with publishing houses whose business it is to attend to the commas, semicolons, and split infinitives. On the other hand, no amount of grammatical or literary polish can make a writer if he neglects the cultivation of a literary personality. As Buffon says,"The style is the man."Style is not a method, a system or even a decoration for one's writing; it is but the total impression that the reader gets of the quality of the writer's mind, his depth or superficiality, his insight or lack of insight and other qualities like wit, humor, biting sarcasm, genial understanding, tenderness, delicacy of understanding, kindly cynicism or cynical kindliness, hardheadedness, practical common sense, and general attitude toward things. It is clear that there can be no handbook for developing a"humorous technique"or a"three-hour course in cynical kindliness,"or"fifteen rules for practical common sense" and"eleven rules for delicacy of feeling. "

We have to go deeper than the surface of the art of writing, and the moment we do that, we find that the question of the art of writing involves the whole question of literature, of thought, point of view, sentiment and reading and writing. In my literary campaign in China for restoring the School of Self-Expression (hsingling) and for the development of a more lively and personal style in prose, I have been forced to write essay after essay giving my views on literature in general and on the art of writing in particular. I have attempted also to write a series of literary epigrams under the general title"Cigar Ashes." Here are some of the cigar ashes:

(a) Technique and Personality

Professors of composition talk about literature as carpenters talk about art. Critics analyze a literary composition by the technique of writing, as engineers measure the height and structure of Taishan by compasses.

There is no such thing as the technique of writing. All good Chinese writers who to my mind are worth anything have repudiated it.

The technique of writing is to literature as dogmas are to the church—the occupation with trivial things by trivial minds.

A beginner is generally dazzled by the discussion of technique—the technique of the novel, of the drama, of music and of acting on the stage. He doesn't realize that the technique of writing has nothing to do with the birth of an author, and the technique of acting has nothing to do with the birth of a great actor. He doesn't even suspect that there is such a thing as personality,   which is the foundation of all success in art and literature.

(b) The Appreciation of Literature

When one reads a number of good authors and feels that one author describes things very vividly, that another shows great tenderness of delicacy, a third expresses things exquisitely, a fourth has an indescribable charm, a fifth one's writing is like good whiskey, a sixth one's is like mellow wine, he should not be afraid to say that he likes them and appreciates them, if only his appreciation is genuine. After such a wide experience in reading, he has the proper experiential basis for knowing what are mildness, mellowness, strength, power, brilliance, pungency, delicacy, and charm. When he has tasted all these flavors, then he knows what is good literature without reading a single handbook.

The first rule of a student of literature is to learn to sample different flavors. The best flavor is mildness and mellowness, but is most difficult for a writer to attain. Between mildness and mere flatness there is only a very thin margin.

A writer whose thoughts lack depth and originality may try to write a simple style and end up by being insipid. Only fresh fish may be cooked in its own juice; stale fish must be flavored with anchovy sauce and pepper and mustard—the more the better.

A good writer is like the sister of Yang Kueifei, who could go to see the Emperor himself without powder and rouge. All the other beauties in the palace required them. This is the reason why there are so few writers who dare to write in simple English.

(c) Style and Thought

Writing is good or bad, depending on its charm and flavor, or lack of them. For this charm there can be no rules. Charm rises from one's writing as smoke rises from a pipe-bowl, or a cloud rises from a hill-top, not knowing whither it is going. The best style is that of"sailing clouds and flowing water,"like the prose of Su Tungp'o.

Style is a compound of language, thought and personality. Some styles are made exclusively of language.

Very rarely does one find clear thoughts clothed in unclear language. Much more often does one find unclear thoughts expressed clearly. Such a style is clearly unclear.

Clear thoughts expressed in unclear language is the style of a confirmed bachelor. He never has to explain anything to a wife. E.g., Immanuel Kant. Even Samuel Butler often gets so quizzical.

A man's style is always colored by his "literary lover." He grows to be like him more and more in ways of thinking and methods of expression. That is the only way a style can be cultivated by a beginner. In later life, one finds one's own style by finding one's own self.

One never learns anything from a book when he hates the author. Would that school teachers would bear this fact in mind!

A man's character is partly born, and so is his style. The other part is just contamination.

A man without a favorite author is a lost soul. He remains an unimpregnated ovum, an unfertilized pistil. One's favorite author or literary lover is pollen for his soul.

A favorite author exists in the world for every man, only he hasn't taken the trouble to find him.

A book is like a picture of life or of a city. There are readers who look at pictures of New York or Paris, but never see New York or Paris itself. The wise man reads both books and life itself. The universe is one big book, and life is one big school.

A good reader turns an author inside out, like a beggar turning his coat inside out in search of fleas.

Some authors provoke their readers constantly and pleasantly like a beggar's coat full of fleas. An itch is a great thing.

The best way of studying any subject is to begin by reading books taking an unfavorable point of view with regard to it. In that way one is sure of accepting no humbug. After having read an author unfavorable to the subject, he is better prepared to read more favorable authors. That is how a critical mind can be developed.

A writer always has an instinctive interest in words as such. Every word has a life and a personality, usually not recorded by a dictionary, except one like the Concise or Pocket Oxford Dictionary.

A good dictionary is always readable, like the P.O.D.

There are two mines of language, a new one and an old one. The old mine is in the books, and the new one is in the language of common people. Second-rate artists will dig in the old mines, but only first-rate artists can get something out of the new mine. Ores from the old mine are already smelted, but those from the new mine are not.

Wang Ch'ung (A. D. 27-c. 100) distinguished between"specialists" and "scholars,"and again between"writers" and"thinkers."I think a specialist graduates into a scholar when his knowledge broadens, and a writer graduates into a thinker when his wisdom deepens.

A"scholar's"writing consists of borrowings from other scholars, and the more authorities and sources he quotes, the more of a"scholar" he appears. A thinker's writing consists of borrowings from ideas in his own intestines, and the greater thinker a man is, the more he depends on his own intestinal juice.

A scholar is like a raven feeding its young that spits out what it has eaten from the mouth. A thinker is like a silkworm which gives us not mulberry leaves, but silk.

There is a period of gestation of ideas before writing, like the period of gestation of an embryo in its mother's womb before birth. When one's favorite author has kindled the spark in one's soul, and set up a current of live ideas in him, that is the"impregnation."When a man rushes into print before his ideas go through this period of gestation, that is diarrhoea, mistaken for birth pains. When a writer sells his conscience and writes things against his convictions, that is artificial abortion, and the embryo is always stillborn. When a writer feels violent convulsions like an electric storm in his head, and he doesn't feel happy until he gets the ideas out of his system and puts them down on paper and feels an immense relief, that is literary birth. Hence a writer feels a maternal affection toward his literary product as a mother feels toward her baby. Hence a writing is always better when it is one's own, and a woman is always lovelier when she is somebody else's wife.

The pen grows sharper with practice like a cobbler's awl, gradually acquiring the sharpness of an embroidery needle. But one's ideas grow more and more rounded, like the views one sees when mounting from a lower to a higher peak.

When a writer hates a person and is thinking of taking up his pen to write a bitter invective against him, but has not yet seen his good side, he should lay down the pen again, because he is not yet qualified to write a bitter invective against the person.

(d) The School of Self-Expression

The so-called"School of Hsingling"started by the three Yüan brothers at the end of the sixteenth century, or the so-called"Kungan School"(Kungan being the native district of the brothers) is a school of self-expression. Hsing means one's"personal nature,"and ling means one's"soul"or"vital spirit. "

Writing is but the expression of one's own nature or character and the play of his vital spirit. The so-called"divine afflatus"is but the flow of this vital spirit, and is actually caused by an overflow of hormones in the blood.

In looking at an old master or reading an ancient author, we are but watching the flow of his vital spirit. Sometimes when this flow of energy runs dry or one's spirits are low, even the writing of the best calligraphist or writer lacks spirit or vitality.

This"divine afflatus"comes in the morning when one has had a good sleep with sweet dreams and wakes up by himself. Then after his cup of morning tea, he reads the papers and finds no disturbing news and slowly walks into his study and sits before a bright window and a clean desk, while outside there is a pleasant sun and a gentle breeze. At this moment, he can write good essays, good poems, good letters, paint good paintings and write good inscriptions on them.

The thing called"self"or"personality"consists of a bundle of limbs, muscles, nerves, reason, sentiments, culture, understanding, experience, and prejudices. It is partly nature and partly culture, partly born and partly cultivated. One's nature is determined at the time of his birth, or even before it. Some are naturally hardhearted and mean; others are naturally frank and straightforward and chivalrous and big-hearted; and again others are naturally soft and weak in character, or given over to worries. Such things are in one's"marrow bones"and the best teacher or wisest parent cannot change one's type of personality. Again other qualities are acquired after birth through education and experience, but insofar as one's thoughts and ideas and impressions come from the most diverse sources and different streams of influence at different periods of his life, his ideas, prejudices and points of view present a most bewildering inconsistency. One loves dogs and is afraid of cats, while another loves cats and is afraid of dogs. Hence the study of types of human personality is the most complicated of all sciences.

The School of Self-Expression demands that we express in writing only our own thoughts and feelings, our genuine loves, genuine hatreds, genuine fears and genuine hobbies. These will be expressed without any attempt to hide the bad from the good, without fear of being ridiculed by the world, and without fear of contradicting the ancient sages or contemporary authorities.

Writers of the School of Self-Expression like a writer's most characteristic paragraph in an essay, his most characteristic sentence in a paragraph, and his most characteristic expression in a sentence. In describing or narrating a scene, a sentiment or an event, he deals with the scene that he himself sees, the sentiment that he himself feels and the event as he himself understands it. What conforms to this rule is literature and what does not conform to it is not literature.

The girl Lin Taiyü in Red Chamber Dream belonged also to the School of Self-Expression when she said,"When a poet has a good line, never mind whether the musical tones of words fall in with the established pattern or not. "

In its love for genuine feelings, the School of Self-Expression has a natural contempt for decorativeness of style. Hence it always stands for the pure and mild flavor in writing. It accepts the dictum of Mencius that"the sole goal of writing is expressiveness. "

Literary beauty is only expressiveness.

The dangers of this school are that a writer's style may degenerate into plainness (Yüan Chunglang), or he may develop eccentricity of ideas (Chin Shengt'an), or his ideas may differ violently from those of established authorities (Li Chowu). That is why the School of Self-Expression was so hated by the Confucian critics. But as a matter of fact, it is these original writers who saved Chinese thought and literature from absolute uniformity and death. They are bound to come into their own in the next few decades.

Chinese orthodox literature expressly aimed at expressing the minds of the sages and not the minds of the authors and was therefore dead; the hsingling school of literature aims at expressing the minds of the authors and not the minds of the sages, and is therefore alive.

There is a sense of dignity and independence in writers of this school which prevents them from going out of their way to say things to shock people. If Confucius and Mencius happen to agree with them and their conscience approves, they will not go out of their way to disagree with the Sages; but if their conscience disapproves, they will not give Confucius and Mencius the right of way. They can be neither bribed with gold nor threatened with ostracism.

Genuine literature is but a sense of wonder at the universe and at human life.

He who keeps his vision sane and clear will have always this sense of wonder, and therefore has no need to distort the truth in order to make it seem wonderful. The ideas and points of view of writers of this school always seem so new and strange only because readers are so used to the distorted vision.

A writer's weaknesses are what endear him to a hsingling critic. All writers of the hsingling school are against imitation of the ancients or the moderns and against a literary technique of rules. The Yüan brothers believed in "letting one's mouth and wrist go, resulting naturally in good form" and held that "the important thing in literature is genuineness." Li Liweng believed that "the important thing in literature is charm and interest." Yüan Tsets'ai believed that "there is no technique in writing." An early Sung writer, Huang Shanku, believed that "the lines and form of writing come quite accidentally, like the holes in wood eaten by insects."

(e) The Familiar Style

A writer in the familiar style speaks in an unbuttoned mood. He completely exposes his weaknesses, and is therefore disarming.

The relationship between writer and reader should not be one between an austere school master and his pupils, but one between familiar friends.

Only in this way can warmth be generated.

He who is afraid to use an"I"in his writing will never make a good writer.

I love a liar more than a speaker of truth, and an indiscreet liar more than a discreet one. His indiscretions are a sign of his love for his readers.

I trust an indiscreet fool and suspect a lawyer.

The indiscreet fool is a nation's best diplomat. He wins people's hearts.

My idea of a good magazine is a fortnightly, where we bring a group of good talkers together in a small room once in a fortnight and let them chat together. The readers listen to their chats, which last just about two hours. It is like having a good evening chat, and after that the reader goes to bed, and next morning when he gets up to attend to his duties as a bank clerk or accountant or a school principal posting notices to the students, he feels that the flavor of last night's chat still lingers around his cheeks.

There are restaurants for giving grand dinners in a hall with goldframed mirrors, and there are small restaurants designed for a little drink. All I want is to bring together two or three intimate friends and have a little drink, and not go to the dinners of rich and important people. But the pleasure we have in a small restaurant, eating and drinking and chatting and teasing each other and overturning cups and spilling wine on dresses is something which people at the grand dinners don't understand and cannot even"miss."

There are rich men's gardens and mansions, but there are also little lodges in the mountains. Although sometimes these mountain lodges are furnished with taste and refinement, the atmosphere is quite different from the rich men's mansions with vermillion gates and green windows and a platoon of servants and maids standing around. When one enters the door, he does not hear the barking of faithful dogs and he does not see the face of snobbish butlers and gatekeepers, and when he leaves, he doesn't see a pair of "unchaste stone lions" outside its gate. The situation is perfectly described by a writer of the seventeenth century: "It is as if Chou, Ch'eng, Chang and Chu are sitting together and bowing to each other in the Hall of Fuhsi, and suddenly there come Su Tungp'o and Tungfang Su who break into the room half naked and without shoes, and they begin to clap their hands and joke with one another. The onlookers will probably stare in amazement, but these gentlemen look at each other in silent understanding. "

(f) What Is Beauty?

The thing called beauty in literature and beauty in things depends so much on change and movement and is based on life. What lives always has change and movement, and what has change and movement naturally has beauty. How can there be set rules for literature or writing, when we see that mountain cliffs and ravines and streams possess a beauty of waywardness and ruggedness far above that of canals, and yet they were formed without the calculations of an architect? The constellations of stars are the wen or literature of the skies, and the famous mountains and great rivers are the wen or literature of the earth. The wind blows and the clouds change and we have the pattern of a brocade; the frost comes and leaves fall and we have the color of autumn. Now do the stars moving around their orbits in the firmament ever think of their appreciation by men on earth? And yet the Heavenly Dog and the Cowherd are perceived by us by an accident. The crust of the earth shrinks and stretches and throws up mountains and forms deep seas. Did the earth consciously create the Five Sacred Mountains for us to worship? And yet the T'aihua and the K'uenluen Mountains dash along with their magnificent rhythm and the Jade Maiden and the Fairy Boy stand around us on awe-inspiring peaks, apparently for our enjoyment. These are but free and easy strokes of the Creator, the great art master. Can clouds, which sail forth from the hill-tops and meet the lashing of furious mountain winds, have time to think of their petticoats and scarves for us to look at? And yet they arrange themselves, now like the scales of fish, now like the pattern of brocade, and now like racing dogs and roaring lions and dancing phoenixes and gamboling unicorns, like a literary masterpiece. Can autumn trees that are feeling the pinch of heat and cold and the devastation of frost, and that are busily occupied in slowing down their breath and conserving their energy, have time to paint and powder themselves for the traveller on the ancient highway to look at? And yet they seem so cool and pure and sad and forlorn, and far superior to the paintings of Wang Wei and Mi Fu.

And so every living thing in the universe has its literary beauty. The beauty of a dried-up vine is greater than the calligraphy of Wang Hsichih, and the austerity of an overhanging cliff is more imposing than the stone inscriptions on Chang Menglung's tomb. Therefore we know that the wen or literary beauty of things arises from their nature, and those that fulfill their nature clothe themselves in wen or beautiful lines. Therefore wen, or beauty of line and form, is intrinsic and not extrinsic. The horse's hoofs are designed for a quick gallop, the tiger's claws are designed for pouncing on its prey; the stork's legs are designed for wading across swamps, and the bear's paws are designed for walking on ice. Does the horse, the tiger, the stork or the bear ever think of its beauty of form and proportions? All it tries to do is function in life and adopt a proper posture for movement. But from our point of view, we see the horse's hoofs, the tiger's claws, the stork's legs and the bear's paws have a striking beauty, either in their fullness of contour and suggestion of power, or in their slenderness and strength of line, or in their clearness of outline, or in the ruggedness of their joints. Again the elephant's paws are like the lishu style of writing, the lion's mane is like the feipo, fighting snakes write wonderful wriggling ts'aoshu ("grass script"), and floating dragons write chuanshu ("seal characters"), the cow's legs resemble pafen (comparatively stout and symmetrical writing), and the deer resembles hsiaok'ai (elegant"small script"). Their beauty comes from their posture or movement, and their bodily shapes are the result of their bodily functions, and this is also the secret of beauty in writing. When the shih or posture of movement requires it, it may not be repressed, and when the posture or movement does not require it, it must stop. Hence a literary masterpiece is like a stretch of nature itself, well-formed in its formlessness, and its charm and beauty come by accident. For this thing we call shih is the beauty of movement, and not the beauty of static proportions. Everything that lives and moves has its shih and therefore has its beauty, force, and wen, or beauty of form and line.

生活 | The
的 | Importance
艺术 | of Living

第十三章

与上帝的关系

一. 宗教的恢复

世上有许多人自以为认识上帝，知道上帝的爱憎。因此一个人在讨论这个题目时，不免被有些人认做亵渎，也被另一些人认做先知。人类以各个分别的说起来，不过是地壳的千百万万分之一，而地壳又不过是宇宙的千百万万分之一，真是极其渺小的物事，竟敢说认识上帝。

然而没有生活的哲学是完备的，没有一个人类精神生活的概念是充分的，除非我们把自己引进到和周遭世界的生活有满意而融洽的关系里边去。人类已很够重要，是我们研究中最重要的题旨，这就是人性主义的要素。然而人类是生活于一个宏大的宇宙中的，这宇宙也和人类一般的奇特。所以凡是忽视了周遭的大世界，忽视了它的起由和结果的人，都不能算做有着一个满意的生命。

正统派宗教的缺点，在于在历史的进展中和一些完全不涉及宗教范围的物事发生了不可分析的关系——物理学、地质学、天文学、犯罪学、性的概念和妇女观念。如若它专自限于良心的范围，则重新定向的工作便不必像目下这般的困难了。毁灭"天堂"和"地狱"的观念，较易于毁灭上帝的观念。

反之，科学把宇宙的神秘的一种更新更深的意义和物质之为动力的一种别称的新概念，展开于现代基督徒的眼前。对于上帝本身，詹姆士·金斯（James Jeans）曾说，宇宙实好似一个伟大的思想，而不似一具伟大的机器。计算的本身，证明宇宙中实在有所谓算术上不能加以计算的东西。宗教须往后退，它不应该像以前涉及自然科学范围中的许多物事，而应承认它们乃是不属于宗教的物事。宗教也不应该让神灵的阅历去倚赖着完全不相干的说法，如：人类的历史已有四千余年，或一百万年，或地球的形状是扁的、圆的，或是像折叠桌子一般，或是由印度的象或中国的龟所擎着。宗教应该限于道德的范围，限于良心的范围，它自有和花木鱼星的研究的一般的尊严。圣保罗是首先动手术割治犹太教的人。他把饮食（吃有蹄的动物）和宗教分拆开来，使宗教得益不浅。宗教不但从分拆饮食之中可以得到益处，也可从分拆地质学和解剖学之中得到益处。宗教不必再去做一个天文学和地质学的涉猎者和一个古代传说的保存者。宗教尽可以在生物教师讲课时闭口不言，就不像向来那么愚蠢，而易于得到人类的崇敬了。

照现代所有的宗教而论，每个人将不得不把自己从所信仰的宗教中拯救出来，不论我们对于神学信条的意见如何。我们未尝不能在跪在地上默默作礼、眼望着彩色玻璃的教礼和崇拜的氛围下投身于上帝的门下。在这种意义上，崇拜成为真正的美术经验，真是出于本心的美术经验，犹如我们看着太阳向山林的背后落下去一般。在这个人的心目中，宗教是良心的最后事实，因为这个美术经验是非常近于诗意的。

但他对于现代的教会必然蔑视，因为他崇拜的上帝，并不是一个花些小钱即能买得动的上帝。他不能在乘船向北行驶时，叫风向北吹；不能在向南行驶时，叫风向南吹。为着顺风而感谢上帝，乃是绝对的无礼，也是自私。因为这包含着上帝这个特别人向北行驶时，便不顾及另有许多向南行驶的人了。宗教应该是一种灵的交流，当中不能含有此造对彼造有所求的交换情事。他必不能够了解教会的真义，他对于宗教所经过的转变必觉得奇怪。他如将宗教照目前的形式下定义必会愕然无措。宗教是它的现状加了神秘情感的赞颂吗？抑是某种已经成为非常神秘，已经雕饰，已经遮掩的道德真理，庶使教士之流可以从而得到生活吗？启示对宗教的关系，是否也是如"秘方"对用广告宣传的"秘制药品"之关系一般的吗？它抑或是一种利用不能见的、不可思议的事物在那里变戏法，因为不能见的不可思议的事物乃是最便于变戏法的事物吗？信仰是否应该以知识为基础，还是信仰乃是开始的知识的终点吗？它抑或像一个棒球，可以由爱梅·麦克弗逊（Aimee McPherson，二十世纪初福音传教士和创建者）向观众打去——是一种乔（Joe）可以用接棒球的法子去接过来的事物吗？……基督是否必须在托尔斯泰被希腊的正统教会除名之后，于大风雪中将他抱在怀中吗？或是基督将要立在曼宁主教（Bishop Manning）的窗外，招呼那些坐在长椅中的富家孩子，一再做他的请求说"让这些孩子到我这里"吗？

所以宗教在我们的心中所留下的是：一种令人不舒服的——然而在我是异常满意的——感觉，觉得宗教所在于我们的生命中的，将是一种对生活的美，生活的伟大和生活的神秘更简单化的感觉。当中虽也有一种责任，但已撤去了神学所堆积于表面的自以为准确的东西。在这个形式中的宗教是简单的，它于现代的人类已是够好。中古时代的神灵神权统治思想已渐渐退化。至于个人的永生问题——即宗教用以打动人心的第二个大理由——现在有许多人都已是抱着要死便死，而并无不满意的态度了。

我们对永生的成见，当中略带一些病理性质。人类的期望长生是可以谅解的，但如若没有基督徒从中推波助澜，则必不至于被人类重视到这般畸形的地步。它已不是一种微妙的回想、一种崇高的幻想、位于虚无和事实中间的诗意境界，而已成为一种十分一本正经的事实。尤其是在修道士的心目中，死亡的意识或死后的生活，已成为生活中主要的关怀事件了。事实上，五十岁以上的人们，不论是异教徒或基督徒，大多并不怕死。这就是他们为什么不为死亡所威吓，并不把天堂和地狱十分放在心上的理由。我们常常听见他们很高兴地讨论自己身后的碑铭和坟墓的式样，以及火葬的好处等。我这话并不单说凡是自知必升天堂者是如此，也是指着对死亡抱一种现实见解，以为人死不过似灯烛的火焰熄灭一般者而说的。目下识见高超的名人当中，有许多个都表示不相信有所谓个人的长生而并不在意——如威尔斯（H.G.Wells，英国著名科幻作家）、阿尔伯特·爱因斯坦（Albert Einstein）、阿瑟·凯兹爵士等人——但我以为并不一定需识见十分高超的人们方能克服死亡的恐怖。

有许多人已将别种更有意义的永生代替了这种个人的永生——如种族的永生、功绩和影响的永生。当我们去世之后，倘若所遗留的功绩依旧继续影响我们自己社会中的人生——不论这影响是怎样微小——而在其中活动，便已够了。我们可以将花朵摘下来，将花瓣丢在地上，然而它的香味依旧存留于空气中。这是一种更好的、更合理的、更为公的永生。在这种真实的意义上，我们可以说路易·巴斯德（Louis Pasteur）、路德·伯班克（Luther Burbank，美国植物育种家）和托马斯·爱迪生（Thomas Edison）至今还在我们之中活着。他们的身体虽然已死，但这又有什么关系？因为所谓"身体"者，无非是许多化学的组成分子不断有变化组合状态的一个抽象的综合罢了。人们开始了悟自己的生命不过是像永流的大河中的一滴水，因此对于这生命之流乐于做一些贡献。倘若他能少怀一些自私心，他自会觉得满足了。

I. The Restoration of Religion

SO many people presume to know God and what God approves and God disapproves that it is impossible to take up this subject without opening oneself to attack as sacrilegious by some and as a prophet by others. We human creatures who individually are less than a billionth part of the earth's crust, which is less than a billionth part of the great universe, presume to know God!

Yet no philosophy of life is complete, no conception of man's spiritual life is adequate, unless we bring ourselves into a satisfactory and harmonious relation with the life of the universe around us. Man is important enough; he is the most important topic of our studies: that is the essence of humanism. Yet man lives in a magnificent universe, quite as wonderful as the man himself, and he who ignores the greater world around him, its origin and its destiny, cannot be said to have a truly satisfying life.

The trouble with orthodox religion is that, in its process of historical development, it got mixed up with a number of things strictly outside religion's moral realm—physics, geology, astronomy, criminology, the conception of sex and woman. If it had confined itself to the realm of the moral conscience, the work of re-orientation would not be so enormous today. It is easier to destroy a pet notion of "Heaven" and "Hell" than to destroy the notion of God.

On the other hand, science opens up to the modern Christian a newer and deeper sense of the mystery of the universe and a new conception of matter as a convertible term with energy, and as for God Himself, in the words of Sir James Jeans, "The universe seems to be nearer to a great thought than to a great machine." Mathematical calculation itself proves the existence of the mathematically incalculable. Religion will have to retreat and instead of saying so many things in the realm of natural sciences as it used to do, simply acknowledge that they are none of religion's business; much less should it allow the validity of spiritual experience to depend on totally irrelevant topics, like whether the age of man is 4,000 odd years or a million, or whether the earth is flat, or round, or shaped like a collapsible tea-table, or borne aloft by Hindu elephants or Chinese turtles. Religion should, and will, confine itself to the moral realm, the realm of the moral conscience, which has a dignity of its own comparable in every sense to the study of flowers, the fishes and the stars. St. Paul performed the first surgical operation upon Judaism and by separating cuisine (eating hoofed animals) from religion, immensely benefited it. Religion stands to gain immensely by being separated not only from cuisine, but also from geology and comparative anatomy. Religion must cease to be a dabbler in astronomy and geology and a preserver of ancient folkways. Let religion respectfully keep its mouth shut when teachers of biology are talking, and it will seem infinitely less silly and gain immeasurably in the respect of mankind.

Such religion as there can be in modern life, every individual will have to salvage from the churches for himself. There is always a possibility of surrendering ourselves to the Great Spirit in an atmosphere of ritual and worship as one kneels praying without words and looking at the stained-glass windows, in spite of all that one may think of the theological dogmas. In this sense, worship becomes a true aesthetic experience, an aesthetic experience that is one's own, very similar in fact to the experience of viewing a sun setting behind an outline of trees on hills. For that man, religion is a final fact of consciousness, for it will be an aesthetic experience very much akin to poetry.

But what contempt he must have for the churches, as they are at present. For the God that he worships will not be one that can be beseeched for daily small presents. He will not command the wind to blow north when he sails north, and command the wind to blow south when he sails south. To thank God for a good wind is sheer impudence, and selfishness also, for it implies that God does not love the people sailing south when HE, the important individual, is sailing north. It will be a communion of spirits without one party trying to beg a favor of the other. He will not be able to comprehend the meaning of the churches as they are. He will wonder at the strange metamorphosis that religion has gone through. He will be puzzled when he tries to define religions in their present forms. Is religion a glorification of the status quo with mystic emotion? Or is it certain moral truths so mystified and decorated and camouflaged as to make it possible for a priestcraft to make a living? Doesn't revelation stand in the same relation to religion as "a secret patented process" stands in relation to certain advertised nostrums? Or is religion a juggling with the invisible and the unknowable because the invisible and the unknowable lend themselves so conveniently to juggling? Is faith to be based on knowledge, or does faith only begin where knowledge ends? Or is religion a baseball that Sister Aimée McPherson can hit with a baseball bat right into an audience—something that Joe can catch and "get" in the way that he catches a baseball?... Did Christ really have to receive Tolstoy in his arms in a blazing snowstorm after he was excommunicated from the Greek Orthodox Church? Or is Jesus going to stand outside Bishop Manning's cathedral window and beckon to the rich men's children in their pews and repeat His gentle request, "Suffer little children to come unto me?"

So we are left with the uncomfortable and yet, for me, strangely satisfying feeling that what religion is left in our lives will be a very much more simplified feeling of reverence for the beauty and grandeur and mystery of life, with its responsibilities, but will be deprived of the good old, glad certainties and accretions which theology has accumulated and laid over its surface. Religion in this form is simple and, for many modern men, sufficient. The spiritual theocracy of the Middle Ages is definitely receding and as for personal immortality, which is the second greatest reason for the appeal of religion, many men today are quite content to be just dead when they die.

Our preoccupation with immortality has something pathological about it. That man desires immortality is understandable, but were it not for the influence of the Christian religion, it should never have assumed such a disproportionately large share of our attention. Instead of being a fine reflection, a noble fancy, lying in the poetic realm between fiction and fact, it has become a deadly earnest matter, and in the case of monks, the thought of death, or life after it, has become the main occupation of this life. As a matter of fact, most people on the other side of fifty, whether pagans or Christians, are not afraid of death, which is the reason why they can't be scared by, and are thinking less of, Heaven and Hell. We find them very often chattering glibly about their epitaphs and tomb designs and the comparative merits of cremation. By that I do not mean only those who are sure that they are going to heaven, but also many who take the realistic view of the situation that when they die, life is extinguished like light from a candle. Many of the finest minds of today have expressed their disbelief in personal immortality and are quite unconcerned about it–H. G. Wells, Albert Einstein, Sir Arthur Keith and a host of others–but I do not think it requires first-class minds to conquer this fear of death.

Many people have substituted for this personal immortality, immortality of other kinds, much more convincing–the immortality of the race, and the immortality of work and influence. It is sufficient that when we die, the work we leave behind us continues to influence others and play a part, however small, in the life of the community in which we live. We can pluck the flower and throw its petals to the ground, and yet its subtle fragrance remains in the air. It is a better, more reasonable and more unselfish kind of immortality. In this very real sense, we may say that Louis Pasteur, Luther Burbank and Thomas Edison are still living among us. What if their bodies are dead, since"body"is nothing but an abstract generalization for a constantly changing combination of chemical constituents! Man begins to see his own life as a drop in an ever flowing river and is glad to contribute his part to the great stream of life. If he were only a little less selfish, he should be quite contented with that.

二．我为什么是一个异教徒

　　宗教终是一桩属于个人的事件，每个人都必须由他自己去探讨出自己的宗教见解。只要他是出于诚意的，则不论所探讨得到的是什么东西，上帝绝不会见怪他。每个人的宗教经验都是对他本人有效的，因为我已说过它是一种不容争论的东西。但是，如若一个诚实的人将他对于宗教问题的心得用诚恳的态度讲出来，也必是有益于他人的。我在提到宗教时，每每避开它的普泛性而专讲个人的经验，就是这个缘故。

　　我是一个异教徒。这句话或许可以作为一种对基督教的叛逆，但叛逆这个名词似乎略嫌过火，还不能准确地描写出一个人怎样在他的心理演变中，逐渐地背离基督教。他怎样很热诚地极力想紧抱住基督教的许多信条，而这些信条仍会渐渐地溜了开去。因为其中从来没有什么仇恨，所以也谈不到什么叛逆。

　　因为我生长在一个牧师的家庭中，有一个时期也预备去做传道工作，所以在意旨的交战之中，我的天然感情实在是向着基督教方面，而并不是反对它。在这个情感和意识交战的当中，我渐渐地达到了一个确定地否认"赎罪说"的地位。这个地位照简单的说法，实在不能不称之为一个异教徒的地位我始终觉得只有处在有关生命和宇宙的状态的信仰时，我方是自然自在而无所交战于心。这个程度的演变极其自然，正如儿童的奶牙脱落，或已熟的苹果从树头掉落一般。我对这种脱落当然是不加以干涉的。照道家的说法，这就是生活于道里边。照西方的说法，这不过是依据自己的见解，对自己和宇宙抱一种诚恳的态度罢了。我相信一个人除非对自己抱着一种理智上的诚恳态度，他便不能自在和快乐。一个人若能自在，便已登上天堂了。在我个人，做一个异教徒也无非是求自在罢了。

"是一个异教徒"这句话，其实和"是一个基督徒"在意义上有什么高下之分？这不过是一句反面的话，因为在一般的读者心目中，"是一个异教徒"这话的意义，无非说他不是一个基督徒罢了。而且"是一个基督徒"也是一句很广泛很含混的说法，而"不是一个基督徒"这句话也同样意义不很分明。最不合理者，是将"异教徒"这名词的意义定为一个不信宗教或上帝的人。因为根本上，我们对"上帝"或"对于生命的宗教"的态度还没有能够定出确切的意义。伟大的异教徒大都对大自然抱着一种深切的诚敬态度，所以我们对异教徒这个名词，只可取其通俗的意义，将它作为不过是一个不到礼拜堂里去的人（除为了一次审美的行动外，我确不大到礼拜堂里去）。是一个不属于基督教群，而并不承认寻常的正统教义的人的解说。

在正的方面，中国的异教徒（只有这一种是为我所深知而敢于讨论的）就是一个以任心委运的态度去度这尘世生活的人。他禀着生命的久长，脚踏实地，很快乐地生活着。时常对于这个生命觉到一种深愁，但仍很快地应付着。凡遇到人生的美点和优点时，必会很深切地领略着，而视良好行为的本身即是一种报酬。不过我也承认他们因想升到天堂去，才做良好的行为，反之，如若没有天堂在那里诱引，或没有地狱在那里威吓，即不做良好行为的"宗教的"人物，自有一些怜悯和鄙视的心思。倘若我这句话是对的，则此间有很多的异教徒，不过自己不觉得罢了。现在的开明基督徒和异教徒其实是很相近的，不过在谈到"上帝"时，双方才显出他们的歧异点。

我以为我已经知道宗教经验的深度，因为一个人不必一定像纽孟枢机主教（Cardinal Newman，英国罗马天主教领袖与作家）一般的大神学家才能获得这种经验——否则基督教便失去了它的价值，或已经被人误解了。在我眼前看来，一个基督徒和一个异教徒之间的灵的生活，其区别之点不过是崇信基督教者是生活于一个由上帝所统治和监视的世界中，他和这个上帝有着不断的个人关系。所以也可说他是生活在一个由一位仁慈的父亲所主持的世界中，他的行为水准须谐合于他以一个上帝之子的地位所应达到的标准。这个行为水准显然是一个普通人难于在一生中，或甚至在一个星期中，或甚至在一天之中毫无间断地达到的。他的实际生活实是游移于人类的生活水准和真正的宗教生活水准之间的。

在另一方面，这异教徒住在这世界上像一个孤儿一般，他不能期望天上有一个人在那里照顾他，在用祈祷方式树立灵的关系时即会降福于他的安慰。这就显然是一个较为不快乐的世界，但也自有益处和尊严，因为他也如其他孤儿一般不得不学习自立，不得不自己照顾自己，并更易于成熟。我在转变为异教徒之中，始终使我害怕的，并不是什么灵的信仰问题，而就是这个突然掉落到没有上帝照顾我的世界里边去的感念，这个害怕直到最后的一刹方才消灭。因为当时我也如从小即是基督徒的人一般，觉得如若一个人的上帝其实并不存在，这个宇宙的托底便好似脱落了。

然而，有时一个异教徒也会将这个更为和暖的、更为快乐的世界，同时看成一个更为稚气的、更像尚在生长中的世界；一个人如若能够长久保持着这个幻想，确是一件好而有益的事情，他的观念将和佛教徒对生命的观念相近似；这个世界将因此好似更为彩色华丽，不过也将因此成为一个不十分实在的，所以价值较低的世界。在我个人说起来，凡是不十分实在的和彩色过重的事物都是要不得的。一个人如要得到一种真理，必须付一笔代价，不论它的后果如何，我们终是需要真理的。这个境地在心理上，正和一个杀人者所处的境地相同：如若一个人犯了一次杀案，最好办法就是自首。我就是因了这个理由鼓起勇气转变为一个异教徒的。但一个人在承认一切之后，自会没有惧怕。心里安适就是一个人在承认一切之后的心境（这里我觉得我已受了佛教和道家思想的影响）。

我或者也可以将基督徒的和异教徒的境地用下列的说法加以区别：我个人的异教思想同时为了自傲心和自卑心弃绝了基督教，是情感上的自傲心和理智上的自卑心，但笼统地说起来，自卑的成分比自傲居多。我是为了情感上的自傲心，因为我深不愿见除了我们是人类的理由，所以应该做和蔼合礼的男女人之外，还有别的理由；在理论上，如若你喜欢将思想分类，这个当可归入可做代表的人性主义思想。但大半我是为了自卑心，为了理智上的自卑心，因为在现代天文学的面前，我不能再相信一个寻常人类会被大创造者视为一个重要的分子，因为一个人类不过是地球上一个极其微渺的分子，地球也不过是太阳系中一个极其微渺的分子，而太阳系更不过是大宇宙中一个极其微渺的分子罢了。人们的大胆和傲然夸张，实是所以使我倾跌的东西。我们对于那个"超人"所做的工作，所知道者只不过是几千万分之一，怎能够说已经知道了他的性质？怎可以对他的能耐做假定之说呢？

人类个人的重要，无疑是基督教的基本教义之一。但我们试看在基督徒的日常生活中，这条教义已将他们引向怎样可笑的夸张。

在我母丧后出殡的四天之前，忽然大雨倾盆，这雨如若长此下去（这在漳州，秋天是时常如此的），城内的街道都将被水淹没，而出殡也将因此被阻。我们都是特地从上海赶回去的，如若过于耽搁日子，于我们都是很不便的。我的一个亲戚（她是一个极端的但也并不是不常见的中国笃信基督者的榜样）向我说，她向来信任上帝，上帝是必会代他的子女设法的。她即刻做祈祷，雨竟停止了，显然是为了这样便可以让我们这个小小的基督徒家庭举行我们的出殡礼。但这件事里边所含的意义是：倘若没有我们这件事夹在当中，上帝便将听任全漳州的万千人民遭受大水之灾，如以往所常遭到的一般；或也可说是：上帝不是为了漳州万千的人民，而只是为了我家这少数几个人要赴着晴天出殡，所以特地将雨停止。我觉得实是一种最不可思议的自私自利，我不能相信上帝是会替如此自私的子女想什么法子的。

还有一个基督教牧师写了一篇自传，其中述说在他的一生中上帝许多次照顾他的故事，希望因此荣归于上帝。其中有一件上帝照应他的事件是：当他筹集了六百元去购买到美国的船票的那一天，上帝特地将汇兑率降低一些，以便这位重要人物在购买美金船票时可以便宜一些。以六百元所能购买的美金而言，高低的相差至多不过一二十元，难道上帝单单为了使他这个儿子可以得到一二十元的便宜，竟肯使巴黎、伦敦和纽约的交易所经历一次金融风潮吗？我们应记得这种荣归于上帝的说法，在基督徒群中是并非罕见的。

人们的寿限大都不过七十岁，他们竟会这般厚颜自傲。人类以其集合体而言，也许已有一部很动人的历史，但以各个而言，在宇宙中正如苏东坡所说，不过是沧海之一粟，或如朝生暮死的蜉蝣罢了。基督徒不肯谦卑。他们对于这股他们自己也是其中一分子的生命巨流（这股巨流永远向无穷无尽处流去，如一条大河之流向海中，永远变迁，而也是永远不变的）的集体的永存，从来不知道感觉满足。瓦器将向窑工问："你为什么将我烧成这个模式，为什么将我烧成这般脆法？"瓦器因为易于破碎，所以感觉不满足。人类有了这样一具奇异的身体，几乎近于神圣的身体，也仍感觉不满足。他还要长生不老！他不肯让上帝安宁。每天还要做祈祷，他每天还要从这个万物之源那里讨些个人的赏赐。他为什么不让上帝得些安宁呢？

从前有一个中国学者，他不信佛教，但他的母亲很相信。她极其虔诚，整天不停地念"阿弥陀佛"时，她的儿子即在旁边唤一声"妈妈"，她恼了。"这样看起来，"她的儿子向她说，"菩萨如果也听得见您这般唤他时，他不也要发恼吗？"

我的父母都是极虔诚的基督徒。每晚听我父亲领头做晚祷，便可以知道他的虔诚程度。我是一个对宗教感觉很敏锐的孩子。我以一个牧师儿子的地位，受到教会教育的便利，我从其长处获得益处，但也从其短处获得痛苦。对它的长处，我是始终感谢的，对于它的短处，则将它转变成我的力量。因为依照中国哲学的说法，一个人的生命是没有所谓好运或厄运的。

我是不许到中国戏院里边看戏的，不许听说书的，完全和中国的民间神话故事隔绝。当我踏进教会学校之后，我父亲所教我读过的一些《四书》完全荒废了。这或许于我是一种益处——因为这一来，我在从未受过西方教育之后，能以一个西方小孩走到东方新奇世界里的愉快心境再回去研究这些旧学。当我在学校读书时，我完全抛弃毛笔而专用自来水笔，是于我最有益的事情，因为这使我在心理上始终觉得东方是一个完全新的事情，在心理上始终觉得东方是一个完全新鲜的世界，直到我已有了做研究它的准备的时候。如若维苏威火山不将庞贝城掩没，则庞贝的古迹必不能保存得这样完备，那地方石板街上所留下的车辙必不能保存到今日，教会学校的教育就是我的"维苏威火山"。

思想这件事总是危险的。而且，思想总是和魔鬼有联系的。当我在学校受教育时，也就是我最虔信宗教的时代，我心中对于基督教生活的美丽感觉和一种对任何物事都想探求其理由的念头已渐渐地发生冲突。但很奇怪，当时我并没有感觉到那种几乎使托尔斯泰因之而自杀的痛苦和失望。在每一个阶段中，我仍觉得自己还是一个统一的基督徒，在信念上仍很融洽，不过比上一个阶段开通一些，在盲从教条上次数略少一些。无论如何，我终究还随时想到"山上的教训"，圣诗中如"看那些田中的百合花啊！"这种句子太好了，相信它不会是假的。我就因了这些，因了意识到内心的基督教生活，所以生出了新的力量。

但教义很可怕地从我的心头渐渐地溜了出去，许多浅近的事情渐渐地使我觉得不自在。"肉体的复活"这一条，当基督未能在第一世纪中如人所期望的第二次降临里边实现，诸圣徒没有从他们的坟墓里边肉身走出来时，即已证明是不成立的，这一条现在依然存在于圣徒的信条中。这就是很浅近的事情中之一端。

后来，我又加入了神学班，以求深造，于是又发现了教义中的另一条也使我起疑的地方。那一条就是"处女生儿"，美国各神道学院的主任教授对于这一条都各抱着不同的见解。最使我动恼的是：中国信徒在受洗礼之前，必须将这一条囫囵承认，不许稍生疑问，而同一教会里边的神学家不许公然认为是一件疑问。这好似有些虚伪，而且似乎是不公允的。

我读到高级的神学，研究到"水门"究竟在哪里那种细微问题时，便觉得责任已经解除，因而对于神学不肯认真，结果是我学科的成绩渐渐低落。我的教师即以为我的性情根本不适于做一个教会牧师，因此主教也以为我不如从此脱离，不愿再在我的身上耗费徒然的教诲了。这在我现在看来，也好似一种不露相的好运。因为我很疑惑如若我当时依旧读下去而终身穿上了一件牧师的长袍之后，是否真能够心口如一！这种对于神学家和一般的教徒所信仰的信条的反抗意念，在我看来，实在近于我所谓的"背叛"了。

这个时候，我已达到深信基督教的神学家实是基督教大敌的地步。他们有着两个我最不能了解的矛盾点：第一，他们将基督教信仰的整个结构完全系在一个苹果上。如若亚当没有吃苹果，世上即不会有原始的罪恶；如若世上并没有原始的罪恶，世上便不需要什么救赎。不论那个苹果在象征上有怎样的价值，但这一点终是极显明的。基督本人从来没有提起过原始的罪恶或救赎这件事情，所以它其实并不符合基督的训诲。总而言之，我从研究文学之后，也如现代的美国人一般，不能意识到我有着什么罪恶，而且绝不相信我有罪恶。我所能意识到的就是：上帝只要能如我的母亲爱我的一半，便绝不会将我打到地狱里边去。这是我内心意识里边的最后一次行为，不论为了哪一种宗教，都不能不承认其为事实。

还有一个问题，在我看是尤其不合理的。这就是：当亚当和夏娃在蜜月中吃了一个苹果时，上帝即异常大怒，罚他们的子孙世世代代为了这一件小小的罪过而受罪，但是，当同是这班子孙将上帝的独子害死时，上帝异常快活，将他们一起赦免。不论人们对这件事有怎样巧妙的解释相佐证，我总认它是极不合理的。这也就是使我不自在的末了一件事情。

我在毕业之后，依旧是一个很热心的基督徒，会自动地在北京的清华学校（非教会学校）里组织了一个主日《圣经》班，这事曾使当时的许多同事教员心里很不高兴。这《圣经》班的圣诞日集会使我最受痛苦，因为我是在拿一件自己所不相信的伪事在那里告诉给中国的儿童听。自从我将一切都借着理智破解之后，留在我心中的就只剩了爱心和恐惧两件事，一种渴望能依赖一个全智的上帝，庶使我可以觉得快乐的爱心，如若没有了这个一再抚慰的爱心，我便不能如此快乐和安宁——和堕落到孤儿世界中去的恐惧心。最后我居然获救了。我和一位同事辩论说："如若没有上帝，人民便不肯行善，而世界必将颠倒了。"

"不然。"我的孔教同事回说，"因为我们都是懂道理的人类，所以我们应该能够过一种合于道理的人类生活。"

这个令人崇尚人类生活尊严的说法，割断了我和基督教的最后一丝关系，从此之后，我便成为一个异教徒了。

现在我已完全明白了。异教的信仰是一种更为简单的信仰。它没有什么假定之说，也无须做什么假定之说。它专就生活事实而立论，所以使良好的生活更为人所崇尚。它在不责善之中使人自然知道行善。它并不借着种种假定的说法，如：罪恶、得救，十字架、存款于天上，人类因了上天第三者的关系，所以彼此之间有一种彼此应尽的义务，等——都是一些曲折难解，难于直接证明的事情——去劝诱人们做一件善事。如若一个人承认行善的本身即是一件好事，他即会自然而然将宗教的引人行善的饵诱视做赘物，并将视之为足以掩罩道德真理的彩色的东西。人类之间的互爱应该就是一件终结和绝对的事实。我们应该不必借着上天第三者的关系而彼此相爱。基督教在我看来，好似已使道德成为一件非常困难、非常复杂的事情。罪恶反而是一件极动人、极自然和极可悦的东西。在另一方面，异教主义倒好似能够将宗教从神学里拯救出来，而恢复了它的信仰的简单性和感觉的尊严。

其实，我颇已看出有许多神学的谬说怎样从第一、第二、第三世纪中渐渐地产生，将"山上训诲"的简单真理歪曲成一种严厉、不合人情、自以为是的结构，以供一个祭司阶级自私地利用。从"启示"这个名词即能看出其中的隐情。这启示就是一种授予一个先知的特别秘密或神圣计划，由这先知以师生授受的方式世代传袭下去；这启示也是各种宗教中从回教和摩门教到活佛的喇嘛教和爱迪夫人的基督教科学所都具有的，以便他们可以各自握着，当做一种得救的特有注册专利品。凡是祭司阶级都是依赖这个启示为他们的日常食粮而获得生活。"山上训诲"这个简单真理必须修饰起来，必须将上帝所重视的百合花镀上金子。于是我们有了"第一个亚当""第二个亚当"，如此类推下去。圣保罗的逻辑在基督教的早年时代似乎很能动人听闻，令人很难责难，但在现在较为乖滑、较富于意识的人的心目中，便十分勉强、缺乏力量了。而崇尚启示的弱点即在这种亚洲式演绎逻辑和现代对真理较为乖滑的领悟之间，显露于现代人的眼前了。所以，只有借着回到异教主义和不承认启示，一个人方能回到原始式的（在我看来是较为满意的）基督教。

所以说一个异教徒为不信宗教的人是错误的，其实他所不信的不过是各式各样的启示罢了。一个异教徒是必然信仰"上帝"的，不过他因恐旁人误会，所以不肯说出来。中国的异教徒都是信仰"上帝"的，文学中所用以表示这个"上帝"的名词，其最常见者就是"造物者"。唯一的不同点就在：中国的异教徒很诚实地听任这位"造物者"隐处在一个神秘的光环中，不过对他表示着一种尊畏和虔敬即以为足够了。对于这个宇宙的美丽，对于万物的巧妙，对于星辰的神秘，对于上天的奇伟，对于人类灵魂的尊严，他都是能领会的。他接受死亡，他接受痛苦，视之不过为生命所不可免的东西，视之如旷野的阵风，如山间的明月，而从无怨言。他以为"委心任运"乃是最虔敬的态度和宗教信仰，而称之为"生于道"。如若"造物者"要他在七十岁死亡，他便坦然在那时去世。他又相信"天理循环"，所以世界绝不会永远没有公道。此外，他便无所求了。

II. Why I Am a Pagan

RELIGION is always an individual, personal thing. Every person must work out his own views of religion, and if he is sincere, God will not blame him, however it turns out. Every man's religious experience is valid for himself, for, as I have said, it is not something that can be argued about. But the story of an honest soul struggling with religious problems, told in a sincere manner, will always be of benefit to other people. That is why, in speaking about religion, I must get away from generalities, and tell my personal story.

I am a pagan. The statement may be taken to imply a revolt against Christianity; and yet revolt seems a harsh word and does not correctly describe the state of mind of a man who has passed through a very gradual evolution, step by step, away from Christianity, during which he clung desperately, with love and piety, to a series of tenets which against his will were slipping away from him. Because there was never any hatred, therefore, it is impossible to speak of a rebellion.

As I was born in a pastor's family and at one time prepared for the Christian ministry, my natural emotions were on the side of religion during the entire struggle rather than against it. In this conflict of emotions and understanding, I gradually arrived at a position where I had, for instance, definitely renounced the doctrine of redemption, a position which could most simply be described as that of a pagan. It was, and still is, a condition of belief concerning life and the universe in which I feel natural and at ease, without having to be at war with myself. The process came as naturally as the weaning of a child or the dropping of a ripe apple on the ground; and when the time came for the apple to drop, I would not interfere with its dropping. In Taoistic phraseology, this is but to live in the Tao, and in Western phraseology it is but being sincere with oneself and with the universe, according to one's lights. I believe no one can be natural and happy unless he is intellectually sincere with himself, and to be natural is to be in heaven. To me, being a pagan is just being natural.

"To be a pagan"is no more than a phrase, like"to be a Christian."It is no more than a negative statement, for to the average reader, to be a pagan means only that one is not a Christian; and, since"being a Christian"is a very broad and ambiguous term, the meaning of"not being a Christian"is equally ill-defined. It is all the worse when one defines a pagan as one who does not believe in religion or in God, for we have yet to define what is meant by"God"or by the"religious attitude toward life."Great pagans have always had a deeply reverent attitude toward nature. We shall therefore have to take the word in its conventional sense and mean by it simply a man who does not go to church (except for an aesthetic inspiration, of which I am still capable), does not belong to the Christian fold, and does not accept its usual, orthodox tenets.

On the positive side, a Chinese pagan, the only kind of which I can speak with any feeling of intimacy, is one who starts out with this earthly life as all we can or need to bother about, wishes to live intently and happily as long as his life lasts, often has a sense of the poignant sadness of this life and faces it cheerily, has a keen appreciation of the beautiful and the good in human life wherever he finds them, and regards doing good as its own satisfactory reward. I admit, however, he feels a slight pity or contempt for the"religious"man who does good in order to get to heaven and who, by implication, would not do good if he were not lured by heaven or threatened with hell. If this statement is correct, I believe there are a great many more pagans in this country than are themselves aware of it. The modern liberal Christian and the pagan are really close, differing only when they start out to talk about God.

I think I know the depths of religious experience, for I believe one can have this experience without being a great theologian like Cardinal Newman—otherwise Christianity would not be worth having or must already have been horribly misinterpreted. As I look at it at present, the difference in spiritual life between a Christian believer and a pagan is simply this: the Christian believer lives in a world governed and watched over by God, to whom he has a constant personal relationship, and therefore in a world presided over by a kindly father; his conduct is also often uplifted to a level consonant with this consciousness of being a child of God, no doubt a level which is difficult for a human mortal to maintain consistently at all periods of his life or of the week or even of the day; his actual life varies between living on the human and the truly religious levels.

On the other hand, the pagan lives in this world like an orphan, without the benefit of that consoling feeling that there is always someone in heaven who cares and who will, when that spiritual relationship called prayer is established, attend to his private personal welfare. It is no doubt a less cheery world; but there is the benefit and dignity of being an orphan who by necessity has learned to be independent, to take care of himself, and to be more mature, as all orphans are. It was this feeling rather than any intellectual belief—this feeling of dropping into a world without the love of God—that really scared me till the very last moment of my conversion to paganism; I felt, like many born Christians, that if a personal God did not exist the bottom would be knocked out of this universe.

And yet a pagan can come to the point where he looks on that perhaps warmer and cheerier world as at the same time a more childish, I am tempted to say a more adolescent, world; useful and workable, if one keep the illusion unspoiled, but no more and no less justifiable than a truly Buddhist way of life; also a more beautifully colored world but consequently less solidly true and therefore of less worth. For me personally, the suspicion that anything is colored or not solidly true is fatal. There is a price one must be willing to pay for truth; whatever the consequences, let us have it. This position is comparable to and psychologically the same as that of a murderer: if one has committed a murder, the best thing he can do next is to confess it. That is why I say it takes a little courage to become a pagan. But, after one has accepted the worst, one is also without fear. Peace of mind is that mental condition in which you have accepted the worst. (Here I see for myself the influence of Buddhist or Taoist thought.)

Or I might put the difference between a Christian and a pagan world like this: the pagan in me renounced Christianity out of both pride and humility, emotional pride and intellectual humility, but perhaps on the whole less out of pride than of humility. Out of emotional pride because I hated the idea that there should be any other reason for our behaving as nice, decent men and women than the simple fact that we are human beings; theoretically and if you want to go in for classifications, classify this as a typically humanist thought. But more out of humility, of intellectual humility, simply because I can no longer, with our astronomical knowledge, believe that an individual human being is so terribly important in the eyes of that Great Creator, living as the individual does, an infinitesimal speck on this earth, which is an infinitesimal speck of the solar system, which is again an infinitesimal speck of the universe of solar systems. The audacity of man and his presumptuous arrogance are what stagger me. What right have we to conceive of the character of a supreme being, of whose work we can see only a millionth part, and to postulate about his attributes?

The importance of the human individual is undoubtedly one of the basic tenets of Christianity. But let us see what ridiculous arrogance that leads to in the usual practice of Christian daily life.

Four days before my mother's funeral, there was a pouring rain, and if it continued, as was usual in July in Changchow, the city would be flooded, and there could be no funeral. As most of us came from Shanghai, the delay would have meant some inconvenience. One of my relatives–a rather extreme but not an unusual example of a Christian believer in China–told me that she had faith in God, Who would always provide for His children. She prayed, and the rain stopped, apparently in order that a tiny family of Christians might have their funeral without delay. But the implied idea that, but for us, God would willingly subject the tens of thousands of Changchow inhabitants to a devastating flood, as was often the case, or that He did not stop the rain because of them but because of us who wanted to have a conveniently dry funeral, struck me as an unbelievable type of selfishness. I cannot imagine God providing for such selfish children.

There was also a Christian pastor who wrote the story of his life, attesting to many evidences of the hand of God in his life, for the purpose of glorifying God. One of the evidences adduced was that, when he had got together 600 silver dollars to buy his passage to America, God lowered the rate of exchange on the day this so very important individual was to buy his passage. The difference in the rate of exchange for 600 silver dollars could have been at most ten or twenty dollars, and God was willing to rock the bourses in Paris, London, and New York in order that this curious child of His might save ten or twenty dollars. Let us remind ourselves that this way of glorifying God is not at all unusual in any part of Christendom.

Oh, the impudence and conceit of man, whose span of life is but threescore and ten! Mankind as an aggregate may have a significant history, but man as an individual, in the words of Su Tungp'o, is no more than a grain of millet in an ocean or an insect fuyu born in the morning and dying at eve, as compared with the universe. The Christian will not be humble. He will not be satisfied with the aggregate immortality of his great stream of life, of which he is already a part, flowing on to eternity, like a mighty stream which empties into the great sea and changes and yet does not change. The clay vessel will ask of the potter, "Why hast thou cast me into this shape and why hast thou made me so brittle?" The clay vessel is not satisfied that it can leave little vessels of its own kind when it cracks up. Man is not satisfied that he has received this marvelous body, this almost divine body. He wants to live forever! And he will not let God alone. He must say his prayers and he must pray daily for small personal gifts from the Source of All Things. Why can't he let God alone?

There was once a Chinese scholar who did not believe in Buddhism, and his mother who did. She was devout and would acquire merit for herself by mumbling, "Namu omitabha!"a thousand times day and night. But every time she started to call Buddha's name, her son would call, "Mamma!"The mother became annoyed. "Well"said the son, "don't you think Buddha would be equally annoyed, if he could hear you?"

My father and mother were devout Christians. To hear my father conduct the evening family prayers was enough. And I was a sensitively religious child. As a pastor's son I received the facilities of missionary education, profited from its benefits, and suffered from its weaknesses. For its benefits I was always grateful and its weaknesses I turned into my strength. For according to Chinese philosophy there are no such things in life as good and bad luck.

I was forbidden to attend Chinese theaters, never allowed to listen to Chinese minstrel singers, and entirely cut apart from the great Chinese folk tradition and mythology. When I entered a missionary college, the little foundation in classical Chinese given me by my father was completely neglected. Perhaps it was just as well—so that later, after a completely Westernized education, I could go back to it with the freshness and vigorous delight of a child of the West in an Eastern wonderland. The complete substitution of the fountain pen for the writing brush during my college and adolescent period was the greatest luck I ever had and preserved for me the freshness of the Oriental mental world unspoiled, until I should become ready for it. If Vesuvius had not covered up Pompeii, Pompeii would not be so well preserved, and the imprints of carriage wheels on her stone pavements would not be so clearly marked today. The missionary college education was my Vesuvius.

Thinking was always dangerous. More than that, thinking was always allied with the devil. The conflict during the collegiate-adolescent period, which, as usual, was my most religious period, between a heart which felt the beauty of the Christian life and a head which had a tendency to reason everything away, was taking place. Curiously enough, I can remember no moments of torment or despair, of the kind that drove Tolstoy almost to suicide. At every stage, I felt myself a unified Christian, harmonious in my belief, only a little more liberal than the last, and accepting some fewer Christian doctrines. Anyway, I could always go back to the Sermon on the Mount. The poetry of a saying like"Consider the lilies of the field" was too good to be untrue. It was that and the consciousness of the inner Christian life that gave me strength.

But the doctrines were slipping away terribly. Superficial things first began to annoy me. The"resurrection of the flesh,"long disproved when the expected second coming of Christ in the first century did not come off and the Apostles did not rise bodily from their graves, was still there in the Apostles' Creed. This was one of those things.

Then, enrolling in a theological class and initiated into the holy of holies, I learned that another article in the creed, the virgin birth, was open to question, different deans in American theological seminaries holding different views. It enraged me that Chinese believers should be required to believe categorically in this article before they could be baptized, while the theologians of the same church regarded it as an open question. It did not seem sincere and somehow it did not seem right.

Further schooling in meaningless commentary scholarship as to the whereabouts of the"water gate"and such minutiae completely relieved me of responsibility to take such theological studies seriously, and I made a poor showing in my grades. My professors considered that I was not cut out for the Christian ministry, and the bishop thought I might as well leave. They would not waste their instruction on me. Again this seems to me now a blessing in disguise. I doubt, if I had gone on with it and put on the clerical garb, whether it would have been so easy for me to be honest with myself later on. But this feeling of rebellion against the discrepancy of the beliefs required of the theologian and of the average convert was the nearest kind of feeling to what I may call a"revolt. "

By this time I had already arrived at the position that the Christian theologians were the greatest enemies of the Christian religion. I could never get over two great contradictions. The first was that the theologians had made the entire structure of the Christian belief hang upon the existence of an apple. If Adam had not eaten an apple, there would be no original sin, and, if there were no original sin, there would be no need of redemption. That was plain to me, whatever the symbolic value of the apple might be. This seemed to me preposterously unfair to the teachings of Christ, who never said a word about the original sin or the redemption. Anyway, from pursuing literary studies, I feel, like all modern Americans, no consciousness of sin and simply do not believe in it. All I know is that if God loves me only half as much as my mother does, he·will not send me to Hell. That is a final fact of my inner consciousness, and for no religion could I deny its truth.

Still more preposterous another proposition seemed to me. This was the argument that, when Adam and Eve ate an apple during their honeymoon, God was so angry that He condemned their posterity to suffer from generation to generation for that little offense but that, when the same posterity murdered the same God's only Son, God was so delighted that He forgave them all. No matter how people explain and argue, I cannot get over this simple untruth. This was the last of the things that troubled me.

Still, even after my graduation, I was a zealous Christian and voluntarily conducted a Sunday school at Tsing Hua, a non-Christian college at Peking, to the dismay of many faculty members. The Christmas meeting of the Sunday school was a torture to me, for here I was passing on to the Chinese children the tale of herald angels singing upon a midnight clear when I did not believe it myself. Everything had been reasoned away, and only love and fear remained: a kind of clinging love for an all-wise God which made me feel happy and peaceful and suspect that I should not have been so happy and peaceful without that reassuring love—and fear of entering into a world of orphans. Finally my salvation came. "Why," I reasoned with a colleague, "if there were no God, people would not do good and the world would go topsy-turvy. "

"Why?" replied my Confucian colleague. "We should lead a decent human life simply because we are decent human beings," he said.

This appeal to the dignity of human life cut off my last tie to Christianity, and from then on I was a pagan.

It is all so clear to me now. The world of pagan belief is a simpler belief. It postulates nothing, and is obliged to postulate nothing. It seems to make the good life more immediately appealing by appealing to the good life alone. It better justifies doing good by making it unnecessary for doing good to justify itself. It does not encourage men to do, for instance, a simple act of charity by dragging in a series of hypothetical postulates—sin, redemption, the cross, laying up treasure in heaven, mutual obligation among men on account of a third-party relationship in heaven—all so unnecessarily complicated and roundabout, and none capable of direct proof. If one accepts the statement that doing good is its own justification, one cannot help regarding all theological baits to right living as redundant and tending to cloud the luster of a moral truth. Love among men should be a final, absolute fact. We should be able just to look at each other and love each other without being reminded of a third party in heaven. Christianity seems to me to make morality appear unnecessarily difficult and complicated and sin appear tempting, natural, and desirable. Paganism, on the other hand, seems alone to be able to rescue religion from theology and restore it to its beautiful simplicity of belief and dignity of feeling.

In fact, I seem to be able to see how many theological complications arose in the first, second, and third centuries and turned the simple truths of the Sermon on the Mount into a rigid, self-contained structure to support a priestcraft as an endowed institution. The reason was contained in the word revelation–the revelation of a special mystery or divine scheme given to a prophet and kept by an apostolic succession, which was found necessary in all religions, from Mohammedanism and Mormonism to the Living Buddha's Lamaism and Mrs. Eddy's Christian Science, in order for each of them to handle exclusively a special, patented monopoly of salvation. All priestcraft lives on the common staple food of revelation. The simple truths of Christ's teaching on the Mount must be adorned, and the lily He so marveled at must be gilded. Hence we have the"first Adam"and the"second Adam,"and so on and so forth. But Pauline logic, which seemed so convincing and unanswerable in the early days of the Christian era, seems weak and unconvincing to the more subtle modern critical consciousness; and in this discrepancy between the rigorous Asiatic deductive logic and the more pliable, more subtle appreciation of truth of the modern man, lies the weakness of the appeal to the Christian revelation or any revelation for the modern man. Therefore only by a return to paganism and renouncing the revelation can one return to primitive (and for me more satisfying) Christianity.

It is wrong therefore to speak of a pagan as an irreligious man: irreligious he is only as one who refuses to believe in any special variety of revelation. A pagan always believes in God but would not like to say so, for fear of being misunderstood. All Chinese pagans believe in God, the most commonly met–with designation in Chinese literature being the term chaowu, or the Creator of Things. The only difference is that the Chinese pagan is honest enough to leave the Creator of Things in a halo of mystery, toward whom he feels a kind of awed piety and reverence. What is more, that feeling suffices for him. Of the beauty of this universe, the clever artistry of the myriad things of this creation, the mystery of the stars, the grandeur of heaven, and the dignity of the human soul he is equally aware. But that again suffices for him. He accepts death as he accepts pain and suffering and weighs them against the gift of life and the fresh country breeze and the clear mountain moon and he does not complain. He regards bending to the will of heaven as the truly religious and pious attitude and calls it"living in the Tao."If the Creator of Things wants him to die at seventy, he gladly dies at seventy. He also believes that"heaven's way always goes round"and that there is no permanent injustice in this world. He does not ask for more.

一. 合于人情的思想之必要

思想是一种艺术，而不是一种科学。中国和西方的学问之间，最大的对比就是：西方太多专门知识，而太少近于人情的知识；至于中国则富于对生活问题的关切，而欠于专门的科学。我们眼见西方科学思想侵入了近于人情的知识的区域，其中的特点就是：十分专门化和无处不引用科学的与半科学的名词。这里，我所谓"科学"的思想，是指它在一般的意义上而言，而尚不是真正的科学思想，因为真正的科学思想是不能从常识和幻想分析开来的。在一般的意义上，这种科学思想是严格的、合于逻辑的、客观的、十分专门化的，并在方式和幻想的景物中是"原子式"的。这东西两种形式的学问，其对比终还是归结于逻辑和常识的冲突。逻辑如若剥去了常识，它便成为不近人情；而常识如若剥去了逻辑，它便不能够深入大自然的神秘境界。

当一个人检视中国的文学和哲学界时，他将得到一些什么东西呢？他会察觉那里边没有科学，没有极端的理论，没有假说，而且没有真正的性质十分不同的哲学。例如中国诗人王维，他不过借儒道以正行为，借佛教以净心胸，并借历史、画、山、河、酒、音乐和歌曲以慰精神罢了。他生活在世界中，但也是出世的。

所以，中国即成为一个人人不很致力于思想，而人人只知道尽力去生活的区域。在这里，哲学本身不过是一件很简单而且属于常识的事情，可以很容易地用一两句诗词包括一切。这区域里面没有什么哲学系，广泛地说起来，没有逻辑，没有形而上学，没有学院式的胡说；没有学院式专重假定主义，较少智力的和实际的疯狂主义，较少抽象的和冗长的字句。机械式的唯理主义在这里是永远不可能的，而且对逻辑的必须概念都抱着一种憎恶的态度。这里的事业生活中没有律师，而哲学生活中也没有逻辑家。这里只有着一种对生活的亲切感觉，而没有什么设计精密的哲学系，这里没有一个康德或一个黑格尔，而只有文章家、警语作家、佛家禅语和道家譬喻的拟议者。

中国的文学，以其全面而言，粗看似乎只见大量短诗和短文，在不爱好的人们看起来，似乎是多得可厌，但其中实有种种的类别和种种的美点，正如一幅野外景色一般。这里面有文章家和尺牍家，他们只需用五六百个字，便能将生活的感觉表示于一篇短文或短札中，其篇幅比美国低级学校儿童所做的论说更短。在这种随手写作的书札、日记、笔记和文章中，我们所看到的大概是对一次人生遭遇的评论，对邻村中一个女子自尽的记载，或对一次春游、一次雪宴、一次月夜荡桨、一次晚间在寺院里躲雨的记载，再加上一些这种时节各人谈话的记录。这里有许多散文家同时即是诗人，有许多诗人同时即是散文家，所有的著作每篇至多不过五七百字，有时单用一句诗文即能表出整个生活哲学。这里有许多警喻、警语和家信的作家，他们写作时都是乘兴所至，随后写去，并不讲究什么严格的系统。这使派系难于产生。理智阶级常被合于情理的精神压伏，尤其是被作家的艺术的感觉性压伏，而无从活动。事实上，理智阶级在这里是最为人所不信任的。

我无须指出逻辑本能乃是人类灵心的一种最有力的利器，因而科学的成就成为可能。我也知道西方的人类进步至今还在根本上是由常识和批判精神所统制着，这常识和批判精神是比逻辑精神更为伟大的东西，我以为实在是代表着西方思想的最高形式。我也无须明说西方的批判精神比在中国更为发展。在指出逻辑的思想的弱点中，我不过是指着某一种特别的缺点而说的，即如他们的政治中也有着这一种的弱点，如德国人和日本人的机械式政治即属于此类。逻辑自有它的动人之处，我认为侦探小说的发展就是一种逻辑灵心的最令人感兴趣的产品，这种文字在中国完全没有发展过。但是过度耽于逻辑思想也自有其不利之处。

西方学问杰出的特质就是专门化和分割知识，将它们归入各式各类的门类。逻辑思想和专门化的过于发展，再加上好用专门的名词，造成了现代文明的一个奇特事实，即哲学已和它的背景分隔得如此遥远，已远落在政治学和经济学的后面，以致一般的人们都会走过它的旁边而竟觉着好似没有这样一件东西。在一个平常人的心目中，甚至在某些有教养的人的心目中，都觉得哲学实在是一种最好不必加以过问的学科。这显然是现代文化中的一种奇特的反常现象，因为哲学本应是最贴近人们的脑怀和事业的物事，现在反倒远在千里之外。希腊和罗马的古典型的文化便不是如此，中国的文化也不是如此。也许是现代人对于生活问题——其实是哲学中的正常题旨——不感兴趣，或许是我们已经走离哲学的原始概念太遥远了。我们的知识范围已经推广到如此广大，由各类专家所热心守卫的知识门类已经如此众多，以至于哲学这一门其实应是人们所宜最先研究的学问，反而被打入没有人愿意做专门研究的场地里边去了。美国某大学的布告可以作为现代教育状况的一个典型，这布告说："心理学科现在已经开放，凡是经济学科的学生，愿意者都可以加入。"所以经济学科的教授已将自己一科里学生的友爱和幸福托付给心理学科的教授，同时为了答谢好意，他又容许心理学科的学生踏进经济学科的围场以表示友谊。同时，"知识之王"的哲学则如战国时的君王一般，不但已不能从他的学科附庸各国收取贡礼，而且觉得他的权力和国土日渐减缩，只剩较少的食粮和不足的人民效忠于他了。

因为现在我们已达到一个只有着知识门类而并没有着知识本身的人类文化梯阶；只是专门化，但没有完成其整体；只有专门家，而没有人类知识的哲学家。这种知识的过分专门化，实和中国皇宫中尚膳房的过分专门化没有什么分别。当某一个朝代倾覆的时节，有一位贵官居然得到了一个从尚膳房里逃出来的宫女。他得意极了，特地在某天邀请了许多朋友来尝尝这位御厨高手所做的菜肴。当设宴的日期快到时，他即吩咐这宫女去预备一桌最丰盛的御用式酒席。这宫女回说，她不会做这样的一席菜。

"那么，你在宫中时，做些什么呢？"主人问。
"噢，我是专做席面上所用的糕饼的。"她回答。
"很好，那么你就替我做些上好的糕饼吧。"
宫女的答语使他几乎跳起来，因为她回说："不，我不会做糕饼，我是专切糕饼馅子里边所用的葱的。"

现在的人类知识和学院式学问的场地里边，情形就和这个相仿。我们有着一位略晓得一些生命和人类性质的生物学家；有着一位略晓得一些同一题目的另一部分的精神病学家；有着一位通晓人类早年历史的地质学家；有着一位知悉野蛮人种的心性的人类学家；有着一位如若偶然是个心胸开通者，可以教给我们一些人类过去历史所反映出来的人类知识和人类愚行的历史学家；有着一个有时也能帮助我们认识我们的行为，但仍是多偏于告诉我们一种学院式的呆话——如：刘易斯·卡罗尔乃是一个忧郁主义者，或从他的用鸡为试验的实验室里走出来而宣布说，巨响对于一只鸡的影响是使它们的心跳跃的心理学家——有些以教授为业的心理学家，在我看来，当他们错误时，他们是使人昏迷的，而在不错误时更令人昏迷。但在专门化的程序中，同时并没有应该并进地完成整个切要程序，即将这类知识的多方面综合成一个整体，以达到它们所拟达到的最高目的——生命的知识——的程序。现在我们或许已经做了将知识完成整个的预备，耶鲁大学中的人类关系学会，和哈佛大学立校三百年纪念会中的演讲词都可以做这一点的证明。不过，除非西方的科学家能用一种较简单的、较不合逻辑的思想方法去从事于这件工作，完成整体这件事简直没有成功的日子。人类知识不单是将专门知识一件一件地加上去而成的，也不能单从统计式平均数的研究中去获得它，只能借着洞察而获得成就，只能借着更普通的常识、更多的智能和更清楚但更锐敏的直觉方能获得成就。

逻辑的思想和合理的思想之间，或可称为学院式思想和诗意的思想之间，有着一种很明显的区别。学院式的思想，我们所有的已很多了，但是诗意的思想现代中尚还稀见。亚里士多德和柏拉图其实是很摩登的，他们所以如此，不但因为希腊人很近似现代人，而且因为他们实在是——严格的说法——现代思想的祖先。亚里士多德虽也有他的人性主义见解和中庸之道的学说，但他确是现代教科书作家的祖师，他实在是首创将知识分割成许多门类者——从物理学和植物学直到伦理学和政治学。他显然就是首创为普通人所不能了解的不相干的学院式胡说者，而后来现代美国社会学家和心理学家更助纣为虐，又比他更为厉害。柏拉图虽有着真正的人类洞察力，但在某种意义上，他实在应负如新柏拉图主义学派所崇尚的对于概念和抽象观念崇拜的责任，这个传统的思想不但没有被加些更多的洞察力以为调和，反而被现代专讲概念和主义的作者熟习，而将它视做好似实有一个独立的存在性。最近的现代化心理学实是剥削了我们的"理智""意旨"和"情感"部门，并帮助杀害那个和中古时代的神学家在一起时尚还是一个整体的"灵魂"……

很明显，现在所需要的似乎是一种需经过改造的思想方式，一种更富有诗意的思想，方能更稳定地观察生命和观察它的整体。正如已过世的詹姆斯·哈维·罗宾逊（James Harvey Robinson）所警告我们的："有些谨慎的观察家很坦白地表示他的真诚意见说，除非将思想提升到比目下更高的平面之上，文明必然将要受到某种绝大的阻碍。"罗宾逊教授很智慧地指出，"良心的驱使和洞察力似乎是在彼此猜忌。但其实它们很可以成为朋友。"现代的经济学家和心理学家似乎有着太多的良心驱使而缺乏洞察力。对世事施用逻辑的危险这一点是不应该过分重视的，但因科学思想的力量和尊严在现代如此巨大，以致虽有人曾做种种的警告，这一类的学院式思想依旧不断地侵入哲学的区域，深信人类的灵心可以如一组沟渠一般加以研究，人类的思想浪潮也可以如无线电波一般加以测量。它的后果是逐渐在那里扰乱我们的思想，同时于实用的政治学上有着极恶劣的影响。

I. The Need of Humanized Thinking

THINKING is an art, and not a science. One of the greatest contrasts between Chinese and Western scholarship is the fact that in the West there is so much specialized knowledge, and so little humanized knowledge, while in China there is so much more concern with the problems of living, while there are no specialized sciences. We see an invasion of scientific thinking into the proper realm of humanized knowledge in the West, characterized by high specialization and its profuse use of scientific or semi-scientific terminology. I am speaking of "scientific" thinking in its everyday sense, and not of true scientific thinking, which cannot be divorced from common sense on the one hand, and imagination on the other. In its everyday sense, this "scientific" thinking is strictly logical, objective, highly specialized and "atomic" in its method and vision. The contrast in the two types of scholarship, Oriental and Occidental, ultimately goes back to the opposition between logic and common sense. Logic, deprived of common sense, becomes inhuman, and common sense, deprived of logic, is incapable of penetrating into nature's mysteries.

What does one find as he goes through the field of Chinese literature and philosophy? One finds there are no sciences, no extreme theories, no dogmas, and really no great divergent schools of philosophy. Common sense and the reasonable spirit have crushed out all theories and all dogmas. Like the poet Po Chüyi, the Chinese scholar"utilized Confucianism to order his conduct, utilized Buddhism to cleanse his mind, and then utilized history, paintings, mountains, rivers, wine, music and song to soothe his spirit. "He lived in the world and yet was out of it.

China, therefore, becomes a land where no one is trying very hard to think and everyone is trying very hard to live. It becomes a land where philosophy itself is a pretty simple and common sense affair that can be as conveniently put in two lines of verse as in a heavy volume. It becomes a land where there is no system of philosophy, broadly speaking, no logic, no metaphysics, no academic jargon; where there is much less academic dogmatism, less intellectual or practical fanaticism, and fewer abstract terms and long words. No sort of mechanistic rationalism is ever possible and there is a strong hatred of the idea of logical necessity. It becomes also a land where there are no lawyers in business life, as there are no logicians in philosophy. In place of well thought out systems of philosophy, they have only an intimate feeling of life, and instead of a Kant or a Hegel, they have only essayists, epigram writers and propounders of Buddhist conundrums and Taoist parables.

The literature of China as a whole presents us with a desert of short poems and short essays, seemingly interminable for one who does not appreciate them, and yet as full of variety and inexhaustible beauty as a wild landscape itself. We have only essayists and letter-writers who try to put their feeling of life in a short note or an essay of three or five hundred words, usually much shorter than the school composition of an American schoolboy. In these casual writings, letters, diaries, literary notes and regular essays, one finds here a brief comment on the vicissitudes of fortune, there a record of some woman who committed suicide in a neighboring village, or of an enjoyable spring party, or a feast in snow, or boating on a moonlight night, or an evening spent in a temple with a thunderstorm raging outside, generally including the remarks made during the conversation that made the occasion memorable. We find a host of essayists who are at the same time poets, and poets who are at the same time essayists, writing never more than five or seven hundred words, in which a whole philosophy of life is really expressed by a single line. We find writers of parables and epigrams and family letters who make no attempt to co?rdinate their thoughts into a rigid system. This has prevented the rise of schools and systems. The intellect is always held in abeyance by the spirit of reasonableness, and still more by the writer's artistic sensibility. Actually the intellect is distrusted.

It is hardly necessary to point out that the logical faculty is a very powerful weapon of the human mind, making the conquests of science possible. I am also aware that human progress in the West is still essentially controlled by common sense and by the critical spirit, which is greater than the logical spirit and which I think represents the highest form of thinking in the West. It is unnecessary for me to admit that there is a very much better developed critical spirit in the West than in China. In pointing out the weaknesses of logical thinking, I am only referring to a particular deficiency in Western thought, and sometimes in Western politics also, e.g., the Macht-politik of the Germans and the Japanese. Logic has its charm also, and I regard the development of the detective story as a most interesting product of the logical mind, a form of literature which failed entirely to develop in China. But sheer preoccupation with logical thinking has also its drawbacks.

The outstanding characteristic of Western scholarship is its specialization, and cutting up of knowledge into different departments. The over-development of logical thinking and specialization, with its technical phraseology, has brought about the curious fact of modern civilization, that philosophy has been so far relegated to the background, far behind politics and economics, that the average man can pass it by without a twinge of conscience. The feeling of the average man even of the educated person, is that philosophy is a"subject"which he can best afford to go without. This is certainly a strange anomaly of modern culture, for philosophy, which should lie closest to men's bosom and business, has become most remote from life. It was not so in the classical civilization of the Greeks and Romans, and it was not so in China, where the study of the wisdom of life formed the scholars' chief occupation. Either the modern man is not interested in the problems of living, which are the proper subject of philosophy, or we have gone a long way from the original conception of philosophy. The scope of our knowledge has been so widened, and we have so many"departments"of knowledge zealously guarded over by their respective specialists, that philosophy, instead of being the first of man's studies, has left for it only the field no one is willing to specialize in. Typical of the state of modern education is the announcement of an American university that"the Department of Psychology has kindly thrown open the doors of Psychology 4 to the students of Economics 3." The professor of Economics 3 therefore commits the care of his students to the professor of Psychology 4 with his love and blessings, while as an exchange of courtesy, he allows the students of Psychology 4 to tread in the sacred precincts of Economics 3 with a gesture of friendly hospitality. Meanwhile, Philosophy, the King of Knowledge, is like the Chinese Emperor in the times of the Warring Kingdoms, who instead of drawing tribute from the vassal states, found his authority and domain daily diminishing, and retained the allegiance of only a small population of very fine and loyal, but poorly fed subjects.

For we have now come to a stage of human culture in which we have compartments of knowledge but not knowledge itself; specialization, but no integration; specialists but no philosophers of human wisdom. This over-specialization of knowledge is not very different from the over-specialization in a Chinese imperial kitchen. Once during the collapse of a dynasty, a rich Chinese official was able to secure as his cook a maid who had escaped from the palace kitchen. Proud of her, he issued invitations for his friends to come and taste a dinner prepared by one he thought an imperial cook. As the day was approaching, he asked the maid to prepare a royal dinner. The maid replied that she couldn't prepare a dinner.

"What did you do, then?"asked the official.

"Oh, I helped make the patties for the dinner,"she replied.

"Well, then, go ahead and make some nice patties for my guests. "

To his consternation the maid announced:"Oh, no, I can't make patties. I specialized in chopping up the onions for the stuffing of the patties of the imperial dinner. "

Some such condition obtains today in the field of human knowledge and academic scholarship. We have a biologist who knows a bit of life and human nature; a psychiatrist who knows another bit of it; a geologist who knows mankind's early history; an anthropologist who knows the mind of the savage man; a historian who, if he happens to be a genial mind, can teach us something of human wisdom and human folly as reflected in mankind's past history; a psychologist who often can help us to understand our behavior, but who as often as not tells us a piece of academic imbecility, such as that Lewis Carroll was a sadist, or emerges from his laboratory experiments on chickens and announces that the effect of a loud noise on chickens is that it makes their hearts jump. Some educational psychologists always seem to me stupefying when they are wrong, and still more stupefying when they are right. But along with the process of specialization, there has not been the urgently needed process of integration, the effort to integrate all these aspects of knowledge and make them serve the supreme end, which is the wisdom of life. Perhaps we are ready for some integration of knowledge today, as is evidenced by the Institute of Human Relations at Yale University and in the addresses at the Harvard Tercentenary. Unless, however, the Western scientists proceed about this task by a simpler and less logical way of thinking, that integration cannot be achieved. Human wisdom cannot be merely the adding up of specialized knowledge or obtained by a study of statistical averages; it can be achieved only by insight, by the general prevalance of more common sense, more wit and more plain, but subtle, intuition.

There is clearly a distinction between logical thinking and reasonable thinking, which may be also expressed as the difference between academic thinking and poetic thinking. Of academic thinking we have a great deal, but of poetic thinking we find very little evidence in the modern world. Aristotle and Plato are strikingly modern, and that is so, perhaps not because the Greeks resembled the moderns, but because they were strictly the ancestors of modern thought. In spite of his humanistic point of view and his Doctrine of the Golden Mean, Aristotle was strictly the grandfather of the modern textbook writers, being the first man to cut up knowledge into separate compartments—from physics and botany to ethics and politics. As was quite inevitable, he was the first man also to start the impertinent academic jargon incomprehensible to the common man, which is being outdone by the American sociologists and psychologists of today. And while Plato had real human insight, yet in a sense he was responsible for the worship of ideas and abstractions as such among the Neo-Platonists, a tradition which, instead of being tempered with more insight, is so familiar among us today in writers who talk about ideas and ideologies as if they had an independent existence. Only modern psychology in very recent days is depriving us of the watertight compartments of "reason," "will" and "emotion" and helping to kill the "soul" which was such a real entity with the medieval theologians...

It seems that a regenerated form of thinking, a more poetic thinking, which can see life steadily and see it whole, is eminently desirable. As the late James Harvey Robinson warns us: "Some careful observers express the quite honest conviction that unless thought be raised to a far higher plane than hitherto, some great set-back to civilization is inevitable." Professor Robinson wisely pointed out that, "Conscientiousness and Insight seem suspicious of one another, and yet they might be friends." Modern economists and psychologists seem to me to have an overdose of conscientiousness and not enough of insight. This is a point which perhaps cannot be overemphasized, the danger of applying logic to human affairs. But the force and prestige of scientific thinking have been so great in the modern age, that in spite of all warnings, this species of academic thinking constantly encroaches upon the realm of philosophy, with the jejune belief that the human mind can be studied like a sewerage system and the waves of human thought measured like the waves of radio. The consequences are mildly disturbing in our everyday thinking, but disastrous in practical politics.

二. 回向常识

中国人都憎恶"逻辑的必要"那个名词，因为在中国人的心目中，世事之中无所谓逻辑的必要。中国人对于逻辑的不信任，起点于不信任字眼，进而惧怕界说，最后则对一切系说、一切假说表示天性的憎恨。因为使哲学派成为可能者，都是字眼、界说和系说的罪恶。哲学的腐化起于对字眼的偏见。中国作家龚定庵说，圣人不说话，有能为的人才说话，愚人才会做辩论——其实龚氏本人就是一个最好做辩论的人，但他仍说这句话。

因为这就是哲学的悲惨经过：即哲学家不幸都是好说话的人，而不是肯守缄默的人。所有的哲学家都喜欢听他自己的语声。即如老子，他虽是第一个指点给我们知道"大块"是无言的，但他自己在出函谷关去隐居深山、乐享余年之前，仍免不了听从人劝，遗留下传诸后世的五千言。尤其足以代表这类天才哲学言谈家的就是孔子游遍"七十二国"以说诸国之君；又如苏格拉底，他在雅典的街上走来走去，遇到走路的人即叫住他，问他几句话，以便他自己可以发生聪明的意见给自己听。所以"圣人不多言"这句话乃是相对的说法。不过圣人和才子之间仍有一种区别，因为圣人谈到生活都是以亲身的阅历为中心，才子则只知道研究解释圣人的说话，而笨人更是只知道将才子的说话咬文嚼字地辩论。在希腊的修辞学家当中，我们可以看见这种专以咬文嚼字为尚的纯粹谈论家。哲学本是一种对智慧的爱好，已变成了对字句的爱好，等到修辞学的风尚渐渐滋长，哲学便和生活越离越远了。等到后来，哲学家竟专顾多用字眼，多用长的句子；短短的警语多变成了长句，句子变成了论据，论据变成了专书，专书变成了长篇大论，长篇大论变成了语言学的研究；他们需要更多的字眼以定所用字的界说，并将它们归类，他们需要更多的派别以区别和隔离已经设立的派别。这个程序接连不断地进行着，直到对于生活的直接切己的感觉或知悉完全丧失，致使外行竟敢于诘问："你在那里说些什么？"同时，在后来的思想历史中，少数几个对生活本身感觉到直接撞击的独立思想家——如歌德、萨缪尔·约翰逊、爱默生、威廉·詹姆斯——都拒绝在谈论家的胡言乱语中发言，并始终极固执地反对归类的精神。因为他们是聪明的，他们替我们维持着哲学的真意义，就是生活的智慧。在许多情形中，他们都抛弃了论据，回向警语。一个人在丧失了说出警语的能力时，方去写长篇；而他在论证之中依旧不能明白发表他的意思时，方去著作一本专书。

人的爱好字句，是他走向愚昧之途的第一步，他的爱好界说乃是第二步。他越从事于分析，他越需要界说，他越加定界说，他越是趋向一个不可能的逻辑的完美境界，因为企求逻辑的完美就是愚昧的迹象。因为字句是我们思想的材料，所以定其界说的企图乃是完全可嘉的，于是苏格拉底在欧洲创始了一个定界说狂。其危险在于我们意识到曾由我们定其界说的字眼时，便不能不将用以定界说的字眼也定出它们的界说来，因此，其结果除了用以定生活的界说的字眼以外，我们又有了专用以定别的字眼的界说的字眼，而定字眼的界说这桩事便成了我们的哲学家的主要成见了。忙碌的字眼和空闲的字眼之间显然有一种分别，前者在我们的日常工作生活中尽它们的责任，而后者只存在于哲学家的研究团体中。此外苏格拉底和弗兰西斯·培根的界说和现代大教授的界说之间，也是有着一种分别的。莎士比亚对生活有着最切己的感觉，但他也居然能从容地过去，并没有做什么定界说的企图，也可说是因为他没有做定界说这件事，所以他所用的字眼都有着一种别个作家所缺少的"实体"，而他的文字中也充满着一种现代所缺少的人类悲剧的意味和堂皇的气概。我们无从将他的文字限制到某一个动作效能的范围之内，正如我们无从将他的文字限制到一个对妇女的特别观念里边。因为它们都在有了定界说的性质时方使我们的思想僵硬，因而剥夺了生活本身发光的、幻想的色彩特质。

但如若字眼为了必须的理由分割了我们的在表示程序中的思想，那对于系统的爱好更能损害我们对于生活深切的知悉。系统不过是一种对真理的从旁斜视，因此，这系统越加有着逻辑的发展，那种灵心上的斜视也变得越加可怕。人类只想看见偶然所能看到的真理的片面，并将它发展和提升到一个完善的逻辑系统的地位的欲望，即是我们的哲学会和生活势必越离越远的理由。凡是谈到真理的人，都反而损害了它；凡是企图证明它的人，都反而伤残歪曲了它；凡是替它加上一个标识和定出一个思想派别的人，都反而杀害了它；而凡是自称为信仰它的人，都埋葬了它。所以一个真理，等到被树立成为一个系统时，它已死了三次，并被埋葬了三次了。他们在真理出丧时所唱的挽歌就是："我是完全对的，而你是完全错误的。"他们所埋葬的是哪一种真理在根本上无关紧要，不过总是已经埋葬了它。真理便如此在防护它的人手中受到了虐待，而一切哲学的党派，不论古今，都只是专心致力于证明一点，即"我是完全对的，而你是完全错误的"。德国的哲学家们写了一本挺厚的书，想要证明某一种有限制的真理，结果反而将那真理变了一个胡说，这班人大概可算是最坏的冒犯者了，不过这种思想的疾病在西方的思想界中几乎是随处有的，而只在深浅中有些分别罢了，在他们越趋于抽象时，这个病症也越深。

这种不近人情的逻辑，其结果是造成了一种不近人情的真理。今天我们所有的哲学是一种远离人生的哲学，它差不多已经自认没有教导我们人生意义和生活智慧的意旨，实在早已丧失了我们所认为哲学的精英对人生的切己感觉和对生活的知悉。威廉·詹姆斯即称这种对人生的切己感觉为"经验的要素"。等到日子长久之后，威廉·詹姆斯的哲学和逻辑所加于现代西方思想方式的蹂躏必会一天厉害过一天。但我们如想把西方哲学变成近于人情，则必须先将西方逻辑变成近于人情。我们须回到一种对现实和生活，尤其是对于人性急于接触的思想方式，而不单是求得不错、合于逻辑、没有不符之处便算完事。我们对于笛卡儿（Descartes）著名的发现"我思故我在"这句名言所表率的思想的疾病，应该拿惠特曼所说那句较为近于人性和较为有意义的话"我照现在的地位，我已尽够"去替代。生活或存在无须跪在地上恳求逻辑代它证明世上确有它这样的事物。

威廉·詹姆斯终其身在那里企图证明中国式思想方式，并替它辩护，不过自己没有觉得罢了。当中不过有着下列的分别：他如果真是一个中国人，必不会用这许多字眼去做他的论证，而只用那么三五百个字写一篇短文，或在他的日记中写上短短几句话便算完事了。他将要对着字眼胆怯，恐怕越多用字眼，便越加会引起误会。但威廉·詹姆斯在他对生活的深切感觉，对人类阅历的透彻，对机械式的理智主义的反抗，对于思想切心想保持它的流动状态，并对那些自以为已经发现了一个万分重要的、绝对的、无所不包的真理，而将它纳入一个自以为满足的系统中的人们的不耐

烦当中，简直就是一个中国人。他的坚持在艺术家的意识上，属于知觉的
现实比属于概念的现实更为重要这一点上也像一个中国人。其实所谓哲学
家者，就是一个时常将他的感觉力集中于最高的焦点去观察生活的变动，
随时预备碰到更新的和更奇怪的矛盾事情、前后不符的事情和一切不合于
常例的事情。在他拒绝的一个系统之中，他所拒绝的并非因它不对，而只
因它是一个系统，在这个举动中，他实在破坏了西方的哲学派。照他的说
法，宇宙的一元概念和多元概念之间的分别实是哲学中最重要的分别。他
使哲学有放弃空中楼阁而回到生活本身的可能。

　　孔子说："道不远人，人之为道而远人，不可以为道。"他还有一句聪
明的话，很像詹姆斯的口气："人能弘道，非道弘人。"不，世界并不是
一个三段论法或一个论据，而是一个生物；宇宙不做声，只是生活着；它
并不做什么辨认，只是进行着。某英国天才作家说："理智不过是神秘物
事中的一个节目；而在最高傲的意识国统治的背面，理智和惊奇涨红了脸
相对着。不可避免的事情变成了平凡，而疑惑和希望成了姐妹。宇宙是粗
野的，如鹰的翅膀带着一些竞技的意味，这还算是一件可喜的事。大自然
就是一个神奇之迹，同一的事物不再重回，即使回来也必是已经不同的。"
在我看来，西方的逻辑家所需要的只是一些自谦心，如有人能将他们的黑
格尔派的自以为是医好，他们就能得救了。

II. The Return to Common Sense

THE Chinese hate the phrase"logical necessity" because there is no logical necessity in human affairs. The Chinese distrust of logic begins with the distrust of words, proceeds with the abhorrence of definitions and ends with instinctive hatred for all systems and theories. For only words, definitions and systems have made schools of philosophy possible. The degeneration of philosophy began with the preoccupation with words. A Chinese writer, Kung Tingan, said:"The Sage does not talk, the Talented Ones talk, and the stupid ones argue"–this in spite of the fact that Kung himself loved very much to argue!

For this is the sad story of philosophy: that philosophers belonged to the genus of the Talkers and not the Silent Ones. All philosophers love to hear their own voice. Even Laotse himself, who first taught us that the Creator (the Great Silent One) does not talk, nevertheless was persuaded to leave five thousand words to posterity before he retired outside the Hankukuan Pass to pass the remainder of his life in wise solitude and oblivion. More typical of the philosopher-talker genius was Confucius who visited"seventy-two kingdoms"in order to get a hearing from their kings, or, better still, Socrates who went about the streets of Athens and stopped passers-by to ask them questions for the purpose of hearing himself give wise answers. The statement that the"Sage does not talk"is therefore only a relative one. But still the difference exists between the Sage and the Talented Ones, because the Sage talks about life, as he is directly aware of it; the Talented Ones talk about the Sage's words and the stupid ones argue about the words of the Talented Ones. In the Greek Sophists we have the pure type of Talkers who are interested in the play and interplay of words as such. Philosophy, which was the love of wisdom, became the love of words, and in proportion as this Sophist trend grew, the divorce between philosophy and life became more and more complete. As time went on, the philosophers began to use more and more words and longer and longer sentences; epigrams of life gave place to sentences, sentences to arguments, arguments to treatises, treatises to commentaries, and commentaries to philological research; more and more words were needed to define and classify the words they used and more and more schools were needed to differ and secede from the schools already established; the process continued until the immediate, intimate feeling or the awareness of living has been entirely lost sight of, and the layman has the perfect right to ask,"What are you talking about?"Meanwhile, throughout the subsequent history of thought, the few independent thinkers who have felt the direct impact of life itself–a Goethe, a Samuel Johnson, an Emerson, a William James–have refused to speak in the jargon of the Talkers and always been intractably opposed to the spirit of classification. For they are the wise ones, who have kept for us the true meaning of philosophy, which is the wisdom of life. In most cases, they have forsaken arguments and reverted to the epigram. When man has lost the ability to speak in epigrams, he writes paragraphs; when he is unable to express himself clearly in paragraphs, he develops an argument; and when he still fails to make his meaning clear in an argument, he writes a treatise.

Man's love for words is his first step toward ignorance, and his love for definitions the second. The more he analyzes, the more he has need to define, and the more he defines, the more he aims at an impossible logical perfection, for the effort of aiming at logical perfection is only a sign of ignorance. Since words are the material of our thought, the effort at definition is entirely laudable, and Socrates started the mania for definitions in Europe. The danger is that after being conscious of the words we define, we are further forced to define the defining words, so that in the end, besides the words which define or express life itself, we have a class of words which define other words, which then become the main preoccupation of our philosophers. There is evidently a distinction between busy words and idle words, words that do duty in our workaday life and words that exist only in the philosophers' seminars, and also a distinction between the definitions of Socrates and Francis Bacon, and the definitions of our modern professors. Shakespeare, who had the most intimate feeling of life, certainly got along without trying to define anything, or rather because he did not try to define anything, and for that reason, his words had a"body"which the other writers lacked, and his language was infused with that sense of human tragedy and grandeur that is often missing today. We can no more hold his words down to any one particular function than we can hold him down to any particular conception of woman. For it is in the nature of definitions that they tend to stifle our thought and deprive it of that glowing, imaginative color characteristic of life itself.

But if words by necessity cut up our thoughts in the process of expression, the love of system is even more fatal to a keen awareness of life. A system is but a squint at truth, and the more logically that system is developed, the more horrible that mental squint becomes. The human desire to see only one phase of truth which we happen to perceive, and to develop and elevate it into a perfect logical system, is one reason why our philosophy is bound to grow stranger to life. He who talks about truth injures it thereby; he who tries to prove it thereby maims and distorts it; he who gives it a label and a school of thought kills it; and he who declares himself a believer buries it. Therefore any truth which has been erected into a system is thrice dead and buried. The dirge that they all sing at truth's funeral is,"I am entirely right and you are entirely wrong."It is entirely immaterial what truth they bury, but it is essential that they do the burying. For so truth suffers at the hands of its defenders, and all factions and all schools of philosophy, ancient and modern, are occupied only in proving one point, that"I am entirely right and you are entirely wrong."The Germans, with their Gründluhkeit, writing a heavy volume to prove a limited truth until they have turned it into an absurdity, are perhaps the worst offenders, but the same disease of thinking may be seen or noted more or less in most Western thinkers, becoming worse and worse as more and more abstract they become.

As a result of this dehumanized logic we have dehumanized truth. We have today a philosophy that has become stranger to life itself, that has almost half disclaimed any intention to teach us the meaning of life and the wisdom of living, a philosophy that has lost that intimate feeling of life or awareness of living which we spoke of as the very essence of philosophy. It is that intimate feeling of life which William James has called"the stuff of experience."As time goes on, I feel that the philosophy and logic of William James will become more and more devastating to the modern Western way of thinking. Before we can humanize Western philosophy, we must first humanize Western logic. We have to get back to a way of thinking which is more impatient to be in touch with reality, with life, and above all with human nature, than to be merely correct, logical and consistent. For the disease of thinking typified by Descartes' famous discovery:"I think, therefore I exist,"we have to substitute the more human and more sensible statement of Walt Whitman's:"I am sufficient as I am. "Life or existence does not have to go down on its knees and beg logic to prove that it exists or that it is there.

William James spent his life trying to prove and defend the Chinese way of thinking, without knowing it. Only there is this difference, that if William James had been a Chinese, he would not have written so many words to argue it out, but would have merely stated in an essay of three or five hundred words, or in one of his leisurely diary notes, that he believed it because it is so. He would be shy of the words themselves, for fear that the more words he used, the greater the chances for misunderstanding. But William James was a Chinese in his keen awareness of life and the varieties of human experience, in his rebellion against mechanistic rationalism, his anxiety to keep thought constantly fluid, and his impatience with people who think they have discovered the one all-important, "absolute" and universal truth and have enclosed it in a self-sufficient system. He was Chinese, too, in his insistence on the importance of the artist's sense of perceptual reality over and against conceptual reality. The philosopher is a man who holds his sensibilities at the highest point of focus and watches the flux of life, ready, to be forever surprised by newer and stranger paradoxes, inconsistencies, and inexplicable exceptions to the rule. In his refusal to accept a system not because it is incorrect, but because it is a system, he plays havoc with all the Western schools of philosophy. Truly, as he says, the difference between the monistic conception and the pluralistic conception of the universe is a most pregnant distinction in the history of philosophy. He has made it possible for philosophy to forget its beautiful air-castles and return to life itself.

Confucius said, "Truth may not depart from human nature; if what is regarded as truth departs from human nature, it may not be regarded as truth." Again he says, in a witty line that might have dropped from James's lips, "It is not truth that can make men great, but men that can make truth great." No, the world is not a syllogism or an argument, it is a being; the universe does not talk, it lives; it does not argue, it merely gets there. In the words of a gifted English writer: "Reason is but an item in the mystery; and behind the proudest consciousness that ever reigned, reason and wonder blushed face to face. The inevitable stales, while doubt and hope are sisters. Not unfortunately, the universe is wild, game-flavored as a hawk's wing. Nature is miracle all: the same returns not save to be different." It seems what the Western logicians need is just a little humility; their salvation lies in some one curing them of their Hegelian swelled-heads.

三. 近情

　　和逻辑相对的有常识，或更好一些的说法，还有近情的精神。我以为近情精神实是人类文化最高的、最合理的理想，而近情的人实在是最高形式的有教养的人。世人没有一个人是完美无缺的，他只能力争上游去做一个近乎情理的生物。我正期待着世界上将有一个世人在个人的事件上，并在国家的事件上，都会得着这个近情精神之鼓舞的时期。近情的国家将生活于和平之中，近情的夫妻能生活于快乐之中。在我替我的女儿挑选丈夫时，我将只有一个标准：他是否是一个近情的人？我们当然不能期望世上有终身不吵架的夫妻，我们只能期望他们都是近情的男女，只近情地吵架，并近情地言归于好。只有在世界的人类都是近情的人时，我们才能得到和平和快乐。这近情的时代如果有来临的一天，就是和平时代的来临。在这时代中，近情的精神必会占最大的势力。

　　近情精神是中国所能贡献给西方的一件最好的物事。我并没有说中国那些向人民预征五十年钱粮的军阀是近情的，我的意思只是说，近情的精神乃是中国文明的精华和她最好的方面。我这个发现曾偶然由两位久居中国的美国人所证实。其中一位居住中国已经三十年，他说，中国的一切社会生活乃是以

"讲理"为基础的。在中国人的争论之间，他们最后一句有力的论据必是："这岂是合于情理的吗？"而最严重的、最平常的斥责之词就是：这人是"不讲理"的。一个人如若在争论之中自承不近情理，则他已是输了。

我曾在《吾国与吾民》一书中说过："在一个西方人，一个说法只须合于健全的逻辑，他便认为是已很充足。但在一个中国人，则一个说法虽然在逻辑上已是很对，他也还不肯认为充足，同时还必须求其近于人情。'近情'在实际上比合于逻辑更为人所重视。Reasonableness这个字，中文译做'情理'，其中包括着'人情'和'天理'两个元素。'情'代表着可以活动的人性元素，而'理'代表着宇宙之万古不移的定律。"一个有教养的人就是一个洞悉人心和天理的人。儒家借着和人心及大自然的天然程式的和谐生活，自认可以由此成为圣人者也不过是如孔子一般的一个近情的人，而人所以崇拜他，也无非因为他有着坦白的常识和自然的人性罢了。

人性化的思想其实就是近情的思想。专讲逻辑的人永远自以为是，所以他不近人情，也是不对的；至于近情的人则自己常疑惑自己是错的，所以他永远是对的。近情的人和专尚逻辑的人，不同处可以在他们信札后面的附言中看出来。我最爱读朋友所给我的信后的附言，尤其是那种和信的正文互相矛盾的附言。这种附言里边包括着一切近情的后想，一切疑惑不决之点和忽然而发的聪明说话和常识。一个温和的思想家就是一个企图用长篇大论的论据证明一个说法之后，忽然回到了直觉的地位，由于一阵忽然而发的常识，立刻取消他以前所做的论证而自认错误的人，这就是我所谓人性化的思想。

我们只需拿各人所写的信来看看，专尚逻辑的人必是在信的本文中罄其所欲言，而近情的人，即有着人类精神的人，必是在附言中说他的话。譬如一个人的女儿请求她的父亲许她进大学读书，她的父亲或许在回信之中列出许多极合于逻辑的理由，第一怎样，第二怎样，第三怎样，例如：

已有三个哥哥在大学读书，负担已经很重；她的母亲正在家中患病，需要她在旁服侍；等。他在信末署名之后，又加写了一行附言："不必多说了，一准在秋季开学时入校吧。我总替你想法子。"

或如一个丈夫写信给他的太太，发表离婚的决心，并列出许多似乎毫无驳诘余地的理由，如：第一，太太对他不忠；第二，他每次回家从来吃不到热饭；等。所列的理由很充足极平允，倘若委托给律师办理，则事理上将更为严正，口气将更为有理由。但是他在写完这信时，心中忽然有所感触，便又提起笔来在后面加了一行："算了吧，可爱的苏菲，我真是一个坏坯子。我将带一束鲜花回家了。"

上述的两封信里边，其论据都是极为合理的，不过当时说这些话的是一个心在逻辑的人，而在附言中已变了一个有着真正人类精神的人在那里说话——一个近人情的父亲和一个近人情的丈夫。如此就是人类灵心的责任，人类有灵心，并不是叫它去做愚笨的逻辑的辩论，而应是在互相冲突的冲动、感觉和欲望永远变迁的海洋中企图保持一种合于理智的平衡。这就是人事中的真理，也就是我们所企图达到的地步。无从答复的论据常可由怜悯之情答复它，充足的理由常可由爱情打破它。在人事之中，不合逻辑的行为常是最能动人的。法律本身就承认它未必能处处绝对的平允，也时常不能不迁就人情，所以一国的元首都另有一种特赦权，如林肯所用以赦免那个"母亲的儿子"一样。

近情精神使我们的思想人性化，并且使我们不坚信自己总是对的。它的影响在于刨去我们行为的棱角，并使它调和起来。和近情精神相反，就是思想和行为中，我们的个人生活中，国家生活中，婚姻、宗教与政治中的一切方式的热狂和武断。我以为，在中国热狂和武断是较少的。中国的暴众虽也易于鼓动（例如庚子年的拳匪），但近情的精神确在某种程度上使我们的皇帝专制、我们的宗教和所谓"欺压女性"人性化。近情精神在这些当中当然都是有限制的，不过它确是存在着的。这精神使我们的皇帝、我们的上帝和我们的丈夫都成了单纯的人类。中国的皇帝并不像日本天皇那么半神道，而中国的史家已演绎出一个皇帝受命于天，但他如失德，便将丧失天命的假说。他如失德，我们可以杀他的头，在历代的兴衰中，被人砍去脑袋的皇帝已不知道有多少个，这就破除了我们的皇帝乃是神圣的或半神圣的念头。我们的圣人也没有被人尊奉为神道，而不过始终认他们为聪明的教师，我们的神道也不是完善的模范，而不过是像我们的官府一般唯利是图，很是腐败，可以用甘言和贿赂打动。凡是出乎情理之外的事情，我们一概称之为"不近人情"，太过于矫情的人就是大奸，因为他在心理上是反常的。在政治的区域内，某些欧洲国家人们心中的逻辑和他们的行事实在异常不近人情。

在某种意义上，我们可以说现代的欧洲并不由近情的精神所统治着，也不是由具有理智的精神所统治着，而实在是由疯狂的精神所统治的。

看看欧洲的现象，使人发生一种不宁的感觉，这种不宁，并不是由于看见国家的目标、国界和殖民地要求的冲突而发生，因为这些都是理智的精神所能够应付的，而实在是由于看见欧洲各统治者那种心境而发生的。这就是等于跨上一辆街车，驶到一处陌生的地方，而忽然对司机发生了不信任的心思。这不信任并不是由于疑心司机不认识路，因而疑心他不能将自己载到目的地，而是由于听见司机在那里胡言乱语，前言不搭后语，因而疑心他未必是清醒的。如若司机还有着一支手枪，而坐者并没有离开汽车的方法，他的不宁当然将更为增加了。我敢信这幅人心的调整画并不是人心的本身，而不过是一时失常，不过是暂时疯狂的一个阶段，将来自会像瘟疫一般自己消灭的。我敢证言，人心是终属有能力的，敢信人类不免一死的灵心虽是有限制的，但其智力实仍是远胜于欧洲之不顾一切的司机，而到了最后，我们终能和平地生活，因为到了那时节，我们都已学会怎样做近情的思想了。

III. Be Reasonable

IN contrast to logic, there is common sense, or still better, the Spirit of Reasonableness. I think of the Spirit of Reasonableness as the highest and sanest ideal of human culture, and the reasonable man as the highest type of cultivated human being. No one can be perfect; he can only aim at being a likeable, reasonable being. In fact, I look forward to the time when the people of the world will be informed with this reasonable spirit, both in their personal and their national affairs. Reasonable nations live in peace and reasonable husbands and wives live in happiness. In the selection of husbands for my daughters, I shall have only one standard: is he a reasonable person? We cannot imagine perfect husbands and wives who never quarrel; we can only conceive of reasonable husbands and wives who quarrel reasonably and then patch up reasonably. Only in a world of reasonable beings can we have peace and happiness. The Reasonable Age, if that should ever come about, will be the Age of Peace. It will be the age in which the Spirit of Reasonableness prevails.

The Spirit of Reasonableness is the best thing that China has to offer to the West. I do not mean that Chinese warlords are reasonable when they tax the people fifty years ahead; I mean only that the Spirit of Reasonableness is the essence and best side of Chinese civilization. I had this discovery of mine accidentally confirmed by two Americans who had lived a long time in China. One, who had lived in China for thirty years, said that the foundation of all Chinese social life is based on the word chiangli, or "talk reason." In a Chinese quarrel, the final clinching argument is, "Now is this reasonable?" and the worst and commonest condemnation is that a man "pu chiangli" or "does not talk reason." The man who admits being "unreasonable" is already defeated in the dispute.

I have said in My Country and My People that: "For a Westerner, it is usually sufficient for a proposition to be logically sound. For a Chinese it is not sufficient that a proposition be logically correct, but it must be at the same time in accord with human nature. In fact to be 'in accord with human nature,' to be chinch'ing (i.e., to be human), is a greater consideration than to be logical." "The Chinese word for 'reasonableness' is ch'ingli, which is composed of two elements, ch'ing (jench'ing) or human nature, and li (t'ienli) or eternal reason. Ch'ing represents the flexible, human element, while li represents the immutable law of the universe." A cultured man is one who understands thoroughly the human heart and the laws of things. By living in harmony with the natural ways of the human heart and of nature, the Confucianist claims that he can become a sage. But then the sage is no more than a reasonable person, like Confucius, who is chiefly admired for his plain, common sense and his natural human qualities, i.e., for his great humanness.

Humanized thinking is just reasonable thinking. The logical man is always self-righteous and therefore inhuman and therefore wrong, while the reasonable man suspects that perhaps he is wrong and is therefore always right. The contrast between the reasonable man and the logical man is often shown in the postscripts to letters. I always love the postscripts in my friends' letters, especially those that entirely contradict what has been said in the body of the letter. They contain all the reasonable afterthoughts, the hesitancies, and the flashes of wit and common sense. The genial thinker is one who, after proceeding doggedly to prove a proposition by longwinded arguments, suddenly arrives at intuition, and by a flash of common sense annihilates his preceding arguments and admits he is wrong. That is what I call humanized thinking.

We can imagine a letter, in which the logical man speaks in the body of the letter and the reasonable man, the truly human spirit, speaks in the postscript. A father may be writing to his daughter, who has been begging him for a chance to go to college, and in the letter he proceeds to outline the various perfectly sound reasons,"firstly,""secondly, "" thirdly,"why he cannot send her to college with a kind of consistent, cumulative, unanswerable logic, those reasons being that he has already three sons to support at college, that her sick mother needs someone to keep her company at home, and so on. After signing his name he adds a little line:"Hang it, Julie, prepare to go to college in the fall. Somehow I will do it. "

Or let us imagine a husband writing to his wife, announcing his decision to arrange for a divorce and giving an unimpeachable array of reasons for it: firstly, that his wife has been unfaithful, secondly, that he can never get a hot meal when he comes home, and so on. They are perfectly valid reasons, even righteous reasons, and if he engages a lawyer to do the job, the logic will be still more perfect and the tone more righteous still. But after writing the letter, something happens in his mind, and he scrawls a barely legible note:"Damn it, darling Sophie, I'm a rotten dog myself. I'm coming home with a bunch of flowers. "

While the arguments in both letters are perfectly sound and valid, it is the logical man who speaks there, while in the postscripts there speaks a truly human spirit—a human father and a human husband. For such is the duty of the human mind, that it is not called upon to make a stupidly logical argument, but should try to maintain a sane balance in an ever-changing sea of conflicting impulses, feelings, and desires. And such is truth in human affairs that that is true which we will to make so. The unanswerable argument can always be answered with some compassion, and validity itself invalidated by love. In human affairs, it is often the illogical course of conduct that is the most convincing. Law itself admits the incompleteness of its claim to absolute justice when it often has to fall back on"a reasonable interpretation"of a clause, or when it allows the chief executive the power of pardon, so well exercised by Abraham Lincoln to a mother's son.

The reasonable spirit humanizes all our thinking, and makes us less sure of our own correctness. Its tendency is to round out our ideas and tone down the angularities of our conduct. The opposite of the reasonable spirit is fanaticism and dogmatism of all sorts in thought and behavior, in our individual life, national life, marriage, religion and politics. I claim that we have less intellectual fanaticism and dogmatism in China. While a Chinese mob is quite excitable (witness the Boxers of 1900), the Spirit of Reasonableness has humanized to a large extent our monarchical autocracy, our religion, and the so-called"suppression of women."All this must be taken with certain qualifications, but is nevertheless true. It makes our emperors, our gods, and our husbands merely human beings. The Chinese emperor was not a semi-divine being like the Japanese ruler, and Chinese historians have evolved the theory that the emperor rules by a mandate from Heaven, and when he misrules, he forfeits that"heavenly mandate."When he misrules, we just chop off his head, and we have chopped off the heads of too many kings and emperors in the many rising and falling dynasties for us to believe they are"divine"or"semi-divine."Our sages are not canonized as gods, but always regarded merely as teachers of wisdom, and our gods are not models of perfection, but venal and corrupt and open to cajolement and bribery like our officials. Anything which goes beyond reasonableness is immediately condemned as puchi'n jench'ing ("moving far away from human nature"), and a man who is too saintly or too perfect can be a traitor because he is psychologically abnormal.

In the sphere of politics, there is something terribly inhuman in the logic of the minds of men and conduct of affairs in certain states of Europe...

In a sense we may say that today Europe is not ruled by the reasonable spirit, nor even by the spirit of reason, but rather by the spirit of fanaticism. Looking at the picture of Europe today gives one a feeling of nervousness, a nervousness which comes not so much from the mere presence of conflicts of national aims and state boundaries and colonial claims, which the spirit of reason should be amply able to deal with, but rather from the condition of mind of the men who are the rulers of Europe. It is like getting into a taxicab in a strange city and being suddenly overcome by a distrust of the driver. It is not so bad that the driver doesn't seem to be acquainted with the map of the city and cannot take one to his destination by the proper route; it is more alarming when the passenger in the back seat hears the driver talk incoherently and begins to suspect his sobriety. That nervousness is decidedly heightened when the inebriate driver is armed with a gun and the passenger has no chance of getting out. One has reason to believe that this caricature of the human mind is not the human mind itself, that these are mere aberrations, mere stages of temporary insanity, which will burn themselves out like all waves of pestilence. One has reason to express a reassurance in the capacities of the human mind, to believe that the human mortal mind, limited as it is, is something infinitely higher than the intellect of the reckless drivers of Europe, and that eventually we shall be able to live peaceably because we shall have learned to think reasonably.